AF530531

Food Processing Technologies, Co-Product Utilization and Quality Assurance

Food Processing Technologies, Co-Product Utilization and Quality Assurance

Edited by :

MRIDULA D.

R.T. PATIL

M.R. MANIKANTAN

SATISH SERIAL PUBLISHING HOUSE

403, Express Tower, Commercial Complex, Azadpur, Delhi-110033 (India)

Phone : 011-27672852, Fax : 91-11-27672046

E-mail :info@satishserial.com, hkjain1975@yahoo.com

Website : www.satishserial.com

Published by :

SATISH SERIAL PUBLISHING HOUSE

403, Express Tower, Commercial Complex, Azadpur, Delhi-110033 (INDIA)
Phone : 011-27672852 Fax : 91-11-27672046
E-mail : info@satishserial.com, hkjain1975@yahoo.com

ISBN 978-93-81226-16-2

Typeset at: Harminder Singh Kharb *for* Laxmi Art Creation
Printed at: Salasar Imaging Systems, Delhi-110035

Preface

India produces about 96.43 million tonnes rice, 78.40 million tonnes wheat, 40.73 million tonnes coarse cereals, 15.11 million tonnes pulses and 28.82 million tonnes oilseeds. Thus, as for food security scenario of our country is concerned, we are in a good position to some extent but the nutritional security is still a big challenge in front of the country. Only 5-6% of the produce is processed in India as against 83% in Malaysia, 72% in Thailand and 70% in Brazil. We have to improve a lot in the loss prevention and there is an urgent need to enhance the commercial processing of our agri-produce. As per the available records, commercial processing is limited to only about 10% in our country as against more than 70% in countries like Malaysia, Thailand and Brazil. While demand of processed products in domestic and international market is increasing day by day due to rapid urbanization, changing life style, increased number of working women and due to rising per capita income. It is not only the demand of the processed food products but attitude of consumers is also being changed. They are becoming more health conscious now. Health conscious consumers give first priority to nutritious food products with functional properties because of the relationship between diet and disease. Consumers are becoming aware about nutritional labeling and phytofactors present in the food products. Earlier we were purchasing the food products just by checking only the manufacturing and expiry date but now first we check the nutritional information and complete labeling on the pack and then decide to purchase of a particular item. Hence, there is a need to change the research strategies from the traditional style to the newer directions so that users can get the best quality end products with maximum retention of the naturally occurring phytofactors and other health benefits to meet the demands of health conscious consumers as well as to strengthen India's agri- business.

This book on "Food Processing Techniques, Co-Product Utilization and Quality Assurance" is an effort in this direction to bring out the recent developments on these aspects that will certainly be helpful in planning the research programmes in these areas. This book will be useful to researchers, students and also to knowledge transfer functionaries to promote food processing as well food quality and safety issues.

Editors

Preface

List of Contributors

A.K.Thakur
ICAR Research Complex for
Eastern Region, Research Centre
Plandu, Ranchi-834 010, Jharkhand
akthakur_ciphet@rediffmail.com

Anil Kumar Dixit
Division of Transfer of Technology
Central Institute of Post Harvest
Engineering & Technology (CIPHET)
Ludhiana-141 004, Punjab
anil_dixit7@rediffmail.com

B.C. Sarkar
Department of Food Engineering &
Technology, Sant Longowal Institute of
Engineering and Technology (Deemed to
be University), Longowal-148 106, India

D.B. Singh
Division of Post Harvest & Technology
Central Institute of Temperate
Horticulture, Rangreth-190 007
Jammu and Kashmir, India
deshbsingh@yahoo.co.in

D.N.Yadav
Division of Food Grains & Oilseeds
Processing, Central Institute of
Post Harvest Engineering & Technology
Ludhiana-141 004, Punjab
dnyadav1977@yahoo.co.in

Dattatreya M. Kadam
Division of Food Grains & Oilseeds
Processing, Central Institute of Post
Harvest Engineering & Technology
Ludhiana-141 004, Punjab
kadam1k@yahoo.com
kadam1k@gmail.com

Deepak Raj Rai
Division of Transfer of Technology
Central Institute of Post Harvest
Engineering & Technology
Ludhiana-141 004, Punjab
d_r_rai@yahoo.com

Devinder Dhingra
Central Institute of Post Harvest
Engineering & Technology
Ludhiana-141 004, Punjab
devinder.dhingra@gmail.com

Deepika Goswami
Division of Food Grains & Oilseeds
Processing, Central Institute of Post
Harvest Engineering & Technology
Ludhiana-141 004, Punjab
deepikagoswami83@yahoo.com
deepikagoswami@rediffmail.com

H.K. Sharma
Department of Food Engineering & Technology, Sant Longowal Institute of Engineering & Technology (Deemed to be University), Longowal-148106, Punjab

Harinder Singh Oberoi
Division of Agriculture Structure & Environment Control, Central Institute of Post Harvest Engineering & Technology
Ludhiana-141 004, Punjab
hari_manu@yahoo.com

K.K. Singh
Indian Council of Agriculturte Research
Krishi Bhawan, New Delhi
singh_ciae@yahoo.com

Kiran Bains
Department of Food and Nutrition
Punjab Agricultural University
Ludhiana-141 004, Punjab
kiranbains68@hotmail.com

K. Narsaiah
Division of Agriculture Structure & Environment Control, Central Institute of Post Harvest Engineering & Technology
Ludhiana-141 004, Punjab
knarsan@yahoo.com

Mahesh Kumar
Department of Processing & food Engineering, Punjab Agricultural University
Ludhiana-141 004, Punjab
maheshpau.68@rediffmail.com

Mandeep Kaur
Department of Food Engineering & Technology, Sant Longowal Institute of Engineering and Technology (Deemed to be University), Longowal-148106, India
mandeepnahal@gmail.com

M.R. Manikantan
Division of Food Grains & Oilseeds Processing, Central Institute of Post-Harvest Engineering and Technology
Ludhiana-141 004, Punjab
maniciphet@gmail.com

M.S. Meena
Senior Scientist (Agril. Extension)
Division of Socio-Economic Extension & Training, ICAR Research Complex for Eastern Region
Patna-800 014, Bihar
ms101@rediffmail.com

Monika Sharma
Division of Food Grains & Oilseeds Processing, Central Institute of Post Harvest Engineering and Technology
Ludhiana-141 004, Punjab
sharma.monikaft@gmail.com

Mridula D.
Division of Food Grains & Oilseeds Processing, Central Institute of Post Harvest Engineering and Technology
Ludhiana-141 004, Punjab
mridulads4@gmail.com

Neha Babbar
Central Institute of Post-Harvest Engineering and Technology
Ludhiana-141 004, Punjab
nehababbar84@rocketmail.com

P. Barnwal
Division of Food Grains & Oilseeds Processing, Central Institute of Post Harvest Engineering & Technology
Ludhiana-141 004, Punjab
pbarnwal@rediffmail.com

Pranita Jaiswal
Division of Agriculture Structure & Environment Control, Central Institute of Post Harvest Engineering & Technology
Ludhiana-141 004, Punjab
pranitajaiswal@gmail.com

Prasoon Verma
Agricultural Engineering Section
Indian Institute of Pulses Research
Kanpur-208 024, U.P.
prasoonverma@rediffmail.com

R.K. Vishwakarma
Division of Horticulture Crop Processing
Central Institute of Post Harvest
Engineering & Technology
Abohar-152116, Punjab
vishwakarma_raj@sify.com

R.K. Gupta
Division of Horticulture Crop Processing
Central Institute of Post Harvest
Engineering & Technology
Abohar-152 116, Punjab
rkguptaciphet@gmail.com

R.T. Patil
Central Institute of Post Harvest
Engineering & Technology
Ludhiana-141 004, Punjab
rtpatil@sify.com, ramabhau@yahoo.com

S.N. Bhowmik
Division of Food Grains & Oilseeds
Processing, Central Institute of Post
Harvest Engineering & Technology
Ludhiana-141 004, Punjab
snb70@yahoo.co.in

S.N. Jha
Division of Agriculture Structure &
Environment Control, Central Institute of
Post Harvest Engineering & Technology
Ludhiana-141 004, Punjab
snjha_ciphet@yahoo.co.in

S.K. Nanda
Project Coordinator
AICRP on Post Harvest Technology
Central Institute of Post Harvest
Engineering & Technology
Ludhiana 141 004, Punjab
sk_nanda4578@sify.com

Sangeeta Chopra
Division of Transfer of Technology
Central Institute of Post Harvest
Engineering & Technology
Ludhiana-141 004, Punjab
dhingra.sangeeta@gmail.com

Tejinder Singh
Department of Food Science and
Technology, Punjab Agricultural University
Ludhiana-141 004, Punjab
tejinder108@yahoo.co.in

Uppal Singh
Central Institute of Post Harvest
Engineering and Technology
Ludhiana-141 004, Punjab
uppal_ciphet@yahoo.com

U.S. Shivhare
University Department of Chemical
Engineering & Technology
Panjab University, Chandigarh-160 014, U.T.

V.E. Nambi
Division of Horticulture Crop Processing
Central Institute of Post Harvest
Engineering & Technology
Abohar-152 116, Punjab
eyarkainambi@yahoo.co.in

Vidyadhar Kawalkar
Ramkrishna Mission Ashram Hospital
Bhopal 462 039, M.P.
drkawalkar@gmail.com

Vinod K. Bhargav
Division of Agriculture Structure &
Environment Control
Central Institute of Post Harvest
Engineering & Technology
Ludhiana-141 004, Punjab
vkbciphet@gmail.com

□□□

Contents

1

Overview on Emerging Food Processing Technologies

R.T. PATIL

The technological development are needed to handle the food gently and safely during primary, secondary and tertiary processing so that losses are minimized as well as safe, nutritious and easy to consume food is made available to the consumers. The advancement in the basic sciences have made possible use of higher and emerging technologies through the basic unit operations like cleaning, grading, sorting, pasteurization, blanching, sterilization, cooking, packaging, safe transport and processing for intermediate or value added products. The processes adopted conventionally for these operations were either simple, mechanical or thermal, however they have progressed further to achieve the same effect of cooking or precooking through non thermal and semi thermal operations so that acceptable safe product with greater nutrition and fresh like taste is obtained. For necessity to provide - in addition to gentle processing and fresh like products and nutritional and sensory benefits, it is necessary to integrate the food chain with newer advanced processes which have been explained below.

Pulsed electric fields (PEF)

When exposed to high electric field pulses, cell membranes develop pores which may be permanent or temporary, depending on the intensity and treatment conditions. Pore formation increases membrane permeability which results in the loss of cell content or intrusion of surrounding media. Low intensity treatment has the potential to induce stress reactions in plant cells resulting in the promotion of a defense mechanism by increased production of secondary metabolites. An irreversible perforation of the cell membrane reduces its barrier effect permanently and causes cell death. PEF can be applied in fruits and vegetables processing while pressing, extraction or drying for them to be more effective. In meat, it can be used for brining and pickling for shelf life extension, as the mass transport and micro-diffusion could be enhanced to make processing more effective. It can also be used in pasteurisation of liquid foods because the loss of cell vitality caused by electroporation is capable of inactivation of microorganisms

High pressure processing

Minimally processed, fresh tasting, microbiologically safe and additive-free food with an extended shelf life can be successfully produced using high pressure (HP) technology. The pressures currently used range from tens of MPa in common homogenizers or supercritical fluid extractors to up to 350 or 800 MPa in ultra HP homogenizers or HP pasteurization units, respectively. High Pressure units are used for the inactivation of vegetative microorganisms to extend shelf life of the processed food. India produces speciality fruits like Jamun, bael, custard apple, pomegranate etc the products of which are sought after world over due to their nutritional benefits. Hence keeping the freshness even while long distance transportation and storage can be achieved through high pressure processing.

Ultrasound

High power ultrasound treatments have the potential to improve a large range of key processes in food production such as emulsification and homogenization, crystallization, mass transfer processes and heat transfer. Furthermore, food structure can be altered leading to tissue softening or disruption and migration as well as alteration of proteins. Viscosity reduction can contribute to improved characteristics during heat transfer or extrusion processes. In terms of food safety, ultrasound can contribute to higher hygienic standards in food production. The versatile application of ultrasound in food technology and its characteristics allows the improvement of a large variety of processes. Depending on the application, ultrasound can contribute to savings in energy or water consumption as well as in the use of chemicals and other resources.

Ohmic heat processing

In Ohmic heating, an alternating electrical current is passed through a food sample. This results in internal energy generation in foods. This produces an inside-out heating pattern, which is much faster than conventional outside-in heating. Ohmic heating is somewhat similar to microwave heating but with very different frequencies. The advantage of ohmic heating is that it uniformly heats foods with different densities. Pea puree is one of the natural ingredients used in baby food processing. Blanching of puree is as important as the sterilization process to enhance the quality of baby foods. Enzyme inactivation and the colour improvement are also achieved in blanching. Long blanching time during water blanching of vegetable purees adversely affects the quality and yield of the product. It was found that ohmic blanching applied by using 30 V/cm and above voltage gradient inactivated peroxidase enzyme in less time than the water blanching. The ohmic blanching at 50 V/cm gave the shortest critical inactivation time of 54 s with the best colour quality. Ohmic heating can be used for heating liquid foods containing large particulates, such as soups, stews, and fruit slices in syrups and sauces, and heat sensitive liquids. The technology is useful for the treatment of proteinaceous foods, which tend to denature and coagulate when thermally processed. Juices can be treated to inactivate enzymes without affecting the flavor.

Nanotechnology

Nanotechnology derived food packaging materials is the largest category of current nanotechnology applications for the food sector. Food packaging materials with improved mechanical, barrier and antimicrobial properties are of interest, and also nano-sensors for traceability and monitoring of conditions of food during transport and storage. An effort to extend the shelf-life and enhance food quality while reducing packaging waste has encouraged the exploration of new, nanotechnology-based materials. Chitosan, zein, starch etc. are bio-polymers and these are being exploited to develop new materials for packaging and coating applications in food industry.

Irradiation

The technology of ionizing irradiation consists of irradiating the products with electron energy sources, a type of light capable of eliminating the structures of pests, insects, mites, bacteria, and even viruses. At low dose levels (1 kGy or less), most fresh-cut vegetables show little change in appearance, flavor, color, and texture, although some products can lose firmness. The destruction of spoilage organisms due to irradiation increases the shelf life of most fresh and fresh-cut vegetables.

Functional foods from algae

Algae are amongst the simplest organisms that can use light energy to drive the synthesis of biomass. Research on functional foods (such as astaxanthin) and ingredients (such as omega-3 oils and emulsifiers) from algal sources has lead to recent product commercialization, however the future is likely to see further ingredients (flavours, flavour enhancers etc) becoming available from this source.

Microwave heat processing

The use of microwaves in food processing can be extended for killing insects, blanching of fruits and vegetables, drying of fruits, vegetables and dairy products, stabilization of rice bran and wheat flour and pre-treatment of oilseeds for efficient oil extraction. Volumetric heating by microwaves is always fast, effective and eliminates conduction as the main mode of transferring heat to places where it is needed. This and other comparative advantages offer scope for application in the food and fibre industry. The microwave heating was found considerably more effective for inactivating enzymes, reduced indirect heating requirement and water use for green beans, and green peas. The microwave blanching was less severe heat treatment required to inactivate lipoxygenases and hence may result in improved product flavour, colour, texture and nutritive value. In case of microwave cooking of chickpeas and common beans, the solid losses, released in the cooking water, were significantly less. The blanching snow peas in a heat-sealable microwave bag produced a product of equivalent sensory quality and superior nutritional quality compared with conventional blanching methods. Microwave (MW)-related (MW-assisted or MW-enhanced) combination drying is a rapid dehydration technique that can be applied to specific foods, particularly to fruits and vegetables. The advantages of MW-related combination drying include the following: shorter drying time, improved product quality, and flexibility in producing a wide variety of dried products.

Extrusion processing

Extrusion refers to the forming of products to the desired shape and size by forcing the material through a die opening under pressure. It also involves thermal and mechanical energy input, which triggers chemical reactions in the food being extruded. Extrusion has been used extensively in the food industry to process a variety of raw materials into breakfast cereals, ready to eat expanded snacks mainly from cereals like maize and wheat, pastas, texturized vegetable protein, flat bread, meat products, and pet foods. Advantages of extrusion include continuous processing, enhanced hygiene, lower energy needs, and enhanced control. The combination of drum drying and twin-screw extrusion allows for the production of 100% fruit products.

Some of prioritized applications of emerging technologies suitable for India

1. Utilization of smart and bio-degradable packaging technologies for foods

- Development and optimization of controlled release mechanism technology using volatiles of plant origin for shelf-life extension and microbiological safety of food through incorporation in the packaging materials.
- Development of food specific packaging materials having insect-repelling and water absorbing properties for their longer and safe shelf-life.
- Retention of volatile and flavour in packaged food materials through smart packaging interventions.
- Utilization of time-temperature indicators and colour code schemes to access safe shelf-life of packaged food materials.
- Application of poly-lactic acid based bio-degradable clamshell containers and films/coatings for their utilisation in packaging of food materials.
- Development of Bio-polymer Nano-composites for Packaging of Food Commodities.

2. Development non thermal protocols using pulse electric field (PEF) / HPP for food processing and preservation

- Enhancement of Shelf Life of Milk and RTS Beverages.
- Enhancement of diffusivity of enzymes (A. fumigatus, xylanase, protease) and effectiveness of modified enzymes on dehulling and cooking quality of pulses.
- Development of non thermal extraction processes for obtaining bioactive mixtures/ compounds from variety of by products and process wastes.

3. Nano technological applications for development of controlled release and functional foods

- Micro-Encapsulation and Nano-Encapsulation of Herbal, Medicinal Extracts and Food Ingredients.
- Nano-composites for food packaging with improved barrier or antimicrobial properties.
- Nano-emulsions/delivery systems for increased absorption of nutraceuticals and health supplements.
- Nano-sensors for traceability and monitoring conditions of foods during transport and storage.

- Modification of food structures for improved solubilization and bioavailability of minerals and vitamins using nano-technologies (hyper homogenization, molecular mixing, and multi-emulsion technology).

4. Application of supercritical fluid extraction in food processing
   - Extraction of antioxidants, herbal medicines, and food colors from botanicals, decaffeination of coffee and tea, reduction of fat and cholesterol levels in food products.
   - Extraction of active ingredients (lycopene, β-carotene, capsaicin, gingerol, caffeine, tocopherol, terpenoids, phenolics, pyrethrum, azadirachtins, etc), flavoring materials (aromas and essential oils) from natural products
   - Food processing generates a large variety of byproducts and waste including bran, germ, husk, spent barley/adjuncts (cereals based); and, peel, rind, seeds, pomace (fruits and vegetables based). These byproducts contain valuable antioxidants, vitamins rich essential oils, pigments, flavor compounds, etc. The SCFE can be applied on these byproducts in order to reduce their environment impact and to produce value added products.

5. Use of irradiation technologies for food processing and preservation
   - Standardize the irradiation protocols for retention of medicinal properties and shelf-life enhancement of medicinal extracts
   - Develop irradiation technologies suitable for sprout inhibition, delay in ripening and senescence, insect disinfestation, psteurisation/sterilisation of processed & raw foods, reduction in antinutritional factors and for improved keeping quality of food grains and flours.

❑❑❑

2

Maize Processing Technologies

P. BARNWAL

Maize is one of the most versatile crop and grown all over the world (almost in more than 100 countries). It is also called as "corn" by some researchers. It is one of the major crops in America, Africa and Asia. Many diverse types of maize are found in Central America and Mexico and these are considered to be the primary centre of its origin. In most South-East Asian (SEA) countries, maize is becoming an important feed crop for poultry and swine. In India, its production probably occurred in the beginning of the seventeenth century *i.e.* during the early days of the East India Company.

In Indian agriculture, maize comes next to rice, wheat and jowar for area and production. From 1950-51 to 2008-09, there is significant increase in area, production and yield of maize from 3.16 to 8.17 million hectares, 1.73 to 19.73 million tonnes and 547 to 2414, respectively (Anon 2009). Though maize is grown almost in all states of the country, the major maize growing states are: Andhra Pradesh, Karnataka, Rajasthan, Bihar, Maharashtra, Tamil Nadu, Uttar Pradesh, Madhya Pradesh, Gujarat, Himachal Pradesh, Jammu & Kashmir, Punjab, West Bengal and Jharkhand etc. for year 2008-09 (Anon

2009). The maize productivity/yield (kg/hectare) is higher in Andhra Pradesh, Tamil Nadu, West Bengal, Punjab, Karnataka, Bihar, Maharashtra, Himachal Pradesh, Jammu & Kashmir, Rajasthan, Uttar Pradesh, Gujarat, Jharkhand and Madhya Pradesh etc. while area under maize (million hectares) is higher in Karnataka, Rajasthan, Andhra Pradesh, Madhya Pradesh, Uttar Pradesh, Maharashtra, Bihar, Gujarat, Jammu & Kashmir, Himachal Pradesh, Tamil Nadu, Jharkhand, Punjab and West Bengal.

The production of maize from 1998-99 to 2009-10 is shown in Table 1 (Anon 2009). It is clear that maize production in Kharif season is more than that in Rabi season. The total maize production has increased from 11.15 million tonnes in 1998-99 to 19.73 million tonnes in 2008-09.

Table 1: Production of maize in India from year 1998-99 to 2009-10

Year	Production (Million tonnes)		
	Kharif	Rabi	Total
1998-1999	9.54	1.61	11.15
1999-2000	9.71	1.80	11.51
2000-2001	10.22	1.82	12.04
2001-2002	11.25	1.91	13.16
2002-2003	9.27	1.88	11.15
2003-2004	12.73	2.25	14.98
2004-2005	11.48	2.70	14.18
2005-2006	12.16	2.55	14.71
2006-2007	11.56	3.54	15.10
2007-2008	15.11	3.85	18.96
2008-2009	14.12	5.61	19.73
2009-2010*	12.00	4.68	16.68

*Advance estimates

Indian maize has diversified color kernels e.g. white, red, purple, brown or sometimes multi-colored kernels. Generally, maize is more resistant to drought than either rice or wheat, yields per hectare are good, the crop is free of bird predation, and matures relatively quickly. It has complete enclosure of cobs inside leaves "husks" which ensures easy harvesting.

Types / classification of maize

The genus Zea is a monotypic belongs to tribe Maydeae of family Poaceae (Graminae). On the basis of endosperm characteristics (i.e. kernel size, shape and composition etc.), E. Lewis Sturtevant classified the maize into seven groups and these are described as below:

(i) Zea mays var. indurata or 'Flint corn'

It is predominant in Europe, Asia, Central and South America. Its kernel is soft and starchy in the center and completely enclosed by a very hard outer layer. The kernels are usually rounded but are some times short and flat. Its colour may be white or yellow. Most of the commercial maize, grown in India, belongs to flint corn.

(ii) Zea mays var. indentata or 'Dent corn'

It is referred as dent corn because of dent depression on the top of the kernel. The colour of kernel may be yellow or white. The kernel has hard outer starch and soft inner starch. The hard starch extends on the sides and the soft starch is in the center and extends to the top of the kernels. In the drying and shrinking of the soft starch, various forms and degrees of indentation result. This is the most common type of maize grown in U.S.A. and stores well.

(iii) Zea mays var. everta or 'Pop corn'

It possesses exceptional popping qualities. Size of the kernels is small but the endosperm is hard. When these are heated, the pressure built up suddenly within the kernels results in an explosion and the grain is turned inside out resulting in pops. It has higher percentage of hard starch than flint corn.

(iv) Zea mays var. saccarata or 'Sweet corn'

The endosperm is composed of sweetish starch and characterized by a translucent, horny appearance when immature. After maturity or on drying, these cells collapse, making the kernels / grains shrivelled or wrinkled. It has sweeter taste than other corns. The crop is mainly grown in northern half of the USA. The cobs are picked up green for canning and table purpose.

(v) Zea mays var. amylacea or 'Soft corn' or 'Flour corn'

It resembles to the flint corn in appearance and ear characteristics. Its kernels are of all colours, but white and blue are the most common. The grains are entirely composed of soft starch alone and have a little or not dent. It is one of the oldest types of maize grown in USA and South Africa.

(vi) Zea mays var. tunicata or 'Pod corn'

The pod corns are characterized by having each kernel enclosed within a pod or husk in the ear which in turn enclosed in husk like other types of corn. It is a primitive type of corn, having less importance i.e. it may be oldest domesticated maize but it is not grown commercially at present.

(vii) Zea mays var. ceratina kulesh or 'Waxy Corn'

Its kernels look to have waxy appearance with gummy starch in them because of amylopectin. The kernel, when cut or broken, gives a waxy appearance. It produces the starch similar to tapioca starch for making adhesive for articles. Its origin is supposed to be in China but many waxy hybrids, developed in USA, are producing starch similar to that of tapioca and are grown commercially.

Besides above types or classifications, the unfertilized young ear obtained from sweet corn or other corn is known as baby corn. It is rich in minerals and vitamins and can be harvested within 45-50 days after sowing for marketing. The ears are removed within two days of tasselling and utilized as salad and vegetable or for soup making. It can be preserved in 1.5 percent saline solution and can be used at later period. Indian maize is mainly characterized as dent corn which is useful for starch processing by method of wet milling.

Botanical, structural and nutritional description

The botanical name of Maize is *Zea mays* Linn. It is an annual plant, which usually grows to a height of one to four meters or more in some cases (up to eight meters) (Fig.1). It belongs to family Gramineae and Genus Zea. The length of maize cob may be from 8 to 42 centimetres. In general, there are 300 to 1000 grains (seeds) in one maize cob (Fig. 2). It is normally a monoccious plant having two types of inflorescence, the female inflorescence which develops into an ear (Cob) and male inflorescence which contains the male flowers. The male flowers are borne in a cluster (tassel) on top of end of the stem as a terminal panicle, while the female flowers are borne inside the young cobs which spring from the nodes on the stem usually located about midway on the stalk.

The various parts of maize kernel are shown in Fig. 3. The maize kernels consist of three main parts: (i) pericarp (hull or bran), (ii) germ, and (iii) endosperm. The approximate share of various parts in maize kernel and the fat content throughout the maize kernel is given in Table 2 and 3, respectively (Chakraverty *et al.,* 2003).

Fig. 1 : Maize plant

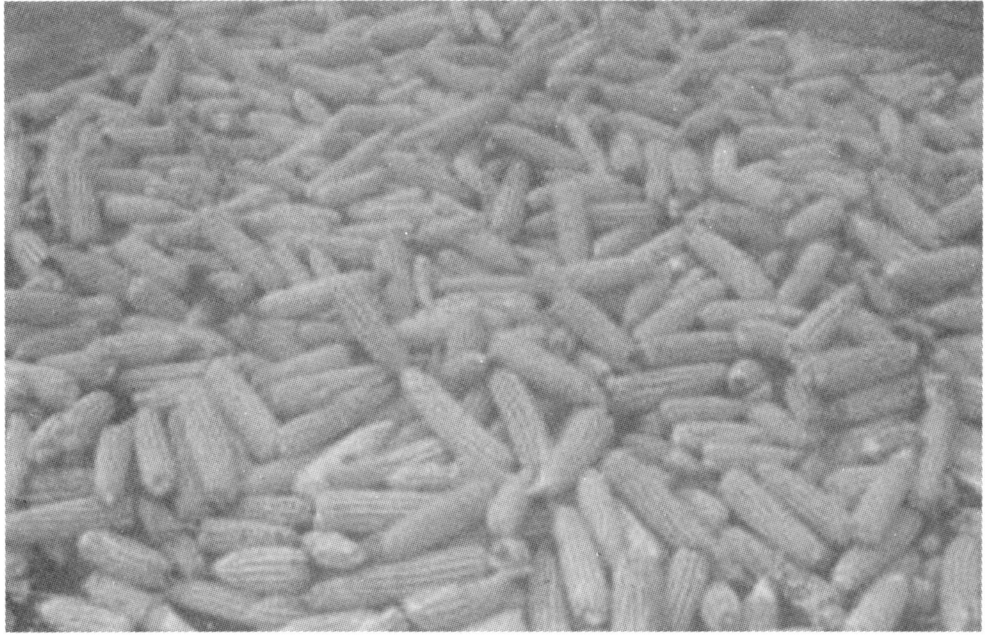

Fig. 2 : Maize cobs

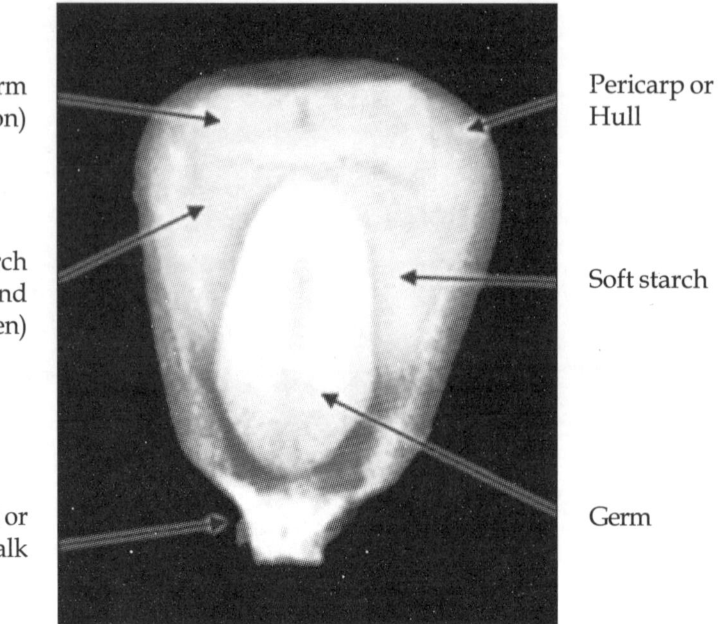

Fig. 3 : Maize kernel and its parts

Thus, maize germ contains a high percentage of fat (up to 30% on dry basis) and germ itself shares about 10-12% of whole maize kernel (Table 2-3). Therefore, if germ fragments remain in the products made from milling, it will adversely affect the shelf life of the processed products.

Table 2: Composition of maize kernel

S. No.	Constituents	% of kernel
1	Endosperm	80-82
2	Germ	10-12
3	Bran	5–6
4	Tip cap	1–1.5

Table 3: Fat content throughout the maize kernel (approx.)

S. No.	Component	%
1	Whole maize kernel	4–6
2	Endosperm	0.5–1.0
3	Germ	25–30
4	Bran /hull	5-6

Maize is one of the nutritious crops and its nutritional composition is shown in Table 4.

Table 4 : Proximate analysis of maize (per 100 g)

S. No.	Component	Value
1	Moisture (%)	14.90
2	Protein (g)	11.10
3	Fat (g)	3.60
4	Mineral (g)	1.50
5	Fiber (g)	2.70
6	Carbohydrate (g)	66.20
7	Calorific value / Energy (kcal)	342.00
8	Calcium (mg)	10.00
9	Phosphorus (mg)	348.00
10	Iron (mg)	2.00
11	Thiamine (mg)	0.42
12	Riboflavin (mg)	0.10

Source: Nutritional value of Indian Foods, NIN Hyderabad, A.P.; Rajoo (1998)

Maize processing

Maize processing is essential to fractionate it into various food and industrial products. It can be used to produce more nutritious, palatable and popular food products from maize. Traditionally, maize is processed into maize flour in mills for making *chapatti* or bread. The immature maize cobs are roasted and eaten all over the country. It is an important raw material for animal and poultry feed and corn flakes manufacturing units. Traditional pounding technique is used for removal of pericarp of the maize. In India, maize grains are dry-milled in electrically or hand operated stone mills (Khetarpaul *et al.,* 2003). The germ contains a major share of fat (up to 30% on dry basis) and germ itself shares 10-12% of whole maize kernel. The maize flour thus produced is screened through domestic sieves to remove 3–4% bran and the flour is high in oil. Thus, the shelf life of maize processed products (maize flour etc.) reduces as rancidity development in maize oil is much faster *e.g.* if germ fragments remain in the products made from processing, it will adversely affect the shelf life of the processed products. Therefore, presence of germ fragments in the maize based processed products should be avoided for its better shelf life (Chakraverty *et al.,* 2003; Khetarpaul *et al.,* 2003).

From Fig. 3, it is clear that germ is very difficult to separate as it is attached firmly inside the maize kernel. Since the degermination process is very complicated, the removal of germ is not critical for situations where maize grits/meals are going to be consumed within a short period of time from the time it is produced/processed.

As mentioned above, maize germ which is relatively high in oil, needs to be removed completely for longer shelf life of maize based processed products as the oil undergoes degradation during storage producing free fatty acids and oxidation products responsible for the bitter taste and rancid smell.

There are two methods of maize processing viz. traditional or non-degerming process and non-traditional or degerming process (Fig.4). Degerming is usually done to separate the germ and the bran from the maize kernel so that the resulting endosperm could produce highly starch-concentrated refined flour after grinding. Thus, degerming involves the separation of bran and germ to yield pure starchy endosperm from maize kernel. The separation of these parts with maximum yields and purity of prime products is the main objective of the processing using a minimum amount of energy.

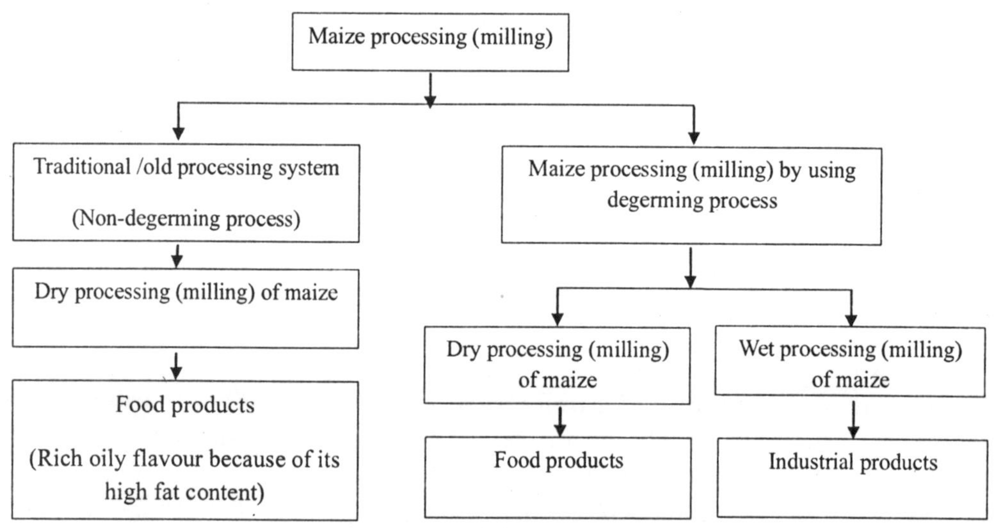

Fig. 4: Types of maize processing

Depending on the end product desired, there are two methods of maize processing *i.e.* dry and wet processing/milling methods.

In general, dry milling is used for producing various food products (flaking grits, flour, meal etc.) while wet milling is used for production of industrial products (starch, corn oil, animal feeds etc. or germ, fiber, protein and starch) (Rajoo, 1998). However, in both dry and wet milling methods, the germ is separated from the grain in order to extract and recover the oil. The degermed maize products, obtained by dry/wet milling methods, have longer shelf life than non-degermed maize processed products.

Dry milling or tempering degerming (T.D.) process

In case of dry milling of maize grains (with less than 17 percent moisture), the desired prime product is the large sized flaking grits because large grits can be reduced to smaller size but not vice-versa. In it, the maize should be broken at particular moisture content, so that it should release the germ without breakage i.e. with whole germ. For proper separation of germ and bran, prior to degerming, the moisture content of the maize kernels is increased to toughen the bran, which makes it easier to remove together with the germ from the endosperm. If maize kernels have cracks, it will break readily and do not respond properly to degerming and thus low yields of large grits and poorer degerming. The stress cracked kernels decreased with increase in water temperature.

Dry milling produces grits, maize flour, semolina and minimum amount of maize meal etc. The grits is the main product of dry milling process, which may be used as porridge by boiling. The processing units use grits for manufacture of products like ready-to-eat snacks (maize flakes), wall paper paste and manufacture of glucose by direct hydrolysis.

The various steps in dry milling of maize are as follows:

a) Dry cleaning, and if necessary, wet cleaning

b) Conditioning/Tempering

c) Degerming (separation of hull, germ and tip cap from the endosperm in the degerminator)

d) Drying and cooling of degermer product

e) Grading and grinding

f) Sifting and classifying

g) Purifying

h) Further drying of product, if necessary

i) Processing of germ fraction for recovery of crude corn oil and

j) Packaging and transportation of products

The most common type of dry milling of maize/corn is Tempering Degerming (T.D.) process (Saunders *et al.*, 2001). In it, the corn is first cleaned to remove dirt and debris and to separate fine and broken kernels from whole corn (Fig. 5). The corn is tempered by increasing moisture, and the hull (pericarp), germ, and tip cap are removed, leaving the endosperm. This endosperm is then converted into flours, meals, and grits through a series of mills, sifters, and gravity tables. Corn germ is the raw material for corn oil and bran is generally an animal feed item but with further processing, used in some specialty foods as a source of fiber.

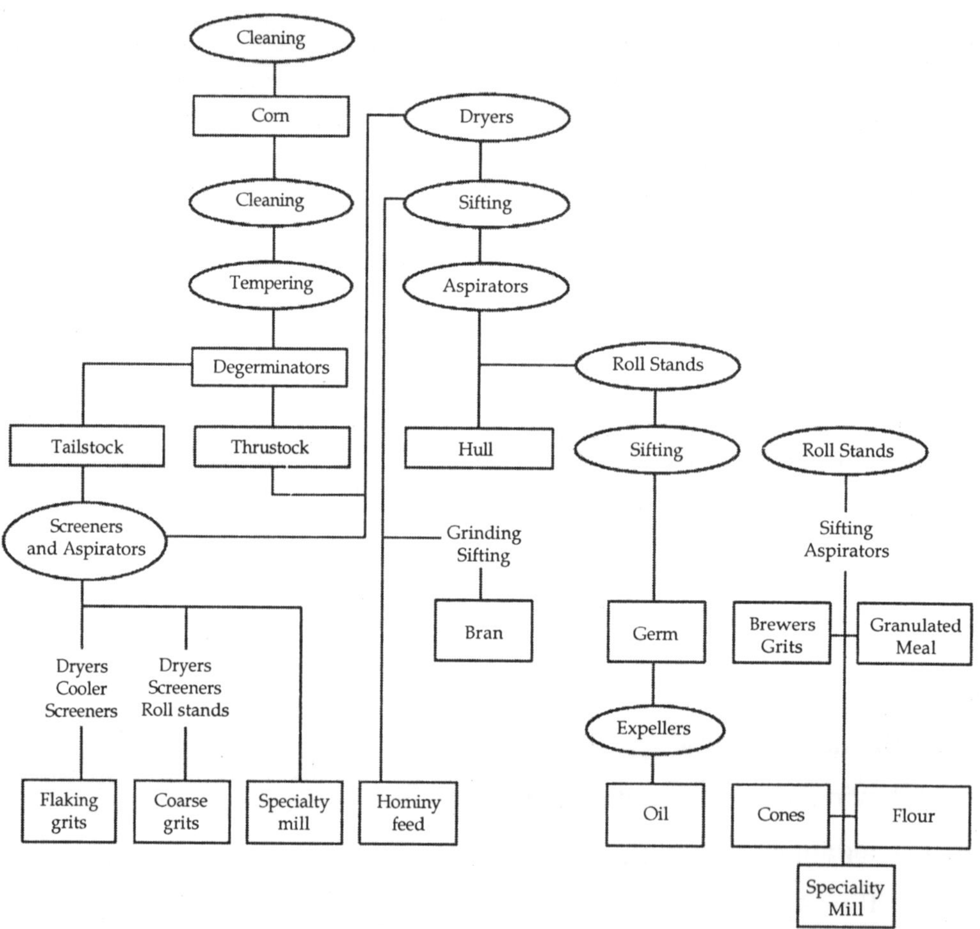

Fig. 5 : Flow diagram of maize (corn) dry milling process (Saunders *et al.*, 2001)

Another similar flow diagram for tempering-degerming process (also known as TD process or partially wet degerming process), is shown in Fig. 6. In this process, water is added to maize grain to raise its moisture content to 20 to 23 percent. After this, it is tempered to hydrate the germ and the germ becomes more resilient during the degermination step. The kernels are fractionated using a degerminator, which breaks the kernel into pericarp, endosperm and germ components, using a combination of grinding and abrasion. Germ and pericarp components are separated from endosperm products using sieving, aspiration and gravity table methods. Endosperm particles are classified by particle size and contain less than 1% fat for its storage stability. The largest endosperm particles, i.e. flaking grits, are premium products typically used in breakfast cereals. Smaller particles are ingredients in maize-based snack foods. Pericarp and endosperm components having high fat content (more than about 1.5% crude fat) are combined to form the co-product hominy feed (protein 12% db, fat 3% db), which is used in animal diets. Germ from dry milling (18 to 20% fat) is sold separately as an animal dietary ingredient or combined with the hominy feed.

Wet milling

By wet milling method, maize is generally processed to manufacture maize starch worldwide. The average recovery of various products and co-products of maize during the wet milling are: starch (60-62%), gluten (8-9%), germ (6-7%) and husk (22-24%). The co-products of starch manufacture include maize oil, maize steep liquor, gluten etc. as the important value added products.

The Flow diagram of maize (corn) wet milling process is shown in Fig. 7. It involves the semi-continuous steeping of maize/corn kernels for 24 to 36 hour at 52°C using a combination of sulfur dioxide (SO_2) and *Lactobacillus* fermentation. Kernels are immersed during steeping to allow hydration in the presence of *Lactobacillus* bacteria, followed by diffusion of sulfur dioxide into the kernel. Steeping prepares kernels for subsequent milling and separation steps to recover germ, fiber, protein and starch. Following a wet grinding step, germ, fiber, protein and starch components are recovered using size and density separation methods. Starch is recovered as high purity (>99.5% db) and used to make modified starches, syrups and ethanol. The starch, the major constituent of the corn kernel, is used in its native form or after chemical or enzymatic modification, in foods and industrial products. The starch is also converted into glucose or fructose for use as food sweeteners. Glucose can be fermented in to ethanol for fuel or beverages or into many other chemicals. Corn gluten feed (protein 23% db; yield 24%) is made from steep water solids and fiber and corn gluten meal (protein 67% db, yield 5%) is produced from the endosperm protein. Germ from wet milling (35 to 40% fat) is used for oil extraction.

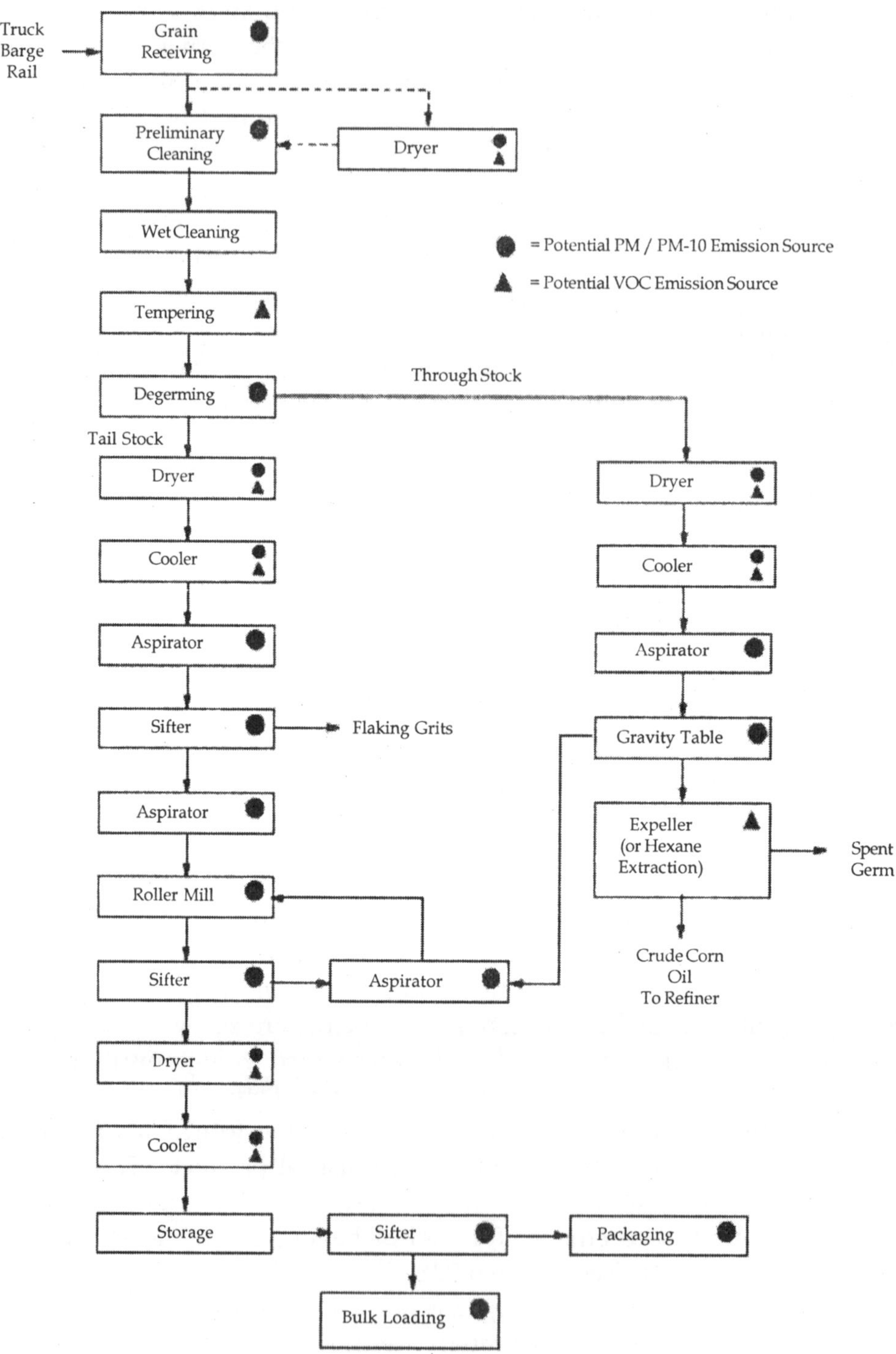

Fig. 6 : Another flow diagram for dry milling of maize (with degerming)

The principal products resulting from the wet-milling process are cornstarch, high-fructose corn syrup (a sweetener), and ethanol from fermentation (Fig. 7). Overall, co-products from the wet milling process contain higher concentrations of fat, starch and protein than those from dry milling.

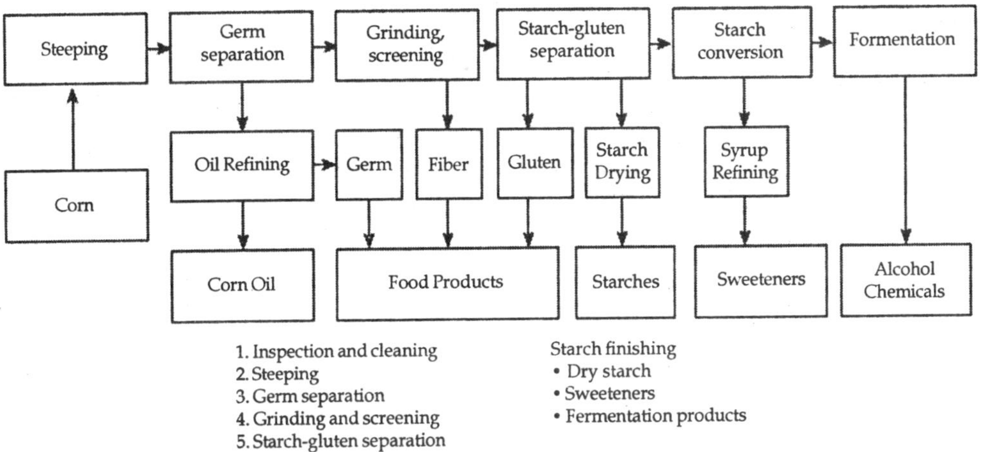

Fig. 7 : Flow diagram of maize (corn) wet milling process (Saunders *et al.*, 2001)

Intermittent milling and dynamic steeping (IMDS) process

The intermittent milling and dynamic steeping (IMDS) process is an alternative method for wet milling of maize (Lopes-Filho and Brandemarte, 2002). Fig. 8 presents a flow diagram of the IMDS method/process. In it, the steeping stage can be reduced to 5 hour by soaking maize in water at 60°C for 2 hour and cracking the kernels to remove solution components diffusional barriers with minimum germ damage.

The various steps of IMDS process involves initial moisture determination of the grains, kernels soaking in water at 60°C for 2 hour, kernel cracking by a roller mill, and dynamic steeping into a solution with different concentrations of lactic acid and SO_2 according to pre-defined treatments. For homogeneous cracking of the kernels, three times are needed to pass them through the roller mill. For the first two times, the gap between the cylinders is 3.2mm while for the last time, it is 3.0mm. Dynamic steeping is performed at 52°C in different solution conditions of lactic acid and SO_2 concentrations. The times of preliminary soaking plus dynamic steeping resulted in a total of 8 hour for the process. The presence of lactic acid in the steeping solution decreased the amount of fibre in germ fraction. It has been found that the smallest difference is 5.5% between IMDS (0.2% SO_2, 0.55% lactic acid) and the conventional 36 hour steeping process. An average of 1.3% of kernels remained un-cracked after IMDS process (Lopes-Filho and Brandemarte, 2002).

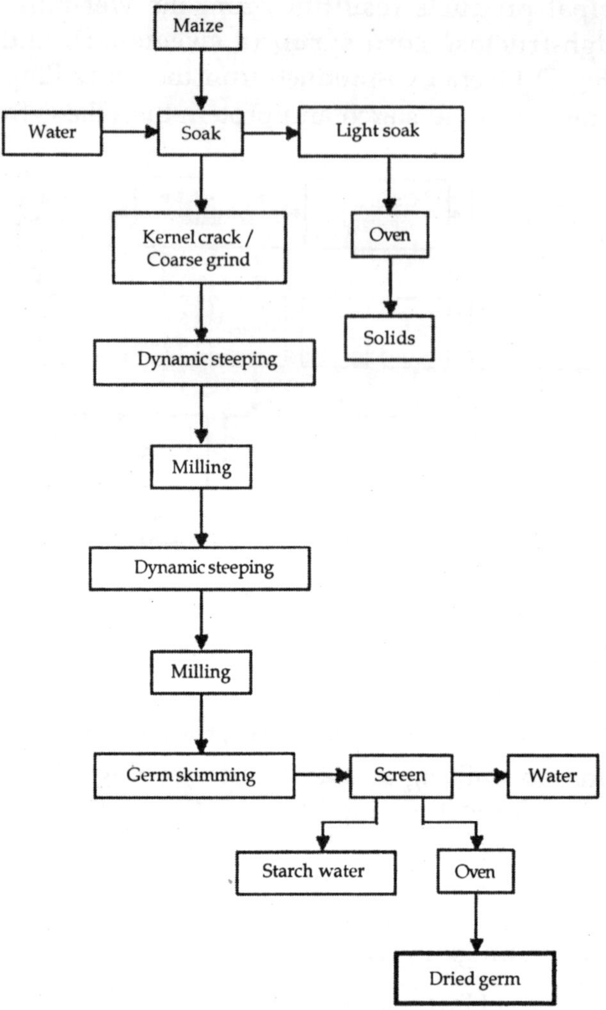

Fig. 8 : Intermittent milling and dynamic steeping (IMDS) process for maize

There are some other methods for maize processing also i.e. alkaline processing and dry-grind ethanol process etc. (Murthy *et al.,* 2009; Dowd, 2003; Noureddini *et al.,* 2009).

Technologies for maize processing

In India, maize is processed by industries using large scale dry/wet milling methods using specially designed and imported degerming equipments like Beall degermer and Entoleter machines. However, some efforts have been made by research and development organizations on the various post harvest machines for processing of maize crop. The various degerming machines are given as under:

1. Beall Degerminator
2. Satake maize degermer
3. Entoleter pin mill
4. Corn Dehuller
5. Roller Mill
6. Impact devices for degermination of maize without adding water i.e. without tempering etc.

Thus, in industries, larger capacity imported degerminating units viz. Beall Degerminator (1.27-3.0 tonne/hr), Satake maize degermer (5.0 tonne/h) and Entoleter pin mill (13.97 tonne/h) etc. are used. These are of higher capacity as well as extremely costly and hence not suitable for small / medium scale entrepreneurs as well as producers at production catchments/areas.

The maize processing machines/mills may be used to produce various milled products i.e. big and medium grits, small grits, bran, meal and germ that could be utilized by subsidiary units for its further value addition. The big and medium grits, having low-fat, low fibre content, may be used to produce maize flakes having better shelf-life. The small grits can be utilized in the brewery industry, and also may be extruded with different formulations to produce or make a variety of processed products. Maize germ may be used for extraction of oil for edible purposes. The deoiled germ and meal, obtained from the maize mill, may be used in the manufacture of cattle feed. Maize bran may be utilized in bakery industries as a source of dietary fibre. Thus, maize processing is suitable for making various processed products for food and industrial purposes.

Utilization of maize

Maize is one of the most diversified crops of all cereal grains in terms of its applications. It may be used at any stage of its growth. It has many uses e.g. human food and nutrition products (*flour, roti/chapatti,* corn flakes, vegetable (baby corn), pop corn etc.), industrial products (starch, oil, and other value-added products) and livestock feed (beef, dairy and poultry feed etc.). In industrialised countries, maize is largely utilized as livestock feed and as raw material for industrial products. In low income countries, it is mainly used for human consumption.

Maize is mainly used as food and feed purpose in India. For snack food purposes, maize is used in the form of corn flakes, popcorn, sweet corn etc. Maize stalks may be used as dry season feed for farm animals and also as soil mulch where it is in abundance.

Maize oil is widely used as a cooking medium and for manufacturing of hydrogenated oil. The oil has the quality of reducing cholesterol in the human blood like sunflower oil. Maize acts as a source in the manufacture of starch, syrup, dextrose, oil, gelatin, lactic acid etc. The maize flour is used as a thickening agent in the preparation of many edibles like soups, sauces and custard powder. Maize syrup is used as an agent in confectionery units. Maize sugar (dextrose), used in pharmaceutical formulations, is a sweetening agent in soft drinks etc. Maize gel on account of its moisture retention character is used as a bonding agent for ice-cream cones, as a dry dusting agent for baking products. The maize may be used extensively for livestock feeds viz., cattle, poultry and piggery in the form of seeds/grains and fodder. The green fodder may be fed to milch cattle to enhance the milk production to a considerable extent.

The maize cob has many important agricultural and industrial uses. Approximately it forms 15 to 18% of the total ear weight and contains 35% cellulose, 40% pentose and 15% lignins. Their uses in agriculture includes as a litter for poultry and as soil conditioner. Based on the physical properties of the cob, the ground/powder of maize cob may be used as fillers for explosives in the manufacture of plastics, glues, adhesives, reyon, resin, vinegar and artificial leather and as diluent and carrier in the formulation of insecticides and pesticides, based on the chemical properties, the processed cobs find their use in the manufacture of furfurol, fermentable sugars, solvents, liquid fuels, charcoal gas and other chemicals, and also in the manufacture of pulp, paper and hard boards.

Maize has diverse end-uses. It is a good source of carbohydrates and can be used as food at various stages of development and every part of the plant has commercial value. The maize starch has polymeric uses for food (jellies; sauce thickener; salad dressing; instant /frozen/canned foods etc.) as well as non-food (textiles, paper, metal refining and packaging film etc.) uses and the biodegradability is an added advantage/ benefit.

Therefore, maize processing technologies (dry/wet milling methods) may be used for processing and better utilization of maize. The good quality maize procured from the farmers/ growers may be processed for its value addition, under one roof, into various products e.g. maize flakes, ready to eat (RTE) flakes, extruded maize products, maize roti flour, maize germ oil extraction, feed formulations, and bakery products. With the diversification of products from maize, it may boost the production of good quality maize in the country directly benefiting the grower/farmer. With minor modifications, the technology and machinery used for industrial processing of maize may be used for value addition to other grains also to ensure the working of the mills all round the year.

References

Anon (2009). Directorate of Economics & Statistics (DES), Department of Agriculture and Cooperation, Ministry of Agriculture, Government of India. http://dacnet.nic.in/eands/

Chakraverty A, Mujumdar AS, Raghavan GSV, Ramaswamy HS, (2003). Handbook of Postharvest Technology : Cereals, Fruits, Vegetables, Tea and Spices, Marcel Dekker, Inc., New York, USA.

Dowd MK, (2003). Improvements to laboratory-scale maize wet-milling Procedures. Industrial Crops and Products, 18(1): 67-76.

Khetarpaul N, Grewal RB, Jood S, Singh U, (2003). Emerging Trends in Post-Harvest Processing & Utilization of Plant Foods, Agrotech Publishing Academy, Udaipur, India.

Lopes-Filho JF, Brandemarte EA, (2002). Maize Germ Recovery and Quality using the Intermittent Milling and Dynamic Steeping Process, Biosystems Engineering 83 (2): 185–190.

Murthy GS, Sall ED, Metz SG, Foster G, Singh V, (2009). Evaluation of a dry corn fractionation process for ethanol production with different hybrids. Industrial Crops and Products, 29(1): 67-72.

Noureddini, H, Malik M, Byun J, Ankeny AJ, (2009). Distribution of phosphorus compounds in corn processing. Bioresource Technology, 100(2): 731-736.

Rajoo RK, (1998). Maize: The Golden Grain of Himachal Pradesh, Kalyani Publishers, Ludhiana, India.

Saunders DS, Meredith FI, Voss KA, (2001). Control of Fumonisin: Effects of Processing, Environmental Health Perspectives, 109(2): 333-336.

Websites

http://agricoop.nic.in; www.agmarknet.nic.in ; www.apeda.com ; www.nabard.org ; www.codexalimentarius.net ; http://dacnet.nic.in/eands/ ; www.fao.org ; http://faostat.fao.org/ ; www.fciweb.nic.in ; www.icar.org.in ; http://www.indiamart.com ; http://www.maizeindia.org/Research/Achievements.htm; http://naasindia.org; www.ncdc.nic.in.

□□□

3

Newer Dimensions in Pulse Processing

PRASOON VERMA

Pulses are essential component as sources of protein to vast vegetarian population of our country. If animal proteins are considered as complete protein, combination of cereal and pulses provides essential amino acids to a vegetarian diet. Proteins present in pulses are complementary to the proteins available in wheat and rice. Pulses are rich sources of methionine whereas cereals have more of lysine. That is why majority of vegetarian Indian cuisines include appropriate proportion of cereals and pulses. Being the cheapest source of proteins, pulses also play an important role in food security of the nation.

Despite being largest producer of pulses, per capita availability of pulses in India is much below the recommended level of 80g, as suggested by World Health Organization and FAO. Considering other sources of proteins, ICMR has lowered the pulse requirement to 47g/day/person, but the pulse availability scenario is still meager and at present it is less than 30g. This is due to almost near stagnant production around 15 million tonnes (mt) for over last four decades and increase in population during the period. To meet domestic demand, India is importing about 1.5mt of pulses annually. This demand and supply gap can either be filled by raising production or by reducing post harvest losses, which are to the tune of 20-30%. Both of these approaches are equally important for improving per capita availability of pulses. Efforts are being made to develop input responsive, high yielding pulse varieties through crop improvement, but with moderate success. A brief account of post harvest losses indicate loss of 25%, and considering 5% grains retained as seed, about 30% of pulse produced goes out of food chain. Storage (5-10%) and milling (15-20%) are the major contributors to the post harvest losses. It is needless to say that curbing losses will make more pulse available for human consumption.

An overview of post harvest chain is described in the following table :

Total Pulse Produced	**100 units**
Threshing, transport and primary processing	2.00
Storage before milling	7.50
Seed purpose	9.05
Consumed as whole grain	13.57
Pulse available for milling	**67.88 units**
Milling losses	10.18
Husk separated	8.83
Pulse available as *dal*	48.87
Pulse available for human consumption	**13.57 + 48.87 = 62.44**

Pulse milling

In India, pulses are mostly consumed in the form of *dal*, after dehusking and splitting. The process of dehusking and splitting is called milling. Thus, milling is the most common post harvest operation for pulses. Removal of outer seed coat is done to improve appearance, texture, cooking quality, palatability and digestibility of pulses. It has also been found that *dal* is less susceptible to insect or pest attack during storage in comparison to whole grain. In India, pulse milling is performed in about 15000 mills scattered all over the country, making it the third largest food processing industry after rice and wheat milling.

The major steps involve in pulse milling are : (i) pre-milling treatment to loosen the seed coat and (ii) removal of husk and (iii) splitting into cotyledons. Behaviour of grains during milling depend on type and variety of the legumes, thickness, texture and wax in husk, thickness of gum layer between seed coat and cotyledons, shape, size, uniformity and hardness of the grain, pre-harvest and pre-milling storage conditions. These characteristics play a major role in influencing pre-treatments, milling performance and *dal* recovery. In case of pulses, seed coat is tightly attached to the cotyledons due to the presence of gummy substances. The nature and quantity of gums differs from pulse to pulse, variety to variety. Depending upon quantity and quality of gums, pulses are categorized in difficult-to-mill and easy-to-mill kind of pulses. Pigeonpea, moongbean, urdbean etc., are considered as difficult to mill kind of pulses, whereas chickpea, lentil, fieldpeas etc., can be milled easily. Processes and equipments for loosening, husk removal and splitting of cotyledons vary from place to place. Basically, dehusking of pulses is a traditional age-old practice, originated from home and later grown to cottage and large-scale commercial industry.

Milling process

Essentially pulse milling process involves cleaning, grading, pitting, pre-treatment, milling and polishing operations. The major steps involved in pulse milling are discussed below:

i. *Cleaning and grading*: Pre-cleaning and grading of raw material is an essential step in milling. It involves removal of dust, dirt and foreign material, off-sized, immature, shriveled, infested and damaged grains.

ii. *Pitting*: This is a common practice in commercial *dal* mills using carborundum/emery-coated roller. Abrasive or refractory action of rotating rollers scratches the seed surface to facilitate entry of oil/water applied to the grain during pre-milling treatment.

iii. *Pre-milling treatments*: Depending upon adherence of seed coat to cotyledons, different pre-milling treatments are given to different pulses. Dry and wet treatments are quite prevalent throughout the country for pigeonpea.

a) Wet treatment

Soaking grains in water followed by sun drying is an age-old practice for loosening of seed coat. Soaking in water dissolves the gum present between the seed coat and cotyledons. Wetting followed by prolong sun drying up to 10-12% moisture, loosens the seed coat and makes it brittle. This help in easy dehusking and splitting when grains are subjected to shearing surfaces. In some areas instead of soaking, which takes longer drying time, few liters (1-5 liters) of water is sprinkled and mixed with a quintal of raw grains, followed by sun drying up to appropriate milling moisture, prior to milling.

b) Dry treatment

At commercial scale the difficult-to-mill pulses are generally given the treatment of edible oil for loosening the husk. Before oil application, cleaned material is scratched/pitted with the help of abrasive rollers, so that oil can penetrate inside the grain quickly. After thorough mixing grains are kept for few hours for uniform penetration of oil into grains. Oil application is followed by alternate sun drying and heaping for few days. In the process the husk gets loosened and starts detaching from the cotyledons.

c) Other methods

Various researchers have developed many pre-milling treatments. At GBPUA&T, Pantnagar, sodium bi-carbonate, vinegar and enzyme treatments were evaluated. But none of the treatments reached up to commercial acceptance. Thermal treatment developed by CFTRI, Mysore has found some takers for large scale milling, but high initial investment and operating costs are major bottlenecks in wider adoption of the technology.

iv. *Tempering*: Once mixing with oil/water treatment is completed, grains are heaped and covered with gunny bags and left for 12 to 18 hrs. This process helps in uniform penetration of oil/water into the seeds.

v. *Drying* : In most of the *dal* mills in India sun drying of pretreated grains is commonly practiced. Mechanical dryers are also used in some mills during rainy seasons for drying of treated grains. For sun drying grains are spread in thin layer and turned frequently with rake for uniform drying. This operation makes *dal* milling a very lengthy process (2-3 days) and also uniform drying is difficult to achieve. The sun-dried grains require more passes and consume more energy in transporting material within the premises. The drying time with the use of dryers ranges between 2-3 hours, which gives tremendous time saving.

vi. *Dehusking and splitting*: At commercial level, *dal* mills by and large use emery rollers for dehusking and splitting. The pulses with thin seed coat like mungbean have tendency to split even before dehusking. More than three passes are required for milling of difficult-to-mill kind of pulses. It also depends on type of pulse crop, pre-milling treatment, processing technique, grain size, variety *etc*. Husk and powder are generally separated with the help of aspirator. Separation of milled fractions *viz.* unhusked whole, dehusked whole, unhusked *dal*, dehusked *dal* and broken is done to achieve quality product. Mixture of husk and powder is used as cattle feed.

vii. *Polishing*: It is done to give uniform appearance and increase consumer appeal to the *dal* produced. It also helps in removal of husk from unhusked *dal*. Consumers prefer both polished and unpolished *dals*. Different kinds of *dal* polishing techniques such as nylon polish, oil/water polish, leather and *makhmal* polish are used. For pigeonpea and mungbean polishing is generally done using oil. At commercial level, polishing with marble dust and non-edible colours is quite common, but it is harmful, if consumed without proper washing, hence, neither desirable nor recommended.

Traditional pulse milling

At home scale, traditionally, hand pounding of grains in a mortar and pestle was practiced for dehusking, after water pre-treatment. Grinding of treated grains in hand-operated wooden and stone *chakkies* was also used quite commonly. Separation of cotyledons and husk was done by manual winnowing. *Dal* prepared from water soaking or wet treatment takes longer time to cook and has poor cooking quality.

Commercial milling

In commercial milling of pulses, usually dry treatment is followed to reduce treatment time. Pitting, oil application, dehusking and splitting operations are performed mechanically. Sun drying is quite prevalent in pulse milling industry, but it makes pulse milling a weather dependent process. Mostly roof tops or open yards are used for grain drying. Adoption of mechanical dryers saves drying time along with reducing dependency on weather conditions. Even now most of the commercial pulse mills completely depend on solar energy. This makes near closer situation in *dal* mills during rainy and winter seasons. Difficult-to-mill legumes, such as pigeonpea, blackgram, and greengram, require more oil or water in treatments and prolonged sun-drying. Pulses like, chickpea, lentil, peas etc., with loosely attached husk to cotyledons, require less oil/water application and shorter duration of drying.

Pulse milling and allied machineries

i. Stone *chakki*: At domestic level, pulse milling i.e. dehusking and splitting is performed in between two horizontal stone discs, stationary lower and rotary upper disc. Normally 2-4 passes are required to achieve de-husking over 90%.

ii. *Abrasive roller*: In commercial mills dehusking is done with the help of emery or carborundum rollers. The abrasive roller rotates inside perforated casing with a fixed clearance between roller and the casing.

Residence time inside the roller is controlled with outlet opening. Different emery grades are used for different pulse grains.

iii. *Abrasive cone*: Like rice polisher, vertically mounted abrasive cone is used for dehusking and splitting of pulses. The cone rotates inside conical perforated casing. Gap between the casing and cone can be increased or decreased depending upon size of the grain to be milled.

iv. *Burr mill*: Burr mills are also being used for dehusking and splitting of pulses. Light impurities from milled fractions are removed with the help of aspiration. In order to reduce breakage some mills use stationary hard rubber and rotating steel discs. In some cases, graded and cleaned raw grains are passed through the disc to impart scratching to the grains prior to water or oil treatment.

v. *Pulse splitter*: Splitter is used to split the dehusked wholes, obtained after passing through abrasive roller, into two halves.

vi. *Rotary grader*: Rotary screen graders give better performance than reciprocating sieves in grading of grains according to size. Grading of raw grains and separation of finished *dal* is done with the help of rotary graders.

vii. *Color sorter*: Colour sorting machine is used to separate grains on the basis of colours. Whole dehusked and unhusked grains are difficult to separate with the help of sieves. Passing whole dehusked grain from abrasive roll results into higher powdering, therefore, *gota* removal prevents powdering loss. This machine with the help of high resolution camera and sensors separates the grains of different colours.

viii. *Destoner*: De-stoner is designed to remove impurities such as grain-sized sand, stones, metals, glasses and impurities present in any grain mass. The machine consists of vibration motors, pressure fans and screen.

ix. *Treatment mixer*: A worm type conveying mechanism is used for thorough mixing of oil or water applied to pitted grains.

x. *Polisher*: To bring uniformity of colour, polishing is done for better market acceptability of product. Leather belt, nylon or jute ropes, makhmal etc kind of surfaces are used to polish split *dal* with help of, water, oil and edible colour.

xi. *Elevators*: In pulse mills, bucket elevators are used to carry material to different machines and floors.

Improved processes and technologies

i. Pre-milling treatments

Krishnamurti *et al.* (1972), tried to substitute *sirka* (vinegar) to replace edible oil in commercial milling of pulses. Results were comparable with traditional oil treatment milling, therefore, the process could not get acceptance among millers. Kurien (1972) exposed the pigeonpea grains to heated air at 300°C for few minutes. Reportedly this helps 99.5% husk removal in single pass. This method reduces loss and eliminates dependency on weather conditions. In 1977, Kurien applied dry heated air of 120-180°C to the grains and obtained 95-99% dehusking in abrasive machine. High energy input is the major bottleneck in adoption of the technology. Saxena *et al.* (1981) tested aqueous solutions of various chemicals, *viz.*, calcium hydroxide, sodium hydroxide, sodium carbonate, sodium bicarbonate and sodium chloride at normalities between 0.3-1.0 N. Cylindrical abrasive rollers were used to evaluate the treatments. Treatment with normal solution of sodium bicarbonate was found most effective. Reddy (1981) treated pitted pigeonpea sample with normal sodium bicarbonate solution and tempered for 12 hours, followed by mixing grains with oil @ 250g/100kg. Thus treated grains were again tempered for 12 hours. 96.2% dehusking was observed with abrasive rolls. Karsolia and Bal (1980) tried sand roasting at 75°C, 100°C and 125°C with roasting time ranging 1-10 mintues. Roasting at 100°C for 5 minutes gave maximum effectiveness of 85.2% shelling. Saxena (1985) extended the work of sodium bicarbonate treatment and soaked pigeonpea grains with the solution at 4, 6, 8 and 10% concentrations for 0.5, 1 and 2 hours. The grains were dried up to 8.5-10% moisture in oven at 65°C. The study revealed 97% dehusking and 80% *dal* recovery in single pass.

To eliminate soaking and reduce drying time, Mayande (1987), mixed 5 liters of 8% sodium bicarbonate solution with 100 kg grain, tempered the grains for 6 hours and dried it to 10% moisture. Milling of such treated grains gave dehusking of 97%. Verma *et al.* (1993), explored feasibility of enzyme pre-treatment for pigeonpea. Incubation temperature, period and moisture content were optimized. Maximum hulling efficiency of 89% was obtained at 0.08g enzyme/260g sample.

ii. *Dal* Mill

Since traditional pulse milling is labour intensive, time consuming and weather dependent, improved processing method and machinery were developed at CFTRI, Mysore. The process essentially involves loosening of husk by incipient roasting, followed by tempering in specially designed bins. Dehusking is done in improved pearling machine to obtain dehusked whole *i.e., gota* in single operation. *Dal* is obtained with the help of impact splitter after moisture treatment and aeration under controlled conditions.

iii. Mini *Dal* Mills

Like any other industry, high capacity pulse processing plants are mostly confined to cities. Raw materials are purchased from farmers at lower rate by middlemen and again sold back in rural areas after value addition, at higher cost through the middlemen. In this process of procurement, processing and sale, the cost of *dal* gets doubled and the maximum profit goes to the pocket of middlemen. Therefore, a need was felt to develop entrepreneurship among the rural mass for processing of pulses. With this view, small-scale pulse milling units were conceived. Considering the prospects of pulse milling as cottage/ village industry, many research organizations have developed different models of small capacity *dal* mills. Basically, these mills are based on various milling techniques. In the attrition type of machines, coaxially mounted pair of abrasive discs (one stationary and other rotating) is used. Gap between the discs can be increased or decreased according to grain size. During milling the grain fed from the center travels to the periphery of the discs. In abrasion type mills, an abrasive roller is fixed on rotating shaft, inside a drum. In such batch type mills the grain resides for specified time. In the third type of machine, the grain is subjected to combination of thrust and abrasion, between the rotating conical roller and perforated screen. Models of low capacity *dal* mills developed at CIAE, Bhopal, CFTRI, Mysore, PKV, Akola and IIPR, Kanpur are some of the popular machines. While CIAE *Dal* Mill and PKV Akola *Dal* Mills use cylindrical emery roller mechanism for milling, the CFTRI mill has conical emery roll for milling. Within capacity range of 75-150 kg/h, all these mini *dal* mills use abrasive surfaces for dehusking and splitting except IIPR Mini *Dal* Mill. IIPR Mini *Dal* Mill uses abrasion for pitting or scratching of grain surface prior to treatment, followed by soft splitting in rubber-steel disc mechanism.

iv. Snack foods

Different snack foods are prepared from different pulses since ages. These are prepared, usually, by heating, roasting, puffing, flaking, grinding *etc.* at domestic and cottage level. With introduction of new processing machineries and fermentation technologies, various new pulse based products are being commercially manufactured and marketed. Extrusion technology has lead to the development of different snacks made from pulse flour. Production and consumption of starch rich products *viz.*, pasta, noodles, vermicelli, flakes *etc.*, have become popular in domestic market. Pulses are being added to ingredient of these extruded products to make it more nutritious. Now fermented products like *khman, dokhla, idli, dosa etc.* are being produced at large scale. These products can also be prepared at home from ready-to-cook kind of materials.

v. Dryers

Traditionally, sun drying is most common method of reducing grain moisture. But it is very time consuming, weather dependent and labour intensive process. In pulse milling moisture conditioning is an essential process, dryers specific to pulse milling industry can be very useful. Continuous or batch drying also takes less space, hence minimize the need of drying yards. This will ensure uniformity of the treated material as well.

vi. *Gota* separator

Gota separator developed by CFTRI, Mysore claims to save power to an extent of 15-20%. Use of this machine reduces powdering loss by 2-3% in subsequent passes, which is ultimately added to *dal* recovery.

vii. Husk aspiration system

During dehusking in abrasive rolls, though major husk portion is removed and gets mixed with cotyledon powder, but some bigger husk particles is collected at outlet with milled fractions. An efficient aspiration system should remove the husk to get clean product, without causing any air pollution. In the aspiration system developed, air is blown at control rate in cross current direction to remove husk. Out coming mass come across air at required pressure and flow in cascade. At junction box low pressure is created to separate quibbling which are carried away with air flow. Finally dust is collected in cyclones. Air locks and filters remove remaining light particles to release clean air.

Future research thrust

i. *Variety improvement*: As pulse production and productivity has not grown at desired rate, main emphasis of breeders is to develop high yielding varieties. Considering high milling losses due to tight adherence of husk with cotyledons, varieties need to be improved from milling point of view.

ii. *Response to pre-treatments*: Response of different pulse cultivars differs for different pre-milling treatments. Study about response to pre-milling treatments will help in identifying varieties for better recovery and variety selection in crossing programmes.

iii. *Drying behaviour*: Drying of pulse grains is an essential process of pre-milling treatments. Therefore, study of drying characteristics of different pulse cultivars is required to study conditioning parameters for each pulse crop.

iv. *Design of dehusking mechanism*: In order to achieve better dehusking with minimum milling loss, there is a need to further improve dehusking machineries. Appropriate grit size of abrasives, roller structure and configuration should be designed and evaluated for higher recoveries.

v. *Improvement in pre-treatments*: Commercially oil treatment is widely adopted throughout the country. Like pulses, edible oil is also being imported to meet domestic requirement. Hence there is a need to substitute oil with other cheap chemicals or non-conventional oils.

vi. *Utilization of milling by-products*: Broken obtained from pulse milling are sold at low prices. Utilizing broken for secondary and tertiary processing can fetch higher returns. Husk and powder mixture, which at present goes for cattle feeding, if utilized properly, can be used to develop high fiber low protein diet supplement.

vii. *Composite pulse milling plant*: Lot of material handling is involved in pulse milling industry. Process and equipments need to be developed to make a fully automated plant, as it is available for rice milling. For this modification in pretreatment methods and milling machineries is required.

viii. *Ready-to-eat/cook pulse products*: Since ages ready-to-consume edible products are being made from pulses. With increasing number of working women, scope for nutritive ready-to-eat kind of products has increased. Many Indian cuisine, including fermented and non-fermented products can be developed for such exclusive market. Use of pre digested *dal* analogue may result into better assimilation of proteins in human body.

Pulses are chief sources of protein for vegetarian population of India but due to near stagnant pulse production since four decades and increase in population, present pulse availability is far below than recommended level. Though India is the largest producer of pulses, but to meet domestic requirement, still pulses are being imported. As per estimates about 25-30% of this precious commodity is lost during various post harvest handling operations, of which storage and milling contribute maximum. By reducing these losses more pulse can be made available for human consumption. With adoption of improved storage and milling techniques, developing new methods for loss reduction, and utilizing by-products, protein availability can be improved to meet nutritive requirement of the country. Cereal based products with incorporation of pulses make a complete food. Primary, secondary and tertiary processing of pulses can be adopted for better employment and income opportunities at rural and cottage scale. When pulse breeders are thriving hard to develop high yielding varieties, reduction in post harvest losses and utilization of milling by-products seems to be the definite way to improve pulse availability scenario. The slogan adopted for Save Grain Campaign, **'Grain Saved is Grain Produced'** is quite valid for pulses as well.

Glossary

- *Pulse milling*: The process of de-husking and splitting pulse grain is called milling
- *Pre-milling treatment*: Pretreatment is given to loosen the adherence of seed coat to the cotyledons for ease of husk removal.
- *Pitting*: Process of scratching of grain surface for easy penetration of oil or water during pretreatment.
- *Tempering*: Conditioning of treated grains for uniformity of treatment.
- *Polishing*: To give uniform appearance to finished product (whole grain or *dal*), polishing is done.

References

Ilyas, SM and Wanjari, OD, (2001). Important issues in post production processing, storage and by-product utilization of pulses and soybean. National seminar on Emerging Trends in Processing, Handling, Storage and by-product Utilization of Pulses and Soybean, G B Pant University of Agric. & Tech., Pantnagar, Jan 18-19.

Karsolia, RP and Bal, Satish, (1980). Dry conditioning of pulses, ISAE Paper No. 80-0420, ISAE, New Delhi.

Krishnamurti, K, Girish, GK, Ramasivan, T, Bose, SK, Singh, K and Tomar, RPS, (1972). A new process for removal of husk of red gram using sirka. Bulletin of Grain Technology 10: 181-186.

Kurien PP, (1984). Dehulling Technology of Pulses. Research and Industry 29: 207-214.

Kurien, PP, Desikachar, HRS, Parpia, HBA, (1972). Processing and utilization of grain legumes in India. Tropical Agriculture Research Series No.6. Symposium on Food Legumes, TARC, Ministry of Agriculture and Forestry, Tokyo, Japan.

Kurien, PP and Parpia, HBA, (1968). Pulse milling in India (1) processing and milling of tur arhar (Cajanus cajan L). J Fd Sci Technol 5: 203.

Lal, RR and Verma, Prasoon, (2007). Post Harvest Management of Pulses, Technical Bulletin, Indian Institute of Pulses Research, Kanpur.

Lal, RR and Verma, Prasoon, (2007). IIPR Mini Dal Mill for Employment Generation, Technical Bulletin, Indian Institute of Pulses Research, Kanpur.

Mayande, VM, (1987). Pigeonpea milling technology: biochemical and engineering studies. Thesis, M Tech (Agril. Engg.), G B Pant University of Agriculture & Technology, Pantnagar.

Reddy, VPR, (1981). Dehusking of arhar grain (Cajanus cajan L). M Tech Thesis. G B Pant University of Agriculture & Technology, Pantnagar.

Saxena RP, (1985). Milling of Pigeonpea Grain and Associated Aspects. Ph.D. Thesis. G B Pant University of Agriculture and Technology, Pantnagar.

Saxena, RP, Singh BPN, and Singh, JK, (1981). Effect of chemical treatment on husk removal of arhar (Cajanus cajan) grain. Paper No. 81-PAS-116, New Delhi, India : Indian Society of Agricultural Engineers.

Verma P, Saxena, RP, Sarkar BC, More PK, (1993). Enzymatic pretreatment of pigeonpea (Cajanus cajan L) grain and its interaction with milling. J Food Sci Technology 30:368-370.

□□□

4

Prospects of Enzyme Conditioning in Dehulling of Pulses

S.N. BHOWMIK

India contributes nearly 25% of the total world production of pulses and consumes 27% of global. Pulses serve as a major sources of proteins to the Indian population. The import is 11% of global. Agro-climatic conditions in the sub-continent are favourable for production of pulses and various processing practices for pulses are in tune with the age-old food habits. More than 75% of pulses are converted to dehulled split cotyledon called '*dhal*' by milling in the country.

Dhal industry is the third largest food processing industry after rice and wheat milling industries. There are about 14000 *dhal* mills in the country. These mills are mostly in unorganized sector and face problems like inadequate efficient machines. The losses during milling are estimated to the level of 7-8%. If these losses are reduced, the import in pulse milling sector can be reduced. Also the pulses industry is not energy efficient due to unorganized sector. The attractive price for good quality *dhal* over the brokens, has revived the interest of entrepreneurs for up grading existing mills and also to establish modern milling equipment and protocols.

Dehulling is one of the most important operations in postharvest handling of pulses and, hence, plays an important role in processing and utilization of pulses. The seed coat (hull) of pulses is often indigestible and may have a bitter taste. Dehulling improved the palatability and taste of chickpea, pigeonpea, and lentil. Removal of hull facilitated a reduction of polyphenols or tannins, antinutritional factors mostly present in seed coat. Dehulling also reduced cooking time by removing the impermeable seed coat of bean, which hindered water uptake during cooking. The true protein digestibility and net protein utilization of dehulled components were significantly higher than those of the whole seed of pigeonpea (Singh, 1993).

Dehulling efficiency (yield of dehulled seeds) is an important quality characteristic for lentil breeders, processors, and exporters. An efficient and improved dehulling of pulses is of vital importance in reducing dehulling losses. Pigeonpea seed is split into *dhal* by a milling process called dehulling. *Dhal* yield of pigeonpea varies between 50 and 80% (with a mean of 62%) in small- scale milling operations and between 60 and 85%(with a mean of 71%) in large-scale milling operations (Singh and Jambunathan, 1981).

Pulse conditioning for loosening the husk from the cotyledons is one of the important steps in dehulling of legumes. The legumes are generally pre-treated to loosen the hulls before they can be separated using mechanical means. It is generally designed to toughen the hull and loosen the gummy bond between the hull and the cotyledon and to harden the cotyledon to reduce damage. Loosing the hulls during dehulling is traditionally achieved either by wet or dry methods (Kurien, 1977). Pre-dehulling treatments may also involve heat treatment alone or soaking in water, chemical solutions, tempering followed by hot dehulling (Phirke and Bhole, 2000; Phirke *et al.*, 1992; Ramakrishnaiah and Kurien, 1983; Srivastava *et al.*, 1988). The limitations of these treatments are either a shape deformation or poor cooking quality of dehulled splits. These treatments are labour intensive and time consuming.

Green gram (*Vigna radiate*), black gram (*Vigna mungo*), red gram (*Cajanus cajan*) and horse gram (*Macrotyloma* uniflorum) are common legumes consumed as whole, dehulled splits, canned, boiled, roasted or ground into flour to make variety of desserts, noodles, snacks and main dishes. These legumes are classified as hard-to-dehull because of the presence of mucilage and gum forming a strong bond between the hulls and cotyledons. The chemical nature of the mucilage and gums present in the interface between the husk and cotyledons play an important role in dehulling of these legumes (Ramakrishnaiah and Kurien, 1983). These mucilage and gums of legume seeds are network of cellulosic microfibrils embedded in the matrix of non starch polysaccharides (NSP) and proteins (Cosgrove, 1997). Partial hydrolysis of these NSP and/or proteins by enzymatic reactions facilitates the easy dehulling of legumes (Verma *et al.*, 1993). The dehulling efficiency of 88.93% was achieved at enzyme concentration of 0.08 g protein per 260g pigeon-pea grain. Protease pre-treatment of green gram and black gram resulted in higher yield of dehulled kernels while xylanase pre-treatment was very effective in improving the dehulling properties of horse gram compared to protease, which resulted in the yield of more undehulled kernels and fines (Sreerama *et al.*, 2009).

Enzyme conditioning versus traditional conditioning

Traditionally two conditioning techniques are followed to loosen the husk from the cotyledon before dehulling viz., wet conditioning and dry conditioning. The major disadvantages of these techniques are: firstly they are labour intensive and time consuming process. Secondly, the cooking quality of the pulses is poor especially with wet technique. Thirdly, on soaking for longer duration leads to leaching of important nutrients from the pulses into the soaked water. Fourthly, indiscriminate utilization of synthetic chemicals is an environmental hazard that dictates bad health. Fifthly, high dehulling loss is incurred due to powdering and breakage especially in dry method. Sixthly, limited quantity of pulses can be processed at a given time. In comparison enzymes from natural sources (ie plant and microbial origin) are more eco-friendly, energy efficient and cost effective process. Moreover, the cooking quality can be enhanced along with nutrient availability by enzymatic process. Enzymes are heat labile and hence with no residual effect in cooked food.

Prospects for enzymatic treatments

The effectiveness of enzymatic process is a function of moisture content of grain, incubation temperature, hydrolysis time, enzyme concentration and enzyme activity during the process. The plant cell wall material of dicotyledons (including pulses) is mainly composed of three major categories of polysaccharides: cellulose, hemicelluloses and pectines. The proportion and composition of these cellwall polysaccharides varies depending on cell type and plant species. For enzymatic hydrolysis it is very much essential to understand the chemical composition of the substrate that needs to be hydrolysed and accordingly enzyme treatment in combination/ individually needs to be designed. Hence characterization of cell wall polysaccharides, gums and mucilage is essential prior to designing any enzyme formulations. Enzymes are specific in action and their concentrations may differ from species to varieties. Commercially available enzymes applied in this regard are viz., protease, xylanase, and cellulase. Attempts to hydrolyze the mucilaginous bond enzymatically though proved advantageous over conventional practices yet confront monetary shortfall.

Table 1: Comparative analysis of yield of milled fractions, dehulling properties, cooking properties of pigeonpea subjected to various culture filtrates pre-dehulling treatment

Dehulling treatment	Population density (log CFU ml^{-1})	Moisture (%)	Dehulled grains (%)	Undehulled grains (%)	Fines (%)	Hulls (%)	Dehulling Index	Degree of dehulling (%)	Dehulling efficiency (%)	Cooking time (min)	Dispersed solids (g/100g)
Control	0.0	5.7	61.8	17.7	9.5	10.9	0.45	82.3	61.9	20	17.32
Trichoderma reesei											
3 days	7.0	5.3	67.4	12.6	8.6	11.3	0.57	87.5	67.7	21	24.00
6 days	7.1	6.0	70.5	9.4	8.2	11.8	0.64	90.7	70.8	21	23.70
9 days	7.2	6.0	71.6	7.9	7.9	12.5	0.68	92.1	71.7	20	20.39
12 days	7.3	5.8	71.4	8.4	8.1	12.0	0.66	91.6	71.8	20	20.82
Aspergillus oryzae											
3 days	7.1	5.0	70.8	9.64	8.0	11.5	0.65	90.4	70.8	19	28.48
6 days	7.3	6.0	71.4	7.8	7.8	12.9	0.69	92.2	71.5	19	21.65
9 days	7.9	6.2	70.8	8.9	7.1	13.1	0.68	91.1	70.9	20	26.09
12 days	7.6	5.9	72.9	6.63	6.5	13.9	0.74	93.4	72.9	21	21.23

Source : Nath Bhowmik et. al., 2010

Utilization of culture filtrates containing extracellular enzymes of gums/ mucilage hydrolysing microorganism for pulse conditioning shall not only be a cost effective technique but also be an innovative approach for pulse milling. Moreover, drudgery in formulating pure enzyme combinations (for specific crop and variety) and their cautious application subsequently can be curtailed by culture filtrate. This is because microorganism spills out variable proportion of vital metabolites including several extracellular enzymes and organic acids, as a function of nature of the growth substrate to which it is exposed in a rather safe manner inherently. This aspect may perhaps not only increase the dehulling efficiency and cooking quality of the pulses but also can improve the nutritional and functional properties. Cellulolytic microorganisms are a heterogenic group of aerobic and anaerobic microorganisms that are able to produce polysaccharide hydrolysing enzymes. *Trichoderma reesei, Aspergillus fumigatus, A. oryzae* produces cellulase and hemicellulase principally and are recognized as vital inoculants for polysaccharide degradation commercially.

In a preliminary experiment as depicted in Table 1, it has been observed that the cellulolytic fungal growth significantly influences the dehulling efficiency upto 72%. Increment of cellulosic enzymes, organic acids in culture filtrates with corresponding fungal population density is the plausible cause (Nath Bhowmik *et al.,* 2010). However, no significant change in the cooking times and solid dispersion of dehulled splits was observed for control and culture filtrate pre-treated legume. This suggest that, the moisture gradient formed in the seed during enzyme reaction, 1:1 (w/v) may not have affected the quality of protein, starch, and perhaps other moisture sensitive components of the grain.

India is the biggest grower and consumer of pulses in the world. Pulses are consumed principally as dehulled splits called *dhal*. However, the *dhal* yield is 62% in small-scale milling operation with a milling loss estimated to be 7-8%. Dehulling is traditionally achieved either by wet or dry conditioning. The decline in *dhal* milling entrepreneurship is mainly attributable to lack of standard pulse-dehulling method, the traditional one being inefficient and time-consuming that result in heavy losses due to breakage and powdering of grains besides shape deformation and poor cooking quality. Enzyme conditioning is prospective to reduce the dehulling loss upon improving the dehulling efficiency. Attempts to condition pulses by cellulosic enzymes though proved advantageous over conventional milling yet confront shortfall due to high cost and poor accessibility. Moreover developing a common protocol of pure enzyme conditioning technique covering wide range of legume grain species and varieties is distant if not impossible. Pulse grain pre-treatment with culture filtrates of effective cellulolytic microorganism alone or in consortium upon culturing on suitable food substrates holds better option due to low cost and user-friendly nature of the approach. Culture filtrate dehulling pre-treatments reduce the amount of fines.Use of culture filtrate of cellulolytic fungi has the potential to increase efficiency and quality output in grain processing operations in developing countries.

References

Cosgrove, DJ, (1997). Assembly and enlargement of the primary cell wall in plants. Annual Review of Cell and Developmental Biology 13: 171-201.

Kurein, PP, (1977). Grain milling technology. Paper presented at FAO experts meeting on grain legume processing. Central Food Technological Research Institute (CFTRI), Mysore.

Nath Bhowmik S, Manikantan MR, Goswami D, Patil RT, (2010). Effect of growth of Trichoderma and Aspergillus on dehulling efficiency of pigeonpea. In: International Symposium on Recent Advances in Cross-disciplinary Microbiology: Avenues & Challenges, December 14-17, Birla Institute of Technology, Mesra, Jharkhand.

Phirke, PS, Bhole, NG, (2000). Pretreatment of pigeon pea grain for improvement of dehulling characteristics. International Journal of Food Science and Technology 35 : 305–313.

Phirke, PS, Pumbarka, S, Tapre, AB, (1992). Evaluation of chemical pretreatment of pigeon pea grains for milling. Indian Journal of Agricultural Engineering 2: 141–142.

Ramakrishnaiah, N, Kurien, PP, (1983). Variability in the dehulling characteristics of pigeon pea (Cajanus cajan) cultivars. Journal of Food Science and Technology 20 : 287–291.

Singh, U, (1993). Protein quality of pigeonpea (Cajanus cajan L.) as influenced by seed polyphenols and cooking process. Plant Foods for Human Nutrition 43: 171-179.

Singh, U, and Jambunathan, R, (1981). A survey of methods of milling and consumer acceptance of pigeonpeas in India. Pages 419-425 in Proceedings of International Workshop.on Pigeonpeas, 15-19 Dec 1980, ICRISAT, India. Vol. 2. Patancheru, A.P. 502 324, India.

Sreerama, YN, Sasikala, VB, Pratape, VM, (2009). Effect of enz,me pre-dehulling treatments on dehulling and cooking properties of legumes. Journal of Food Engineering 92 : 389–395.

Srivastva, V, Mishra, DP, Chand, L, Gupta, RK, Singh, BPN, (1988). Influence of soaking on various biochemical changes and dehusking efficiency in pigeon pea (Cajanus cajan L.) seeds. Journal of Food Science and Technology 25 : 267–271.

Verma, P, Saxena, RP, Sarkar, BC, Omre, PK, (1993). Enzymatic pre-treatment of pigeonpea (Cajanus cajan L.) grain and its interaction with milling. Journal of Food Science and Technology 30 : 368-370.

□□□

5

Oilseed Processing : New Developments and Future Prospects

H.K. SHARMA AND B.C. SARKAR

Oilseeds are mainly composed of oil and protein. Extraction of oils from oilseeds is a key step for their commercialization. Fats and oils are the most concentrated source of energy in our diet and also act as carrier for lipo-soluble vitamins like Vitamin A, D, E and K. The fats/oils also act as an important ingredient for several food stuffs and non-food products and are recognized as one of the richest source of essential fatty acids.

Oilseeds occupy an important place in global agriculture by providing vegetable oils and high-protein meals for food, feed and industrial uses. The leading oilseeds crops by volume of the world production are soybean, cottonseed, peanut, sunflower, rapeseed (canola), copra (coconut) and palm. The combined world production of these major oilseed crops in 2001-2002 was 324 million metric tons. Soybeans constitute the largest share of the world oilseed supply. The United States, Brazil, Argentina, People's Republic of China, and India dominate world soybean production. Together these countries account for approximately 90% of the world soybean crop. Production of other oilseeds is widely distributed throughout the world. The oil content of oilseeds varies considerably among crop species, cultivars within species and the agro climatic conditions under which they are grown. The approximate oil content of soybean, cottonseed, peanut, sunflower, rapeseed, copra and palm kernel are 20%, 23%, 45%, 40%, 35%, 67% and 50%, respectively. The economic exploitation of oilseeds begins with the primary process of separating oil and protein meal. The technology of oil separation has evolved with the progressive development of oilseed production from a subsistence agricultural practice to an extensive commercial agricultural enterprise (Wijeratne *et al.,* 2010).

Oil scenario in India

Oilseeds and animals are the main sources of fat. Though India has the largest number of animals compared to any single country of the world, but the animal fats are not preferred and our efforts to exploit fats of land and marine animals are negligible. The major responsibility of oil production in India, both for edible and industrial usages depends on vegetable oil seeds production. In the global context, India is one of the major producers of oilseeds which are the second major agricultural crop in terms of tonnage and value. In fact, India enjoys a premier position in the world, occupying over 15 percent of its cropped area, accounting 10 percent of the total global output of the vegetable oils and fats. The major oilseeds of India are groundnut, rape seed mustard, linseed, sesame and castor. Groundnut and rape seed mustard account about 85 percent of the total production of oilseeds in the country. Soybean, sunflower, safflower, cotton seed and coconut are the other important oilseeds produced in India. For processing of oilseeds and oil bearing materials, India has a large network of oil mills employing more than 10 million people. Oil milling industries at present, consist of about 0.35 million village ghanies, 70,000 power driven ghanies, 50,000 mechanical expellers, 360 solvent extraction plants and 90 vanaspati units. The installed capacity of oil mills is estimated as 35 million tons per annum besides, there is solvent extraction plants of about 6.8 million tons per annum capacity in the country (Shukla *et al.,* 1992).

The production of oilseed in India during the year 2003-04 was about 25 million tons (Mt). The high oil bearing oilseed (rapeseed, mustard, sesame, groundnut and sunflower) is about 58% of the total production leading to an amount of 14.5 Mt whereas the production of oilseed in India during the year 2009-10 was about 23.11 Mt (Table 1). The total oilseed produced in India is not available for utilisation because of heavy losses starting from harvest to consumer utilisation (Fig. 1). Various options are being tried to reduce the losses and enhancing oil availability. Table 2 shows the persistent gap between demand and supply which can possibly be reduced.

- Development of high yielding variety and wide application
- Improved cultural practices
- Reduction in harvest and post harvest losses
- Improvement in deoiling process.

Out of the above possible options mentioned for overcoming the oilseed crisis, improvement in deoiling process has great potential to alleviate the edible oil problem because about 0.5Mt oil is lost due to inefficient processing of oilseed every year.

Table 1 : Indian Oilseeds Production (In Million Tons) (*Source* : Trade Estimate)

Oilseeds	2009-2010	2008-2009	2007-2008	2006-2007	2005-2006	2004-2005	2003-2004
Groundnut	5.12	5.92	6.89	5.35	6.25	6.07	7
Castor seed	0.93	0.98	0.91	0.78	0.91	0.80	0.67
Sesame seed	0.76	0.58	0.66	0.61	0.60	0.72	0.80
Rapeseed and Mustard	6.32	6.2	4.59	6.02	6.77	6.60	6.25
Linseed	0.16	0.13	0.18	0.19	0.20	0.20	0.23
Niger seed	0.08	0.08	0.07	0.07	0.11	0.15	0.10
Safflower	0.15	0.17	0.17	0.19	0.21	0.21	0.19
Sunflower	0.99	1.15	1.46	1.6	1.62	1.58	1.21
Soybean	8.50	8.9	9.46	7.96	7.05	5.85	6.85
Total	23.01	24.11	24.39	22.77	23.72	22.18	23.30
Grand Total	23.11	24.26	24.59	22.97	23.97	23.90	23.30

(including Toria)

Table 2 : Edible Oil Demand Projection

	2004	2010	2015
Total Demand (Mt)	10.90	15.6	21.3
Total Area under Oilseeds (M Hectares)	23.40	28	32
Yield (Tons/hectare)	1.07	1.2	1.4
Production of Oilseeds (Mt)	25.10	33.6	44.8
Domestic supply of edible oils (Mt)	7	10.1	13.4
Total edible oil imports - (Mt)	4.30	5.9	8.3
Imports as share of demand	39.40%	38.10%	39.50%

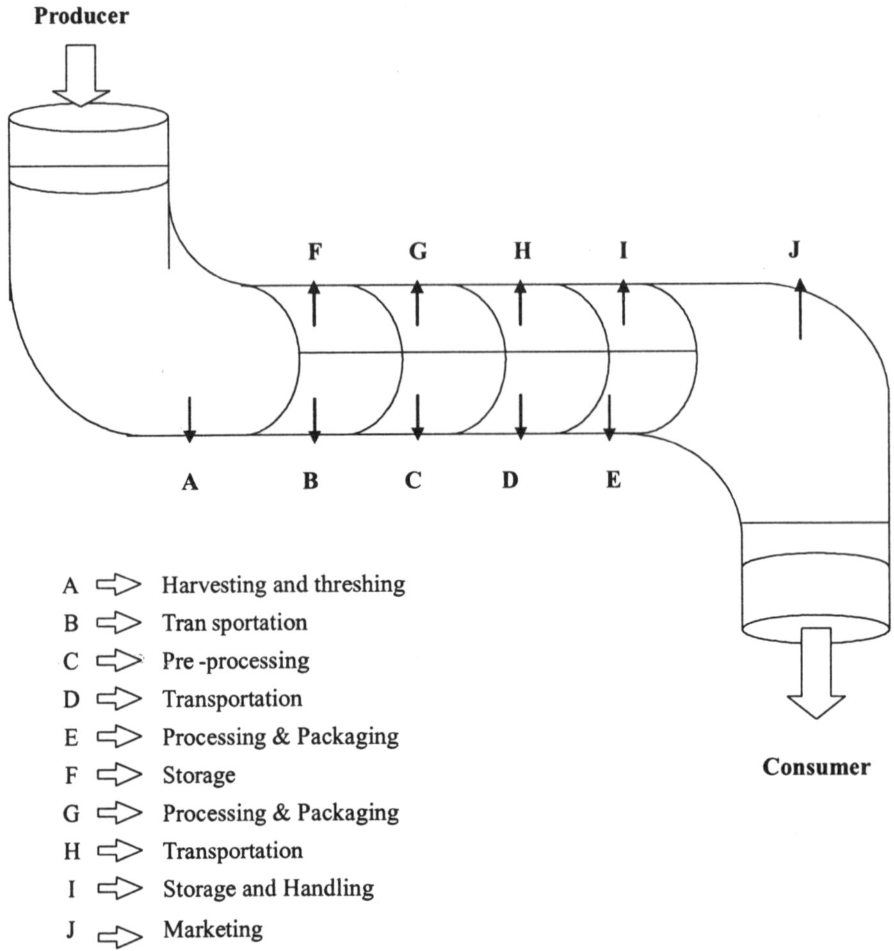

Fig. 1 : The leaky oilseed pipe showing possible losses

Handling and storage of oilseeds

Proper handling and storage of oil-containing materials are very important to minimize deterioration and maintain good quality of both oil and meal. Whole, intact, low-moisture oilseeds (about 8-10 percent moisture) may be stored for an extended time under suitable conditions. Deterioration of oilseeds is accompanied by respiration or carbon dioxide evolution and heat generation by oxidation reactions. Sound, intact seeds may release less than 10 cubic centimeters of carbon dioxide per gram of seed per day, while damaged, high moisture seeds may release 50 cubic centimeters or more of carbon dioxide. Furthermore, heat generation by oxidation reactions may increase the temperature of stored seed, accelerating deterioration even to the point of charring the seeds.

Activity of native enzymes present in oilseeds, infestation by insects and mites along with microbial activity during storage are also important factors affecting oil and meal quality. In general, high moisture content (above 14-15 percent moisture) in seeds has an adverse effect on oil and meal quality. Oil splitting, or acid generation, may be accelerated by microbial growth (mold and/or bacterial growth) and enzyme activity in high-moisture seeds. Oil in mature seeds may contain about 0.5 percent free fatty acids. However, if seeds are damaged mechanically or by frost or become wet during harvest, handling and storage, then the acidity of oil can be much higher.

Sprouting, considered a damage factor, is another important issue during storage of high-moisture seeds. Sprouted seeds may have lower oil and higher free fatty acid content as compared to sound seeds. Today, air dryers are essential components of modern storage facilities to maintain oilseed quality. Most oilseed storage bins are equipped with aeration ducts and ventilation blowers to cool the seeds. Since oxidation reactions are aerobic processes, a low oxygen atmosphere in storage bins helps to slow down oxidation and quality deterioration. Mature seeds can be stored longer than immature seeds because of the lower activity of oil-splitting enzymes in mature seed. Proper storage of harvested oilseeds may also contribute to desirable changes. For example, oil extraction yields from fresh soybeans can be lower than for soybeans stored more than five months. The storage of soybeans also decreases the chlorophyll content of green beans. Chlorophyll is not desirable in edible oils and needs to be removed during the oil refining process. For the better yield during processing, the following steps are recommended.

Cleaning

Oilseeds need to be cleaned to remove plant stems, sticks, leaves and foreign material before storage. Such material may decompose and cause heating in stored seed mass diminishing oil and meal quality. Foreign materials in seeds are typically separated out by a combination of rotating or vibrating coarse screens, reels and aspiration. This process is commonly referred to as scalping. Sand and dirt also are removed by fine screening. Magnetic devices remove metal contaminants from seeds. Plate magnets, drum magnets or electromagnets installed over conveyor belts are commonly used.

Seed drying

The moisture content of oilseeds often needs to be reduced to minimize degradation in storage and to improve the effectiveness of downstream processing. For example, soybeans are often received at 13 percent moisture and need to be dried to 10 percent moisture to facilitate efficient hull removal. Large, vertical, open-flame grain dryers can be used for oilseeds as well. These dryers have multiple columns of oilseeds which slowly migrate downward. The upper portion of the column is used for drying and lower section is for cooling.

Dehulling

The amount of hull on oilseeds varies significantly. The percentage of hulls for cotton seed, sunflower seeds and soybean are 45, 25 and 7 percent, respectively. Hulls and shells of oilseeds do not contain a significant amount of oil (less than 1 percent). Most oilseeds need to be separated from their outer husk or shell prior to oil extraction. Dehulling also is referred to as shelling or decorticating. Dehulling increases oil production efficiency, capacity of the extraction equipment and reduces wear in the expeller as the husks are abrasive. If not removed, hulls reduce the total oil yield by absorbing or retaining oil in the pressed cake. Furthermore, wax and color compounds present in the hulls end up in the extracted oil. These compounds are not desirable in edible oils and need to be removed during the refining process. Dehulling reduces fiber and increases protein content of the meal.

Size reduction and flaking

Most oilseeds are reduced in size to facilitate hull removal, heating, drying and flaking prior to oil extraction. Oilseeds also can be flaked prior to solvent extraction. Flaking ruptures seed cellular structure and reduces the distance that solvent has to travel to reach the oil in the cells. A flaking mill has two large diameter rolls turning in opposite direction and forced together by hydraulic cylinders. As the seeds are pulled through the flaking mill, they are stretched and flattened. Typical flake thickness is in the range of 0.01-0.015 inch or 0.25-0.37 mm. Flaking of oilseeds also increases surface area for increased contact between solvent and seed during the solvent extraction process. Oil from the cracked or flaked seeds should be extracted as quickly as possible (within 24 hours) to minimize meal and oil quality deterioration.

Cooking / Tempering

Oilseeds are cooked or tempered to denature proteins, release oil from the cells and inactivate enzymes. For example, rapeseed contains the enzyme myrosinase. This enzyme catalyzes hydrolysis of glucosinolates which are naturally present in rapeseed. During the hydrolysis process, undesirable compounds such as isothiocyanates and nitriles form. These compounds are soluble in oil and lower the quality of oil. Rapeseed is cooked in multistage cookers to keep the glucosinolates intact and inactivate the myrosinase. Rapeseed is preheated to 68-122° F in less than 5 minutes and contacted with live steam at 248°F. Since canola has much lower levels of glucosinolates than that of conventional rapeseed varieties, the cooking temperature for canola is lower (less than 212°F). Cooked seeds are immediately pressed to separate oil. Cooking also gives seeds proper elasticity for efficient pressing. However, cooking at high temperature is not necessary for sunflower seeds.

Processes for oil extraction from seeds

Oilseed processing and oil extraction processes are designed to obtain high quality oil with minimal undesirable components, achieve high extraction yields and produce high value meal. The common methods for deoiling of oilseed are:

- Animal driven *ghanis*, power operated *ghanis*
- Mechanical pressing (small and large size expellers)
- Solvent Extraction
- Combination of mechanical and solvent extraction

Crushing of high oil bearing seeds in *ghanis* is by far the oldest technology. This technology is still being used in the villages and by the marginal entrepreneurs. Estimate place the total number of village *ghanis* at about 300 thousands and power operated *ghanis* at about 70 thousands. Oil recovery in all these *ghanis* is low and the oil cake produced is unstable. Oil recovery efficiencies of traditional *ghanis* is reported to be 74-82% while that of improved power *ghanis* to be marginally better at 75-85%. In spite of poor oil recovery, *ghanis* are popular in rural areas (Agrawal *et al.,* 1988).

Mechanical pressing (hydraulic and screw) is the oldest and simplest method for oil extraction from oleaginous seeds. No chemical is used for oil extraction and therefore the residue is free of chemical residue. It's a labour intensive technique and its use declined over the years. Continuous screw-presses (expellers) have replaced the hydraulic equipment (Bargale, 1997). It consists of an extruder with a perforated body (slots or holes), a helical screw is used to convey and press out the oil (Balke, 2006). A major drawback of this process is the lower oil extraction. Heat pre-treatment improves the malleability of the seed, lower the shattering and denature some proteins to improve oil extractability (Balke, 2006). The extreme heat generation during this method cause darkening of the oil due to burning of the meal and loss of protein amino acids (Bargale, 1997). The choking and plugging problem results in loss of production capacity and increases in energy, labor and other resource requirements (Rosenthal *et al.,* 1996).

The characteristic rapeseed and mustard flavoured oils are produced due to the presence of water (Achaya, 1994; Head and Swetman, 1999). On crushing the oilseeds, toxic compounds are produced when the enzymes (e.g. myrosinase in rapeseed) contact with substrates (e.g. glucosinolates) (Bargale, 1997). Therefore, after flaking, it is recommended to cook the seeds to inactivate enzymes and reduce the production of free fatty acids (Ward, 1982; Latif, 2009).

Mechanical oil expression of oilseeds is mostly done using screw press. The expellers are available in the capacity range of 5-460 tons/day. Baby oil expellers available in India are of capacities as low as 35kg/h. Oil recovery in expellers is also low (90-93%). Screw presses are suitable for their increased application particularly in decentralised level. The expellers are suitable for oil expelling from high oil bearing materials, which do not lend themselves to direct extraction.

Solvent extraction refers to preferential dissolution of oil by contacting oilseeds with a liquid solvent. This is the most efficient technique to recover oil from oilseeds. The efficiency depends on the oilseed preparation prior to extraction, temperature, mode of operation (batch vs. continuous and co-current vs. counter current operations) and equipment design. It is expected that residual oil in the meal to be less than 1 percent after commercial solvent extraction. The choice of solvent type is based on solubility of oil in the selected solvent, cost and safety. Light paraffinic petroleum fractions, pentane (boiling point 88-97°F), hexane (boiling point 146-156°F), heptanes (boiling point 194-210°F) and octane (boiling point 215-264°C) can be used for oil extraction. Currently hexane is widely used for commodity vegetable oil extraction. The major disadvantage of these solvents is their flammability. Strict precautions have to be taken to avoid fires and minimize explosion risk in the extraction plants. Solvent extraction of oilseeds was developed in Europe in the 1920's. Since then it has been considered important to achieve residual oil contents in the meal of 1% or less, as dictated by economic constraints. In the solvent extraction process mostly hexane is used for relatively low oil content seed (<25%, w/w). The process is efficient in terms of oil recovery (99%), but has limitations in terms of high capital and higher operational cost. Solvent loss increases exponentially with decreased capacity.

Solvents for extraction of foods must have high solvent power, low toxicity, good selectivity and should be safe to operate. The most common type of extractor used is the percolation type in which the solvent flows by gravity through a bed of meal (Anjou, 1972; Beach, 1983). The hexane leaves the extractor with an oil concentration of 25-30% and the defatted meal with a solvent content of 30-35% (Anjou, 1972). The hexane in the meal is removed in a desolventizer-toaster which consists of steam heated trays; the meal is first heated by indirect heat in the upper trays to remove most of the solvent, followed by a direct steam injection in the middle trays, and finally by indirect heat in the lower trays to dry the meal (Anjou, 1972; Beach, 1983). The meal has dark colour with 6-11% moisture content and 300 to 1800 ppm hexane (Beach, 1983; Dahlen and Lindh, 1983). This desolventization of meal is the main source of solvent losses (Dahlen and Lindh, 1983).

Moreover, solvent extraction plant is not suitable for decentralized operation. The industry is also hazard prone due to the perpetual risk of fire and explosion. The neurotoxic properties of hexane are a major concern to food regulatory agencies. Therefore, other solvents like petroleum ether, alcohol, isopropanol, acetone, halogenated solvents, supercritical carbon dioxide etc. have been evaluated. But all other solvents tested have shown certain drawbacks when used in oil seed extraction. The residual solvent in the oil and cake is also a potential hazard. The high temperature used in desolventizing the meal, reduces the nutritional and functional properties of the protein in the meal.

Generally, a two-step process is used for high oil contents (above 35%) materials which consists of a continuous pre-press followed by solvent extraction. The pre-pressing allows solvent extraction to be applied to the oilseeds that are difficult to process by direct extraction methods and it also lowered the solvent requirements (Mustakas, 1980). A single solvent extraction is employed for soybeans (Mustakas, 1980). Despite recent developments in the techniques for minimizing VOC (Volatile organic compounds) emissions, the food industry is still responsible for about 7.5% of the mass emissions from stationary sources (Valentin, 1992). Vegetable oil sector is mainly responsible for the high VOC emission level. According to Mustakas (1980), solvent losses in American soybean oil plants was 4-8 litres/ton of seed processed. On this basis the annual level of hexane emission from the soybean oil industries could be 210-430 million litres.

Even with a two-stage operation, the oil content of the final press cake will be about 5-6%. Extracting the press cake with a solvent offers a way to reduce the oil content to less than 1%. In a pre-press solvent extraction, the press is operated to give a press cake with 15-18% oil and the balance is recovered during solvent extraction. The press cake is first passed through a cake sizer and flaking mill before being conveyed to the solvent extractor. The meal is ground prior to sale as animal feed. Many solvents, both petroleum and non-petroleum types have been used in the laboratory to extract oil from oilseeds.

In the solvent extraction process, oil in whole seeds or large pieces is not extracted to a major extent by non polar solvents in which the lipid phase is otherwise completely soluble, proving the impermeability of at least some intact cells (Othmer and Agrawal, 1955). As long known industrially, cells must be ruptured and membranes denatured to achieve high permeation rates.

Severe micro structural distortion occurs during flaking to conform to the new geometry and increased surface area. This requires extensive rupture and deformation of the cell contents. During solvent oil extraction, the drastic thermal treatment of oilseeds reduces the quality of extracted oil and proteinaceous materials (Yoon *et al.,* 1991). The excessive heating in the conventional process decreased the nutritional availability of some essential amino acids (mainly lysine), which results from the Maillard reaction between free amino groups from proteins and carbonyl groups from reducing sugars (Rosenthal *et al.,* 1996). Bargale (1997) summarized the limitations of the solvent extraction method. The process is not chemically safe for human health. Although desolventization removes most of the solvent, but some traces are always present in the meal. The solvents used for the oil extraction are highly flammable. High initial capital and operating costs are required for handling even smaller quantities of oilseeds. Toxic meals may be produced. Natural flavour cannot be extracted. The energy requirements are high and a lower quality of oil (with high content of phosphatides) is recovered. As a consequence of these drawbacks of solvent extraction method, several new approaches have been tried in order to develop a better technology.

Two-phase solvent extraction method

Two phases in this method consists of polar phase and non-polar phase, the polar phase consists of a lower alkanol containing ammonia (0-14% w/w) and water (0-15% v/v) with hexane as a non-polar phase (Caviedes, 1996). Methanol ammonia was used to facilitate the lipid extraction by breaking the cell wall but it also removed glucosinolates and polyphenols (Rubin *et al.,* 1986). Water addition to the polar phase improved the process because the glucosinolate content in the meal was reduced and the hexane as non-polar phase extracts the oil from the crushed seed (Caviedes, 1996). The canola meals produced by this method have superior functional properties (Naczk *et al.,* 1985). It has 25% more crude protein than the meal produced by hexane extraction. The meal was almost bland flavour and light colour. The water absorption and hydration capacity of the glucosinolate-free meal was similar to the air desolventized meal and higher than the commercial canola meal. A significant increase in fat absorption was observed in methanol-ammonia-water treated meal which was about three times more than the commercial soybean and canola meals. The improvements of fat and water absorption may be due to the unfolding of protein (Naczk *et al.,* 1985; Latif, 2009).

In this process the protein solubility is much lower than the hexane extracted meals (Tzeng *et al.,* 1990; Rubin *et al.,* 1986) which can be increased from 50 to 87% by using sodium dodecyl sulfate (SDS) (Igor *et al.,* 1993). But this process is impractical because it requires high amounts of SDS which can be hard to remove by compatible methods to food and the products obtained will have low purity due to the interactions between SDS and protein (Caviedes, 1996). The involvement of the two solvents in this procedure increases the complexity and energy requirements. The meal desolventizing step is unavoidable and the problems associated with use of solvents are major drawback of this technology (Latif, 2009).

Factors to consider in planning an oil extraction system

Energy and resources

Solvent extraction of vegetable oils should be seen as part of a technological and economic system that includes far more than the extraction plant itself. Factors affecting the operation of a solvent extraction plant include: potential markets; nature, timing, size, and reliability of seed and solvent supply; adequacy and reliability of power, water, and transportation, and of maintenance and storage facilities; and ability to find and train personnel and rigorously enforce safety standards.

Size of operation

The size of the operation is the most important factor in determining which kind of process will be used. For intermediate-scale operations (operations that process up to 200 tons per day), the choice is between batch solvent extraction and expeller (pressure extraction) systems. Batch solvent extraction systems operate more slowly and less efficiently, are more labour intensive and dangerous, and use greater quantities of solvent than properly designed continuous systems do. Because of these drawbacks, expellers are usually preferred for installations. However, there are instances when expeller extraction is not suitable for a small operation; in those cases, batch solvent extraction may be the only practical way to proceed. Continuous solvent extraction should be considered only for systems that will treat 200 tons or more of seed per day.

Site and design

Solvent extraction plants are complex systems that must be carefully engineered for safety because of their special hazards. Because of the danger of explosion, solvent extraction plants need to be located a safe distance away from populous areas, and to be designed by experienced engineers. Installation of a plant without such engineering of details is a dangerous error.

Investment cost

The cost of solvent extraction plants is much higher than the cost of expeller extraction plants, usually about double. However, since a solvent plant recovers a greater proportion of the oil, it may still be the economically wiser choice. For example, solvent extraction should recover about 40 kilograms more oil per ton from dry soybeans than expeller extraction would.

Personnel and safety

It takes less labour but more sophistication to maintain and operate a solvent extraction plant than to maintain and operate an expeller plant. Two people per shift are required for the former, compared to three for the latter. The dangers of solvent explosion make tightly controlled procedures necessary. Workers must be trained to have a wholesome fear of exposure to the solvent and of solvent leakage.

Reliability of throughput

For continuous solvent installations especially, it is essential to be able to depend on a steady throughput. Unanticipated interruptions of seed supply may cause buyers of oil and meal to turn to more reliable sources. Both batch solvent and expeller operations are less vulnerable to the effects of such interruptions as compared to continuous solvent operations (Thijs Adriaans, 2006).

Newer developments

Enzyme-assisted solvent extraction method

Seed rupturing is the primary pre-requisite in solvent extraction of vegetable oils. Industrially, flaking or milling does the rupturing. Complete cell rupturing is necessary for rapid solvent extraction. One way, probably the most efficient way of rupturing the cell wall is breaking it with enzyme. Fullbrook (1983) used crude enzyme, prepared from different sources of microorganisms on rapeseed. He investigated the effect of aqueous enzymatic hydrolysis and hydrolysis in presence of hexane on solvent oil extractability of rapeseed using various sources of organisms for enzyme. Hydrolysis was carried out for 4 h and enzyme concentration used was 0-3% w/w basis. He reported that enzymatic hydrolysis resulted in higher oil yields in the presence of hexane. Enzyme concentration at 3% w/w basis was found to result in maximum extractability. Further, *Aspergillus niger* was found to be better enzyme source than *Bacillus subtilis*. Solvent extraction after hydrolysis of ground samples resulted in low energy and solvent usage and that the oil quality was good which did not require any further processing after extraction. Some studies have been reported regarding the enzyme-assisted solvent extraction. The effect of enzymatic hydrolysis to enhance oil extraction with hexane on canola was demonstrated by Sosulski *et al.* (1988). It was observed that hydrolysis of canola with carbohydrase resulted in increased oil yield and reduced time for oil extraction by solvent extraction. They incorporated flaking, autoclaving and optimized moisture to 30% including 0.12% enzyme concentration and incubation for 12 h at 50°C followed by decreasing moisture content to 4%. Hexane extraction was found to be effective by further grinding of treated flakes. The enzyme-mixtures with mixed activity was found to be superior than individual enzymes to enhance the oil extraction efficiency. Dominguez *et al.* (1995) conducted a similar study on soybean and reported an increase in oil extractability by 8-10% and up to 4% in the case of sunflower oil (Dominguez *et al.*, 1993) with the help of different commercial enzymes. An increased amount of solvent extractable oil was observed by Bhatnagar and Johari (1987) for some oilseeds after treatment with enzymes originating from thermophilus moulds. The cottonseed oil recovery was increased up to 5% with previous enzymatic action while sunflower oil yield was increased by 4.2% after using enzymes. Twenty percent increase was observed in the solvent oil extraction by pre-treating shea tree kernels with a mixture of protease and carbohydrases (Tano-Debrah and Ohta, 1994; 1995a,b). Sarkar *et al.* (1998) used crude enzyme preparation from Aspergillus *fumigatus* for hydrolytic treatment of rapeseed and reported more than five percent increase in oil recovery. Hydrolytic pre-treatment parameters such as moisture content during hydrolysis, enzyme concentration, incubation period and incubation temperature were optimized. The optimum values of four parameters were reported as 23.30% moisture content, 51.61 mg/100g of sample enzyme concentration, 45.5°C incubation temperature and 12.83 h incubation period. The oils obtained with enzymatic technology, stable enough to rancidity, show similar composition and structure to treated ones. This technology was applied on an industrial scale for olive, and a pilot or semi pilot scale for rapeseed and coconut. The improvement in the yield of oil was lower than in laboratory scale trials but other characteristics such as stability; acidity and peroxide value were similar (Dominguez *et al.*, 1994).

The oils obtained by pressing from treated canola samples showed increased levels of free fatty acids, although the other characteristics were maintained (Sosulski and Sosulski, 1990). Peroxide values were low in all samples. Oils from enzyme treated samples contained lower amounts of chlorophyll (Dominguez *et al.*, 1994).

Enzyme-assisted full pressing process

The process has been used for extracting rapeseed/canola oil and can be extended to other plant materials such as coconut, peanut, cottonseed and sunflower seeds. The process uses conventional process flows and equipment. Canola seeds are prepared by flaking and heat treatment to inactivate the inherent myrosinase. The flakes are then rehydrated to 25-50% moisture and the required dosage of enzyme is added. The material is incubated at the optimal temperature, dried to the desired moisture content and pressed to release the oil. Depending on the equipment used, one or more than one pressing may be necessary in order to recover most of the oil (Owusu-Ansah, 1995).

Advantages

The following advantages of the process have been reported:

- Considerable increase in throughput
- Significantly lower level of residual oil
- Can be operated at lower level of temperature
- Better quality of oil

Drawbacks

- Extra energy for drying the seeds after treatment
- Long incubation period (about 12h)

Enzyme-assisted cold pressing (EACP) method

The process takes place under the controlled temperature conditions. The oilseed cell walls have many common features and very similar structure consisting of cellulose fibers to which strands of hemicelluloses are attached (Christensen, 1989; Keegstra *et al.*, 1973; Talmadge *et al.*, 1973). The enzyme preparations capable of attacking cell walls must necessarily contain a mixture of cellulases, hemicellulases, pectinases, and even proteases (Christensen, 1989). Commercial preparations containing these enzymes are often unable to hydrolyze specific components and some more development may be required to degrade, complex polysaccharides in the cell wall of plant material (Christensen, 1989). Similarly, Bouvier and Entressangles (1992) utilized a cellulase preparation to reduce the palm oil losses by 3% and 18% from press fibers and crude juice respectively during clarification as compared with the traditional process (without enzyme).

Moisture free soybean oil yield was increased up to 2.78% which corresponded to an increase of about 11.7% in oil recovery by using mixed-activity crude enzyme (Smith *et al.*, 1993). The enzyme treatment conditions of whole and flaked canola seeds to enhance oil recovery during expeller pressing were optimized by Sosulski and Sosulski (1990). The enzymatic pre-treatment of canola, prior to pressing on the expeller, reduced the pressing time and increased the oil recovery from 75% to 92% of the total oil in single pressing. The residual oil content in the cake was reduced to 6.2% and protein content was slightly increased with high lysine availability. The fiber content was also reduced by 1.1 to 2.6 percentage units and the digestibility of organic matter of treated press cake was improved by 8-21 percentage units over that of the control. The carbohydrase and mixed activity enzyme (Novo Industri SP-249) were used for an expeller process and increased the oil extraction from 72-78% (for the untreated seed) to 90-92% (for the hydrolyzed seeds) (Sosulski and Sosulski, 1993). The expeller performance was improved as the material throughput was increased by 51% and the oil flow by 84%. This process produced better quality oil than the solvent extraction process with lower energy requirements. In another study, Sosulski and Sosulski (1993) investigated the use of carbohydrase enzymes to enhance oil recovery in screw pressing. Enzyme-treated seeds at 6% moisture content were pressed in the expeller at full press conditions. Control seeds were also pressed at the same pressure but without enzyme treatment. The enzyme treatment improved throughput of the expeller, increased oil flow rate and oil recovery. Material throughput was increased by 30-50%, and the oil recovery was increased from 72% (control samples) to 90-93% (enzyme-treated samples). The residual oil content in press cakes from enzyme treated seeds was 7.4%. The oil quality was inferior to the cold pressed control but was much better than that of solvent extracted oil. The feasibility of enzymatically treated soybean and sunflower seed to improve oil extractability was studied by Dominguez *et al.* (1993). They emphasized on particle size, moisture content, enzyme concentration and hydrolysis time parameters. The size reduction was found to be favourable during enzymatic treatment because better accessibility of the enzyme to the cellular wall. By reducing size from 2 mm to 1 mm, the oil extraction efficiency increment was more than 35% for soybean and 50% for sunflower. The moisture content required during the enzyme treatment was 50% for soybean and 20-30% for sunflower. An increase in moisture had little positive effect on oil extractability. The increased oil extractability could clearly be observed even with a very low concentration of enzyme (0.01g/100g of seed). The higher enzyme concentration of up to 1g/100g seed for soybean and 2g/100g seed of sunflower increased the oil extraction and the hydrolysis time of 6 h was found to be optimum for both oilseeds.

Enzyme assisted aqueous extraction method (EAAE)

The solvents used in extraction process have shown certain drawbacks. The residual solvent in the oil and cake is also a potential hazard. Water is also being considered as a solvent. It is non-toxic and non-inflammable. The use of water in oil extraction is based upon a pre-treatment of the milled and boiled seed mass with a solution of cell wall degrading enzymes, followed by a centrifugal step to separate the emulsified oil from the seed residues. Good results of this new oil seed extraction method, with yields of up to 90% were reported for rapeseed (Olsen, 1988).

Aqueous enzymatic oil extraction is an emerging technology in the fats and oil industry, and it may offer many advantages as compared to conventional extraction (Rosenthal *et al.*, 1996). It eliminates solvent consumption, which may lower investment costs (Barrios *et al.*, 1990; Lusas *et al.*, 1982) and energy requirements (Barrios *et al.*, 1990). Simultaneous oil and protein recovery from most oilseeds is possible. Degumming operation can be eliminated and it may allow the removal of some toxins or antinutritional compounds from oilseeds (Caragay, 1983). Several studies have been carried out on aqueous processing of oilseeds (Hagenmaier *et al.*, 1973; Kim, 1990). Several factors make the process particularly appealing to rapeseed-mustard/ canola industry. As such, most of the developmental works on the process has been done with canola/rapeseed. Although seed varieties with low levels of glucosinolates are now in use in the rapeseed/canola industry, other compounds such as tannins, snapping and phytic acids are still significantly abundant in the seeds and remain in the meal produced by conventional extraction processes.

Oil removal in case of aqueous extraction process is based on the principle of insolubility of oil in water. Microbial enzymes break down lipo-proteins and lipo-polysaccharide releasing extra oil. They degrade cell wall which helps in easy extractability of oil. Grinding tears off cell wall which reduces the resistance to oil transfer. Higher moisture content aids in cell wall break down. The dissolution of proteins at the pH far from its isoelectric points destabilizes the matrix and facilitates the easy removal of oil droplets from the cell into the aqueous phase. Followed by centrifugation of the matrix separates oil, aqueous medium and solids. This, in turn appeared in different layers based on their respective density.

Enzyme assisted aqueous extraction of oil from oilseed involves treating the comminuted oilseed with enzymes, keeping it foı some time, extraction of oil droplets in aqueous medium and then separation of oil by centrifugation. Centrifugation separates the mixture into four phases viz, oil, cream, aqueous and cake. The enzymatic-aqueous process yields a final meal with significantly reduced levels of these compounds. The process for canola begins with flaking and heating to inactivate myrosinase and fine milling. Slurry of the milled seeds is prepared and the pH adjusted to the optimum of the enzyme system. The required dosage of enzyme is added and the slurry incubated at the optimum temperature of the enzyme system for a desired residence time. The slurry is then centrifuged to yield a clear crude oil, a protein oil emulsion and an aqueous fraction containing solubilized compounds. Separating the liquid fraction from the solids on a decanter centrifuge before recovering the oil could improve oil yield.

The aqueous extraction process was developed as an alternative to the solvent oil extraction process in the 1950's (Rosenthal *et al.*, 1996). It was thought to be safe and cheap with simultaneous recovery of oil and protein from oil-bearing materials (Cater *et al.*, 1974). The hot water flotation method for oil extraction from oilseed is a traditional process used in the rural areas of many developing countries (Southwell and Harris, 1991). It includes five basic steps: (1) heat conditioning of the seed, (2) grinding, (3) extraction by boiling, (4) oil recovery, and (5) drying (Rosenthal *et al.*, 1996). Aqueous extraction of peanuts was first described by Sugarman (1956) in which peanuts were flaked and wet ground with an addition of water and alkali to maintain a pH of 7.5. The slurry was diluted with water at constant pH, heated to 80 °C and passed through a centrifuge to separate solids, emulsion phase and the aqueous protein solution. The solids were dried and the protein from the aqueous solution was precipitated by lowering the pH and separating the protein precipitate by centrifugation or filtration. The emulsion was broken by heating; controlling its pH and moisture content. High shear and pressure stress can be used to break the emulsion. The stream was then sent to another centrifuge where oil, solids and aqueous phase were obtained.

Subrahmanyan *et al.*, (1959) and Bhatia *et al.*, (1966) extended Sugarman extraction system with slight modification in two points: milling was used to extract most of the oil, and the conditions under which oil was recovered. In this process 86% of oil and 71% of protein was recovered. The low oil recovery was attributed to a low efficiency in the three phase separator where 5% of the oil was not recovered from the aqueous solution. The oil had good quality with low colour and free fatty acid (FFA) contents. The protein product had a composition of 85% protein, 9% fat, 6% moisture and 0.4% ash and the carbohydrate meal had 16.2% protein, 6.1% fat, 12% moisture, 41.6% starch, 10.3% crude fiber and 6% ash contents. Bhatia *et al.* (1966) improved this process and reported 91-94% oil recovery and protein products with 4% fat.

Pandey (2000) studied the effect of pH (2.5-4.5), solid to water ratio (1.2-1:6), extraction time (30-90 min) and temperature (30-70 °C) on aqueous extraction of oil from enzyme treated sesame seed. Mixed activity enzyme prepared from *Aspergillus fumigatus* was used for enzyme treatment. Extraction was carried out on ground sample with continuous stirring at the designed experimental combinations as per the work plan (Fig. 2). The oil was separated from the slurry by centrifugation at 10,035g for 10 min. The protein content and residual oil content of deoiled cake and water soluble protein loss were measured. The oil yield of the process varied from 14.22 to 32.17% (db). The protein content of deoiled cake ranged from 33.05 to 48.7% on oil and moisture free basis. Optimum aqueous extraction parameters were: time = 65.2 min, temperature = 47.6⁰C, pH = 4.46, S: W =1:3.7. Treated seeds extracted under optimal condition found to have enhanced oil yield up to 5.7% over untreated sample. The pH showed highly significant effect on extraction of oil yield from sesame seed.

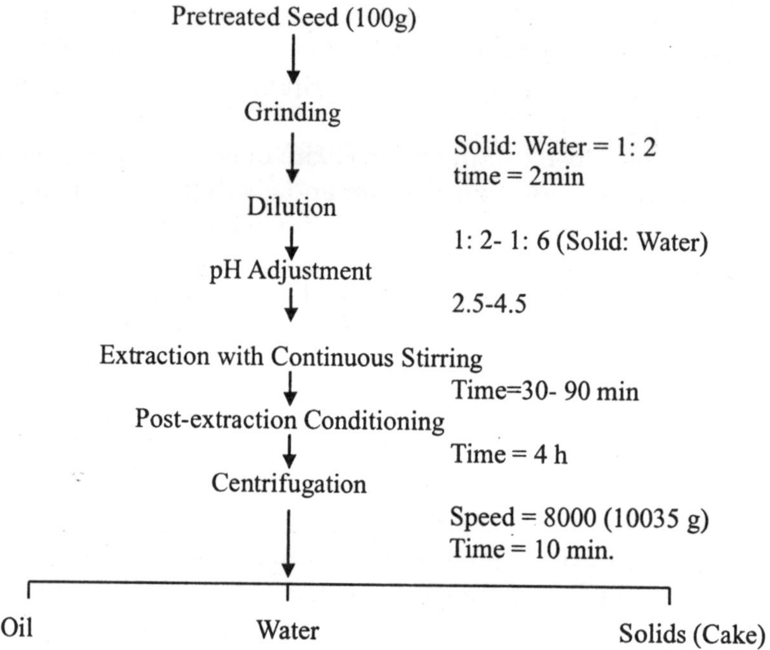

Fig. 2 : Aqueous extraction of enzyme pretreated sesame seed

Hydrolytic enzyme was initially investigated by Sherba *et al.* (1972) for partially releasing oil from full fat, heat-treated soya meal. Fullbrook (1983) hydrolyzed oilseeds in aqueous medium followed by solvent extraction. The hydrolysis was also tried in the presence of a solvent to simultaneously extract the released oil. It was observed that yields could be considerably improved if hydrolysis of the finely ground flour of soybean and rapeseed was carried out in the presence of solvent. By using 3% enzymatic mixture obtained from *Aspergillus niger,* 50% more oil was obtained for rapeseed and 90% of the original extractable oil for soybeans (Fullbrook 1983). Olsen (1988) also performed aqueous hydrolysis of dehulled rapeseed with enzyme preparation including: pectinase, cellulase, and hemicellulase followed by extraction of the residual oil with petroleum ether. In these studies the enzymatic hydrolysis increased the permeability of the seed cell wall for more efficient extraction of the oil. Enzymatic processes had been tried on the seed cell walls to facilitate oil extraction (Cintra *et al.,* 1986; Sosulski *et al.,* 1988; Frevert *et al.,* 1990;

Ho *et al.*, 1992; Ohlson, 1992; Sosulski and Sosulski, 1993; Rosenthal *et al.*, 1995). The use of hydrolytic enzymes such as cellulases, hemicellulases, and pectinases in aqueous enzyme-assisted extraction process break the structure of cotyledon cell walls which makes the structure more permeable. Protease and carbohydrase enzymes are mainly used for this process. The proteins in the cell membranes and inside the cytoplasm are hydrolyzed by the proteolytic enzymes (Rosenthal *et al.*, 1996). Proteolytic enzymes can potentially hydrolyze the lipid body membranes and can also affect the cytoplasmic network thus make the inner structure less tightly bound/compact which facilitates the protein and lipid removal from the cell (Rosenthal *et al.*, 1996). The released oil can be easily separated from the cotyledon cells by an aqueous medium. An enzyme-aided process has also been used for soybeans (Rosenthal *et al.*, 1995) and coconuts (Cintra *et al.*, 1986). The enzyme efficiency based on oil yield enhancement was: mixed activity enzyme > β-glucanase > pectinase > hemicellulase> cellulase (Rosenthal *et al.*, 1996).

For palm oil recovery from the centrifuge sludge of a palm oil mill cellulase, α-amylase, pectinase and protease enzymes were utilized (Ho *et al.*, 1992). A surfactant (SDS) was tried to release the oil into the aqueous phase because the oil remains attached to the plant material even with high speed centrifugation (Ho *et al.*, 1992). Although they optimized conditions for cellulase, pH 4.8 and 30°C for 5 h with and enzyme concentration of 0.5% (w/v), followed by a washing with 0.03 M SDS solution but still it was not economical viable. A mixed solution of α-amylase, protease and poly-galacturonase (PG) was used for coconut oil extraction (Cintra *et al.*, 1986). Under optimal conditions: an enzymatic digestion with 0.1% (w/v) α-amylase, 0.1% (w/v) PG and 0.1% (w/v) protease; (4:1) water to coconut ratio; 80% of oil was extracted with an enzymatic treatment at 40°C for 10 min.; and, centrifugation for 10 min at 12300 x g. An excellent quality of oil was obtained by this process. The hydrolysis was carried out at 50°C for 4 h using 2% (w/w) enzyme and the oil recovered by this process had a very good quality.

Rosenthal *et al.* (1995) used proteolytic and cell-wall degrading enzymes for protein and oil extraction from soybeans but did not observe a significant improvement similar to those reported by Fullbrook (1983) and Ho *et al.* (1992). The particle size and the extraction steps significantly affected the oil and protein extraction from soybeans. The particle size smaller than 150 μm and three or more extraction stages were found to be the optimal conditions for 75% oil and 80% protein. The highest protein and oil extraction yield of both were observed with protease for the hydrolysis of heat treated flour, which depicts that the solubility of protein is very important in the enzymatic extraction of soybean oil.

Enzyme treatment on wet ground sesame seed followed by aqueous extraction increases the oil yield over that of aqueous extraction. The solid-water ratio, pH and extraction period are the most significant parameters that may affect enzyme assisted aqueous extraction process. Enzyme concentration, incubation period and extraction period may also have an interactive effect on oil recovery. The incubation period, solid to water ratio, pH, extraction period and temperature affect the residual oil content in cake. Protein content in cake was affected by solid to water ratio, pH, extraction period and temperature. Protein content in aqueous medium may also be affected by enzyme concentration, solid to water ratio, pH and temperature. The oil obtained is reported to have the quality comparable to that of conventional extracted oil and the phosphorous content of oil in enzymatic process is lesser. But the aqueous extraction has several disadvantages such as:

(a) Yield is low.

(b) The versatile protein, oleoresin, binds with about 20% of the oil in oil bearing leading to lower yield.

(c) Higher use of protease enzyme results in improved oil recovery but produces excessive bitter meals

(d) Effluent disposal problems as large amount of water are used.

(e) Waste water contains dissolved substances such as proteins making the process costly.

Factors affecting oil yield during EACP and EAAE

The unit operations in the aqueous process reported by Cater *et al.* (1974) and Lawhon *et al.* (1981) include grinding, extraction parameters, solid-liquid separation and product recovery.

Grinding of oilseeds

The whole oilseeds or fruits must be ground to fine particles for more efficient extraction of water-soluble components (Cater *et al.*, 1974; Lawhon *et al.*, 1981). Wet or dry grindings are normally used depending on the type of oleaginous materials (Cater *et al.*, 1974). For high moisture content oleaginous materials (coconuts and palm kernel) ,wet grinding is used (Cater *et al.*, 1974; Kim, 1990) and dry grinding is suitable for low moisture oilseeds sunflower seeds, peanut and soybean (Hagenmaier,1974; Rhee *et al.*, 1972; Rustom *et al.*, 1991; Rosenthal *et al.*, 1998). It has been reported that significantly higher protein and oil extraction yields was obtained from smaller particles of non-heat treated soybean flour than the larger sizes (Rosenthal *et al.*, 1998).

Although for peanuts, maximum recovery was observed with fine particle size but a very stable emulsion was also observed in aqueous media (Rhee *et al.*, 1972; Latif, 2009).

Extraction parameters

In enzymatic extraction, the ground seeds are mixed with water, and then agitated to increase extraction. The main parameters that influence the enzymatic extraction yields include enzyme mixture, enzyme concentration, particle size of oilseed, pH, solid-liquid ratio, temperature and time (Cater *et al.*, 1974; Lawhon *et al.*, 1981). Buenrostro and Lopez-Munguia (1986) reported better oil recovery for avocado with 1% (w/w) α-amylase; paste-to-water ratio of 1:5 at 65°C for 1.5 h. The pH during the process depends on the composition of oilseed. In aqueous process, the aim is to obtain high oil and protein yields but it's not possible at a pH value near the isoelectric point of protein, because at this pH, protein can bind the oil in a very stable emulsion (Rosenthal *et al.*, 1996). For peanut, a pH of 4-5 was found to be suitable for oil extraction with no emulsion phase and protein can be extracted easily at a pH of 8.0 (Rhee *et al.*, 1972). Different studies on enzymatic processes were conducted at different temperatures and the optimum temperature range (40-60°C) was reported for several oilseeds (Rhee *et al.*, 1972; Lawhon *et al.*, 1981; Kim, 1990; Rustom *et al.*, 1991; Latif, 2009).

Rosenthal *et al.* (1998) reported that the protein yield was decreased slightly when temperature was higher than 50°C. No significant increase in the palm oil yield was reported when extraction temperature was raised above 45°C (Kim, 1990). For peanut oil extraction, a temperature range of 60-64°C, and a lower temperature of 40-44°C found to be suitable for protein extraction (Rhee *et al.*, 1972). An unfavorable effect of temperatures greater than 60°C was observed on protein extraction yield from peanuts (Rustom *et al.*, 1991). This increased temperature may cause denaturation of protein. The extraction time specially depends not only on the oilseeds type and temperature but also on pH during the extraction process. According to Rhee *et al.* (1972) maximum oil and protein yields from peanut were obtained within 30 min, while Rustom *et al.* (1991) reported that increasing extraction times from 15-40 min had no significant effect on the yields. In another study, it was mentioned that an extraction time of 15 min was sufficient for protein and oil extraction from soybean flour (Rosenthal *et al.*, 1998). Solid-to-liquid ratio is another important parameter that allows the products to be dispersed in the aqueous media. A high amount of water will produce a less stable emulsion and will ultimately simplify the separation process (Rosenthal *et al.*, 1996).

Further research need

Research conducted in India in the area of bioconversion processing of oilseeds is very limited and mainly basic in nature, and the technology is still under developmental stage. Further studies as mentioned below are needed before its application in the oilseed industries.

- Studies on preservation of crude enzyme and enzyme activity on storage conditions.
- Optimisation of process parameters and product yield.
- Electron microscopic study to elucidate the mechanism of enzyme action.
- Toxicological effect on oil cake and oil due to enzyme pre-treatment.
- Cost economics of the process.
- Scale up of the process.

Future prospects

Based on the reports available so far, it may be mentioned that new pre-treatment process can enhance oil recovery in the range of 2-5%. Implementation of newer developments and newer technology can further help to enhance the oil recovery and reduce the persistent gap of supply and demand. This suggests that, if the technology is perfected and applied in the oilseed processing industries, the gap between supply and demand of edible oil can be reduced. However, the efforts are needed to explore the techniques in terms of their economic viability and commercial feasibility.

References

Achaya, KT, (1994). Ghani: a traditional method of oil processing in India. Food, Nutrition and Agriculture. 11: 23-28.

Agrawal, YC, Tikkoo, AK, Singh, BPN and Maharaj Narain, (1988). Post harvest technology of oils: Problems and perspectives. Bulletin of Grain Technology, 26: 99-107.

Anjou, K, (1972). Manufacture of rapeseed oil and meal. In: Rapeseed: cultivation, composition, processing and utilization (edited by L. Appelqvist and R. Ohlson). Elsevier Publishing Co., New York.

Balke, DT, (2006). The production of higher value food ingredients from white mustard via aqueous extraction. Ph.D. Thesis. Department of Chernical Engineering and Applied Chemistry, University of Toronto, Canada.

Bargale, PC, (1997). Mechanical oil expression from selected oilseeds under uniaxial compression. Ph.D. thesis, University of Saskatchewan, Canada.

Barrios, VA, DA Olmos, RA Noyola, and CA Lopez-Munguia, (1990). Optimization of an enzymatic process for coconut oil extraction. Oleagineux. 45:35-42.

Beach, D, (1983). Rapeseed crushing and extraction. In high and low erucic acid rapeseed oils, eds. J.K.G. Kramer, P.D. Sauer and W.J. Pigden, Toronto ON: Academic Press. 181-197.

Bhatia, DS, HAB, Darpra, and BP, Balga, (1966). Peanut protein isolate-production and properties. Journal of Food Science and Technology. 3: 2-5.

Bhatnagar, S, and BN Johari, (1987). Microbial enzymes in the processing of oilseeds. Current Science. 56: 775-776.

Bouvier, F, and B, Entressangles, (1992). Utilization of cellulases and pectinases in the extraction of palm oil. Revue Fruncaise des Corps Gras. 39: 245-252.

Buenrostro, M, and CA, Lopez-Munguia, (1986). Enzymatic extraction of avocado oil. Biotechnology Letters. 8: 505-506.Bulley, N.R. 1984. Supercritical fluid extraction of vegetable oilseeds. Journal of American Oil Chemist's Society. 61: 1362-1365.

Caragay, AB, (1983). Pacing technologies in the fats and oils industry. Journal of American Oil Chemist's Society. 60: 1641-1644.

Cater, CM, KC, Rhee, RD, Hagenmaier, and KF, Mattil, (1974). Aqueous extraction an alternative oilseed milling process. Journal of American Oil Chemist's Society. 51: 137-141.

Caviedes, J, (1996). Aqueous processing of rapeseed (canola). M.Sc. Thesis. Department of Chemical Engineering and Applied Chemistry, University of Toronto, Canada.

Christensen, FM, (1989). Enzyme technology versus engineering technology in the food industry. Biotechnology Applied Biochemistry. 11: 249-265.

Cintra, MO, A, Lopez-Munguia, and J, Vernon. (1986). Coconut oil extraction by a new enzymatic process. Journal of Food Science. 51: 695-697.

Dahlen, JAH, and LA, Lindh, (1983). Mass balance of hexane losses in an extraction plant. Journal of American Oil Chemist's Society. 60: 2009-2010.

Dominguez, H, MJ, Nunez, and JM, Lema, (1993). Oil extractability from enzymatically treated soybean and sunflower: range of operational variables. Food Chemistry. 46: 277-284.

Dominguez, H., M. J. Nunez, and J. M. Lema. (1994). Enzymatic pretreatment to enhance oil extraction from fruits and oilseeds: a review. Food Chemistry. 49: 271- 286.

Dominguez, H, MJ, Nunez, and JM, Lema, (1995). Enzyme-assisted hexane extraction of soya bean oil. Food Chemistry. 54: 223-231.

Frevert, J, R, Frische, J, Hart, and J, Wittkind, (1990). Enzymatic solvent less recovery of oils from plant materials. Ger. Offen. DE 3843027.Fullbrook, P.D. 1983. The use of enzymes in the processing of oilseeds. Journal of American Oil Chemist's Society. 60: 428A-430A.

Fullbrook, PD, (1983). The use of enzymes in the processing of oilseeds. J. Am. Oil Chem. Soc. 60:476-478.

Hagenmaier, RD, CM, Cater, and KF, Mattil, (1973). Aqueous processing of fresh coconuts for recovery of oil and coconut skim milk. Journal of Food Science. 38:516-518.

Hagenmaier, RD, (1974). Aqueous processing of full-fat sunflower seeds; yields of oil and protein. Journal of American Oil Chemist's Society. 51: 470-471.

Head, S, and A, Swetman, (1999).Traditional methods for processing oilseeds. Inform. 10: 151-154.

Ho, CC, MC, Chow, and SH, Ong, (1992). Recovery of residual oil from centrifuge sludge palm oil mill: effect of enzyme digestion and surfactant treatment. Journal of American Oil Chemist's Society. 69: 276-282.170.

Igor, SO, LL, Disoady, and LJ, Rubin, (1993). Catalytic deamidation of canola proteins. Acta Alimentaria, 22: 325-336.

Keegstra, K, KW, Talmadge, WD, Bauer, and P, Albersheim, (1973). The structure of the wall of suspension-cultured sycamore cells based on the interconnections of the macromolecular components. Plant Physiology. 51: 188-196.

Kim, IH, and SH, Yoon, (1990). Effect of extraction solvents on oxidative stability of crude soybean oil. Journal of American Oil Chemist's Society, 67: 165-167.

Latif, Sajid (2009). Analytical Investigations to compare the enzyme-assisted extraction of vegetable oils with conventional methods. Ph.D. Thesis., University of Agriculture, Faisalabad.

Lawhon, JT., KC, Rhee, and EW, Lusas, (1981). Soy protein ingredients prepared by new processes-aqueous processing and industrial membrane isolation. Journal of American Oil Chemist's Society. 58: 377-384.

Lusas, EW, JT, Lawhon and KC, Rhee, (1982). Producing edible oil and protein from oilseeds by aqueous processing. Oil Mill Gazet. 4: 28-34.

Mustakas, GC, (1980). Recovery of oil from soybeans. In: Handbook of Soy Oil Processing and Utilization (Erickson, D. R. Ed.). American Soybean Association and American Oil Chemists' Society, St. Louis. 49-65.

Naczk, M, LL, Diosady, and LJ, Rubin, (1985). Functional properties of canola meals produced by the two-phase solvent extraction system. Journal of Food Science. 50 : 1685-1692.

Ohlson, R, (1992). Modern processing of rapeseed. Journal of American Oil Chemist's Society. 69: 195-198.

Olsen, HS, (1988). Aqueous enzymatic extraction of oil from seeds. In: Asian food conference proceedings Bangkok, Thailand. Reprinted by: Novo Industry A/S, pp A-0604la.

Othmer, DF, and JC, Agrawal, (1955). Extraction of soybeans. Theory and Mechanism. Chem. Eng. Prog. 51: 372-8.

Owusu-Ansah, YJ, (1995). Enzymes in lipid technology and coco butter substitute. In: Technological advances in improved and alternate sources of lipids. Ed. B.S. Kamal and Y. Kakuda, Blackie Academic and Professional.

Pandey, S, (2000). Aqueous oil extraction from enzyme pretreated sesame seed and its optimization.M. Tech Thesis submitted to G. B. Pant University of Agriculture and Technology, Pantnagar.

Rhee, KC, CM, Cater, and KF, Mattil, (1972).Simultaneous recovery of protein and oil from raw peanuts in an aqueous system. Journal of Food Science. 37: 90-93.

Rosenthal, A, DL, Pyle, and K, Niranjan, (1995). Simultaneous aqueous enzymatic extraction of oil and protein from soybean. Personal communication. University of Reading, Reading, England.

Rosenthal, A, DL, Pyle, and Niranjan, K, (1996). Aqueous and enzymatic processes for edible oil extraction. Enzyme Microbial Technology 19: 402–420.

Rosenthal, A, DL, Pyle, and K, Niranjan, (1998). Simultaneous aqueous extraction of oil and protein from soybean: mechanisms for process design. Trans Institute of Chemical Engineers. 76: 224-230.

Rubin, LJ, LL, Diosady, M, Naczk and M, Halfani. (1986). The alkanol-ammonia-Water/ hexane treatment of canola. Canadian Institute of Food Science and Technology Journal. 19: 547-561.

Rustom, IYS, MH, Lopez-Leiva and BM, Nair, (1991). A study of factors affecting extraction of peanut (Arachis hypogaea L.) solids with water. Food Chemistry. 42: 153-165.

Sarkar, BC, Agrawal, YC, Singh, BPN, and Gupta, DK, (1998). Optimization of enzyme pre-treatment of rapeseed for enhanced oil recovery. Journal of Food Science and Technology, 35: 183-186.

Sherba, SE, RB, Steigerwalt, WT, Faith, and CV, Smythe Jr. (1972). Soybean fractionation employing a protease. U.S. patent 3: 640-725.

Shukla, Banshi D, Srivastava, Prabhat K, Gupta, Ram K, (1992). Book on Oilseeds Processing Technology.

Smith, DD, YC, Agrawal, BC, Sarkar, and BNP, Singh (1993). Enzymatic hydrolysis pretreatment for mechanical expelling of soybeans. Journal of the American Oil Chemists' Society.70: 885-890.

Sosulski, K, FW, Sosulski, and E, Coxworth (1988). Carbohydrate hydrolysis of canola to enhance oil extraction with hexane. Journal of American Oil Chemist's Society. 65: 357-361.

Sosulski, K, and FW, Sosulski (1990). Enzyme pre-treatment to enhance oil extractability in canola. In: Canola and Rapeseed: Production, Chemistry, Nutrition and Processing Technology (edited by F. Shahidi), Pp 277-289. New York NY: Van Nostrand Reinhold Publishers.

Sosulski, K, and Sosulski, FW (1993). Enzyme aided vs. two stage processing of canola: Technology, product quality and cost evaluation. J. Am. Oil Chemists Soc., 65: 357-361.

Southwell, KH, and RV, Harris (1991). Extraction of oil from oilseeds using the hot water flotation method. Tropical Science. 32: 217-221.

Subrahmanyan, J, DS, Bhatia, SS, Kalbag, and N, Subramanian (1959). Integrate processing of peanut for the separation of major constituents. Journal of American Oil Chemist's Society. 36: 66-70.

Sugarman, N, (1956). Process for simultaneously extracting oil and protein from oleaginous material. U. S. Patent No. 2 762 820.

Talmadge, KW, K, Keegstra, WD, Bauer and P, Albersheim, (1973). The structure of plant cell walls I. The macromolecular components of the walls of suspension cultured sycamore cells with a detailed analysis of the pectic polysaccharides. Plant Physiology. 51: 158-173.

Tano-Debrah, K, and Y, Ohta (1994). Enzyme-assisted aqueous extraction of fat from kernels of the shea tree Butyrospermum parkii. Journal of American Oil Chemist's Society. 71: 979-983.

Tano-Debrah, K, and Y, Ohta (1995a). Enzyme-assisted aqueous extraction of shea fat: A rural approach. Journal of American Oil Chemist's Society. 72: 251-256.

Tano-Debrah K, and Y, Ohta (1995b). Application of enzyme assisted aqueous fat extraction to cocoa fat. Journal of American Oil Chemist's Society. 72: 1409-1411.

Thijs Adriaans, (2006). Suitability of solvent extraction for jatropha curcas.

Tzeng, YM, LL, Diosady and LJ, Rubin (1990). Production of canola protein materials by alkaline extraction, precipitation and membrane processing. Journal of Food Science. 55: 1147.

Valentin, FHH, (1992). Volatile organic compounds and odors. Environmental Health. Aug. 224-227.

Ward, JA, (1982). Pre pressing of oil from rapeseed and sunflower. Journal of American Oil Chemist's Society. 61: 1358-1361.

Wijeratne, Wilmot B, Wang Tong and Johnson Lawrence A, (2010). Book on Nutritional Enhanced Edible oil Processing. Chapter 4: Extrusion-Based Oilseed Processing Methods.

Yoon, SH, IH, Kim, SH, Kim, and TW, Kwon, (1991). Effects of enzyme treatments and ultrasonification on extraction yields of lipids of protein from soybean by aqueous process. Korean Journal of Food Science and Technology. 23: 673-676.

□□□

6

Novel Products from Groundnut

D.N. YADAV

Groundnut pod profile

There are four botanical types of groundnut namely Virginia runner, Virginia bunch, Valencia and Spanish bunch and they differs in their chemical composition and oil quality. Virginia bunch type seeds are richer in oil and chemical contents followed by Spanish bunch. Protein content is higher in Valencia, while soluble sugars are higher in Virginia runner seeds. Highest oleic acid has been observed in Valencia type. The ratio of oleic to linoleic acid is higher in runner type, with good oil keeping quality. The nutritional quality is higher in Valencia types. In general, it can be stated that Virginia runner types have better chemical composition with balanced oil quality.

Importance and distribution

India along with china accounts for half of the world's groundnut production today. Groundnut is the major oilseed crop in India accounting for 45% of oilseed area and 55% of oilseed production in the country. India is rated as the second largest producer of groundnut in the world with annual production of over 5-6 million tons. Gujarat, Andhra Pradesh, Tamilnadu and Karnataka are the leading producers in the country and accounts for nearly 75% of the total output. Global production of groundnut seed and oil oscillates between 21-24 and 4-5.5 million tons.

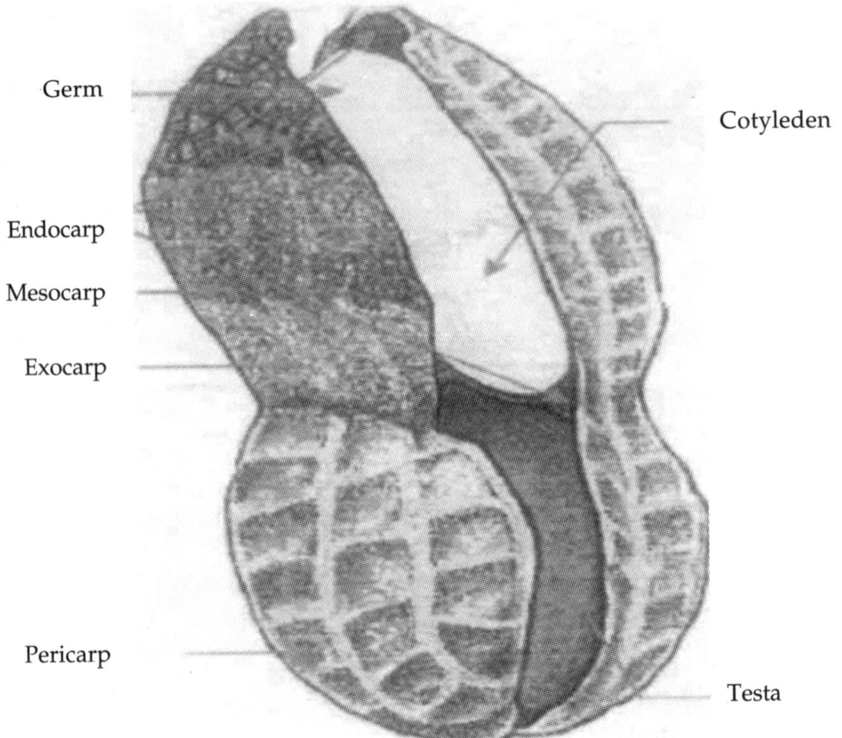

Medicinal value

- Groundnut is one of the most nourishing foods available in the world.
- The U.S. department of agriculture considers it having maximum five important nutrients such as food energy, protein, phosphorus, thiamine and niacin.
- Hence apart from their nutritional value, groundnuts have considerable medicinal value. It contains a good deal of oil which is very easily digested and for this reason they are useful consumptives.
- Eating fresh roasted groundnut with jaggery and goat's milk is a very nutritious food for growing children, pregnant women and nursing mothers.
- It builds a resistance against all infections particularly tuberculosis and hepatitis.

Excessive Bleeding : Several groups of researchers in England have reported using groundnuts or groundnut products in the treatment of haemophilia, an inherited blood disease which causes haemorrhage, beneficial results have also been reported from the use of groundnuts in severe cases of epistaxis or nose bleeding and in case of excessive bleeding during menstruation in women.

Diabetes : Groundnuts are valuable in diabetes. Eating a handful of groundnuts daily by diabetics will not only prevent malnutrition, particularly the deficiency of niacin, but also checks the development of vascular complications.

Diarrhea : Groundnuts are also useful in diarrhea, especially chronic diarrhea which is more frequent immediately after meals. The patients can benefit greatly by drinking goat's milk in which lemon is squeezed with a handful of fresh roasted groundnuts. This type of diarrhea is caused due to nicotinic acid deficiency. The groundnuts, which has required quantity of niacin, are valuable in this disease.

Teeth disorder : Chewing fresh groundnuts with a pinch of salt strengthens the gum, cures stomatitis, kills harmful bacteria and safeguards the enamel of teeth. The mouth should however be washed with water after eating groundnuts.

Beauty aid : Groundnut oil can serve as a beauty aid. A teaspoon of refined groundnut oil, mixed with equal quantity of lime juice, may be applied daily on the face once before going to bed, it keeps the face fresh. Its regular use nourishes the skin and prevents *acne.*

The groundnuts are consumed in many ways and various forms. They may be eaten raw, boiled, steamed or roasted. They are sometimes eaten as a sweetmeat by coating them with sugar. In the west, they form important constituents of confections. They are also taken in the form of nut chocolates, candies and sandwiches.

Nutritional value of edible portion of groundnut per 100 g.

Type of Groundnut	Energy (Kcol)	Protein (g)	Fat (g)	Ca (mg)	Fe (mg)	Thiamin (mg)	Riboflavin (mg)	Niacin (mg)	Vit. A (µg)
Groundnut	567	25.3	40.1	90	2.8	0.90	0.13	19.9	37
Roasted Groundnut	570	26.2	39.8	77	3.1	0.39	0.13	22.1	0

Source: Gopalan C (1971)

Top 10 producers of peanuts (2008)

Country	Production (Million Tonnes)
China	13.09
India	6.60
Nigeria	3.84
United states	1.70
Indonesia	1.48
Myanmar	1.0
Argentina	0.71
Vietnam	0.49
Sudan	0.46
Chad	0.45
World	34.86

Source : www.faostat.org

Utilization: Present scenario

Total in shells	7.29 million tonnes
Equivalents in kernels (70%)	5.1 million tonnes
Retained for showing, direct consumption and export	1.9 million tonnes
Available kernels for crushing	3.2 million tonnes
Total oil availability (40%)	1.28 million tonnes
Deoiled cake	1.92 million tonnes
Protein (34.0%)	0.653 million tonnse

6.53 lakh tonnes high quality protein is presently being utilized as animal feed. In the country like India, where we are facing the problem of protein deficiency particularly in rural areas, utilization of this high quality peanut protein into some popular Indian products will be justifiable.

Importance of the project

Groundnut is the largest oilseed in India in terms of production. It accounted for 35.99 per cent of the oilseeds production of the country during 2007-08. Groundnut contains on an average 40.1 per cent of fat and 25.3 per cent of protein and is a rich source of calcium, iron and vitamin B complex like thiamine, riboflavin, niacin and vitamin A. The quality characteristics and uses of groundnut vary among the developed and developing countries. In developed countries, groundnut is used for the preparation of peanut butter and confectionary products. In developing countries, it is mainly used for oil extraction and its byproduct is utilized for animal feed purposes. Animal milk, in India at any rate, has a venerated place as a food for both children and adults. A milk-like beverage, or a re-constitutable powder from which to generate it by adding water, represents an acceptable way of furnishing proteins and other nutrients to all age groups, often at reasonable cost.

Milk supply in most countries of the world specially, in the developing countries is limited. Keeping in view of the shortage of milk, scientist all over the world worked extensively to find out some substitute for milk which looks like milk but not from the animal source. They have succeeded in their endeavors and the outcome of their strenuous efforts is the manufacture of imitation milks. Imitation milks are those milks, which are made from food materials of vegetable origin such as groundnut. Imitation milks may serve as a boon for the countries where the supply of milk is inadequate. In developed countries, imitation milks can save the masses as heart ailments, which are considered safer than saturated fatty acid. Groundnut and soybean are two major raw material used for preparation of imitation milks.

Groundnut has a potential role to play in combating malnutrition, the present low level in its consumption, especially in the developing countries, should be increased. Preparation of peanut extract (milk like product) and its utilization in the development of dairy analogues such as flavoured beverage, curd, yoghurt, *lassi* and paneer may be one such area of utilization.

Objectives

- Value addition of locally available groundnut through scientific small scale processing
- Providing dairy analogue to the consumers at reasonable prices so as to reduce malnutrition
- To generate self employment and employment to rural skilled labour.

Activities to be undertaken

- Flavoured beverage
- Peanut curd
- Paneer
- Utilization of peanut okra (by-product)

Assumptions

- Capacity of the plant: 300 lt/ day
- Capacity utilization: 80 percent
- Space requirement: 15-20 m^2
- Land and building will be obtained on rent
- Working days in a year: 300 days
- Working hours per day: 8 h
- Recovery @ 7 liter per kg of peanut kernel
- Life span of the machineries: 10 years
- Interest rate @ 11 percent per annum
- Electricity charges @ Rs. 5/ kwh
- Repair and maintenance charges @ 1 percent of the cost of the machine
- Salvage value: 10 percent of the present cost
- Fixed cost (i.e. cost of machinery and equipments) has been annualized.

Proposed marketing channels

- Factory outlet – consumers
- Firm – salesman – consumers
- Firm – hotel and restaurants (paneer and curd) – consumers
- Firm – salesman – food corners (educational institutes and public places) – consumers
- Firm own outlets (thela) at public places – consumers
- Firm – government agencies (through Anganwari) – consumers

Peanut (Full fat) flavoured beverage

It is milky white product prepared from peanut kernels. It is economical, highly digestible, anti nutrients free, flatulence free and nutritious. The proximate composition of the most acceptable beverage prepared by following the given method (Fig. 1&4) was found to be moisture 81.2 ± 0.5%, ash 0.16 ± 0.05%, fat 3.15 ± 0.51%, protein 3.27 ± 0.34%, carbohydrate 12.2% and total solids 18.8 ± 0.5% on fresh weight basis.

Fig. 1 : Peanut flavoured beverage

Peanut (Full fat) curd

Fermented milk product i.e. curd was developed by using peanut milk (Fig. 2&5) and it is sweet in taste, whitish in colour and hasall the nutritional benefits of peanut consumption. The developed curd sampleshad moisture 80.77 ± 1.5%, protein 3.45 ± 0.5%, fat, 3.5 ± 0.4%, ash 0.5 ± 0.1carbohydrate 11.78% and total solids 19.23 ± 0.5% on fresh weight basis.

Fig. 2 : Peanut curd

Peanut (Full fat) paneer

Fig. 3 : Peanut paneer

Peanut paneer is tasty and very nutritious product made by coagulating hot peanut milk with food grade coagulants (Fig. 3 & 6). It is a very versatile food and can be converted into variety of value-added products such as paneer *pakoda*, paneer *tikki*, *kadhai* paneer etc. The paneer had moisture 62.1 ± 0.8%, Protein 13.6 ± 0.5%, Fat 19.2 ± 1.5%, ash, 1.9 ± 0.2% and carbohydrate 3.2 ± 0.1%. Technology for making peanut paneer with variable fat content is also available.

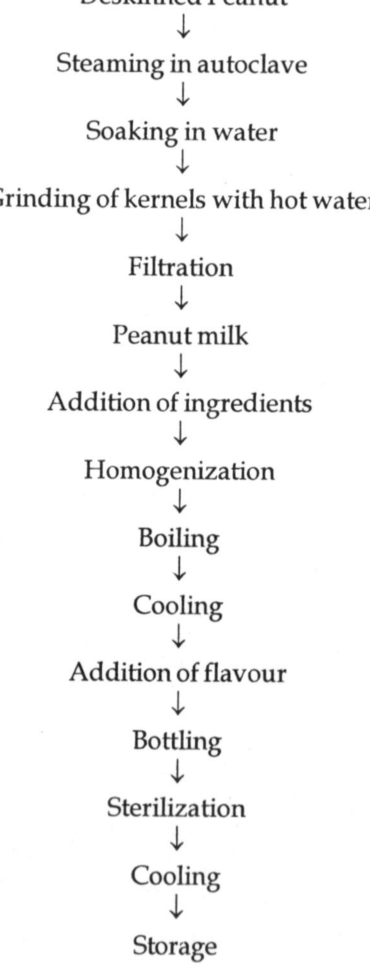

Fig. 4 : Preparation of Peanut Flavoured Beverage

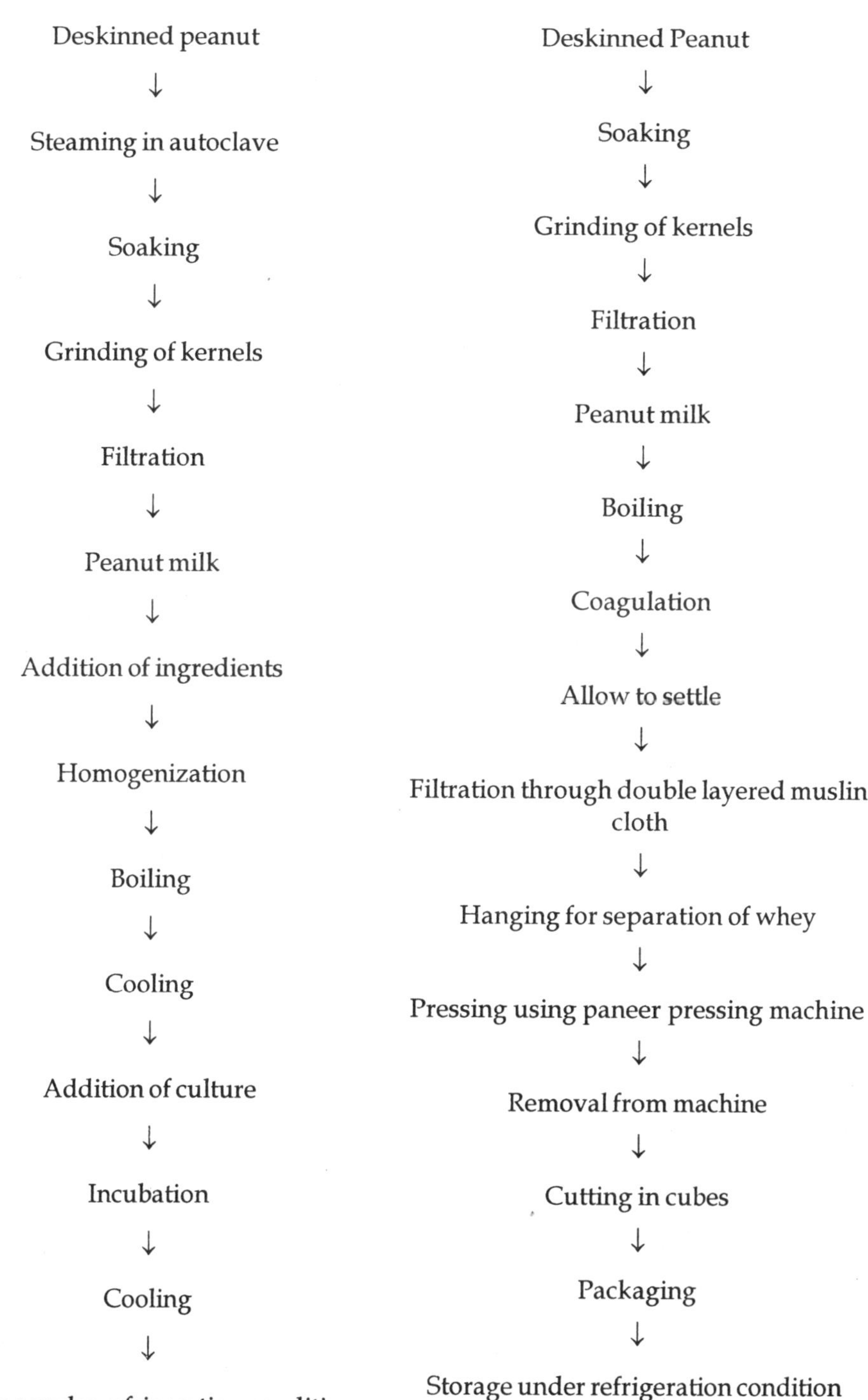

Deskinned peanut
↓
Steaming in autoclave
↓
Soaking
↓
Grinding of kernels
↓
Filtration
↓
Peanut milk
↓
Addition of ingredients
↓
Homogenization
↓
Boiling
↓
Cooling
↓
Addition of culture
↓
Incubation
↓
Cooling
↓
Storage under refrigeration condition

Fig. 5 : Flow diagram of Curd production

Deskinned Peanut
↓
Soaking
↓
Grinding of kernels
↓
Filtration
↓
Peanut milk
↓
Boiling
↓
Coagulation
↓
Allow to settle
↓
Filtration through double layered muslin cloth
↓
Hanging for separation of whey
↓
Pressing using paneer pressing machine
↓
Removal from machine
↓
Cutting in cubes
↓
Packaging
↓
Storage under refrigeration condition

Fig. 6 : Flow diagram for preparation of peanut paneer

Budget estimates

A. Plant Specifications and requirements along with cost

Name of equipment/ product	Capacity	Purpose	Approximate Price (Rs/Unit)
1. Flavoured Beverage			
Groundnut decorticator	100 kg /h	For decortications of groundnut pods	10000
Tray dryer	48 tray	Drying of peanut kernels to remove red skin	70000
Vessels	30 lt capacity, 4 nos.	For soaking of peanut kernels and collection of milk	10000
Autoclave	152 lt 550 x 750 mm	For cooking of peanut kernels and sterilization of milk bottles	180000
Soy / Peanut Processing Plant with deodorizer and filtration unit	300 lt/ day	For making peanut milk from kernels	350000
Homogenizer	50 lt/hr	To homogenize the peanut milk beverage	200000
Weighing Balance	-	Weighing	15000
Capping Machine	Small size	Sealing of bottles	25000
Chest freezer	280 lt		74600
2. Peanut Paneer			
Paneer Pressing Machine	15 kg		10000
Sealing Machine	Small size		5000
3. Peanut curd			
Cup filling machine	250 cup / h	Filling of milk in cups (200ml)	20000
Incubator	336 lt	Curdling of peanut milk	70000
Total amount			**1039600**

a. Annualized fixed Cost = Rs. 169190
b. Rental value of land and building = Rs. 45000
c. Interest on fixed capital (excluding land and buildings) = Rs. 18611

A. Total fixed cost (a+b+c) = Rs. 232801

B. Working capital

Product/ Material	Quantity (per day)	Annual (300 days)	Rate (Rs./ unit)	Cost/ Profit (Rs.)
I. Peanut Flavoured Beverage	100 lt.			
i) Cost of the raw material (peanut)	14.3 kg	4287 kg	50/ kg	214500
ii) Skin Milk Powder	3 kg	900 kg	160/ kg	144000
iii) Sugar	6.5 kg	1950 kg	35/ kg	68400
iv) Flavour	250 ml	75 lt	800/ lt	60000
Sub total (I)				**486900**
Final product	100 lt	30000 lt.		
Cost per bottle 200 ml. (excluding packaging cost)	500	150000		3.25
II. Curd	40 lt.			
Peanut	5.7 kg	1710 kg	50/ kg	85500
Sugar	2.6 kg	780 kg	35/kg	27300
CMC	80 g	24 kg	800/ kg	19200
SMP	1.6 kg	480 kg	160/ kg	76800
Sub total (II)				**208800**
Cost per cup (200 ml)	200	60000		
(excluding packaging cost)				3.48
III. Paneer	15 kg			
Peanut	14.3 kg	4290 kg	50/ kg	214500
Coagulant	180 g	54 kg	220/ kg	11880
Sub total (III)				**226380**
Paneer cost (Rs. / kg)	15	4500		50.31
IV. Total Cost of raw material (I+II+III) (excluding packaging cost)				**922080**
V. Packaging and bottling cost				
Empty Bottle	500	150000	1.0/ bottle	150000

Empty Cup	200	60000	0.50/ cup	30000
Packaging material for paneer	15	4500	2.5/ kg	11250
Sub total (V)				**191250**
VI. Labour, electricity and other charges				
Skilled labour	1	300	200	60000
Unskilled labour	2	600	125	75000
Electricity	20	6000	4	24000
LPG/ fuel	-	40 cylinder	350/ cylinder	14000
Telephone	-	12 months	500/ month	6000
Rent of building and land		12 months	2500/ month	30000
Repair and maintenance	-	-		10000
Transportation			2000/ month	24000
Miscellaneous			1000	12000
Sub total (VI)				**255000**
B. Grand total (IV+V+VI)				**1368330**

C. Total cost per annum (i.e. fixed cost and working cost= A+B) = Rs. 1601131

D. Annual production of beverage, curd and paneer

Production	Quantity/ day	Total quantity (300 days)	Rates/ unit	Returns (Rs.)
Beverage (bottle @ 200 ml)	500	150000	8	1200000
Curd Cups (@200 ml)	200	60000	10	600000
Paneer (kg)	15	4500	75	337500
By product (kg)	9	2700	10	27000
Total				**2164500**

E. Net Profit (D-C) = Rs. 563369

Sensitivity analysis (Profit under risk situation)

	Existing situation-1	Situation-2	Situation-3	Situation-4
Reduction in output		10%	15%	20%
Gross profit	2164500	1948050	1839825	1731600
Net profit	563369	346919	238694	130469
B/C	1.35	1.22	1.15	1.08
Pay back period	1.85	3.00	4.36	7.97
BEP	64.85	74.98	81.33	88.85

The proposed agro processing centre will be an integrated unit for processing of peanut in the production catchment for value addition and better utilization which in turn generate income and employment for the entrepreneur as well as to the associated rural poor and traders. The project is seems to be economically viable with higher benefit-cost ratio 1.35. The sensitivity analysis further indicated that the proposed project is on sound footing, as it can with stand with 20% risk. The payback period for the proposed project is 1.85 years under normal situation, when the plant capacity utilization and other factors operate at optimal level.

Equipments manufacturer / Supplier

M/s SSP (Pvt.) Limited

13, Mile Stone, Mathura Road, Faridabad-121 003

Reference

Gopalan, C. (1971). Nutritive value of Indian Foods. Indian Council of Medical Research Publication, 1971, p 60-114.

http://www.faostat.org/

□□□

7

Postharvest Processing and Value Addition of Fruits and Vegetables

R.K. GUPTA

India being a home of wide variety of fruits and vegetables holds a unique position in production figures among other countries. The diverse agro climatic zones of the country make it possible to grow almost all varieties of fruits and vegetables in India. Fruits and vegetables are more prone to spoilage than cereals due to their nature and composition, and this spoilage occurs at the time of harvesting, handling, transportation, storage, marketing and processing. Horticulture has emerged as the best option for diversification in agriculture for nutritional security, employment generation and enhancing farm income. For increasing the marketability of horticultural produce, profitability and reducing losses, there is an urgent need to create network of infrastructural facilities for storage, packaging, grading, marketing and processed product development. The growing economic importance of horticultural crops, especially of fruits and vegetables could be attributed to an increase demand arising from domestic as well as overseas market. While the increasing domestic demand could be attributed to the increase in income, population growth, changing consumption pattern and growing awareness for better nutrition, the increase in international demand has many reasons including the changing global perspective on trade related to agriculture. Thus task before us is to harness the opportunity by adopting novel techniques in Postharvest Processing and Value addition of Fruits and Vegetables.

Agro-processing is now regarded as the sunrise sector of the Indian economy in view of its large potential for growth and likely socio-economic impact specifically on employment and income generation. Some estimates suggest that in developed countries, up to 14 per cent of the total work force is engaged in agro-processing sector directly or indirectly. However, in India, only about 3-4 per cent of the work force finds employment in this sector revealing its underdeveloped state and vast untapped potential for employment. Properly developed, agro-processing sector can make India a major player at the global level for marketing and supply of processed foods.

Development of value added products from fruits and vegetables

Processing and product development from Aonla

Aonla fruit is nutritionally and medicinally important due to its high vitamin C content and stabilizing tannins. However, it is not consumed in the fresh form due to its high acidic and astringent taste. A number of processed products, therefore, are prepared from aonla fruits to improve their acceptability and exploit nutritional qualities. The products due to their high therapeutic value, can serve as health foods. Besides, the use of aonla in more than 50 products and ayurvedic formulations aonla pulp, RTS, nectar, squash, syrup, jam, preserve, candy, pickle, sauce, *chutney* and dehydrated shreds are more valuable and popular. Diversification of products to suit the requirements of consumer is essential to capture the emerging market. To capture the market, awareness about health aspect of aonla has to be created. These products can easily be made by either women or in Self Help Groups (SHGs).

Value added products of Aonla

Aonla leather

Aonla leather can be prepared by slow heating of aonla pulp, mango pulp and papaya pulp. At the time of or during slow heating, sugar syrup (50° brix) is added. The mixture is then heated slowly till the mixture reached 35° brix. Potassium meta bisulphite preservative is added for increasing the shelf life. The mixture so obtained is spread in trays and kept at 60° C for 36 to 48 h. The leather can be obtained by cutting of the product in to pieces of the desired shape.

Aonla jelly

Aonla jelly can be prepared by the use of aonla juice, pulp and other ingredients like mango pulp, papaya pulp, sugar and pectin added into it. Aonla juice and pulp is mixed with mango pulp and papaya pulp and the mixture is allowed to slow heating. At the time of or during slow heating, sugar is added along with pectin and potassium meta bisulphate. The final product (68° brix) as Aonla jelly is prepared within 10 to 15 minute.

Similarly other fruits and vegetables can be processed into various valuable and nutritious products like fruit beverages, jams, jellies, sauce, ketcup etc. Fruit beverages are easily digestible, highly refreshing, thirst quenching, appetizing and nutritionally far superior to many synthetic and aerated drinks. The process for preparation of such value added products is described as under.

Ready-to-serve (RTS) products from fruits

This is a type of fruit beverage which contains at least 10 percent fruit juice and 10 percent total soluble solids besides about 0.3 percent acid. It is not diluted before serving, hence it is known as ready – to- serve (RTS). RTS can be prepared from different fruits like bael, guava, jamun, aonla, lemon, karonda, phalsa, plum, papaya, ginger etc. The process for preparation of RTS is given below;

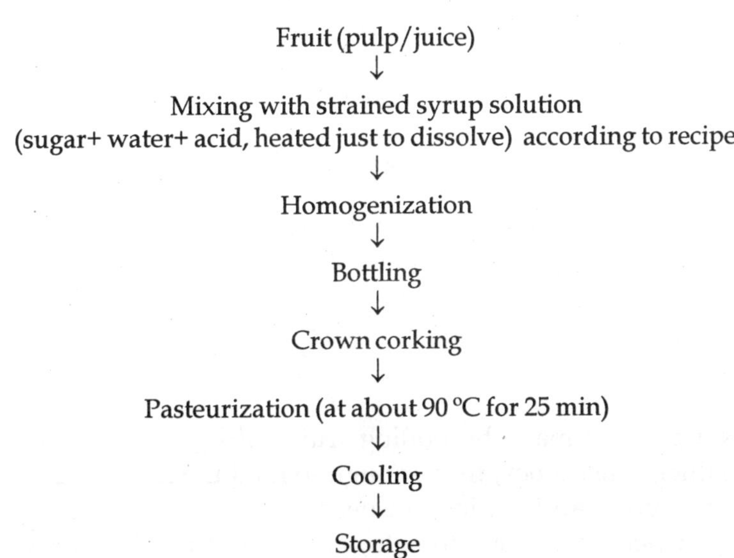

Squash

This is a type of fruit beverage containing at least 25 percent fruit juice or pulp and 40 to 50 percent total soluble solids, commercially. It also contains about 1.0 percent acid and 350 ppm sulphur dioxide or 600 ppm sodium benzoate. It is diluted before serving. Mango, orange and pineapple are used for making squash commercially. It can also be prepared from lemon, lime, bael, guava, litchi, pear, apricot, muskmelon, papaya etc., using potassium metabisulphite (KMS) as preservative, or from jamun, passion fruit, peach, phalsa, plum, mulberry, raspberry, strawberry, grapefruit, etc., with sodium benzoate as preservative. The flow chart for preparation of squash is mentioned below:

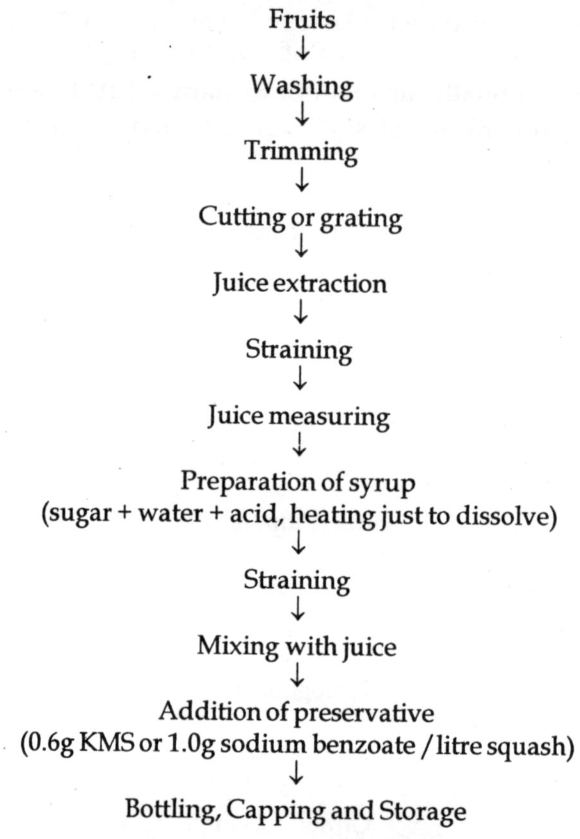

Jam

Jam is a product made by boiling fruit pulp with sufficient sugar to a reasonable thick consistency, firm enough to hold the fruit tissues in position. Apple, pear, sapota, apricot, loquat, peach, papaya, karonda, carrot, plum, strawberry, raspberry, mango, tomato, grapes and muskmelon are used for preparation of jams. It can be prepared from one kind of fruit or from two or more kinds. Commercial jams such as tutti frutti can be prepared from pieces of fruit, fruit scraping and pulp adhering to cores of fruits which are available in plenty in canning factories.

Jam contains 0.5-0.6 percent acid and invert sugar should not be more than 40 percent. This is a type of fruit product containing at least 45 percent fruit juice or pulp and 68 percent total soluble solids, commercially.

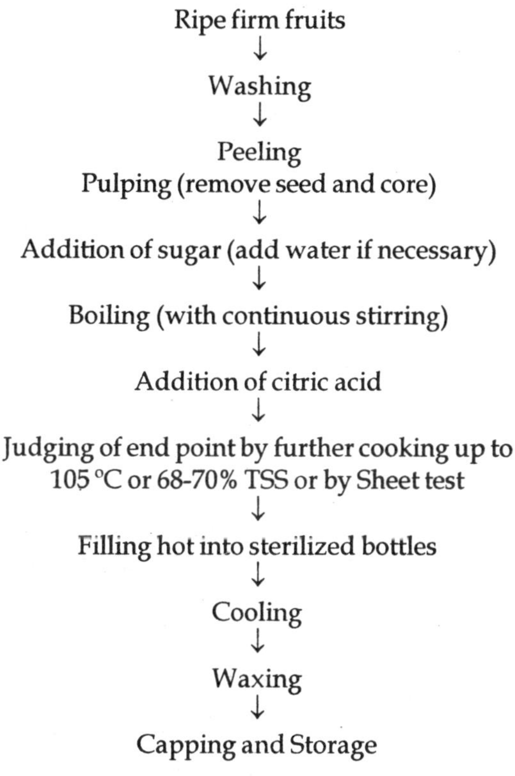

Jelly

Jelly is a semi solid product prepared by boiling a clear, strained solution of pectin containing fruit extract, free from pulp, after the addition of sugar and acid. A perfect jelly should be transparent, well set, but not too stiff, and should have the original flavor of the fruit. It should be of attractive color and keep its shape when removed from the mould. It should be firm enough to retain a sharp edge but tender enough to quiver when pressed. It should not be gummy, sticky or syrupy or have crystallized sugar. Guava, sour apple, plum, karonda, wood apple, loquat, papaya and gooseberry are generally used for preparation of jelly. Apricot, pineapple, strawberry, raspberry, etc., can be used only after addition of pectin powder, because these fruits have low pectin content.

Jelly is a type of fruit product containing at least 45 percent fruit juice and 65 percent total soluble solids, commercially. Its acid content should be 0.5-0.75 percent.

The process for making jelly is described below:

Fruits (ripe, firm, not over ripe)
↓
Washing
↓
Cutting into thick slices
↓
Boiling with water (1.5 times the weight of fruit for about 20-30 min)
↓
Addition of citric acid during boiling (2 g/kg fruit)
↓
Straining of extract
↓
Pectin test (for addition of sugar)
↓
Addition of sugar
↓
Boiling
↓
Judging of end point (Sheet/Drop/Temperature test)
↓
Removal of scum or foam (one teaspoonful edible oil added for 45 kg sugar)
↓
Color and remaining citric acid added
↓
Filling hot into clean sterilized bottles
↓
Waxing (paraffin wax)
↓
Capping
↓
Storage at ambient temperature

Marmalade

This is a fruit jelly in which slices of the fruit or its peel are suspended. The term is generally used for products made from citrus fruits like oranges and lemons in which shredded peel is used as the suspended material. Citrus marmalades are classified into (i) jelly marmalade (ii) jam marmalade.

The procedure for preparation of jelly marmalade is mentioned below, however the procedure for making the jam marmalade is same as that for jelly marmalade. In this case the pectin extract of fruit is not clarified and the whole pulp is used. Sugar is added according to the weight of fruit, generally in the proportion of 1:1. The pulp-sugar mixture is cooked till the TSS content reaches 65 percent.

Marmalade is a type of fruit product containing at least 45 percent fruit juice and 65 percent total soluble solids, commercially. The process for making jelly *marmalade* is described below:

Ripe fruits
↓
Washing
↓
Peeling outer yellow portion (flavedo) thinly
↓
Cutting yellow portion into fine shreds (1.9-2.5 cm long and 0.8-0.12 cm thick)
↓
Cutting into 0.3-0.45 cm thick slices of peeled fruit or crushing into pulp using a grater
↓
Boiling (in 2-3 times its weight of water for 40-60 min)
↓
Straining the extract
↓
Testing for pectin content (Alcohol test)
↓
Addition of sugar (as required)
↓
Cooking to 103-105°C (continuous stirring)

Addition of prepared shreds (shredded peel boiled for 10-15 min in several changes of water for softening and removing bitterness and added @ 62g per kg of extract)

Boiling till jellying point (continuous stirring)
↓
Testing of end point (Sheet/Drop/Temperature test)
↓
Cooling (to 82-88°C with continuous stirring)
↓
Flavoring (orange oil)
↓
Filling in sterilized bottles
↓
Sealing
↓
Storage at ambient temperature

Preserve and Candy

A mature fruit/vegetable or its pieces impregnated with heavy sugar syrup till it becomes tender and transparent is known as a preserve. Aonla, bael, apple, pear, mango, cherry, karonda, strawberry, pineapple, papaya,

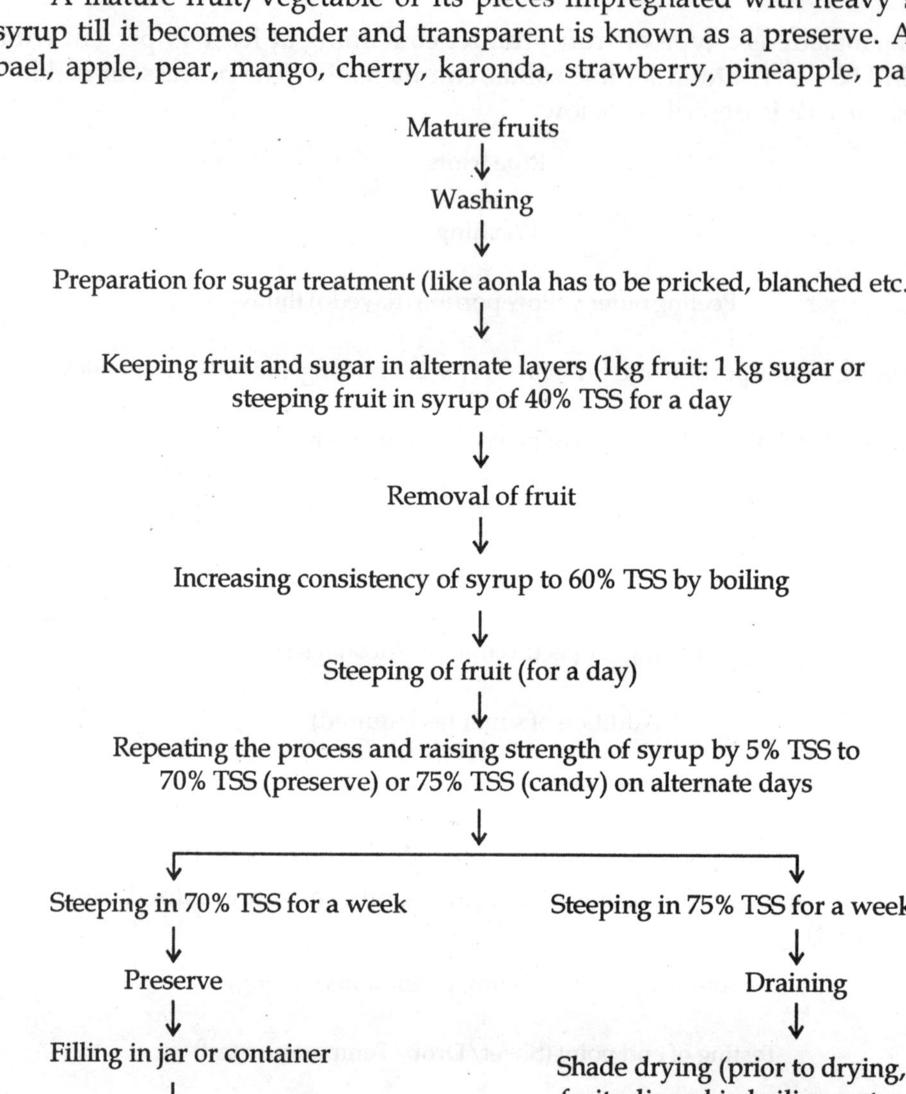

Process flow chart for making preserve and candy

etc., can be used for making preserves. Preserve is a type of fruit product containing at least 55 percent fruit portion and 68 percent total soluble solids, commercially. A fruit/vegetable impregnated with cane sugar or glucose syrup and subsequently drained free of syrup and dried, is known as candied fruit/vegetable. The most suitable fruits for candying are aonla, karonda, pineapple, cherry, papaya, apple, peach and peels of orange, lemon, grapefruit, ginger etc.

Candy is a type of fruit product containing sugar not less than 70 percent and reducing sugar as percentage of total sugar not less than 25 percent. The process is described as under.

Tomato sauce / ketchup

It is made from strained tomato juice or pulp and spices, salt, sugar and vinegar, with or without onion and garlic, and contains not less than 12 percent tomato solids and 25 percent total solids. The process for preparation of tomato sauce / ketchups is described here under.

Tomatoes (fully ripe, red)
↓
Washing
↓
Sorting and trimming
↓
Cutting and chopping
↓
Heating at 70-90°C for 3-5 minutes (to soften)
↓
Pulping or extraction of juice/pulp (mechanically or by sieving)
↓
Straining tomato pulp/juice
↓
Cooking pulp with one third quantity of sugar putting spice bag in pulp and pressing occasionally
↓
Cooking to one third of original volume of pulp/juice
↓
Removal of spice bag (after squeezing in pulp)
↓
Addition of remaining sugar and salt
↓
Cooking
↓
Judging of end point (tomato solids by hand refractometer/ volume by measuring stick *i.e* one third to its original volume)

↓
Addition of vinegar/acetic acid and preservative
↓
Filling hot into bottles at about 88°C
↓
Crown corking
Pasteurization (at 85-90°C for 30 min)
↓
Cooling
↓
Storage at ambient temperature (in cool and dry place)

Some novel products development from Guava

Guava can be processed into a number of products but the development of cheaper and quality product is of paramount importance in present market scenario. It can be processed into some common products like jam, jelly, squash, beverages and slices in sugar syrup. Introduction of new and diversified product is necessary for the survival and growth of processing industry to meet varying taste preferences and demand for internal as well as external market.

1. *Pure fruit juice*: This is the natural juice pressed out of the fruit and remains practically unaltered to its composition during preparation and storage.
2. *Fruit juice concentrate*: A fruit juice from which most of the water has been removed by heating or freezing is known as juice concentrate.
3. *RTS beverage*: This type of fruit juice is not altered in its composition before consumption and is served as such without any dilution.
4. *Fruit squashes*: This is a type of fruit beverages containing at least 25% fruit juice or pulp, 40-50% sugar content, 1.0% acid and 350 ppm sulphur dioxide. This required dilution before serving.
5. *Fruit juice cordial*: It is also a type of squash but with a sparkling, clear, sweetened fruit juice from which the pulp and other insoluble substances have been completely removed.
6. *Fruit syrup*: It also contains 25% juice or pulp but sugar and acid content are in quite higher (65% sugar and 1.3-1.5% acid) as compared to squash. This is also diluted before serving.
7. *Fruit leather*: This is a novel product. It is prepared by dehydrating the guava.

Value added products from pomegranate

During seasonal glut, the prices of fruits are fairly low. Cracking of fruits hampers their quality. Such fruits can be utilized for processing. The following products can be prepared by processing of pomegranate:

1. Juice

Pomegranate fruit with large juicy grains can give an attractive coloured (purplish red) juice with mild acid sweet taste with low tannin content (<0.25 percent). Pomegranate juice makes a delicious drink.

Pomegranate juice can be extracted by following methods.

1. Processing fruit quarters in sack and cloth hydraulic press. It yields about 42 percent juice.
2. Extracting in a spiral type screw press.
3. Extracting by pressing arils in a basket press. This yields 70 percent juice with low tannin (0.12 percent).

Excess tannin from juice can be precipitated by gelatin and is clarified by pectinase enzyme or by heating in a flash pasteurizer at 79-82°C followed by cooling, setting up 24 hours racking of a fettering. Sodium benzoate (600 ppm) is added as a preservative.

The use of sulphur di-oxide or potassium metabisulfite is not recommended for pomegranate because of loss of its natural color due to the bleaching sulphur di-oxide.

2. Concentrate

Fruit juice concentrate, either from single fruit or in blended form, is a product with great potential in account of their wide appreciability in fruit based beverages.

3. Pomegranate rub / Jam

Equal parts of juice and sugar are mixed and the syrup is concentrated on slow fire to a thick consistency (70-75° brix).

4. Pomegranate jelly

An excellent jelly can be prepared from the juice by mixing it with sugar and preservatives. 500 ml of pomegranate juice, 800 g sugar and 2-4 teaspoons of limejuice are mixed and the mixture is boiled. Add 40-50 g of pectin, stirring the content constantly. Bring to rolling boil and boil hard for about half a minute. Remove from heat, skim and pour hot in jars and seal.

Value added products from Ber

Delicious preserve, osmo air-dried products, *chhuhara* and RTS can be made from mature *ber* fruits. Fruits of Umran, Banarsi Karaka and Kaithli cultivars are suitable for the preparation of preserve. Only fully mature fruits at the hard stage are used for making preserve and osmo dehydrated products. Ripe fruits do not make a good product. The fruits are pricked and softened so that sugar impregnation is uniform and effective.

Value added products from Banana

Pulp / Puree

Fully ripe fruits are washed, peeled and pulped in a homogenizer. To avoid browning of pulp, it is de-aerated and held under vacuum or acidified followed by sulphur di-oxide (SO_2) treatment. The pulp is heated, cooled and filled aseptically or with preservative into containers. The complete process is shown hereunder.

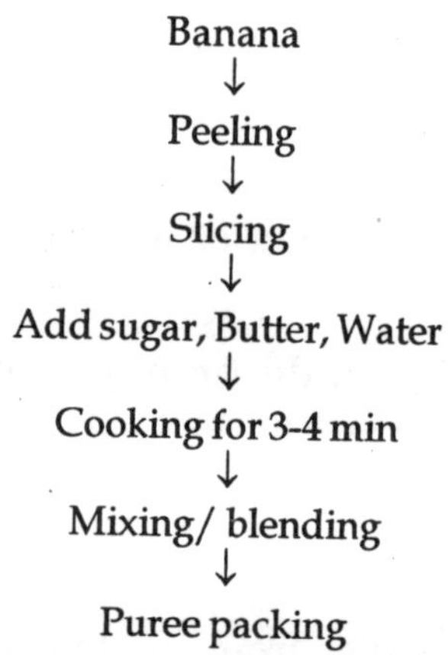

Process for making Pulp/puree of Banana

Banana chips

Raw mature bananas with low sugar content are used for making banana chips. Peel and slice the bananas (2 mm thick slices) with stainless steel knives. Dip the peeled fruit slices in a solution of 0.1% citric acid for 15 minutes to avoid browning. Remove the slices from the solution. Wash them in cold water, drain and spread on flat trays (2.5-5.0 kg/m^2 of tray surface). Trays made of wooden or bamboo strips are suitable. Transfer the trays to a sulphuring box and expose them for an hour to fumes of sulphur dioxide. Burn about 3g fumes of sulphur for a batch of ten trays. Dry the sulphured slices either in the sun or in a drier at 60-63°C temperature. Stop drying when the slices become brittle. The dried slices are fried in edible oil to yield banana chips.

Fried chips are treated with salt and spices and cooled at room temperature before packing. The chips are packed in high-density polyethylene bags or in suitable packing materials and stored in airtight containers. The complete process is shown hereunder.

Peeling of bananas
↓
Putting the peeled bananas in salted iced water
↓
Chopping of bananas in water and adding turmeric
↓
Keep in water for 10 minutes and then drain out water completely
↓
Transfer to a kitchen cloth to remove the moisture
↓
Heat oil till it starts fuming
↓
Deep-fry the slices till almost crisp
↓
Add few slices at a time.
↓
Prepare paste of 1/2 tsp water and 1/4 tsp salt
↓
Add this paste to oil. This will make the slices crispier
↓
Repeat the procedure for remaining slices
↓
Drain the chips on an absorbent paper
↓
Cool and store them in an airtight container

Process for making Banana chips

Some other dried products from fruits and vegetables

Apple: Washing, peeling, coring, trimming and slicing into 5 mm thickness, followed by sulphuring for 30 min or immersion in 1-2% KMS solution for 30 min. This follows the drying at 60-65°C in a dehydrator or sun drying.

Apricot: Washing, cutting into halves, destoning, sulphuring for 30 min, followed by drying at 50-60°C in a dehydrator or sun drying.

Banana: Washing, peeling, cutting into halves lengthwise or crosswise (12 mm thickness), sulphuring for 30 min followed by drying at 55-60°C in a dehydrator or sun drying.

Date: Washing, dipping in boiling solution of 0.5% caustic soda followed by rinsing and drying in a dehydrator at 45-50°C or sun drying.

Fig: Washing, sulphuring of only *Adriatic* varieties for 1 hr followed by drying at 55 to 60°C in a dehydrator or sun drying.

Grapes: Raisins prepared from grapes are the most important dried fruit product. *Muscat* and wine varieties are dipped in boiling solution of 0.5% caustic soda followed by rinsing and sulphuring for 1 hr and then, drying at 55-60°C in a dehydrator or by sun drying. Ripe bunches of grapes are hung inside the dark rooms known as *Kishmish khanas* till berries acquire a greenish or light amber tint. *Monucca* or Raisin is prepared from large seeded *Haitha* grapes which are lye dipped prior to sun drying. In California, the *Sultanina* or *Thompson* seedless grapes are dried.

Mango: Washing, peeling, cutting into 12 mm thick slices, sulphuring for 2 hrs followed by drying at 45-50°C in a dehydrator or sun drying. Ripe mangoes are taken and juicy pulp squeezed by hand and dried as *Aam pappad.*

Papaya: Washing, peeling, removal of seeds, cutting into 6 mm thick slices, sulphuring for 2 hrs followed by drying at 60-65°C in a dehydrator or by sun drying.

Pear: Washing, peeling, cutting into halves, holding in 1-2% salt solution, sulphuring for 30 min or immersion for 20 min in 1-2% KMS solution followed by drying at 60-65°C in a dehydrator or by sun drying.

Peach: Free-stone peaches like *July Elberta* cultivar is used for drying. The drying process includes washing, removal of pits, cutting into halves, sulphuring for 30 min followed by drying at 60-65°C, in a dehydrator or by sun drying.

Prunes: Washing/cleaning, pitting, perforation and drying.

Other fruits: Pomegranate seeds are dried into *Anardana,* cherries and berries are also dried.

Dried products from vegetables

Beans: Washing, peeling, and cutting into 10 mm thick slices, steam blanching for 10 min followed by drying at 60-65°C in sun or dehydrator.

Beetroot: Washing, peeling and cutting into 10 mm thick slices, steam blanching for 10 min and drying at 60-65°C in sun drying or dehydrator.

Bittergourd: Washing, removal of both ends cutting into 10 mm thick slices, blanching for 7-8 min and drying at 65-75°C in sun or dehydrator.

Brinjal: Washing, cutting lengthwise into 10 mm thick slices, blanching for 4-5 min and then, immersion for 1 hr in 1% KMS solution followed by drying at 50-52°C in sun or dehydrator.

Cabbage: Washing, removal of outer leaves and cutting into fine shreds, blanching for 5-6 min, immersion for 10 min in 0.5% KMS solution and draining, followed by drying at 55-60°C in sun or dehydrator.

Cauliflower: Washing, removal of stalks, covering leaves and stems, breaking of cauliflower into suitable size pieces, blanching for 4-5 min. immersion for 1 h in 1% KMS solution and draining, followed by drying at 55-60°C in sun or dehydrator.

Carrot: Washing, scrapping stalks and tips, cutting into 10 mm thick slices, blanching for 2-4 min in boiling solution of 2% common salt, followed by drying at 60-65°C in sun or dehydrator.

Chillies (red): Mature dark red pods are tied to a string and hanged in sun or dried at 50-55°C, in a dehydrator.

Green peas: Washing, removal of shell, blanching grains in boiling water or steam for 3-4 min, immersion in 0.5% KMS solution and draining followed by drying at 60-65°C in sun or dehydrator.

Palak, methi and other leafy green vegetables: Sorting, washing, trimming-off rough stems and stalks and shred making followed by blanching for 2 min in boiling water or steam and then drying at 60-65°C in sun or dehydrator.

Potato: Washing, peeling and cutting into 10 mm thick slices, blanching in boiling water or steam for 3-4 min and immersion in 0.5% KMS solution followed by drying at 60-65°C in sun or a dehydrator.

Tomato: Washing, blanching for 30-60 sec., peeling and slicing into 1 cm thick slices followed by drying at 60-65°C in dehydrator or sun.

Turnip: Washing, removal of stalks, peel, cutting into 5 mm thick slices, blanching for 2-4 min in boiling water and immersion for 1-2 h in 1% KMS solution, followed by drying at 50-55°C in sun or dehydrator.

Mushroom slices: Mushrooms are one of the most important vegetable prepared by dehydration. Steps include trimming, bleaching with SO_2 @ 500-100 ppm for 5-10 min, slicing followed by drying.

Recommendations

- National plan for improvement and extension of agro-processing technology at farm, traditional small industry and modern industry levels should be prepared.

- Value addition technologies pertaining to horticultural crops should be popularized among up-coming small entrepreneurs including women entrepreneurs in terms of minimal processing, intermediate processing, and ready to eat convenience foods and health foods.
- Entrepreneurship Development Programme should be organized for up-coming entrepreneurs in Fruits and Vegetables Processing.

References

CIPHET, Show Case/Annual Report (2007). CIPHET/AICRP on PHT reports/other R&D Institutions reports.

Verma, LR and VK Joshi (2000). Post Harvest Technology of Fruits and Vegetables. 2: 870-873.

□□□

8

Novel Techniques in Post Harvest Processing of Pomegranate

A.K. THAKUR, R.T. PATIL, D.B. SINGH AND R.K. GUPTA

The pomegranate (*Punica granatum* L.) is a highly seasonal fruit with a high level of production as a commercial horticulture in the Indian states of Maharashtra, Karnataka, Gujarat, Andhra Pradesh, Tamil Nadu and Rajasthan. The area under this fruit is expanding day by day and its cultivation is now gaining high importance in northern states like Punjab, Uttar Pradesh, Rajasthan and Himachal Pradesh. Annual production of pomegranate in the country is about 8 lakh tones per year (Table 1). Maharashtra and Karnataka stand first and second, respectively in the production of pomegranate (Table 2). Cultivars like Ganesh, Mridula, Aaraktha, Sindhura are popular varieties grown in the country. Large quantities of fruits are consumed in fresh form, however many products like pomegranate juice, RTS beverage, squash, jelly and anardana (dried arils/seeds) are prepared on commercial scale in some parts of the country. The pomegranate has been regarded as a food medicine of great importance. All parts of the tree, the roots, and the reddish brown bark, leaves, flowers, rind and seeds have featured in medicine for thousands of years. Mature fruits should be immediately removed from the plants as delay in harvesting leads to fruit cracking. Pomegranate is a non-climacteric fruit and requires maturity on the tree for its better post-harvest quality attributes. The fruits are ready to harvest between 125 to 165 days after anthesis. At maturity, the fruit colour changes to dark yellow in summer and dark red in winter. However varieties like Arka Ruby, Mridula and Bhagwa do not show difference in colour change between fruit development and maturity period. In most of the cultivars, the sepals at the anterior end of the fruit curve inwards and turn into dry, black, hard structures at maturity. The fruits are harvested at maturity and their quality is found to little bit improve during storage (Chadha *et al.*, 2004). The fruit are generally harvested fully ripe with a waxy shining surface of reddish yellow or greenish red peel colour, depending on the cultivar. The major storage problem is desiccation of the fruit resulting in a brownish coloured tough peel and browning of arils. Although the peel appears to be thick, it has numerous minute openings that permit free movement of water vapour, making the fruit highly susceptible to water loss (Kader *et al.*, 1984).

Pomegranate is an important fruit crop of many tropical and subtropical regions of the world, grown especially in the moderate climates of Mediterranean countries. As production has increased, proper storage, processing and marketing of pomegranate is needed to meet the demand both in domestic and export markets. Pomegranate is primarily consumed in two different ways. First, the juice contained in arils is extracted by crushing the arils and collecting the juice. This juice can be consumed in pure form, mixed with other juices, or is also converted into concentrated and bottled form. The most common form of such pomegranate concentrate is often referred to as grenadine syrup. A second way is to consume fresh arils as such to enjoy the refreshing juice which releases during the eating. Such arils may be consumed alone or it can be used as a topping on a salad or other food preparations. Sugar contents of Pomegranate juice are about 12 to 16%. This sugar is light for digestion. Pomegranate peels are used for medicines and for colouring cloths. Peels of Pomegranate is used to prepare toothpastes/ toothpowders. Pomegranate juice has iron, phosphorus, calcium, magnesium etc which helps in growth of blood in the body. Processing of Pomegranate provides an opportunity to create a new and innovative market for fresh arils (juice-enclosed seeds), dried arils, juices and wines; and health and nutraceutical products.

Processing of Pomegranate for fresh fruit marketing

Sorting and Grading

Inspection roller-cum-conveyor are used for sorting of pomegranate. Fruits showing greenish colour, fruits with sunspots, affected by borer, disease spot, rusting and rough surface should be eliminated. The pomegranate fruits after proper harvesting at required maturity and sorted properly are graded as per their colour, size and weight. The different grades and their characteristics as suggested by National Horticulture Board (NHB) are presented for domestic and export purpose (Table 3). During the process of grading, mixing of different cultivars should be avoided.

Table 1: Area Production and Productivity of Pomegranate in India

Year	Area (000 ha)	Production (000 Ton)	Productivity, Ton/ha
2006-07	117	840	7.2
2007-08	124	884	7.2
2008-09	109	807	7.4

Source: NHB Horticulture Database-2009

Table 2: States contribution in Pomegranate Production

State	% contribution in production
Maharashtra	68.1
Karnataka	17.1
AP	8.0
Gujarat	4.9
Tamil Nadu	1.2
Rajasthan	0.4
Others	0.2

Source: NHB Horticulture Database-2009

Table 3: Pomegranate grading based on size

Grades	Characteristics
Super size	Attractive, very large, dark red in colour, without blemish weighing >750 g
King size	Attractive and large fruits without blemish; weighing between 500-750 g
Queen size	Fruit are large, attractive without blemish and weighing between 400-500 g
Prince	Fruits are attractive, blemish free and weighing between 300-400 g
12-A	Fruits having 1-2 spots and weighing between 250-300 g
12-B	Fruits weighing <250 g

Packaging

All the graded fruits are wrapped in paper and packed in corrugated fiberboard (CFB) boxes. Brown coloured 3 fold CFB boxes are used for local market while; white coloured 5 fold CFB boxes are used for distant markets. Paper shreds are used as padding material. For 'Super' and 'King' grades, boxes of size 32.5 x 22.5 x 10 cm are used. 'Queen' grade fruits are packed in 37.5 x 27.5 x 10 cm size boxes while 'Prince' and '12-A' grades are packed in 35 x 25 x 10 cm size boxes. Pomegranate fruits have a low rate of respiration and ethylene production. They are susceptible to moisture loss and need to be stored at high humidity. After harvest, the fruits are graded according to size, wrapped in paper, and packed in bamboo baskets or corrugated boxes. In bulk storage, fruits are packed in layers in wooden crates, each containing about 16-18 kg of fruits. Dry grass, rice straw, or papers are used as cushioning material.

Waxing of fruits, alone or in combination with fungicides, has been reported to extend the shelf life of pomegranates (Waskar *et al.*, 1999). The effects of individual shrink film wrapping with two polyolefin films (BDF-2001 and D-955) and skin coating with a sucrose polyester (SPE) Semperfresh™ on the shelf life and quality of soft-seeded 'Ganesh' pomegranates stored at 8, 15 and 25°C were studied (Nanda *et al.* 2001). The shrink-wrapped pomegranates could be stored for 12, 9 and 4 weeks as compared to 8, 6 and 2 weeks by SPE coating at 8, 15 and 25°C respectively, whereas non-wrapped fruits could be kept for 7, 5 and 1 week under similar storage conditions. Peel thickness, freshness and firmness of the fruit were retained and weight loss greatly reduced by shrink wrapping. The weight loss in shrink-wrapped fruits was 1.2–1.3% after 12 weeks of storage at 8°C and 2.2–3.7% after 10 weeks at 15°C. During the same period non-wrapped fruits lost 20.4 and 30.7% at 8 and 15°C, respectively. Changes in acidity, sugars and vitamin C of the shrink-wrapped fruits were lower than that of non-wrapped fruits during 12 weeks of storage at 8°C. Shrink wrapping also reduced the respiration rate of the fruit. No detectable levels of ethylene were produced during storage of pomegranates.

Storage

Pomegranate fruit has better keeping quality than other tropical fruits such as mango, grape, and banana. The storage life of pomegranate is comparable to that of apple. The storage temperature recommended for pomegranate has varied from 0 to 10°C with a shelf life ranging from 2 weeks to 7 months depending on the cultivar. Generally, fruits can be stored at 5°C with 90-95% relative humidity for 2 months. In case of storage beyond two months, temperature should be maintained at 10°C to avoid chilling injury

(Gil *et al.,* 1996). Pomegranates are very susceptible to water loss resulting in shriveling of the skins. Storing fruit in plastic liners and waxing can reduce water loss, especially under conditions of lower relative humidity.

Cool chain

Cool chain is essential during the transport of export quality commodity all the way from the farm to the customer. This helps in maintaining the temperature inside the box at the same low level as in the cold storage. The various stages of the cool chain are; cold store at the farm where harvested fruits can be kept in fresh condition for a short period, refrigerated truck from farm or from grading packing centre to the airport, cold store at the airport, building up of the pallet in a cold store at the airport, loading the aircrafts directly from the cold store in a short time, cargo aircraft maintains cold store temperature in hold, off loading direct into a cold store in the receiving country and refrigerated truck to the customers.

Processing of Pomegranate for value added products

The first step in processing of pomegranate is to extract or separate the arils (juice-enclosed seeds) from its peel and other non-edible parts of the fruit. It has been felt that fresh arils may have a promising market either for direct consumption or further processing into various products. First class pomegranate fruits fetch a fairly good price in the market and are not much used for processing purposes. Hence limited information is available on the utilization of pomegranate fruits for processed products. During seasonal glut, the prices of fruits are fairly low. In some cases cracking of pomegranate fruits hampers their market quality. Such fruits can be utilized for processing. Pomegranate fruits with large juicy grains which can give an attractive colored (purplish red) juice with mild acid, sweet taste and low tannin content (less than 0.25%) are the desirable characteristics for processing of pomegranates.

Separation techniques of arils from pomegranate

The common method for extraction of juice from pomegranate involves cutting the fruit in pieces, aril separation by hand fingers and pressing of arils in screw press or basket press. The juice is then clarified and about 45% juice yield is obtained (http://mofpi.nic.in). For easy extraction of juice, the fruit can be prepared by rolling it on a hard surface to weaken the seed sacks within the fruit. The foremost challenge in juice extraction is peeling of fruits or separation of arils from the peel. In addition, during hand peeling, due to the high content of polyphenols and oxidative enzymes in the fruit rind, the skin of the hands is stained brown, making this process an even more irritating task (Artes *et al.,* 1995). Alternative method for juice extraction is that the fruits are quartered and crushed, or the whole fruits may be pressed in hydraulic press and the juice is strained out. Hydraulic extraction of juice should be done at a pressure less than 100 psi to avoid undue yield of tannin from the rind. The juice from crushed whole fruits contains excess tannin from the rind and this is precipitated out by a gelatin process. After filtering, the juice may be preserved by adding sodium benzoate or it may be pasteurized, allowed to settle for two days, then strained and bottled (Morton, 1987).

Investigation revealed a number of methods employed to separate arils. Tumbling the fruit in a machine like a grape destemmer reportedly works in India. In Georgia (ex-USSR) the fruit was squeezed and lightly crushed between rollers, prior to removing a patch of skin so that the arils could be expelled by hand squeezing. One juicer slices the fruit in half and then uses a hemispherical reamer, much like juicing oranges. Similarly, the FMC orange juice extractor has been used, with limited results so far. In addition, experiments have been run with vibratory screens and flotation (arils sink but the membrane floats).

Khazaei *et al.* (2008) tested an impingement method for extracting pomegranate arils. The method is based on opening the fruit, separating it into two halves, and extracting the arils with the help of pressurized air-jets. Safa *et al.* (2009) used a pneumatic method to extract the pomegranate arils. The method is based on opening the fruit (with a cut) in two halves and dislodging the arils by a scanning air jet. The air under pressure is infused into the halves of the fruit so that the arils substantially disintegrate over a predetermined time period. A novel method and a system were developed, which enable opening the fruit without cutting, extracting the arils with minimum damage, separating the arils from extraneous materials and delivering clean arils to a packaging machine. A Beta-site machine has been constructed and has undergone extensive testing. The results obtained so far indicate that it is possible to extract 1.5 to 2 tons of arils a day in a one-lane machine operated by 1-2 persons with an extraction efficacy of 95% and <5% mechanical damage. (Schmilovitch *et al.*, 2009).

Balasco *et al.* (2009) developed a machine for automatic sorting of pomegranate arils based on computer vision. This work describes the development of a computer vision-based machine to inspect the raw material coming from the extraction process and classify it in four categories. The machine is capable of detecting and removing unwanted material and sorting the arils by colour. The prototype is composed of three units, which are designed to singulate the objects to allow them to be inspected individually and sorted.

Singh *et al.* (2006) studied the manual separation techniques of wild pomegranate for production of anardana. For easy separation of arils, fruits were subjected to sand roasting, hot water dipping and soaking in cold water. Hot water dip for 2 min saved time required for separation of arils from fruits (80% excluding treatment time and 74% including treatment time). Due to the high temperature during sand roasting, quality of arils and *anardana* was inferior when compared to other treatments. The desirable color and chemical composition of the product *anardana* was retained in the treatment hot water dipping for 2 min.

The Juran method for Extracting Arils from Pomegranates (EAP) is fully automated and highly efficient (Juran Metal Works Ltd., Israel: http://www.juran.co.il/) as per innovative process developed by Agricultural Engineering Institute, Israel. Instead of cutting through the fruit and causing a high percentage of damaged arils, the EAP mechanism only scores the peel, and uses a combination of gentle pressure and oscillating air jets to open the fruit and retrieve the arils. The process involves: (a) the fruit is fed inverted, either manually or mechanically, onto a conveyor of open-bottomed circular holders (b) crowns are cut off and grips lift the fruit onto the next level conveyor, pressing it down onto corolla-opening knives that longitudinally score the skin (c) flexible drums press down on the fruit causing it to open in segments and eventually to flatten (d) as the fruit is flattened, the arils fall out with the help of oscillating air jets (e) the untouched and undamaged arils are gravitationally water-separated from other fruit parts such as peel or membrane, and ready to be conveyed for packing.

Hand tool for easy separation of arils from Pomegranate (Thakur *et al.* 2008[a], 2009[c])

The hand tool for easy separation of arils from pomegranate consists of fruit holders having knife arrangement in such a way that it only penetrate into the peel (Fig. 1). Pomegranate fruit has to be held between the pair of holder, and the holders are turned by hand in opposite direction of each other. Due to rotating action of the holders, fruit break into two irregular halves as a simultaneous effect of tension with twist on the peel. During this action the whole fruit experiences a shearing effect; and due to this effect the insides arils become loosen which provides an opportunity for further easy separation of arils. About 30-45% of arils are already separated in the process of irregular breaking performed by the tool due to shearing action on the inner sheath and outer peel. However the separation efficiency depends on the varieties and condition of the pomegranate fruit. The separation efficiency was higher in case of fresh fruit in comparison to the cold stored fruits. However the difference was not significant. The rest of the arils becomes loosen that could be further easily removed by the finger tips as in case of conventional separation. This is a helping tool that facilitates in easy separation of arils from the whole pomegranate fruit without any damage to arils. The hand tool is designed considering the medium size of the fruit because the maximum distribution of medium sized fruit as observed on a particular pomegranate plant. The result of the hand tool is found encouraging and it makes the task of safe separation of arils from the peel and sheath effortless. The unit is primarily made of wood which is tough and light in weight. The tool is capable of breaking the pomegranate fruit safely, it is simple in design and fabrication, cost effective and environment friendly.

Fig. 1 : Operation of hand tool for easy separation of arils

Table 4 : Physical characteristics of Pomegranate fruit and Aril separation efficiency of hand tool

Parameters	Mridula	Ganesh	Cultivar Sinduri	G137	W. Muscut
Major axis dia (L), mm	73.34 ± 3.81	76.43 ± 2.82	78.67 ± 1.77	80.68 ± 6.25	85.76 ± 3.67
Minor axis dia (W), mm	70.23 ± 2.51	73.60 ± 3.12	76.95 ± 3.12	77.01 ± 6.58	83.07 ± 4.30
Axis normal to major and minor (T), mm	66.29 ± 2.69	72.19 ± 3.35	76.37 ± 2.96	74.93 ± 5.71	82.05 ± 5.21
Weight of fruit, g	198.50 ± 24.74	242.60 ± 49.27	251.92 ± 11.25	269.04 ± 47.13	333.05 ± 43.1
Peel thickness, mm	2.97 ± 0.13	3.76 ± 0.27	3.86 ± 0.48	3.48 ± 0.44	3.12 ± 0.18
Rag thickness, mm	0.15 ± 0.04	0.23 ± 0.03	0.27 ± 0.02	0.22 ± 0.02	0.25 ± 0.02
Aril, %	66.42 ± 3.27	62.28 ± 4.37	56.70 ± 4.48	61.97 ± 2.93	60.0 ± 1.66
Size (Geometric mean dia, mm)	69.88 ± 2.71	74.05 ± 2.98	77.32 ± 2.52	77.48±5.78	83.61 ± 4.29
Sphericity	0.953 ± 0.017	0.969 ± 0.016	1.00	0.96 ± 0.039	0.97 ± 0.02
Volume of fruit, cc	178.75	212.70	242.14	243.64	306.0
Density, g/cc	1.11	1.14	1.04	1.06	1.05
Aril separation efficiency by hand tool, by weight	35.6 ± 4.25	32.5 ± 2.84	40.0 ± 6.86	30.1± 3.66	32.0 ± 2.81
Aril separation efficiency by hand tool, by arils no.	36.4 ± 2.32	35.0 ± 3.15	31.32 ± 5.66	32.1 ± 3.09	33.11 ± 4.04

Table 5 : Physical characteristics of arils of different cultivars

Parameters	Mridula	Ganesh	Cultivar Sinduri	G137	W. Muscut
Major axis dia (L), mm	9.91 ± 0.76	9.40 ± 0.49	11.35 ± 0.25	10.62 ± 1.11	9.76 ± 1.05
Minor axis dia (W), mm	6.15 ± 0.77	6.79 ± 0.32	8.12 ± 0.44	6.56 ± 0.59	6.08 ± 0.89
Axis normal to major and minor (T), mm	5.19 ± 0.41	5.62 ± 0.35	7.25 ± 0.46	5.21 ± 0.47	4.67 ± 0.47
Weight of single aril, g	0.34 ± 0.05	0.36 ± 0.05	0.38 ± 0.04	0.22 ± 0.05	0.27 ± 0.05
Size (Geometric mean dia, mm)	6.81 ± 0.58	7.10 ± 0.34	8.74 ± 0.32	7.12 ± 0.51	6.50 ± 0.54
Sphericity	0.687 ± 0.029	0.756 ± 0.014	0.770 ± 0.028	0.67 ± 0.05	0.67 ±0.05

CIPHET-Pomegranate aril extractor (motorized version for processing industry)

The machine namely "CIPHET-Pomegranate Aril Extractor" relates to a mechanical and continuous system for processing of whole Pomegranate fruit of any size, shape and variety; and for recovery of clean, whole and undamaged arils (Thakur *et al.* 2008[b], 2009[c]). The system comprises the main components such as Pomegranate Fruit Breaking unit, Drive unit, Collection Trays where separated arils and extraneous matters are received, Vibrating Sieve unit; and the clean arils and extraneous matter Collection Troughs. The fruit breaking unit consists of an innovative mechanical system arrangement for continuously breaking the fruit in a manner that the major portion (85-92%) of arils is safely separated out. The rest of the arils that is about 8-15% remain attached with the broken peels that are further getting separated over the vibrating screen. The provision for washing and cleaning of the fruit contact parts of machine is also provided so that the machine can be immediately cleaned after the use. The machine is capable to process the whole pomegranate at the rate of approximately 5.0 q per hour with extraction capacity of approximately 90% and with little damage of arils that is less than 5.0%. The developed machine and the separated arils which are almost clean are shown in the Fig. 2. The clean arils can directly be used for eating or further pressing for juice extraction or for *anardana* preparation and for making various value added products. The main advantages of this novel machine "CIPHET-Pomegranate Aril Extractor" is:

- The Pomegranate breaking and arils separation mechanism is highly efficient with little damage and waste
- Continuous operation for rapid processing of large quantities in any shape, size and variety of the Pomegranate
- Peels and other extraneous matter are separated out and yielding a clean arils that can be used for further processing or for fresh eating.

The technical specification of the "CIPHET-Pomegranate Aril Extractor" is:

- Higher aril extraction capacity that is around 5.0 quintal per hour (approximately 35-40 fruit per minute)
- Aril extraction/separation efficiency is in the range of 85-92% depending on variety and Pomegranate fruit characteristics
- Mechanical damage received by separated arils is only 2-3%
- Man power required for safe and smooth running of the machine is 2-3 persons

- Electrical power required- 0.75 kW (1.0hp)
- Over all dimension of machine; Length-1480 mm, Width-660 mm, Height- 1710 mm
- Weight of the CIPHET-Pomegranate Aril Extractor is approximately 250 kg.

Fig. 2: Mechanized version of CIPHET-Pomegranate Aril Extractor

Minimal processing and value added products of pomegranate

Processing of Pomegranate provides an opportunity to create a new and innovative market for fresh arils (juice-enclosed seeds), dried arils, juices and wines; and health and pharmaceutical products. Sugar contents of Pomegranate juice are about 12 to 16%. This sugar is light for digestion. Pomegranate peels are used for medicines and for colouring cloths. Peels of Pomegranate is used to prepare toothpastes/toothpowders. Pomegranate juice has iron, phosphorus, calcium, magnesium etc which helps in growth of blood in the body.

Minimally processed arils

Minimally processed pomegranate arils packed in a convenient package are in good demand in urban market. Some of the processing units are engaged in providing frozen arils in the Indian market. The research and development on the minimally processed arils are going on with the objectives to increase the marketable shelf life of fresh arils. The chemical and organoleptic characteristics of minimally processed seeds of pomegranate 'Primosole' were evaluated after packaging in polypropylene trays (150 g each), sealing with a 40 μm thick polypropylene film and subsequent storage at 5°C for 10 days (Palma et al, 2009). Chemical analysis of centrifuged and filtered juice revealed no significant changes after 10 days of storage. The overall quality, anthocyanin content and antioxidant activity of minimally fresh processed arils of the sweet 'Mollar of Elche' pomegranate cv. (*Punica granatum*, Punicaceae), harvested at two different dates, and stored under modified atmosphere packaging (MAP) at 5 °C was assessed. Hand extracted, chlorine disinfected, rinsed and dried pomegranate arils were exposed to 0.56, 1.13, 2.27, 4.54, 9.08 or 13.62 kJ/m^2 UV-C radiation doses. Minimally fresh processed arils were packed in polypropylene baskets (125 g each), sealed on the top with bioriented polypropylene to generate a passive MAP, and stored up to 13 or 15 days at 5 °C. The respiration rate of fresh processed arils was higher in the late harvested than in earlier harvested fruit. The UV-C radiation did not significantly affect the respiration rate of fresh processed arils. The limit of visual quality acceptance to consumers was reached after 14 days in arils from earlier harvested fruit and after 10 days in late harvested (Violeta López-Rubira *et al.,* 2005).

The effect of different types of semi-permeable films and antioxidant solutions on the quality of minimally processed pomegranate arils was evaluated during storage at 4°C±0.5 for 14 days. Colour changes were not observed and the browning was slight in all the treatments studied but highest on those without antioxidants. pH and acidity values remained constant and the soluble solids content increased only in the arils packaged in perforated polyethylene films due to the high dehydration. The CO_2 concentration inside the BB4 films reached 22%. The aerobics mesophiles count, and moulds and yeast count were low in all the treatments and in the semi-permeable films a reduction was observed. The uses of semi-permeable films in minimally processed pomegranate arils allowed storage for 14 days, at 4°C±0.5, with good chemical, physical and microbiological quality.

Once the arils are extracted from the fruits, the arils/seeds have to be washed with solutions that guarantee the sanitation of the product reducing the microbial load, and that ensure their quality characteristics. These solution include chlorine (100 mg/kg), ascorbic acid (5 g/l), citric acid (10 g/l), and combinations of these (Gil *et al.*, 1996a). The best results with the cultivar 'Mollar de Elche' were obtained washing the seeds with the chlorine solution, followed by a mixture of ascorbic and citric acids, and storing the seeds at 1°C in polypropylene films that allowed the formation of a modified atmosphere appropriate for the conservation of these seeds. The preparation of seeds under very clean conditions and at temperatures close to 0°C prolonged the life of this product and maintained its quality (Gil *et al.*, 1996a). Storage at higher temperatures (4-8°C) produced seeds with lower quality and a shorter commercial life.

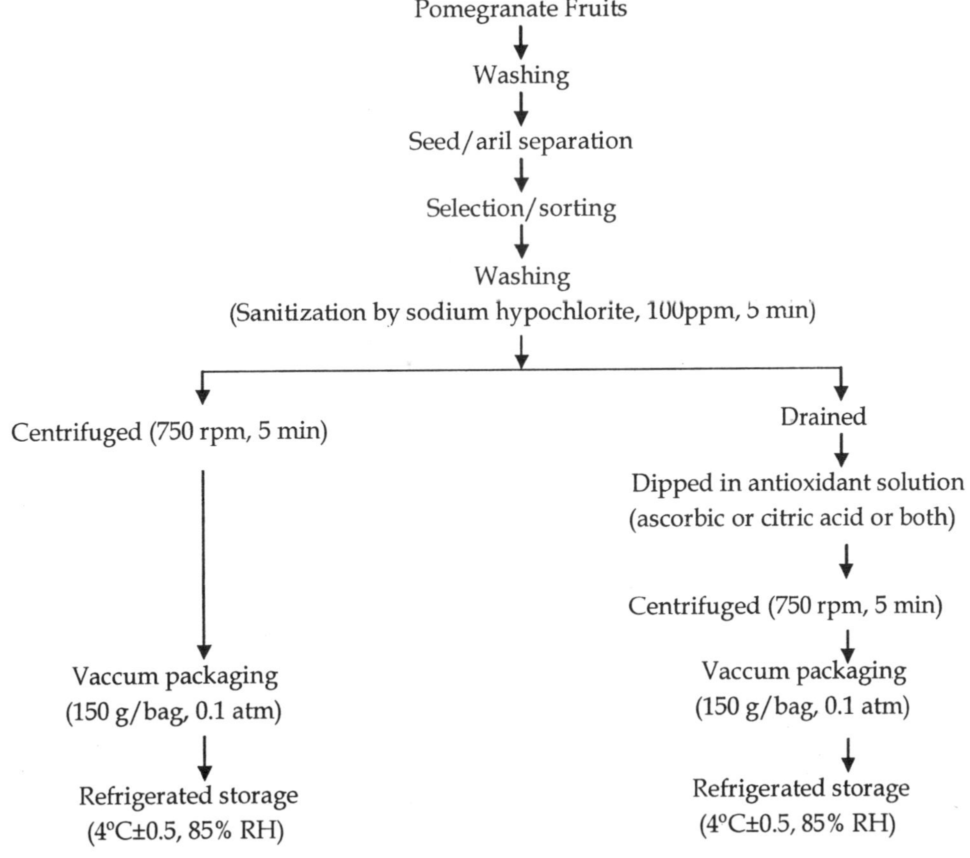

Juice

Pomegranate juice makes a delicious drink. The juice may be extracted from whole fruits or from the grains/arils using cloth hydraulic press which yields about 42%, spiral type screw press yielding about 50% juice and basket press which gives about 70% juice recovery with low tannin (0.12%) content than that obtained from whole fruits (0.175%). Excess tannin from juice can be precipitated by gelatin and is clarified by pectinase enzyme or by heating in a flash pasteurizer at 79-82°C followed by cooling, setting up 24 hours racking of a fettering. Sodium benzoate (600 ppm) is added as a preservative. The use of SO_2 or potassium metabisulfite is not recommended for pomegranate because of loss of its natural color due to the bleaching of SO_2. Shelf life of the juice can be extended by storing at low temperature (10°C). At room temperature (26-30°C) the quality starts deteriorating from 3 to 4 day onwards. It has been observed that there was no significant difference in quality parameters among different cultivars, when stored at low temperature (8-10°C). The juice of pomegranate can be stored up to 4 days under room temperature and up to 10 days at low temperature without any change in quality.

In the case of pomegranate juice, it has been shown that both the extraction and clarification process and the pH of the obtained juice (juices added with organic acids to decrease the pH), have marked effects on juice stability. Storage of frozen, juice permits prolonging the juice life considerably, although it does not completely prevent pigment degradation and browning.

Concentrate

Fruit juice concentrates of pomegranate, singly or in blended form with other promising juices are products having great potential on account of their wide appreciability in fruit based beverages. The concentrate can be stored at low temperature for further processing it for making various products as and when required.

Beverage

A technology for obtaining canned pomegranate beverage is described by Gabuniya *et al.* (1984). The use of bentonite at 1.0-1.5 g/liter and gelatin at 0.05 g/liter as clarifying agents gave better results than milk, charcoal, or casein.

Grenadine

Grenadine is a very appealing product from pomegranate. The quality grenadine can be prepared by mixing juice and sugar in ratio of 60:40 and can be stored up to 60 days at low temperature (5-8°C) and 20 days at room temperature (25-26°C) without any significant change in TSS, acidity, vitamin C and enzymatic browning. Storage in different coloured bottles does not have significant effect on quality of grenadine up to 60 days of storage.

Jelly

An attractive jelly can be prepared from pomegranate juice. The pomegranate jelly of excellent quality can be prepared by mixing juice and sugar in ratio of 2:1 and can be stored up to 180 days at low temperature (8-10°C) and 60-75 days at room temperature without any significant change in TSS, acidity, vitamin C and enzymatic browning.

A related product that has been recently studied has been pomegranate jelly. In this case, the percentage of sugar added is smaller than that of jams and marmalades, and addition of jellying agents (pectins, agar, etc.) is necessary for the preparation of jellies, that generally maintain very well the colour characteristics of the original juice. The effect of the pH is essential for pigment stability and the colour of the prepared jellies. Therefore, it has been shown that jellies prepared with juices from sour pomegranates, have better colour and pigment stability than those produced with sweet pomegranates of cultivar 'Mollar'.

Jam

Equal parts of juice and sugar are mixed and the syrup in concentrated on slow fire to make thick consistency (70-75° Brix). It is also known as anar-rub in Iran.

Syrup

A syrup of about 60°Brix with an added acidity of 1.5% as citric acid had a bright purplish-red color and a delightful taste and flavor. It was preserved by pasteurization or by addition of sodium benzoate.

Wine

The whole pomegranate fruits are pressed without crushing for preparation of wine. Sugar is added to the juice to bring it to 22-23°Brix. Potassium metabisulfite is added to the juice to prevent the growth of undesirable microorganisms. The juice is fermented with starter wine yeast. The wine is aged and finished in the same manner as red grape wine. If a sweet table wine is desired, sugar is added to 8-10°Brix after aging. The wine is flash pasteurized at 60°C, bottled hot, and sealed. The bottles are cooled in sprays of cold water.

Anardana and its powder

Anardana is the major product prepared from wild pomegranate fruits (sour type with high acidity 7.8 to 15.4%) by drying/dehydration of arils. The dehydrated arils of wild pomegranate are acidic and help in improving mouthfeel and digestion. For making anardana, traditionally, the arils are separated manually, sun dried in open and the drying process takes 10 to 15 days for completion. Dry product is packed in gunny bags. The traditionally processed anardana was found to have lot of contamination and microbial load and overall product is of inferior quality lacking desirable colour, nutritional composition and storage capacity. Improved technology to process the wild pomegranate into quality anardana having maximum shelf life, technology is developed at CIPHET, Ludhiana.

Other products such as frozen seeds, seeds in syrup, pomegranate spirits, etc. have also been studied, but not very intensely, and offer a very wide range of pomegranate derived products that will allow the use of low quality fruits that are presently discarded.

Coproducts utilization

Pomegranate waste like rind, leaves and flowers can be processed for value added products having medicinal, industrial and cosmetic value. Technology for preparation of Rind Powder has also been developed. The recovery of rind powder was found to be 15.5% on a whole-fruit basis and 34% on a rind-weight basis. Pomegranate rind powder has potential for use in medicines, leather, dying industries and in tooth powder.

It is to mention the possible use of pomegranate industrialization wastes and residues for the extraction of phytochemicals that have an interest for nutrition and pharmacy. Pomegranate rinds are very rich in polyphenolic compounds (ellagitannins) which, as naturally occurring compounds, or as compounds obtained by transformation of the natural compounds, have a very intense antioxidant activity, and these could be used as ingredients in dietetic formulations. In addition, the rind and the seeds are very rich in fibre, whose nutritional quality has to be evaluated. These residues are also rich in alkaloids, aromatic compounds, and are an interesting source of enzymes. For these reasons, the field of pomegranate by-products is an interesting field to be exploited, once pomegranate industrialization has become a reality.

Quality assurance and future prospects

Quality assurances for processed and minimally processed products are the key to achieve success in the market. An effective quality assurance (QA) system throughout the handling steps between harvest and retail display is essential to provide a consistently good-quality supply of fresh horticultural crops to the consumers and to protect the reputation of a given marketing label. QA starts with the selection of the genotype and its proper time to harvest for the best appearance, textural, flavor (taste and aroma), and nutritional (including phytonutrients) quality. Careful harvesting and handling are required to minimize physical injuries. Each post harvest handling step has the potential to either maintain or reduce quality and in a few cases (such as ripening of climacteric fruits) improve eating quality (Kader, 2001). The availability of low-cost microcomputers and solid-state imaging systems have resulted in increased use of computer-aided video inspection to sort many products into two or more quality grades before marketing. Objective and non-destructive methods of differentiating horticultural perishables on the basis of their flavor and nutritional quality are being tested and will become excellent QA tools as they become more reliable and efficient. Safety assurance can be part of QA and its focus is on minimizing chemical and microbial contamination during production, harvesting, and post harvest handling of intact/whole and fresh-cut fruits and vegetables. Future research and development efforts should focus on developing better methods of monitoring quality and safety attributes of fresh produce as part of a QA system.

The Pomegranate processing industry may become a sunrise industry in India if the quality is maintained throughout to meet the international requirement. There is a great prospect in the field of pomegranate processing particularly in marketing of fresh aril to the domestic and international consumers.

References

Artes F, Gil MI and Martinex JA (1995). Procedimiento para la conservacion de semillas de Granada en fresco. Patent No. 9502362.

Blasco J, Cubero S, Gomez-Sanchis J, Mira P and Molto E (2009). Development of a machine for the automatic sorting of pomegranate (Punica granatum) arils based on computer vision, J Food Engineering, 90(1): 27-34.

Chadha KL, Singh SK, Ram Asrey, Patel VB and Shruti Sethi (2004). Maturity standards in fruits and vegetables. Post harvest management, marketing and trade in horticultural crops (vol. II), Proceedings of the first Indian Horticultural Congress, New Delhi.

Gabuniya, NE, ID Landkipanidze and MV, Magerammov (1984). Technology of obtaining a beverage from pomegranates, Konsernayai Oveschesushilnaya Promyshlennost 7:27.

Gil MI, Artes F and Tomas-Barberan FA (1996). Minimal processing and modified atmosphere packaging effects on pigmentation of pomegranate seeds. J Food Sci., 61: 161-164.

Gil, MI, Martínez, JA and Artés, F (1996). Minimally processed pomegranate seeds. Lebensm Wiss u. Technol., 29: 708-713.

http://mofpi.nic.in

Juran Metal Works Ltd., Israel (Web: http://www.juran.co.il/)

Kader AA, Chardas A and Elyatem, S (1984) Responses of pomegranate to ethylene treatment and storage temperature. Calif. Agr. 38: 14–15.

Kader AA, (2001). Quality assurance of harvested horticultural perishables. ISHS Acta Horticulturae 553: IV International Conference on Postharvest Science.

Khazaie J, Ekrami-Rad N, Safa M and Nosrati S (2008). Effect of air-jet impingement parameters on the extraction of pomegranate arils. Biosystem Engineering, 100(2): 214-226.

Morton J (1987). Pomegranate. In Fruits of warm climates. Ed. Juli F. Morton, Miami, FL, pp. 352-355.

Nanda S, Sudhakar Rao DV and Krishnamurthy S (2001). Effects of shrink film wrapping and storage temperature on the shelf life and quality of pomegranate fruits cv. Ganesh. Post Harvest Biol Technol 22: 61-69.

Palma A, Schirra M, D'Aquino S, La Malfa S and Continella G (2009). Chemical properties changes in pomegranate seeds packaged in polyproplylene trays. ISHS Acta Horticulture 818 (http://www.actahort.org/books/818/818_48 htm).

Safa M, Khazaei J, Kianmehr MH (2009). Extracting the pomegranate arils (seeds) using a pneumatic method. ISHS Acta Horticulture 818 (http://www.actahort.org/books/818/818_52 htm).

Schmilovitch Z, Sarig Y, Ronen B, Hoffman A, Egozi H, Beres H, Bar-Lev E, Grosz F (2009). Development of a method and a system for extracting the seeds (arils) from pomegranate fruits - from concept to commercial utilization. ISHS Acta Horticulture 818 (http://www.actahort.org/books/818/818_53 htm).

Sepulveda E, Galletti L, Saenz C and Tapia M (2000). Minimal processing of pomegranate var. Wonderful. CIHEAM Options Mediterranean's 42:237-242.

Singh DB, Kingsly ARP and Jain RK (2006) Studies on separation techniques of pomegranate arils and their effect on quality of anardana. J Food Engineering (Online published during 2006).

Thakur AK, Patil RT, Singh DB and Gupta RK (2008a). Hand tool for easy separation of arils from Pomegranate (Patent application No.: 479/DEL/2008).

Thakur AK, Patil RT, DB Singh and Gupta RK (2008b). CIPHET-Pomegranate Aril Extractor (Patent application No. 2584/DEL/2008).

Thakur AK, Patil RT, Singh DB and Gupta RK (2009c) Development of tool and machine for safe separation of arils from pomegranate. Paper presented in 2nd International symposium on "Pomegranate and minor including Mediterranean fruits" (ISPMMF-2009) at UAS Dharwad, Karnataka (India), 23-27 June, 2009.

Violeta López-Rubira, Andrés Conesa, Ana Allende and Francisco Artés (2005). Shelf life and overall quality of minimally processed pomegranate arils modified atmosphere packaged and treated with UV-C. Postharvest Biology and Technology 37(2): 174-185.

Waskar DP, Khedkar PB and Garande VK (1999). Effect of postharvest treatments on storage behaviour of pomegranate fruits under room temperature and cold storage. Indian Food Packer 53: 11–15.

□□□

9

Drying of Liquid Foods

DATTATREYA M. KADAM

Among the agricultural processing technologies, drying has a great importance as a post-harvest technology. Drying is the most common form of food preservation and extends the food shelf life; the major objective in drying agricultural products is the reduction of the moisture content to a level, which allows safe storage over an extended period. In addition, it brings about substantial reduction in weight and volume, minimizing packaging, storage and transportation costs (Kadam *et al.*, 2010; 2011). Foaming of liquid and semisolid materials has long been recognized as one of the efficient methods to shorten drying time (Kadam and Balasubramanian 2011).

There are several basic drying methods and a far greater number of modifications of the basic methods. The method of choice depends on the type of food to be dried, the quality level that must be achieved, and the cost that can be justified. Some of the more common drying methods include foam mat drying, drum drying, spray drying, vacuum shelf drying, vacuum belt drying, atmospheric belt drying, freeze drying and these are particularly suited to liquid foods.

Foam-Mat drying of liquid foods

Foam- mat drying is a promising new development in the field of drying aqueous foods. This method offers a wide scope for application in vegetable and fruit juice processing industry as it enables the dehydration of heat sensitive foods or which are difficult-to-dry, sticky, and viscous under relatively mild conditions without undue quality change. This technology is finding an increasing application and importance on commercial scale mostly in the drying of liquids that tender a high quality concentrate such as milk, fruit juices, soluble coffee, etc (Kadam *et al.*, 2010; 2011).

The foam-mat drying is a process in which the transformation of products from liquid to stable foam follows air drying at relatively low temperatures to form a thin porous honey-comb sheet or mat which is disintegrated to yield a free-flowing powder. The dried product obtained from foam-mat drying (Fig. 1) is of better quality, porous and can be easily reconstituted. Concentration of the material prior to conversion into stable foam may or may not be an essential requirement and will depend on its surface tension and consistency. In spite of the fact that a large volume of gas present in the foamed mass impedes the higher rate of heat transfer, drying rate is comparatively high because of enormous increase in liquid-gas interface. The advantages of the foam-mat drying process includes relatively fast drying, easy reconstitution and better quality product. The renewed process is of potential interest in foam-mat drying for developed and developing countries for its simplicity, rapid drying at lower temperature, suitable for all type of juices, retention of nutritional losses and cost effective for producing easily reconsitituble juice powders.

Principle of process

The essential steps in foam-mat drying include as follows (Morgan *et al.*, 1961; Morgan, 1966):

(a) Pre-treatment of the raw material and preparation of liquid concentrate;

(b) Conversion of the concentrate into a stable foam by incorporating air or other gases and a suitable foaming agent;

(c) Exposing the foam in the form of a thin sheet to a current of hot air until it is dehydrated; or hot water conduction method; and

(d) Conditioning and grinding of the dried porous mass into an easily reconstitutble powder in water.

(e) Energy consumption

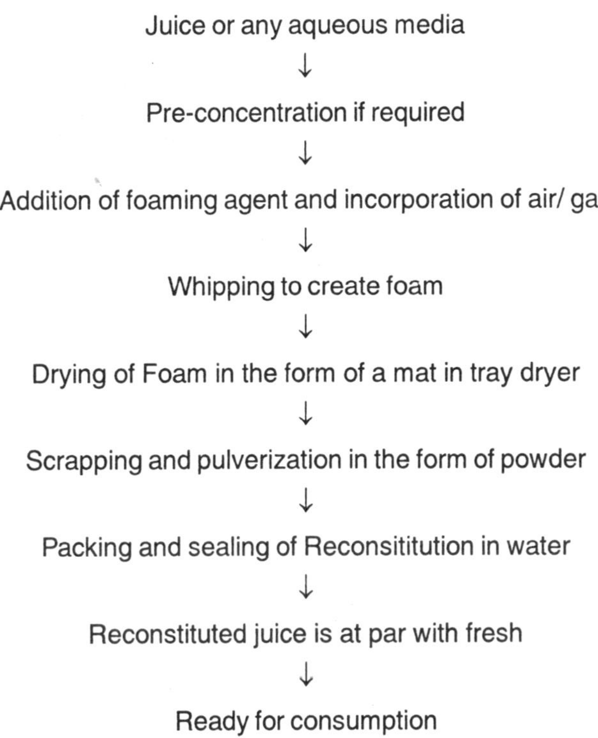

Fig. 1 : Process flow chart of Foam-Mat drying

Types of foaming agents

Foam-mat drying requires stiff stable foams which are not readily made from many food concentrates especially food with low protein content, unless there is an addition of a small quantity of foaming agents or foam stabilizers (Akintoye and Oguntunde, 1991). Foam stabilizers are added at different levels to varying composition of the food concentrate, and their performances could be measured by their foaming, reconstitution, and sensory attributes.

Many types of agents are used for production of foams and each of them offers certain distinct advantages. One of the most commonly used is the monoglyceride, or fatty-ester type. This agent has the capacity to produce extremely fine foams which enable thorough and efficient moisture removal under milder conditions than could be used with some other agents. Hence this agent is particularly useful with the more heat-sensitive materials. It is probably for this reason that most commercial foam-mat dried products have utilized monoglyceride agents.

Different foaming agents have different characteristics and hence impart different foaming properties to the aqueous material to be dried. To dry the food efficiently without effecting its quality different foaming agents are evaluated for their compatibility with a particular food product. Foaming properties such as foam expansion, foam stability and foam density are determined at different concentrates of various foaming agents and based on foaming properties optimum level of best-suited foaming agent is identified (Kadam, *et al.,* 2010a).

Foam expansion

$$\text{Foam Expansion } (\%) = \frac{V_1 - Vo}{Vo} \times 100$$

Where

V_1 is Final volume of foamed product in cm^3

and Vo is the initial volume of the product in cm^3

Foam stability

Foam stability is determined by measuring the reduction in foam volume at definite intervals of time by using the following relationship

$$\text{Foam stability} = Vo \frac{\Delta t}{\Delta V}$$

Where ΔV is the change in volume of foam occurred during the time interval.

Foam density

The density of the foamed material (Juice/ pulp) can be determined in terms of mass by volume and represented as g/cm^3.

$$\text{Foamdensity} = \frac{\rho V_0}{V_1}$$

Where, ρ is the density of the material (Juice/ pulp)

Commercial applications

Over the years, foam-mat drying has been applied to many fruits including tomato juice (Kadam and Balasubramanian, 2011), Citrus/ Kinnow (Kadam *et al.,* 2011) (Fig. 2), mango (Kadam *et al.,* 2010), soymilk, star fruit, bananas, pineapple, lemon, sapota, apple etc.

Fig. 2 : Foam mat dried tomato and kinnow

Addition of soy protein in a solid form at the 10 g/100 g level is very effective foam inducer for foam mat drying of banana to produce free flowing powder reducing the density of the banana puree to 0.5 g/ml after 12 min of whipping. The drying of foamed syrups of different fruits like orange, lemon, grape fruit, tomatoes, strawberry, guava and apple were carried out in a solar energy augmented drying unit. The results showed that foaming decreased the time of drying by 30.8–41.5% in the case of direct sun drying and by 47.1–73.2% in the case of solar drying compared to the drying time of unfoamed syrup. So it was concluded that solar-augmented foam-mat drying technique proves to be of sufficient efficiency and acceptability (Anjaria and Chivate, 1966).

Advantages

1. Foam-mat dried vegetables or fruit powders have less heat induced changes in color and flavor as compared to conventional spray-dried or drum-dried products.
2. The dehydrated powder/flakes are superior to drum dried and spray dried products because of its honey comb structure and better reconstitution properties.
3. The drying of foamed materials is faster than that of non-foamed ones.
4. Foam-mat drying allows processing of hard to-dry materials such as tomato paste and a variety of fruit pulps and juices.

Spray drying of liquid foods

Spray drying is a method of producing a dry powder from a liquid or slurry by rapidly drying with a hot gas. This is the preferred method of drying of many thermally-sensitive materials such as foods and pharmaceuticals. A consistent particle size distribution is a reason for spray drying some industrial products such as catalysts. Air is the heated drying media; however, if the liquid is a flammable solvent such as ethanol or the product is oxygen-sensitive then nitrogen is used (Torrey, 1974).

All spray dryers use some type of atomizer or spray nozzle to disperse the liquid or slurry into a controlled drop size spray. The most common of these are rotary nozzles and single-fluid pressure swirl nozzles. Alternatively, for some applications two-fluid or ultrasonic nozzle are used. Depending on the process needs drop sizes from 10 to 500 micron can be achieved with the appropriate choices. The most common applications are in the 100 to 200 micron diameter range. The dry powder is often free-flowing.

The hot drying gas can be passed as a co-current or counter-current flow to the atomiser direction. The co-current flow enables the particles to have a lower residence time within the system and the particle separator (typically a cyclone device) operates more efficiently. The counter-current flow method enables a greater residence time of the particles in the chamber and usually is paired with a fluidized bed system.

Spray dryer

A spray dryer is a device used in spray drying of liquid foods. It takes a liquid stream and separates the solute or suspension as a solid and the solvent into a vapor. The solid is usually collected in a drum or cyclone. The liquid input stream is sprayed through a nozzle into a hot vapor stream and vaporised. Solids form as moisture quickly leaves the droplets. A nozzle is usually used to make the droplets as small as possible, maximising heat transfer and the rate of water vaporisation. Droplet sizes can range from 20 to 180 μm depending on the nozzle (Fig. 3). Spray dryers can dry a product very quickly compared to other methods of drying. They also turn a solution, or slurry into a dried powder in a single step, which can be advantageous for profit maximization and process simplification (Fig. 4).

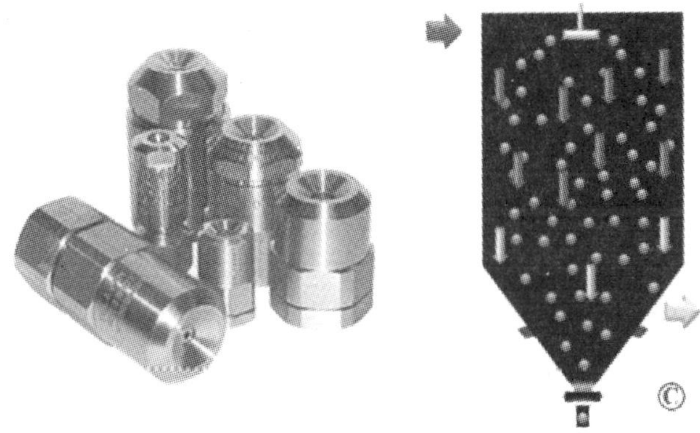

Fig. 3 : Spray drying nozzles and Schematic illustration of spray drying process.

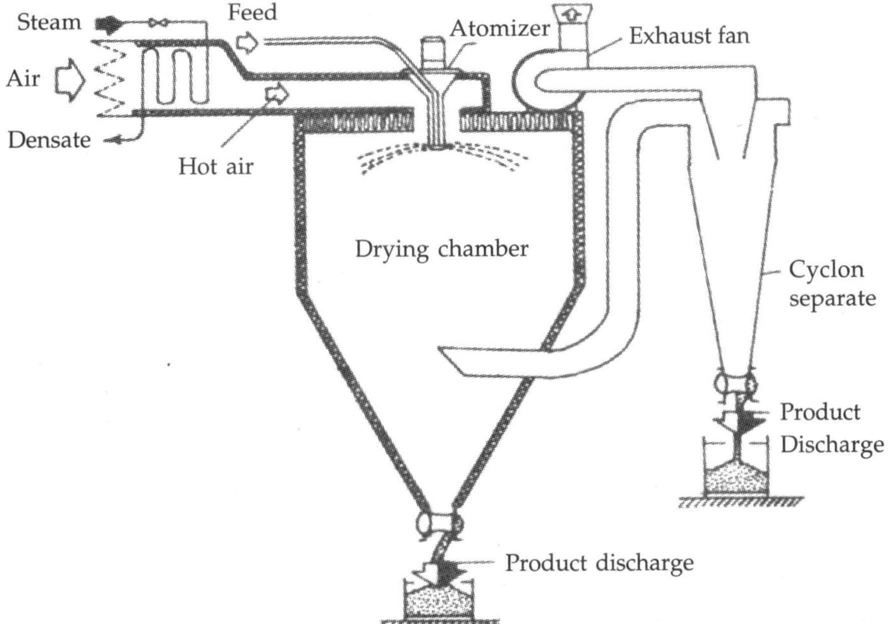

Fig. 4 : Schematic view of Spray Dryer

Micro-encapsulation

Spray drying often is used as an encapsulation technique by the food and other industries. A substance to be encapsulated (the load) and an amphipathic carrier (usually some sort of modified starch) is homogenized as a suspension in water (the slurry). The slurry is then fed into a spray drier, usually a tower heated to temperatures well over the boiling point of water.

As the slurry enters the tower, it is atomized. Partly because of the high surface tension of water and partly because of the hydrophobic/hydrophilic interactions between the amphipathic carrier, the water, and the load, the atomized slurry forms micelles. The small size of the drops (averaging 100 micrometers in diameter) results in a relatively large surface area which dries quickly. As the water dries, the carrier forms a hardened shell around the load.

Load loss is usually a function of molecular weight. That is, lighter molecules tend to boil off in larger quantities at the processing temperatures. Loss is minimized industrially by spraying into taller towers. A larger volume of air has a lower average humidity as the process proceeds. By the osmosis principle, water will be encouraged by its difference in fugacities in the vapor and liquid phases to leave the micelles and enter the air. Therefore, the same percentage of water can be dried out of the particles at lower temperatures if larger towers are used. Alternatively, the slurry can be sprayed into a partial vacuum. Since the boiling point of a solvent is the temperature at which the vapor pressure of the solvent is equal to the ambient pressure, reducing pressure in the tower has the effect of lowering the boiling point of the solvent.

The application of the spray drying encapsulation technique is to prepare "dehydrated" powders of substances which do not have any water to dehydrate. For example, instant drink mixes are spray dries of the various chemicals which make up the beverage (Beristain *et al.*, 1993). The technique was once used to remove water from food products; for instance, in the preparation of dehydrated milk. Because the milk was not being encapsulated and because spray drying causes thermal degradation, milk dehydration and similar processes have been replaced by other dehydration techniques. Skim milk powders are still widely produced using spray drying technology around the world, typically at high solids concentration for maximum drying efficiency. Thermal degradation of products can be overcome by using lower operating temperatures and larger chamber sizes for increased residence times.

Recent research is now suggesting that the use of spray-drying techniques may be an alternative method for crystallization of amorphous powders during the drying process since the temperature effects on the amorphous powders may be significant depending on drying residence times.

Spray drying applications

Food: milk powder, beet root powder, coffee, tea, eggs, cereal, spices, flavorings.

Pharmaceutical: antibiotics, medical ingredients, additives

Industrial: paint pigments, ceramic materials, catalyst supports

Nano spray dryer: The nano spray dryer offers new possibilities in the field of spray drying. It allows to produce particles in the range of 300 nm to 5 μm with a narrow size distribution. High yields are produced up to 90% and the minimal sample amount is 1 mL.

Drum drying of liquid foods

One of the most economical drying methods is drum drying. In this operation, food slurry is contacted with a hot, revolving drum to form a thin layer on the surface. After sufficient residence time to allow the evaporation of water the product is removed from the drum by a scraper device (called a doctor knife) located usually 1/2 to 3/4 of a revolution from the point of application (Torrey, 1974).

Factors affecting the rate of drying and final moisture content are:

1. Residence time on the drum
2. Surface temperature
3. Film thickness

The method of dehydration can only be applied to food slurries or liquid food systems and the product must be able to withstand high temperature-short time exposures without undergoing severe quality changes. It has been most successfully applied to drying of potatoes and has been used to dry milk, soup mixes, baby foods and dried fruit purees (e.g. apple sauce) (Torrey, 1974).

Compared to spray drying, drum drying is a more intense heat treatment which results in more denatured proteins. The milk powder prepared by this technique is less soluble as a result. The temperature uniformity of the heated roller/drum is poor so spray drying results in better quality milk powder. Other products where drum drying can be used are for example starches, breakfast cereals, baby food, instant mashed potatoes to make them cold water soluble (Morgan and Ginnette, 1960).

Drum dryer system description

A drum dryer consists of one or two horizontally mounted hollow cylinder(s) made of high-grade cast iron or stainless steel, a supporting frame, a product feeding system, a scraper, and auxiliaries. The diameter of typical drums ranges from 0.5m to 6m and the length from 1m to 6m.

In operation, steam at temperature up to 200°C heats the inner surface of the drum. The moist material is uniformly applied in a thin layer (0.5mm–2 mm) onto the outer drum surface. Most of the moisture is removed at water boiling temperature. The residence time of the product on the drum ranges from a few seconds to dozens of seconds to reach final moisture contents of often less than 5% (wet basis). The energy consumption in a drum dryer may range between 1.1 kg steam per kg of evaporated water and 1.6 kg steam per kg of evaporated water, corresponding to energy efficiencies of about 60%–90%.

A drum dryer can produce products at a rate between 5 kg/h.m^2 and 50 kg/h. m^2, depending upon type of foods, initial and final moisture content, and other operation conditions.

Types of drum dryers and their applications

Drum dryers are classified into single drum dryer, double drum dryer and twin drum dryer (Fig 5). A double drum dryer has two drums that revolve toward each other at the top (Fig 6). The spacing between the two drums controls the thickness of the feed layer applied to the drum surfaces. A twin drum dryer also has two drums, but they rotate away from each other at the top. Among the three types, single and double drum dryers are most commonly used for fruits and vegetables. For example, large quantities of mashed potato flakes are produced using single drum dryers with specially designed roll feeding system. Double drum dryers are used in California to dry tomato paste. Twin drum dryers are used only for drying materials yielding dusty products. For materials sensitive to heat damage, a vacuum drum dryer may be used to reduce drying temperature. A vacuum drum dryer is similar to other drum dryers except that the drums are enclosed in a vacuum chamber. In continuous vacuum drum dryers, receivers and air locks are designed to provide appropriate seal. Equipment and operation of vacuum drum dryers are relatively expensive, which limits vacuum drum drying to only high-value products or products that cannot be produced more economically by other means (Mujumdar).

Classification and selection of industrial dryers

Perforated (suction) drum dryer is another variation from ordinary drum dryers. It utilizes heated air to heat the drum inside surface and the product is sucked to the perforated drum surface during drying. Several perforated drums can be linked together so that product can be transferred from one drum surface to another to achieve high production rates.

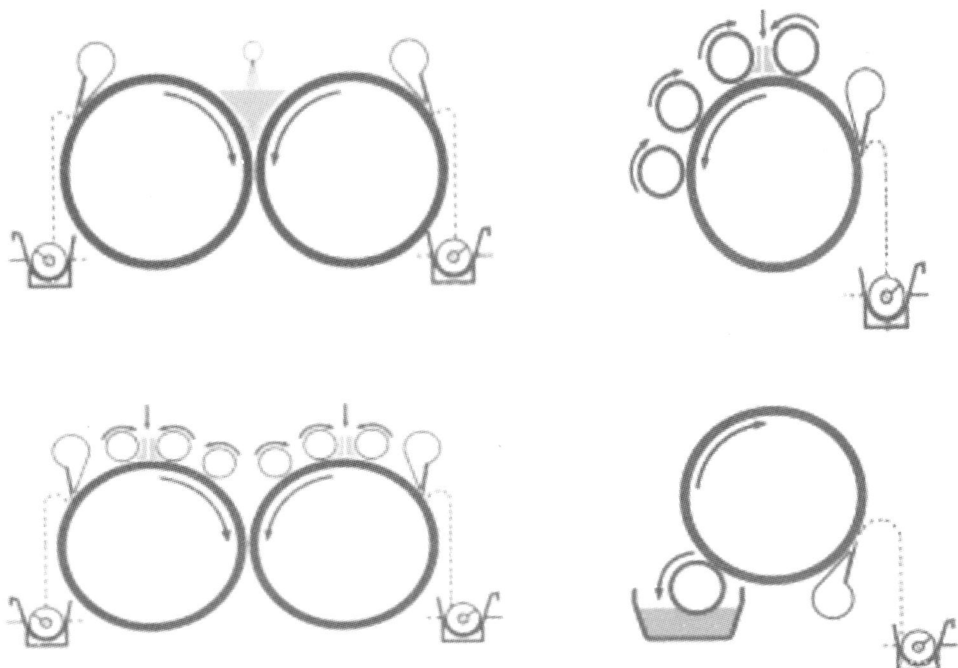

Fig. 5 : View of Different types of drum dryers

Fig. 6 : View of Double drum dryer

The advantages of drum drying include

- The products have good porosity and hence good rehydration due to boiling and evaporation.
- Drum dryers can dry high/more viscous foods, such as pastes and gelatinized or cooked starch, which cannot be easily dried with other methods.
- Drum dryers normally have high energy efficiency.
- Drum drying can be clean and hygienic.
- Drum dryers are easy to operate and maintain.
- The dryers are flexible and suitable for multiple but small quantity production.

The disadvantages of drum drying are the following

- Some products may not form a good film on the drum surface and are not suitable for drum drying.
- Some products, especially those with high sugar content, may not be easily scrapped off from the drum.
- Relatively low throughput compared to spray drying.
- High cost of changing drum surface because of the precision machining that is required.
- Possible scorching of the product to impart cooked flavor and off-color due to direct contact with high temperature drum surface.
- Not able to process salty or other corrosive materials due to potential pitting of drum surface.

Fluidized bed drying

Fluid bed dryers are used for gentle and rapid drying of powders and particles. The raw material/ ingredient to be dried are fluidized by passing hot air through it, setting the particles in motion. The fluid bed is imparted with unique properties as air travels through it, allowing for improved mixing (Torrey, 1974). Fluidized bed technology has been widely used in the pharmaceutical industry for drying, granulating, and coating. A fluid bed dryer significantly reduces drying time compared with a tray dryer or vacuum dryer. A fluid bed dryer exposes the entire product surface area to the high volume air stream and heat is transferred to the product surface by convection (Fig. 7). The advantages of fluid bed dryer are 1) this multi functional machine can be used for drying, granulating, coating and pelleting and 2) the machine dries up to 10-20 times faster and more efficiently than other conventional dryers.

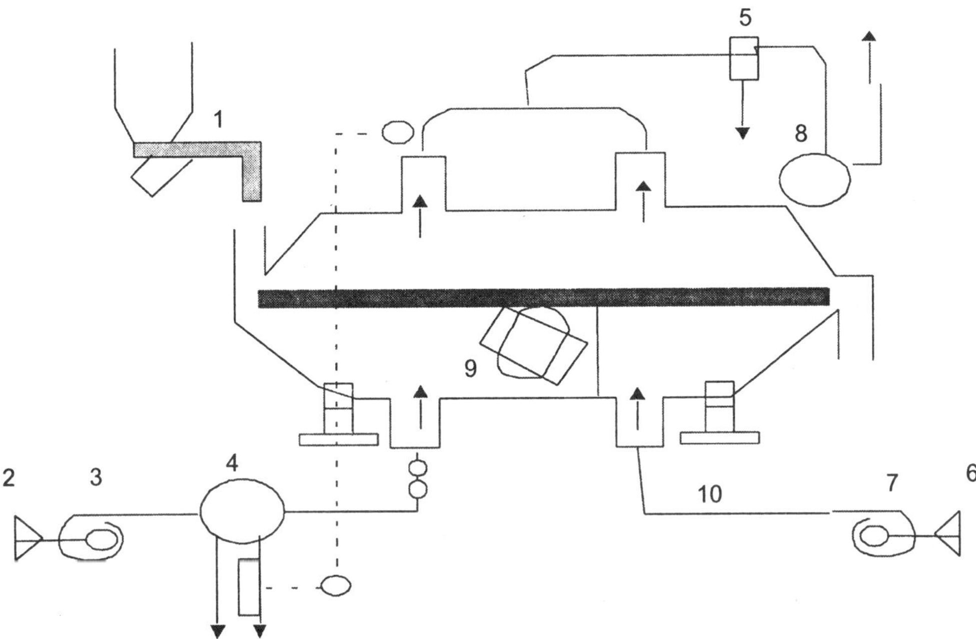

1. Feeding Hopper; 2. Inlet Air Filter; 3. Hot air Blower; 4. Stream Heater; 5. Cyclone Separator; 6. Fan; 7. Cooling Fan; 8. Air Filter; 9. Vibration Motor; 10. Vibrating Separator

Fig. 7 : Schematic view of commercial fluidized bed dryer

In fluid bed drying, heat is supplied by the fluidization gas, but the gas flow need not be the only source. Heat may be effectively introduced by heating surfaces (panels or tubes) immersed in the fluidized layer. Fluid bed drying offers important advantages over other methods of drying particulate materials. Particle fluidization gives easy material transport, high rates of heat exchange at high thermal efficiency while preventing overheating of individual particles. Fluid bed dryers are designed to dry powder particles, where the residual moisture content is higher than what is required in the final powder. Hot drying air is distributed through specially designed perforated plates, on which the powder particles/agglomerates are resting. The velocity of the drying air makes the powder 'fluidize', i.e. the particles get airborne; however, to the extent that only the finest/smallest particles leave the fluid bed with the air, while the bigger particles/agglomerates are conveyed/transported with the upward and forward directed flow of air (due to the perforation technique of the plate) into the subsequent section. Here cold and/or dehumidified air is introduced to cool the powder to the required temperature. Fluid bed drying is suited for powders, granules, agglomerates and pellets with an average particle size between 50 microns and 5 mm (5,000 microns). Very fine, light powders or highly elongated particles may require vibration for successful fluid bed drying.

A *fluidized bed* is formed when a quantity of a solid particulate substance (usually present in a holding vessel) is placed under appropriate conditions to cause the solid/fluid mixture to behave as a fluid. This is usually achieved by the introduction of pressurized fluid through the particulate medium. This results in the medium then having many properties and characteristics of normal fluids; such as the ability to free-flow under gravity, or to be pumped using fluid type technologies. The resulting phenomenon is called fluidization.

Raw materials are fed into the machine from the inlet and moves forward continuously along with the level of fluidized-bed under the action of vibration. The hot air passes through fluidized-bed and carries out hot heat exchange with damp raw material. Then the damp air is exhausted out through cyclone separator and duster and the dry raw material is discharged through discharge outlet. The vibration is created by motor.

Fluid bed dryers consist of a hollow vertical chamber where dry, heated air enters the bottom of the chamber and exhaust air exits through the top or rear of the chamber. There is a perforated diffuser plate at the bottom to diffuse the upward flowing air which helps create a stable and uniform fluidized bed (Fig 8). The diffuser plate can also support the bed when there is no airflow. The heated, dry air acts as both a fluidizing and drying medium for the solids. Fluid bed drying is a batch process so the amount of material that can be dried depends on the size of the fluid bed dryer and nature of the material.

The stationary fluid bed dryer has no moving parts. Fluidization enables the product to flow continuously through the fluid bed, overflowing at the opposite end. Multiple zones in the fluid bed dryer allow heating and cooling in the same unit. Each zone has independent control for temperature, dew point and fluidizing velocity. Residence time can vary up to four fold in the unit by adjusting the weir height for each zone. The fluid bed dryer's gentle but efficient drying action is well suited for friable or soft agglomerated particles. It works best drying products within a narrow particle size range. It has enhanced heat and mass transfer through intimate contact and mixing with the fluidizing gas. In this dryer heat sensitive materials can be dried at temperatures lower than the boiling point due to reduced vapor pressure and indirect heat transfer surfaces can be added to increase capacity or reduce bed area.

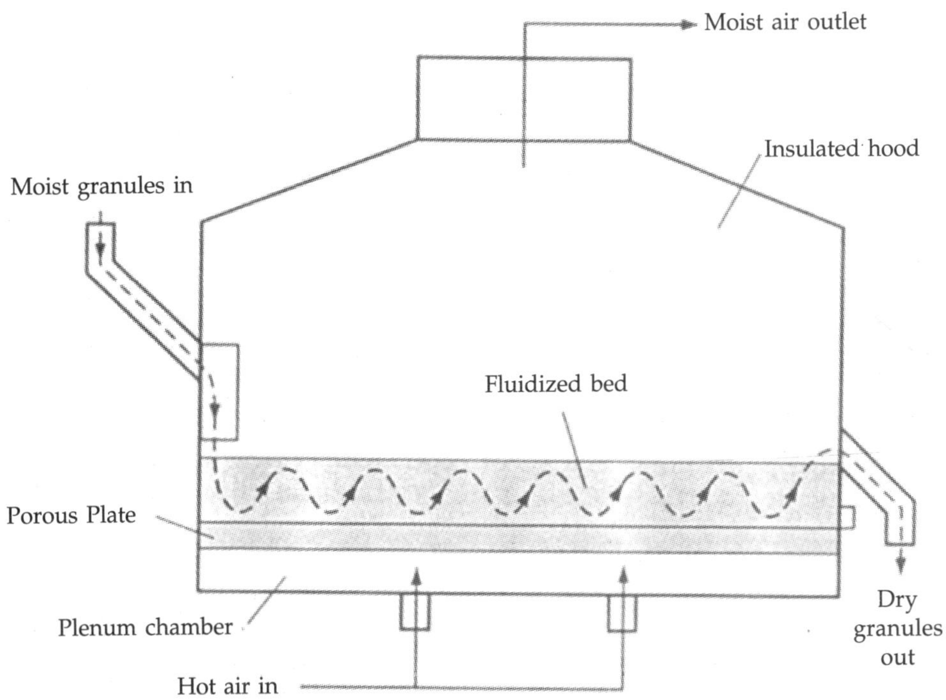

Fig. 8 : Schematic illustration of a fluidized bed dryer

Fluid bed processing involves drying, cooling, agglomeration, granulation, and coating of particulate materials. It is ideal for a wide range of both heat sensitive and non-heat sensitive products. Uniform processing conditions are achieved by passing a gas (usually air) through a product layer under controlled velocity conditions to create a fluidized state (http://www.niroinc.com/drying_dairy_food/fluid_bed_technology.asp).

In fluid bed cooling, cold gas (usually ambient or conditioned air) is used. Conditioning of the gas may be required to achieve sufficient product cooling in an economically sized plant and to prevent pick up of volatiles (usually moisture). Heat may also be removed by cooling surfaces immersed in the fluidized layer. Agglomeration and granulation may be performed in a number of ways depending upon the feed to be processed and the product properties to be achieved. Fluid bed coating of powders, granules, or tablets involves the spraying of a liquid on the fluidized powder under strictly controlled conditions.

There are different types of systems available for fluid bed drying. Continuous and batch dryers, coolers, agglomeration, coating, pelletizing and granulation systems are designed to operate in open cycle (involving water evaporation) or closed cycle (involving mostly organic solvent evaporation). For products posing a risk of dust explosion during processing, pressure shock resistant designs, self-inertized and closed cycle systems are available.

References

Akintoye, OA and Oguntunde, AO (1991). Preliminary investigation on the effect of foam stabilizers on the physical characteristics and reconstitution properties of foam-mat dried soymilk. Drying Technology, 9(1): 245-262.

Anjaria, BV and Chivate, MR (1966). Studies in foam-mat drying. Indian Chemical Engineer, Transactions, 8(3): 64-68.

Beristain, CI, Garica, HS and Vazquez, A (1993). Foam-mat dehydration of Jamaica) hibiscus sabdariffa L.) Instant drink. Drying Technology, 11(1): 221-228.

Berry, RE, Wanger, CJ, Jr., Bissett, OW and Veldhuis, MK (1972). Preparation of instant orange juice by foam-mat drying. Journal of food Science, 37(4): 803-808.

http://serve.me.nus.edu.sg/arun/E_books.htm.

http://www.niroinc.com/drying_dairy_food/fluid_bed_technology.asp.

Jitendra Chauhan J and Yadav P, (2011). Study of Scale up parameters of Fluidized bed Coating, Der Pharmacia Sinica, 2 (1): 228-238 http://www.pelagiaresea rchlibrary.com/der-pharmacia-sinica/vol2-iss1/DPS-2011-2-1-228-238.pdf

Kadam, DM and Balasubramanian S (2011). Foam mat drying of tomato juice. Journal of Food Processing and Preservation. doi:10.1111/j.1745-4549.2010.00492.x

Kadam, DM Wilson, RA and Kaur S (2010). Determination of biochemical properties of foam mat dried mango powder. International Journal of Food Science and Technology. 45 (8); 1626-1632 doi: 10.1111/j.1365-2621.2010.02308.x

Kadam, DM, Patil, RT and Kaushik, P, (2010a). Foam Mat Drying of Fruit and Vegetable Products (Chapter 5). E book entitled "Drying of Food, Vegetables and Fruits" Volume 'A': Fundamental Topics. Edited by Dr. Sachin Jangam (National University of Singapore), Dr. Chung Lim Law (University of Nottingham, Malaysia) and Professor Arun Mujumdar (National University of Singapore).http://serve.me.nus.edu.sg/arun/file/Publications/books/Drying%20o%20Foods%20Vegetables%20and%20Fruits%20Volume%201.pdf pp 111-124.

Kadam, DM, Rai, DR, Patil, RT Wilson, RA, Kaur, S and Kumar R, (2011). Quality of fresh and stored foam mat dried Mandarin powder. International Journal of Food Science and Technology. 46 (4); 793--799 doi:10.1111/j.1365-2621.2011.02559.x

Morgan, AI, Jr. (1966). Foam-mat drying. ASHRAE Journal, 8(11): 58-59.

Morgan, AI, Jr., Ginnette, JM (1960). Continuous process of dehydrating fruit and vegetable juices. US Patent 2,955,943. Cited in:Torrey, M. (ed.), Dehydration of Fruits and Vegetables. Food Technology Review No.13, Noyes Data Corporation, New Jersey, 1974.

Morgan, AI, Jr., Graham, RP, Ginnette, LF and Williams, GS (1961). Recent developments in foam-mat drying. Food Technology, 15(1): 37-39.

Mujumdar AS Classification and Selection of Dryers (Chapter 2). http://serve.me.nus.edu.sg/arun/file/teaching/ME5202/Chapter2%20and%203.pdf. accessed on 01.12.2011.

Torrey, M, (1974). Dehydration of Fruits and Vegetables. Food Technology Review No. 13. Noyes Data Corpn. New Jersey.

□□□

10

Processing of Guar (*Cyamopsis tetragonoloba*) in India

R.K. VISHWAKARMA, U.S. SHIVHARE, S.K. NANDA

The cluster bean (*Cyamopsis tetragonoloba*) is called Guar in India. The word "Guar" is derived from Sanskrit word "*Gau-āhār*" which means the cow fodder. It is a leguminous crop and up to 1953 it was being used as cattle feed (Bhan and Prasad, 1967). It has originated from Indian sub-continent that can be successfully grown in arid regions. The plant is thin but strong and reaches one-meter height at its full growth (Fig. 1). The optimum sowing time of guar is first fortnight of July (Kumar and Singh, 2002). It is harvested in the end of October. The harvesting is done manually in India. Green pods of guar are used as a vegetable for human consumption. It is also used as a green manure crop. However, the principal use and economic importance of guar lies in its industrial processing into guar gum powder.

Unlike the seeds of other legumes, the guar seed has a large endosperm. The concave shaped endosperm contains significant amount of galactomannan gum, which forms viscous gel in cold water. Refined guar gum powder has various food and industrial applications in ice cream, as stabilizer for cheese, instant puddings and whipped cream substitutes, as meat binder, in paper manufacturing, oil well drilling mud, explosives, ore floatation, *etc.*

Fig. 1 : Guar Plant

India is the world's largest producer of guar and is grown in the north-western region encompassing states of Rajasthan (Churu, Nagaur, Banner, Sikar, Jodhpur, Ganganagar, Sirohi, Dausa, Bikaner, Hanumangarh, and Jhunjhunu districts); Gujrat (Kutch, Banaskantha, Mehsana, Sabarkantha, Vadodara, and Ahmedabad districts); Harayana (Bhiwani, Gurgaon, Mahendragrh, Hisar, and Rewari districts); and, Punjab (Bhatinda, Ferozpur, Muktsar, and Mansa districts). India accounts for the 80% of total guar produced in the world followed by Pakistan at second place with 15% of total world production. The guar crop was estimated to be grown on about 2 Mha in India in 2007-08 (Anon, 2010). The state of Rajasthan alone comprises almost 78% area under guar (Kumar and Singh, 2002). Estimated production of guar was 0.8 million tonnes in 2007-08 (Anon, 2010). Rajasthan ranked first in the production of guar with a share of approximately 70% of total production in 2007-08; while, Haryana ranked second with a production of 0.59 lakh tones (Fig. 2). Productivity of guar in India was 322.7 kg/ha in 2007-08.

Nutritional composition

The seeds of guar are light pale to brown in color. The seed is oval to round in shape. It is a dicotyledonous seed, rich in protein, lipids and minerals. Major constituents are: hull (14-17%), endosperm (25-42%), and germ (43-47%). The endosperm is rich in galactomannan gum. The average composition of guar seed is given in Table 1 (Das and Arora, 1978).

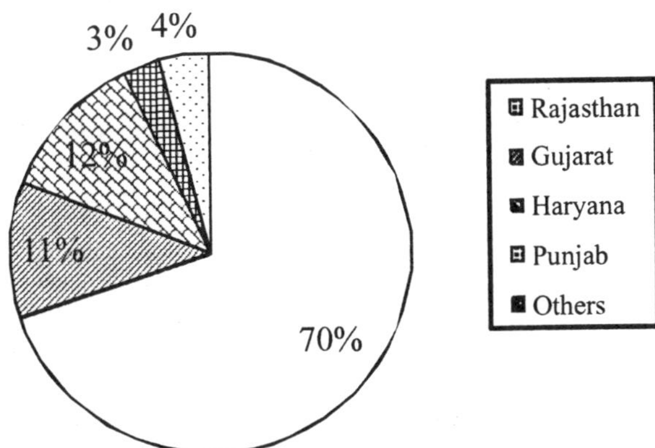

Fig. 2 : State-wise Production Share of Guar in India

Table 1: Average composition of different components of guar seed at 10% moisture content

Seed part	Protein, %	Ether extract, %	Ash, %	Crude fiber, %	Type of carbohydrate
Hull (14-17%)	5.0	0.3	4.0	35.0	D-Glucose
Endosperm (35-42%)	5.0	0.6	0.6	1.5	Galactomannan
Germ (43-47%)	55.3	5.2	4.6	18.0	D-Glucose

The protein content of guar seeds ranges from 24.5 to 34.0% at 10% moisture content (w.b.). Contrary to the other legumes, the germ portion of the guar seed is quite high (43-47%) and constitutes mainly protein. The seed also contains 3.1-7.5% oil (Joshi *et al*, 1990) that is mainly located in germ.

A considerable amount of carbohydrate (78-82%), in the form of galactomannan, is present in endosperm of guar seed (Das and Arora, 1978). It is the most important constituent of the guar and has many food as well as industrial applications. It is composed of D-galactopyranose and D-mannopyranose units. Joshi and Arora (1993) reported the analysis of the endosperm, popularly known as guar gum (Table 2). The molecular weight of the galactomannan varied from 1.5 to 2.0×10^4.

The germ and hull of guar seed together constitute guar meal after processing. It is a rich source of protein and compares well with other vegetable protein source like oilseed cake. The amino acids composition of guar proteins compares well to that of other legume proteins (Ramakrishnan, 1957). Hull and germ contain considerable amount of crude fiber and anti-nutritional factors, which make them, unfit for human consumption. Anti-nutritional factors namely trypsin inhibitors, lectin, and saponins have been reported in guar seed (Gupta, *et al*, 1978; Vidyasagar *et al*, 1978; Arora and Joshi, 1980) in addition to several toxic compounds (gallotanons, gallic acid and its derivatives, myricetin-7, glucocide-3, kaempferol-7 glucocide 3-glucocide, kaempferol-3-rutinoside, kaempferol-3 glucocide, chlorogenic acid, caffeic acid, and ellagic acid) (Kaushal and Bhatia, 1982). Several treatments have been reported to detoxify the guar meal for animal feed uses. Kawatra, *et al* (1969) proposed acid and hot water treatment to detoixify guar meal. Kaur *et al*, (1981) found that the protein concentrate prepared from guar meal is free from toxic constituents and beany flavour and hence they used protein concentrate to fortify the biscuits (Kumar and Singh, 2002).

Table 2: Typical composition of guar gum powder.

S. No	Constituent	Content, %	Metals	Content (ppm)
1	Galactomannan	78.0	As	0.20
2	Water	12.0	Pb	0.03
3	Acid insoluble residue	3.0	Cu	2.50
4	Protein	5.0	Zn	5.60
5	Ash	0.6	Cd	0.05
6	pH (1% solution)	6.5		

Processing of guar seeds

Guar seeds are processed to get endosperm free of germ and hull (popularly known as refined guar gum splits, Fig 3). The refined guar gum splits are further ground to powder and modified as per its uses. The process involves splitting the seed, separation of germ, dehulling, separation of hull, grinding of the endosperm to different particle sizes, and refining of the gum powder. Splitting and dehulling of the seed is done by two processes i.e. dry grinding and wet milling (Gunjal and Kadam, 1991).

Fig. 3: Guar seeds and refined guar gum splits

Wet milling process

For wet milling, Patel (1958) obtained patent for guar gum separation in which guar hull is loosened using chemical agent and also by cooking the seeds in water. Process flow diagram (Gunjal and Kadam, 1991) of this process is shown in Fig. 4, although no details of the specific unit operations have been reported.

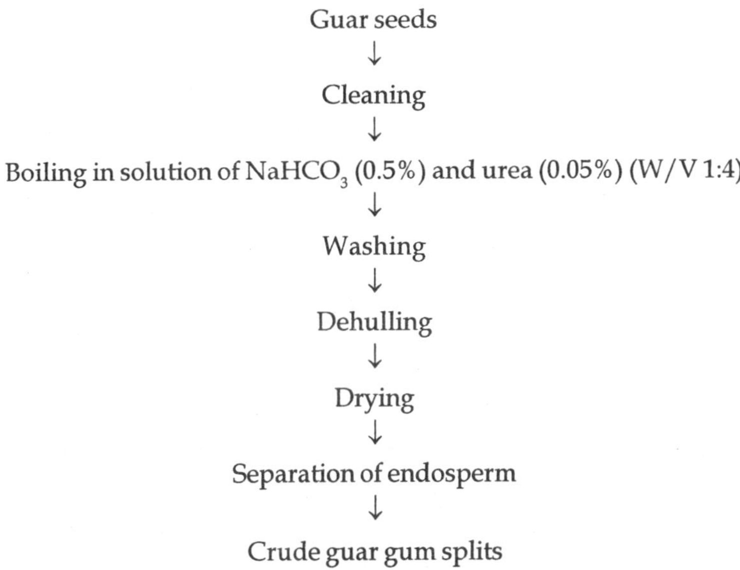

Fig. 4: Flow diagram for milling of guar gum splits by wet process.

Dry milling process

For dry processing, Jiwaji Industrial Research Laboratory, Gwalior obtained Indian patent "The process to obtain refined guar gum in pin-mill without using chemicals" in 1960. As reported in this process, the guar seed are split into two halves using mill having two grinding surfaces rotating at different speeds. The mixture of endosperm and hull is heated to soften the hull and then fed into hammer mill where the hull is removed by abrasion. Then endosperm is pulverized to remove residual hull and germ. Gunjal and Kadam (1991) also mentioned dry milling process for processing guar seed as shown in Fig. 5.

Recovery in wet processing methods is 8 to 10% higher than that in dry processing method, but the quality of the guar gum produced is not good (Gunjal and Kadam, 1991). However, information on type of machines used, processing conditions, efficiency, etc. are not reported in the literature. At present none of the industries use these methods for processing of guar seed.

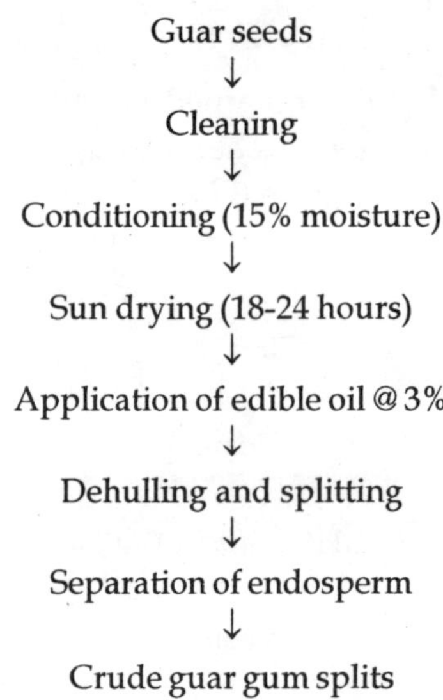

Fig. 5: Flow Diagram for Milling of Guar Gum Splits by Dry Process.

Conventional methods of industrial processing

Jodhpur district of Rajasthan state is the major center for processing of guar. About 125 guar splits manufacturing units out of about 150 units in India are located in the industrial area of Jodhpur. In Bhiwani, only 4-10 industries are located. All industries are using similar type of machine and process for dehulling and splitting of guar seed. The capacity of these plants varies between 3000-5000 tonnes per year. Bigger industries usually set up two or three plants each of 5000 tonnes per year capacity. Usually plants run 24 h from November to April (crop season) and 8-12 h during the remaining period, but bigger industries run their plants 24 h throughout the year (Vishwakarma *et al.,* 2009).

Guar seed is cleaned, graded into two to three sizes using cylindrical grader, and fed at the center of horizontal burr (stone) mill. The upper plate of the burr mill rotates at 500-600 rpm whereas the lower plate is stationary. The clearance between stones is based on size of grain to be fed. Diameter of the plate is usually 61 cm and the coarse emery (14-18 grades) stone is used. Two to six burr mills are used in series depending upon the capacity of plant and dimensions of grain. The guar seed is split in to two halves in this machine. Some portion of germ is also removed during splitting. The seed is not treated prior to splitting. After splitting, split seeds and germ are separated using cylindrical grader.

The splits are then passed through a germ separator (pin mill), where germ is removed from splits. It consists of two mild steel plates of 53 cm diameter. One plate is stationary, whereas the other plate rotates at 3000 rpm. Three rows of pins of high carbon steel of rectangular rod shape (7.5 cm length, 2.5 cm width, and 2.0 cm thickness) are mounted on moving plate. Total 26 pins are placed on the plate. Two rows of pins are mounted on the fixed plate. The material is fed through the center of the machine. The germ is removed from splits through shear, which comes out from the bottom of screen and carried away using blower. It also removes the hull from the edges of splits and makes the splits suitable for handling. Pin mill is used for gum containing splits, because it is a self-wiping type of machine and does not clog during operation.

Thereafter the splits are heated in a kiln to loosen the bond between hull and cotyledon so that dehulling of splits becomes easy. It consists of a 3 m long hollow mild steel cylinder operating at 30 rpm. Diameter of horizontally placed cylinder towards the entry side of splits is 61 cm and towards discharge side is 66 cm, thus a slope of 1°C from horizontal is formed and hence material is conveyed towards discharge. The splits are fed inside of the cylinder. Outer surface of the cylinder is heated using oil fired furnace or firewood. The splits inside the cylinder are heated due to conduction for about 10-15 min. Temperature of the cylinder is usually not controlled; however in some industries, temperature of splits at the outlet is measured and varies from 120-140°C. Heating is considered complete when change in surface colour of splits starts.

The hot splits are fed to dehulling machine from the top. It consists of a mild steel cylinder of 37 cm diameter on which blades of high carbon steel are fixed along the length of cylinder. The capacity of the machine varies between 300-500 kg/h. Usually 3 teeth per inch blades are used. Height of teeth is 0.6 cm. The blades are fixed at a distance of 5 cm all along the periphery of the cylinder. A wire mesh screen (made from wire of 3-4 mm thick galvanized iron) covers the cylinder. Clearance between tip of the blade and screen is about 0.4 cm. The cylinder with blades rotates at 1400-1800 rpm. Since the bond between hull and cotyledon is loosened during heating, subsequent cutting action, impact and friction between splits and blade as well as friction between splits results in dehulling. About 75% part of cotyledon is dehulled in single pass. Usually 2 to 3 dehulling machines in series are therefore used.

The dehulled splits are rapidly cooled and cleaned in air screen cleaner where brokens and hull are separated and cleaned guar gum splits are conveyed for packaging. A process flow diagram for industrial processing of guar is shown in Fig 6.

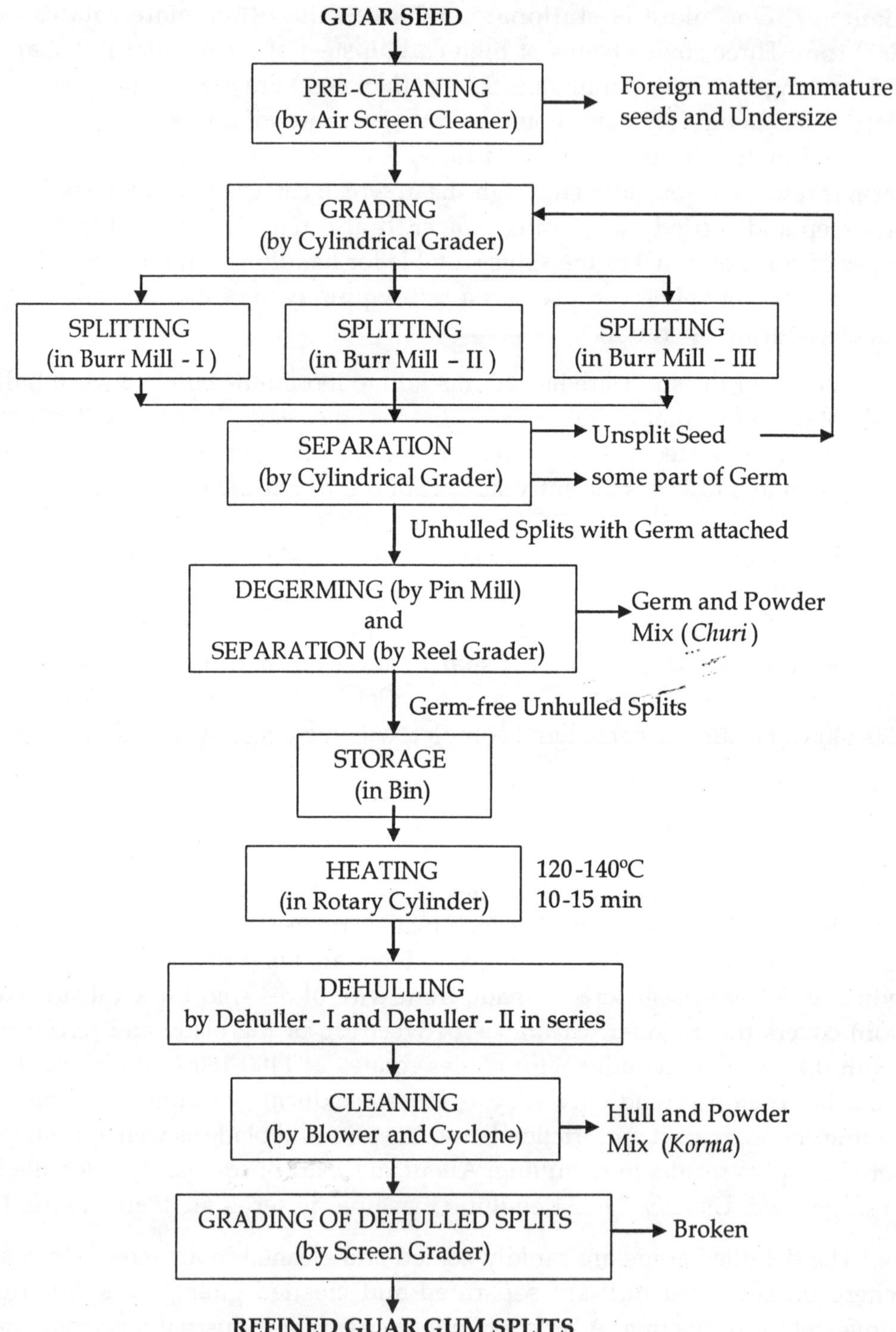

Fig. 6: Flow diagram for industrial processing of guar seed

About 18-25% endosperm is converted into powder or broken and gets mixed with hull during dehulling process. The broken endosperm and powdered hull is used as cattle feed. Thus a considerable loss during dehulling takes place. The following products are obtained through processing of guar seed (Vishwakarma *et al.,* 2009).

1. *Guar gum splits (Guar gum):* Endosperms obtained after dehulling and separation operation. It is of high commercial importance.
2. *Guar by-product (Churi):* This is mixture of hull powder, and germ obtained during splitting and separation of germ. It is rich in protein and used as cattle feed.
3. *Guar meal (Korma):* Hull powder and powdered endosperm obtained during dehulling of the splits is called guar meal. It is used as cattle feed.
4. *Broken endosperm:* This is broken endosperms separated during grading of guar gum splits. It is considered as the low-grade splits.

The guar gum splits are usually of crescent or slightly concave shape. The splits are eventually ground to fine powder of different particle sizes depending upon their uses. Attrition mills, hammer mills, or ultra fine grinders are used in the industry for this purpose. The moisture content of guar gum splits is usually adjusted to 10% (wet basis) after grinding. Some times wet grinding of the splits is also done to get specialized products.

Refining of guar gum

Refining of the guar gum powder is done to remove traces of germ, hull, and compounds other than galactomannan. It is generally done by repeated precipitation with alcohol. Isopropyl alcohol is commonly used in industries. Das and Arora (1978) stated that gum could be purified by mixing 1 kg of crude gum with 10 kg of 1:2 α-picoline-CCl_4 mixtures and drying the filtrate at more than 50°C under reduced pressure to give 30 g of purified gum. Relan, *et al* (1991) purified the guar gum by repeated washing with 80% alcohol at low temperature. The process flow diagram given by Gunjal and Kadam (1991) for refining guar gum powder is given in Fig. 7.

In some industrial processes, the guar gum powder is not washed using isopropanol. The guar gum splits are first screened and then flaked to reduce the thickness. The flaked splits are dried and ground in the ultra fine grinder at uniform rate. The powder is separated using screens. Turbo screens are used some times as additional control. This gives the guar gum powder of good quality.

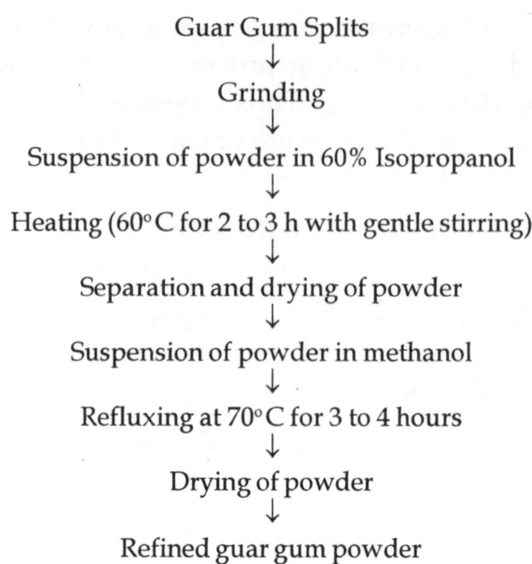

Fig. 7: Refining process of guar gum powder

Properties of guar gum powder

Guar gum powder and its solution is practically odorless and has a bland taste. Its color varies from off white to very light yellow tan. Guar gum is a polysaccharide known as galactomannan, which consists of a mannose backbone with a galactose side chain. It contains 34.6% *D-galactopyranose* units and 63.4% *D-mannopyanose* units. These are high molecular weight hydrocolloided polysaccharide units combined through glycosidic linkages. It is a straight chain compound of *D-mannopyrannose* units joined by β-(1-4) linkages with a side branching unit of every other mannose unit by (1-6) linkage. It has a polymeric structure containing numerous hydroxyl groups, which are treated to produce various derivatives used for special applications in various industries. The active sites present in the D-galactopyranose and D-mannopyranose bind the water molecules and hence galactomannan acts as very good water binder. Its molecular weight is 220000±10%. The greater side branching accounts for its cold water hydration as well as its greater hydrogen bonding activity.

Solution of 1% guar gum powder has a pH between 5.4 and 7.7. The optimum hydration of the powder takes place between pH 7.5 and 9. It has a buffering action and is stable in the solutions of pH 4 to 10.5.

The most important characteristic of guar gum powder is its ability to disperse in water to form viscous colloidal dispersions or sols. The viscosity attained by the solution is dependent on time, temperature, concentration, pH, degree and type of agitation, and particle size of the powder (Greenberg,

2000). A 1% aqueous dispersion of a good quality guar gum has a viscosity of 3-6 Pa.s and about 2 h hydration time is required to reach maximum viscosity (Gunjal and Pawar, 1987). Very fine mesh guar gum powder can be used when there is a need of an immediate high initial viscosity.

Viscosity of the guar gum powder varies with the particle size. Fine powders produce solution of high viscosity. The variation of viscosity with particle size is given in Table 3. It also varies with moisture content of the guar gum at the time of grinding.

Table 3 : Relationship of viscosity with particle size of guar gum powder

S. No	Particle Size (Mesh)	Viscosity (cps)
1.	80	Less than 3000
2.	100	3000-4000
3.	200	4000-6000
4.	300	3000-5000

Hydration rates, water binding properties, and viscosity of the solution are affected by processing techniques used in making the gum powder and particle size. Dry processed guar has better hydration rate and water binding property.

Applications of guar gum

Green guar bean pods are used as vegetable for human consumption. This crop is also grown for cattle feed and as a green manure crop. Guar gum is used for many industrial; food and other applications. Uses depend upon the properties provided by very large molecule in various states of hydration, mostly in solution. Guar gum powder hydrates very easily to produce aqueous solutions because they prevent oil droplets from coalescing. They may also have the property of interfacial binding and thereby act as true emulsifiers.

About 1500 patents on uses of guar gum in different applications have been registered throughout the world (Yadav and Ray, 2002). Guar gum is more than 6 times as effective as starch in thickening power and is used for upgrading starches. Commercially important derivatives of guar gum are:

(i) Hydroxy alkylated guar gum

(ii) Carboxy methylated guar gum

(iii) Oxidised guar gum

(iv) Acetates of guar gum

(v) Cationic derivatives of guar gum

(vi) Sulphated guar gum
(vii) Guar gum formate
(viii) Guar gum acrylamide
(ix) Borate cross linked Guar gum
(x) Reticulated guar gum
(xi) Carboxy methyl hydroxypropyl guar gum
(xii) Depolymerised guar gum

Industrial applications

(i) *Binder applications*: Bailing, palletizing, and briquestting of fine materials to produce large dust-free agglomerates require addition of agents capable of binding fine materials. Choice of binder entails balancing of glue strength, lubricity, and its effect on the final product. Often more than one agent is required to attain all desired characteristics. Guar gum with other additives performs binding functions in several industries.

(ii) *Clay and talc depressant*: Guar gum powders are widely used as clay and talc depressants in froth flotation. A solution of guar gum is added to the conditioning cell upstream of the float cells. During the conditioning process, molecules of guar gum coat the material to be depressed making it water-avid and allowing it to remain in the suspension. Thus, the suspension can be pumped out of the bottom of the float cell.

(iii) *Drilling mud in dewatering wells*: Dewatering wells require clear, biodegradable drilling fluids. Guar gum powder used at 0.5 to 1% in water is an ideal fluid drilling medium. The pseudo-plastic nature of the fluid provides many benefits to the driller. Lubricity under high shear reduces pump power required. High viscosity in the annulus easily suspends cutting preventing to easily separate from the fluid and keep the fluid clear. Guar gum powder biodegrades quickly, leaving no residue. Spilled drilling fluid quickly degrades under attack by soil and air-borne microorganisms. Guar gum left in the well or producing formation breaks down with addition of chlorine bleach or enzyme breaker.

(iv) *Flocculation*: Flocculation is a process of removal of fine particles from a slurry suspension. Guar gum provides bridges between individual particles, thus forming aggregates called *flocs*. The *flocs* grow in size until their density is high enough to settle out of the slurry due to gravity. Guar gum powder are made down in dilute solutions and added to the slurry stream just prior to addition to the clarifier or thickener.

(v) *Fracturing gel well stimulation*: Oil and gas wells produce additional petroleum product after proper stimulation. The process of stimulation increases the producing formation open surface area. Hydraulic piston pumps deliver the water down the casing and into the formation. When water pressure exceeds formation pressure the fractures begin to form. Special fracturing sand delivered with the water into the formation keeps the fractures from collapsing on release of hydraulic pressure. Delivery of sand mixed with water would not be possible without gelling action from guar gum. Water mixed with guar gum becomes viscous and capable of keeping sand in suspension during hydraulic fracturing. Additional components added to guar gum produce thick gels capable of carrying very large amounts of sand using the cross linking process. Breakdown of the gel viscosity after the fracturing process restores formation permeability and allows increased production flow of petroleum products. Enzyme breakers slowly break the bonds in the long polymer chains of guar gum. The gel viscosity returns to that of water and formation permeability returns.

(vi) *Gels*: Guar gum powder produce usable water gels for use in single and multiple use ice packs. Medical and food shippers as well as recreationalists use ice packs for clean, dry cooling of material. Guar gum is particularly valuable for use in freeze point depressed ice packs.

(vii) *Hydro mulch (Tackifier):* A large percentage of gross, seeding for lawns, mining, and highway reclamation is done by hydro mulching. The process of hydro mulching uses slurry of water, fiber mulch, tackifier, seed and fertilizer mixed in a tank. The slurry is thoroughly mixed and applied to the prepared seedbed by spraying. Guar gum powder has been used for the tackfier portion of the slurry in many applications. The tackifier may be added to the slurry in the mixer or is available pre-blended with fiber. The tackifier acts as a "glue" to hold the fiber mulch in place over the seedbed until the seed has sprouted and become established.

Many area of the earth require substantial de-watering. Examples include dams and roadways (for stabilization); landfills and areas containing contaminated ground water. Other geological situations require greater permeability and volume of influence. These situations call for extraction trenches. Injection trenches are constructed for the purpose of returning groundwater to the water table. Construction of extraction or injection trenches utilizes a liquid shoring fluid as support during the construction and backfilling phase. Guar gum powder biopolymer finds extensive use for production of liquid shoring fluid. Guar gum is mixed with fresh water and stabilizers producing a viscous, pseudoplastic fluid. Fluid viscosity reduces readily and is biodegradable. Liquid shoring fluid properly designed by geo-technical engineering provides adequate support to protect excavations from collapse or sloughing during construction.

(viii) *Paper manufacture (Fiber aid):* Guar specialty gums have functionality as adhesive in the manufacture of paperboard. Guar gums in proper aqueous solution will adsorb directly on fibers during the paper making process. On reaching the screen and consolidating into the paper sheet the guar gum acts as an inter-fiber adhesive and strength-building agent. Bonding strength increases by 10 to 20% through the addition of guar gum. Sheet uniformity is increased and printability is improved. Modified guar gums have utility in treatment of waste from paper plants. Fine, colloidal fibers may be flocculated by the use of properly charged modified gum.

(ix) *Waterproofing and gallant (Explosives):* Guar gums find several uses for improving the wet performance of explosives. The addition of guar gum to stick dynamite imparts water-blocking ability to the stick. In the event the stick wrapping is punctured and comes into contact with water, guar gum will from a waterproof block in combination with incoming water, thus preventing further loss of the explosive ability. Guar gum added to ANFO prills provides protection against water invasion into blast holes loaded with the treated ANFO. Guar gum has also been used as a gallant for super saturated nitrate solutions. These solutions are nearly plastic in consistency, yet pumpable and are water-resistant. They may be used for packing in tubes or direct loading of blast hole.

(x) *Textile industry*: Guar Gum shows excellent thickening properties when used for textile sizing, finishing, and printing. It reduces breakage of yarns while sizing and gives better efficiency in loom shed.

In addition, guar gum powder may be used in many industrial applications like carpets, water treatment, paints, tobacco, building and construction products, foundries and ceramics, industrial cleaners and related formulations, etc. It can be used where there is a need of viscous solution, water-binding property, etc.

Food processing applications

Guar gums find wide applications in many of the most common aspects of our lives. Many foods utilize guar gum as a special additive to improve handling, smoothness, shelf life, and other characteristics. Pet foods are improved by the use of guar gum in its formulation. The largest market for guar gum is food industry. Guar gum powder is used as a thickener and a binder of free water in sauces, salad dressing, etc. Some of the food applications of guar gum are described below:

1. *Bakery*: Guar gum is used in the preparation of breads. Small quantity of guar powder when added to the dough increases the yield, improves texture, and gives longer shelf life. Addition of guar gum solutions to dough while kneading gives increased yield of baked goods as well as giving the dough a greater resiliency and a drier, less flabby appearance.

2. *Dairy*: In this field guar gum is used as an excellent binder and stabilizer. It is used in the production of ice creams, sherbets, cheese, liquid milk products, etc. In soft cheeses, the use of guar gum increases the yield of curd solids. The curd is of a soft, compact, and tender texture with separated whey that is very limpid. Guar gum powder binds the free water without disturbing the viscosity characteristics of the mixture. Guar imparts body, texture, chewiness, and heat-stock (as an emulsifier), gives dry and stiff ice cream with retarded melting. It is also used with sodium hexa-meta-phosphate and sodium citrate as a stabilizer in ice cream mixes. It is included in the ice cream and the food chemicals codex. It retards color migration and is very stable in acidic media also.

3. *Meat*: Guar is used as a lubricant and excellent binder for various meat products. It allows storing with less loss of weight and can decrease the filling time for cans. It binds the free water and thus retards shrinkage when used at 0.2-0.5% of meat weight in sausage and pet foods. It acts as lubricant in the sausage stuffing operation.

4. *Dressing and sauces*: Guar gum powder is used as thickener to improve the stability and appearance of salad dressings, relishes, ketchup and others. It stabilizes salad dressing by increasing the viscosity of the emulsion and decreasing the separation rate when applied at 0.2-0.8%. It has replaced many water binders used in pickle and relish sauces where rapid cooling is used.

Besides these, guar gum has important uses in frozen foods because it provides hydration and water binding properties. In cakes, guar gum allows single-step mixing, reduces mixing time, less crumbling tendency in finished cake, easy application of frostings, and greater moisture retention with prolonged shelf life. In canned meat products and pets foods, guar gum allows reduction in the bumping tendency during the cooking of the meat, increase in the case of pumping of the cooked product between the kettle and filling vat, and reduction of splashing and spilling of the product while filling. In dressings and sauces, it is used as thickener because of its ability to disperse in cold water, compatibility with highly acidic emulsions and ability to add viscosity. It is used as a thickening agent in noodles, jams, jellies, etc. It can be used in many food applications and development of new products. It may increase the shelf life of onion and garlic paste by binding the free water and hence reducing the water activity.

Use in pharmaceuticals and cosmetics

In compressed tablets guar gum is used as binder and disintegrator. Several studies have shown significant decrease in cholesterol level after administrating guar gum in human consumption.

In cosmetics and personal care products, guar gum is used as thickener and suspending agent. It imparts good viscosity and gives better foam stability in shampoos. It is also used in toothpastes. Guar Gum is also widely used in tobacco, leather, insecticides and pesticides, crayons, and adhesives depending on viscosity, pH, and to suit their industry specification.

Thus guar gum powder has many industrial, food, pharmaceuticals, and cosmetics applications. It can also be used to enhance shelf life of pastes of perishable foods by reducing the water activity. Extensive research in this aspect is needed.

Status of Indian industry

Indian guar gum industries are manufacturing either guar gum split or guar gum powder. Some industries, e.g. Lucid Gums and Chemicals Ltd., Vikas WSP Ltd., etc. also prepare specialized gums for specific industrial uses. There are more than 150 guar gum split manufacturing units in India and total installed capacity is more than 0.6 million tonnes per annum (*www.dsir.nic.in*). All the split producing units have indigenous plant and machinery. Splits are available in various grades, like in terms of purity as 90%, 92%, 95% and 97%. There are more than 25 units, which manufacture textile grade powder from splits. The total guar gum industry is estimated at about 0.225 to 0.250 million tonnes per annum capacity. The products manufactured in India are mainly food grade guar gum powder and some modified guar gum derivatives for textile printing industry.

There are a few large manufacturers like IGI, HICHEM, Dabur, Vikas WSP, Lucid, Hindustan Gums and Chemicals etc. who produce value added derivatives for export as well as domestic market. The total domestic consumption of guar gum powder inclusive of splits from various end-use sectors is given in Table 4 (*www.dsir.nic.in*). The consumption reached over 50,000 tonnes in 2005.

Table 4 : Consumption of guar gum powder in Indian industries

S. No	Type of industry	Consumption including splits (tonnes)
1.	Textiles	5000
2.	Paper	2500
3.	Explosives	1500
4.	Food and Pharmaceuticals	800
5.	Oil well drilling	130
6.	Cosmetics	150

International markets

India is major producer of guar in the world and guar gum splits and powders are being exported from India to more than 60 countries. Total export of guar gum was 0.13 MT worth 150 million US$ in 2004-05 (Fig. 8). The major share of the guar processed in India is exported either in form of semi processed product i.e. refined splits or in form of guar powder. In recent years export of modified/derivatives gums has also started. USA is the major importing country followed by Germany. Both of these countries account for more than 50% of India's exports of guar products. The other important importing countries from India are Netherlands, U.K., Japan, and Italy.

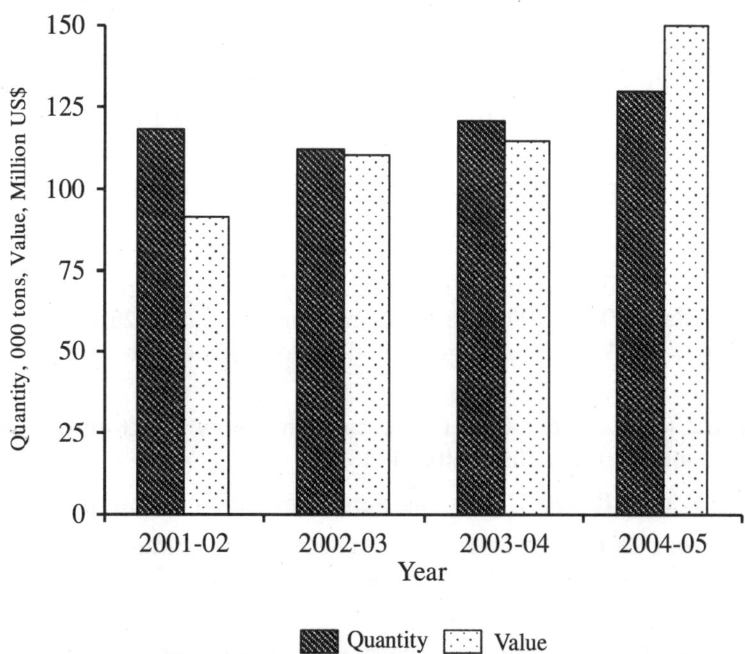

Fig. 8 : Export of guar gum from India.
(*Source* : APEDA, 2006)

Future research

Guar gum is one of the most important agricultural produce of India on account of its high export potential. It is extracted from the endosperm of guar seed and constitutes mainly of galactomannan. The guar gum powder is a very good water binder and emulsifier. It is used in many foods, industrial, and pharmaceutical industries.

The traditional process prevalent in the industry for dehulling, through heating and scratching, does not remove the hull efficiently. Consequently, substantial portion of the endosperm is lost in the form of brokens and powder into the powdered hull, which is sold at a much lower price as cattle feed. Improved processes and machinery therefore need to be developed to minimize such quantitative and financial loss.

While the germ reportedly contains about 55% of protein, it is not regarded suitable for human consumption. More intensive research and development efforts are needed to eliminate anti-nutritional factors and toxic compounds present in it in order to utilize this rich source of protein.

There is a good potential for marketing of value added guar based health foods and medicines as dietary fiber, slimming-aid, fat replacers, etc. in the international markets.

For the long-term survival of guar processing industry, it is very important to develop domestic demand. The associations of guar gum processors may collaborate with Ministry of Food Processing and PFA to encourage utilization of guar in processed foods and prescribe permissible limits. There should be comprehensive research programs to be undertaken by the R&D institutions to promote the applications of guar products as food ingredients.

References

Anon, (2010). Guar Seed Investment Report. (http://www.jrg.co.in/cms/research report/Special-Report04-11-2009Guar-seed-Investment-report—2010.pdf). Commodities Research, JRG Wealth Management, JRG House, Kochi, India. Visited on June 04, 2010.

Arora, SK, and Joshi, UN (1980). A thermostable haemolytic factor in guar (Cyamopsis tetragonoloba). Guar Newsletter, 1 : 13-14.

Bhan, S, and Prasad, R (1967). Guar has many uses. Indian Farming, 17.

Das, B, and Arora, SK (1978). Guar seeds, its chemistry and industrial utilization of gum, in Guar: Its improvement and management, Paroda, R.S. and Arora, S.K. (Eds), India Society of Forage Research, Hissar, 80.

Greenberg, NA (2000). The chemical, physical and physiological properties of partially hydrolysed guar gum. AACC Annual Meeting. November 5-9, 2000. Kansas city, Missouri.

Gunjal, BB, and Kadam, SS (1991). CRC Hand Book of World Food Legumes, Vol. 1 : 289-299.

Gunjal, BB, and Pawar, SG (1987). Physico-chemical, rheological and functional properties of guar gum. Unpublished report. MAU Parbhani.

Gupta, PC, Sagar, V and Pradhan, K (1978). Guar as feed and fodder crop. In: Guar- its improvement and management (eds. Paroda, R. S. and Arora, S. K.). Forage Research. 4A : 109-122.

Joshi, UN, Arora, SK, and Arora, RN (1990). Differential chemical composition of guar species. Guar Research Annals, 6 : 38-40.

Joshi, UN, and Arora, SK (1993). Recent advances in guar gum chemistry and utilization- A review. Forage Research, 19 (3 and 4): 310-328.

Kaur, A, Arora, SK, and Joshi, UN (1981). Amino acid compositions of protein isolate and concentrate. Guar News Letter, 2 : 41-43.

Kaushal, GP, and Bhatia, IS (1982). A study of polyphenols in the seeds and leaves of guar (Cyamopsis tetragonoloba L. Taub.). Journal of Science Food and Agriculture, 33 (5) : 461-470.

Kawatra, BL, Garcha, JS, and Wagle, DS (1969). Studies on guar (Cyamopsis tetragonoloba) protein. Indian Journal of Nutrition Dietetic, 6 : 91.

Kumar, D, and Singh, NB (2002). Guar-An Introduction: in Guar in India. Scientific Publishers (India), Jodhpur.

Patel, RR (1958). Gum. Indian Patent. 61:005. (Cf. Chemical Abstracts 52: 21190c, 1958).

Ramakrishanan, CV (1957). Amino acid composition of crude and germinated guar seed flour protein (Cyamopsis tetragonoloba). Experimentia, 13 : 78.

Relan, PS, Kumar, S, Garg, V, and Arora, SK (1991). Rheological studies of guar gum by viscometric method. Guar Research Annals, 7 : 25-29.

Vidyasagar, Prasad, D, Thakur, RS, and Pradhan, K (1978). Nutritional evaluation of processed guar (Cyamopsis tetragonoloba) meal for broilers. Indian Journal of Poultry Science. 13 (3): 155-160.

Vishwakarma, RK, Nanda, SK, Shivhare, US & Patil, RT (2009). Status on post harvest technology of guar (Cyamopsis tetragonoloba) in India. Agricultural Mechanisation in Asia, Africa and Latin America, 40 (1): 65-72.

Yadav, NS, and Ray, RL (2002). Industrial Aspects in: Guar in India. Scientific Publishers, India, Jodhpur.

□□□

11

Advances in Cereal Processing for Malt Production and its Application in Food Processing

TEJINDER SINGH

Malting and brewing is an important industry in the world and both are associated together. Malt is a raw material for the brewing and distilling industry. With the advent of scientific knowledge and consumer demand for nutritious health products, malt also found place in the food industry. This led to the use of cereal grains for preparation of food malt. Among cereals including wheat, barley, rice, maize, sorghum, oat and finger millet; barley is preferred to make malt for its unique properties, especially for brewing. The malt so produced is commercially called brewer's or distiller's malt. Distillers have the liberty to use malts from others cereals along with barley malt. The most prominent are wheat, rice, sorghum or finger millet malts. Oat malt is used for the manufacture of stout. Wheat malt gives good head retention to beer. Production and use of such malts depends upon demand and the local availability of the cereal grains. Use of food malts in products is preferred for its desirable taste, colour, flavour and nutritional attributes. Food malt which is prepared from cereals other than barely may or may not adhere to the rigid and strict specifications required for brewing. These are diastase activity, modification index and soluble nitrogen content, though yield of extract is an important criterion for any kind of malt, economically. Pharmaceutical industry and night drinks manufacturers are the largest consumers of food malt.

Characteristics of cereal grains for malting

The grain should be sound, free from any discoloration, thermal or mechanical damage; should have more starch and less protein; must show desirable structural strength upon attaining moisture content of 42–44% for adequate handling and transportation. The roots and shoot of the germinating grain must remain intact to obtain malt of suitable quality. Barley has the ability to exhibit such qualities and is preferred for malting. This is due to presence of hull on it, which provide structural support and protect growing acrospires, hence easy to transport from steeping tank to germination beds and finally to the kiln. Though, oat and rice also have hull but the starch protein ratio is not desirable. The hull content in the oat and rice is considerably higher than that of barley, and also the starch content is lower. Consequently both these grains do not provide adequate extract yield and are not economical. Due to large germ in maize, it has the problem of rancidity which develops due to high moisture content and activation of lipase in it. It also contains horny starch at the tip of the grain which is difficult to gelatinize and do not undergo enzymatic hydrolysis. As such these attributes diminish the possibility of using maize for malting despite its high starch content. In addition, maize has only one cell deep aleurone layer, whereas rice has four to six and oat has one to two cell deep layers. Due to this, the ratio between α-amylase and β-amylase is not adequate as is prevalent in barley. Sorghum, however, is malted in African countries for distilling purposes. Excessive presence of tannins in sorghum recede the activity of yeast and cause chill haze problem and thus cannot be used for brewing. Wheat, however, is used for preparation of food malt with little modifications in the malting procedure.

Role of various grain parts in malt production

A cereal grain is generally composed of germ, pericarp, testa and aleurone layers enclosing the starchy endosperm. Some of the grains have ventral furrow. Barley has lemma and palea cemented on the bran of the grain. In the case of rice and oat, hull can be removed during processing for food uses.

Germ

It has a very important and pivotal role in the malting process. Without germ, malt cannot be prepared. It has a cotyledon made of coleorhizae and coleoptile. Both are joined together with an embryonic node. It also has a micropyle area at the tip. Water finds its entry into the grain through the mircopyle area. Testa being semi-permeable in nature, do not allow water to enter into the grain. Coleoptile is a food reserve in the form of reducing sugars and has the ability to grow as embryo. Embryonic node is the site for the synthesis of endogenous gibberellic acid, a growth hormone. Gibberellic acid moves to the adjoining aleurone layer, where it helps to synthesize and release of α-amylase and other hydrolytic enzymes. These hydrolytic enzymes are either not synthesized or released without the presence of gibberellic acid. It is at this stage when exogenous gibberellic acid is provided to produce desirable amount of hydrolytic enzymes. Varieties of barley for malting are selected which produce adequate amount of gibberellic acid and eventually the hydrolytic enzymes. Coleorhizae emerge out of the micropyle area and develop into roots. It provide easy access to steep water to enter into the grain. Excessive root growth always associated with more respiration is a loss of soluble solids which otherwise could find way into wort/ extract. Efforts are made to curtail the root growth and respiration rate without effecting the synthesis of enzymes and progress of modification. For this purpose potassium bromate or combination of ascorbic acid and potassium bromate is applied to the germinating grain. This also helps to control proteolysis and thus soluble nitrogen in the wort.

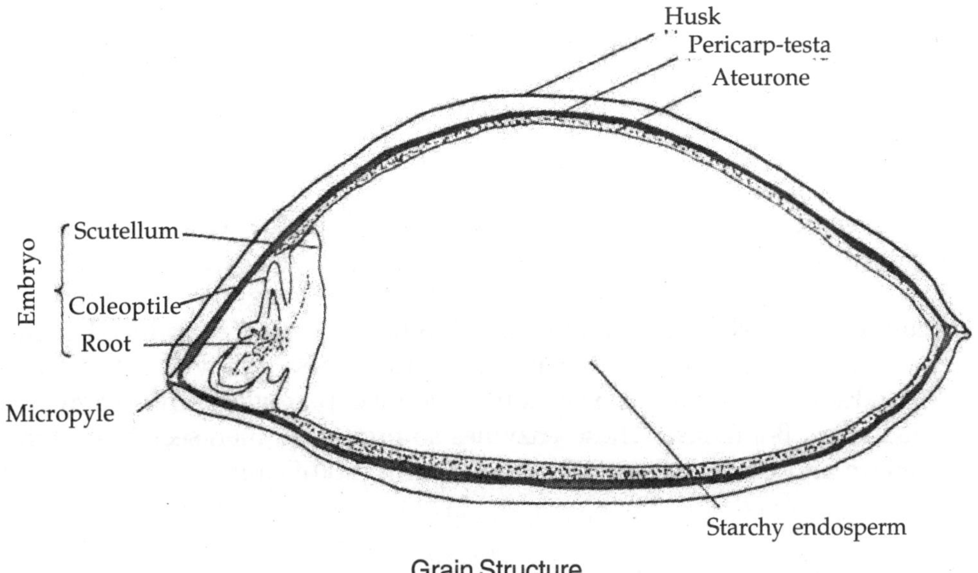

Grain Structure

Aleurone layer

This is the site for synthesis of hydrolytic enzymes. Organelles in the aleurone cells synthesize enzymes. Nucleus of the aleurone cell provides information for the synthesis of the particular enzyme. The enzymes are produced in a very systematic way. Upon release of gibberellic acid from the embryonic nod small amount of α-amylase is produced in the embryo. When the food reserve of the germ is exhausted due to the growth of the plumule, more and more energy is required for germination. This energy is then drawn from the endosperm, which contains starch embedded in the protein matrix. Both are present in a cell, and cell has a cell wall. The cell wall is composed of proteins, cellulose and hemicelluloses.

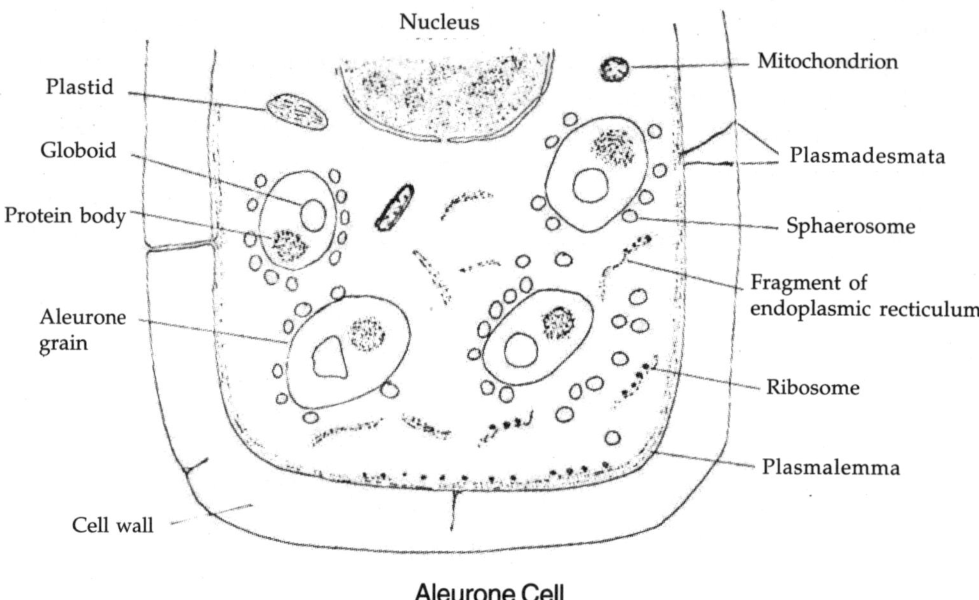

Aleurone Cell

As such, for the release of starch and protein from the cells, hydrolysis of cell wall is required. Therefore, the aleurone cells produce cell wall degrading enzymes. Then proteolytic enzymes are produced for degradation of reserve proteins to release starch. Thereafter, amylolytic enzymes are produced. Some of the grains and legumes have natural gift to possess β-amylase. These are barley and soybean. Wheat also has some amount of β-amylase. This β-amylase is called latent β-amylase. These enzymes so produced, then secrete into the endosperm and cause cell wall and endosperm modification.

Endosperm

It is a dead tissue of the grain and do not respire. However, this is an essential part of the grain that contains starch and reserve proteins. This is required to fulfil the main objective of malting *i.e.* cell wall modification and to provide energy to the growing embryo. The hydrolytic enzymes produced by the aleurone cells penetrate into the endosperm and start hydrolysis. The modification of the endosperm starts from the proximal end and advances to the distal end through the dorsal side of the grain. Hydrolysates, the glucose and amino acids move through the scutellum to the growing embryo. The small starch granules lying under the aleurone layer and on the periphery of the endosperm are hydrolysed to the great extent and then the large starch granules are hydrolysed. The small starch granules though less in weight and more in number are generally considered resistant to gelatinization and enzymatic hydrolysis are easily metabolized during the germination process.

Production of malt

Preparation of malt from a cereal grain involves steeping of grains followed by germination then kilning and removal of roots. During these processes several physical and biochemical changes takes place within the grain affecting the quality of the finished product.

Steeping

This is the first stage to manipulate the malting process to achieve the desirable quality malt. Steeping of grains is done under water at controlled temperature. For barley and wheat, steeping is done at 7°–15°C till the desirable moisture content of 42–44% is achieved. Normally it takes about 48-72 hr. Maize is steeped at 25°C. Finger millet achieves the maximum moisture content up to 30% only. During the initial few hours uptake of water is very fast and thereafter it becomes slow.

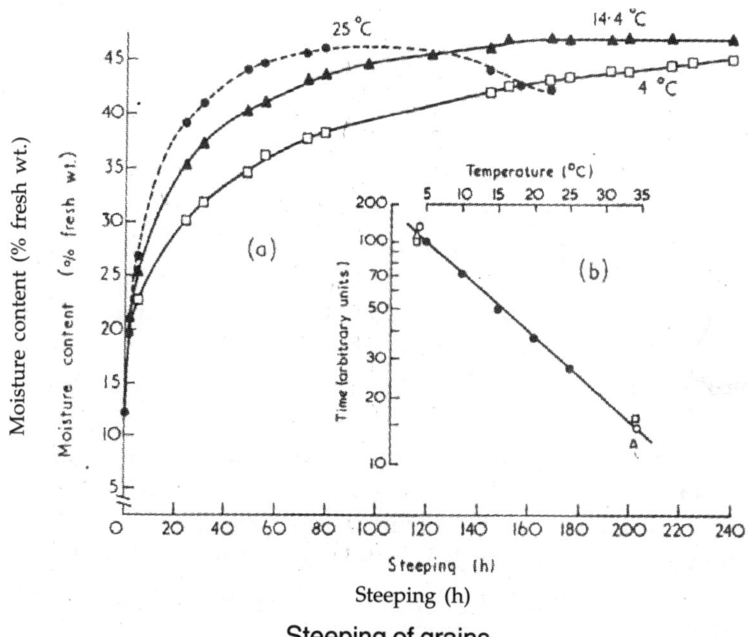

Steeping (h)

Steeping of grains

Increase in moisture content increase the rate of respiration of the steeping grain. Resultantly, the grain takes up oxygen present in the water and produce carbon dioxide. This causes anaerobic conditions in the steeping water. The microflora of the grain also finds its way into the steeping water. Some of the soluble solids including reducing sugars, proteins and polyphenols leach into the steeping water. These tend to produce testinic acid, alcohol etc. Development of such conditions in the steeping water tends to increase the temperature of steeping water and inhibit the growth of the grain by damaging the germ. To overcome these problems, steep water is changed frequently but this causes loss of water and other solids in the steep liquor, which can be used as media for pharmaceutical purposes. The other option is passing of air through the steeping water for the removal of carbon dioxide and increase the availability of oxygen. However, this causes uneven germination. Air-rest and change of water have been founds the best way to avoid any damage to the grain as well as to obtain uniform germination. During this process chitting i.e onset of germination takes place due to elongation of cells in coleorhizae. This opens the micropyle area and allows easy asses of water into the grain. Maltsters take the benefit of this process. They steep the grain for a short time at low temperature such that the moisture content of the grain is near to 30%. Give long air-rest to the steeped grain such that it may start chitting out of water, maintain the temperature by passing the chilled air through the grain and again briefly steep it so that the desirable moisture content is achieved, without which adequate modification of endosperm is not possible. It is at this stage when exogenous gibberellic acid and other additive can be applied to the grain. Hydrogen peroxide along with the phyto-hormone is applied to break dormancy of barley. Water sprayed on the germinating grain for the purpose of increasing the moisture content of the grain is held by the growing roots and rather causes excessive root loss rather than desirable extent of modification.

Germination

Like steeping, germination is also carried out at low temperature between 15°-20°C. Temperature of germination is selected according to the germinating conditions in the field. Maize is germinated at 25°C. The steeped grain is transferred from the steeping tank to the germinating boxes containing false bottom. Temperature is maintained by passing chilled air through the false bottom. This also helps to dissipate heat produced during germination and respiration of the germinating grain. Relative humidity of the germinating chambers is maintained near to 95%. The grain is turned upside down with the turners to prevent matting of the germinating grain due to emergence of roots. Germination is carried out for 4-5 days or till the length of the acrospire is reached to the ¾th of the length of the grain. At this stage the grain becomes mellow, soft and sweet and can be rubbed easily.

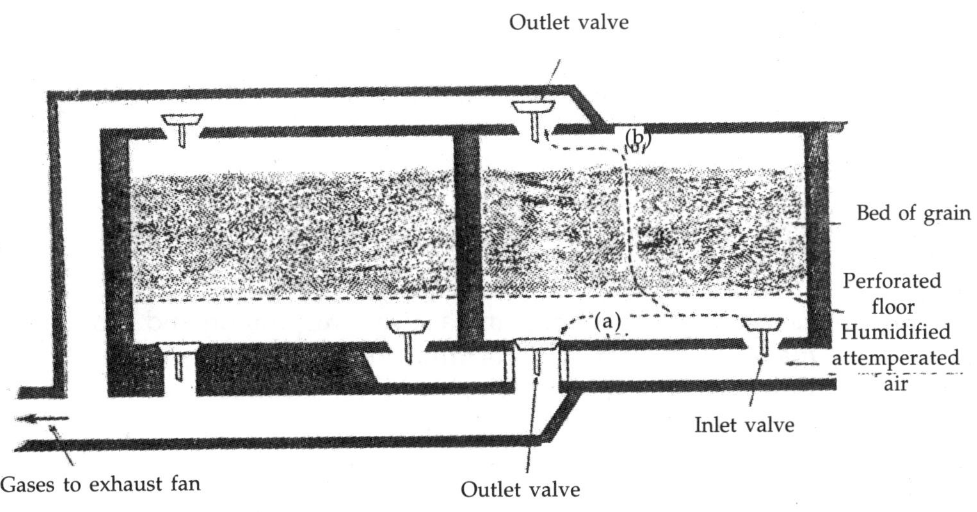

Germination Box

During the germination process, the cell wall which is mainly composed of β-glucan and pentosans is hydrolysed by endo- and exo- β-glucanases, whereas pentosans are hydrolysed by pentosanases, respectively. Though, this is true in the case of barley. For other grains such as wheat pentosanases are more prominent. Both these are linked with proteins. Unless proteins in

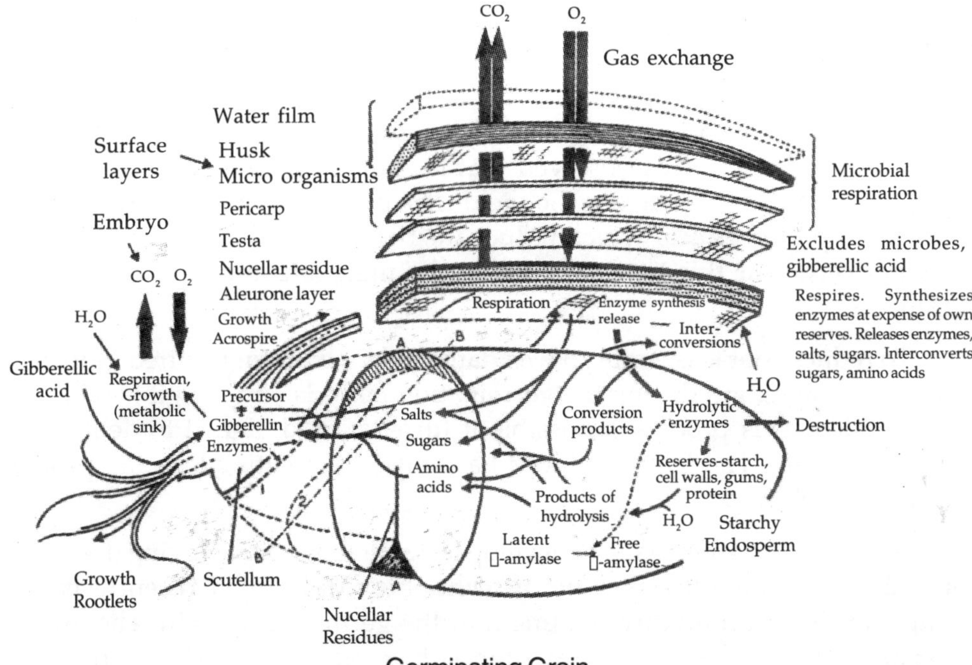

Germinating Grain

the cell wall are hydrolysed, β-glucan or pentosans cannot be assessed by the enzymes. This link between hemicelluloses and protein is hydrolysed by the carboxypeptidases. Thereafter, the reserve proteins in the endosperm are hydrolysed by the proteolytic enzymes and the starch becomes assessable to the amylases. During this time the latent β-amylase is also released. All the carbohydrases hydrolyse starch to maltose, glucose, dextrins and oligosaccharides. This indeed is a normal process for germination of grain in the field, which is curtailed in-between to get malt.

Malting is a long process and causes severe respiration and root losses. Efforts are made to shorten the germination process, decrease root or respiration loss. Use of exogenous giberrellic acid enhances the rate of modification by inducing more hydrolytic enzymes and thus the time germination is reduced and so are the malting losses. Application of potassium bromate reduces root loss. This also controls the excessive proteolysis of the reserve proteins and thus regulates the formation of soluble nitrogen to the desirable limits.

To enhance the penetration of gibberellic acid into the grain, abrasion of the outer surface of the barley grain was attempted. However, this treatment damaged the grain structure and permitted more oxygen to reach the aleurone cells. These cells respond quickly to the exogenous gibberellic acid and release more enzymes rapidly. Though the pattern of modification remains the same as that of normal grain, the endosperm degradation advances more rapidly, extract yield and soluble nitrogen levels are elevated. But handling of grain is somewhat difficult and chances of microbial growth on the surface of the grain are more.

In an attempt to increase extract yield, glucose syrup is applied to the germinating grain. It was observed that the sugar applied finds its way into the extract and hence increase the yield. The only benefit found was that the microflora produces acid which reduce the pH of the mash to a useful extent. It is not certain if this treatment alters the metabolism in the germinating grain.

To reduce root loss, embryo was damaged by battering chitted grain with mechanical agitation during the germination periods and the gibberellic acid was applied. This process is possible if turners are designed to deliberately bruise the grain. The malt produced has only 3.5% malting loss and produce more yield of extract.

Attempts are being made to develop technologies for improving the quality of malt and reduce losses. One of technology developed recently is high temperature high humidity treatment to the germinating grain. The grain is germinated for three days and then subjected high temperature (45°C)

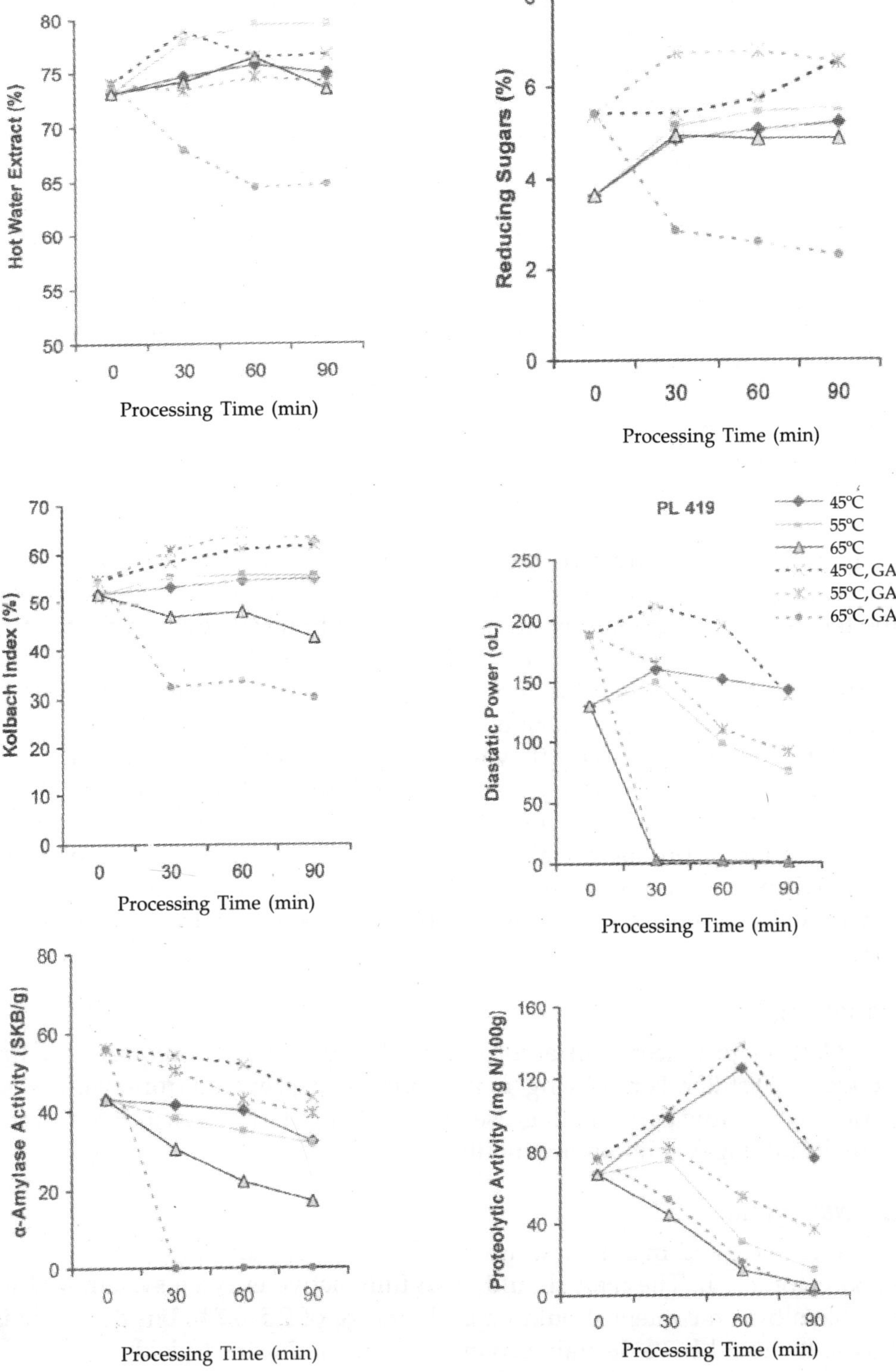

Hot Water Extract (%)
80
75
70
65
60
55
50
0
30
60
90
Processing Time (min)
Reducing Sugars (%)
8
6
4
2
0
0
30
60
90
Processing Time (min)
Kolbach Index (%)
70
60
50
40
30
20
10
0
0
30
60
90
Processing Time (min)
PL 419
45°C
55°C
65°C
45°C, GA
55°C, GA
65°C, GA
Diastatic Power (oL)
250
200
150
100
50
0
0
30
60
90
Processing Time (min)
α-Amylase Activity (SKB/g)
80
60
40
20
0
0
30
60
90
Processing Time (min)
Proteolytic Avtivity (mg N/100g)
160
120
80
40
0
0
30
60
90
Processing Time (min)

treatment under high humidity conditions for 45–60 min. This increases activity of the hydrolytic enzymes produced by the aleurone cells and already present in the endosperm and working on their substrates. The enzymes quickly hydrolyse starch and proteins and advance the modification to the desirable level. This treatment increases the level of reducing sugars, extract yield and even enzyme activity but reduces root and malting losses. There is no need to apply gibberellic acid or any other additive to the germinating grain. The grains difficult to malt can be easily handled using this technology.

Kilning

This is the last process involved in the malting process. Kilning is essential to remove undesirable green malt flavour and to develop desirable colour and flavour. Kilning also restrict the activity of the hydrolytic enzymes. Activity of the enzymes has to be retained because these are required during mashing of malt to produce clear wort. Therefore, green malt is dried in three stages, being at high moisture content originally. In the first stage it is dried at low temperature (45°C) with sufficient flow of air to carry away the water vapours. Evaporation of moisture gives cooling effect to the grain and the enzymes are not damaged. When the moisture content is reduced from 42% to nearly 25%, the temperature of the kiln is slightly increased to reach between 55° to 60°C and the flow of air is also adjusted. Temperature of the kiln is increased to about 70°C to 80°C when the moisture content is reduced to about 8–10%. Air flow rate is also adjusted. Finally curing of the dried malt is carried out at 85°C for 4 h with minimal air flow to get well dried malt with 3–4% moisture content. Failure of following the stepwise kilning will cause stewing, dark colour formation and destruction of enzyme activity. Kilning of green malt at low temperature produces malt with highest enzyme activity. The roots are removed and malt is packed in air-tight containers, being highly hygroscopic in nature. Nevertheless, kilning process is manipulated to get variety of special malts.

Variety malts

Maltster produces malt according to the specifications required by the Brewer or Distiller. For brewing, malt is preferably prepared from two-rowed barley and six-row barley malt is used for distilling. To prepare different types of malts kilning conditions are modified.

Brewer's malt

Well modified malt is first dried at low temperature (55°C) and later on at 85°C for 2–4 h. The malt should have fully active enzyme system and its soluble nitrogen content should be in the range of 0.5–0.7%, but the value is 0.54%. It should have desirable colour, flavour and extract yield.

Distiller malt

Distillers want malt which may have high nitrogen content, made from six-row barley to provide maximum development and survival of enzyme activity to mash adjuncts. For distilling purpose as much carbohydrates as possible is fermented to alcohol. The grain is steeped to 48–50% moisture and germinated at low temperature for a long time. The green malt is dried at 50°C–55°C. For making Scotch whisky, peat is burned to kiln malt. The peat smoke condenses settles on the malt, increase its phenol content and give characteristic flavour to the product.

Pale ale malt

Malt is prepared without using potassium bromate to develop dimethyl sulphide precursor. The green malt is dried at 60°C to 95°C.

Lager malt

Malt is kilned at 55°C to 70°C.

Coloured and special malt

The green malt is drastically modified, stewed and kilned at high temperature or finally roasted. Such malt is used in small quantity to impart colour, special flavour and aroma, high anti-oxidants and palate fullness. To preserve their special characteristics, they should be used fresh. Such malts have little or no enzyme activity and generally mashed along with enzymatically active malt.

Chocolate or black malt

Green malt germinated for 4–6 days is kiln-dried to about 5% moisture. Thereafter, the malt is roasted at 220 to 255°C for 2–2.5 h. During heating, acid fumes are produced and there is a loss of 15% dry matter. The fumes must not be allowed to escape, since the grain is being converted into charcoal, there are chances of catching fire. When adequate colour is obtained, roasting is stopped by sprinkling water on the roasted grain. The water is instantly converted into steam cooling the grain and check the roasting process. This causes swelling of the grain which gives uniformity to the product. Similarly roasted barley is prepared for use in some dark stout.

European brumalt

Grain is steeped at a high moisture content of more than 45% and then geminated for 8 days. The germinated grain is placed in a bed (20 cm deep) for 24 h so that the temperature of the grain may rise to about 50°C and then it may decrease to room temperature with the depletion of oxygen and accumulation of carbon dioxide in the bed. The grain kilned at 90–100°C is very dark in colour and have high reducing sugars.

Crystal or caramel malt

A fully modified malt rich in nitrogen and reducing sugars is rewetted after light kilning, to a moisture content of nearly 50%. The wet malt is heated in the kiln at 65°C–75°C for 1.5–2 h with minimal ventilation, so that the evaporation is restricted and stewing may occur. The interior of the grain is hydrolysed to form syrup. The inverted grain is then dried at 120°C. Liquefied interior of the cool dry malt becomes pale-brown hard caramel like mass which impart very good colour and delicious flavour.

Amber malt

The well modified malt with 3–4% moisture is roasted in the kiln or in a rotating drum at 95°C for 15–20 min and then gradually to 135°C–150°C until the correct colour is attained.

Diastatic malt

The well modified malt is dried at 45°C–55°C for 24 h to get moisture content of 5–8%. The dried malt has full hydrolytic enzymes activity.

Acid malt or Dixon's patent malt

This malt is used to reduce pH in the mash tun. The green malt is sprayed or steeped with 10% solution of lactic acid before kilning. This malt also provide high amount of permanent soluble nitrogen and formal nitrogen for supporting the growth of yeast.

Application of malt in food processing

Depending upon the local availability of the grain, malt prepared from any cereal can be used for food purposes. Malt used for food purposes is called food malt. Soluble nitrogen is of little or no concern. Rather high soluble nitrogen is desirable for many food applications. Nevertheless, maximum yield of extract and quantity of reducing sugars are desirable. Malt is mostly used in the form of extract concentrate for number of food products.

Extrusion cooking

Wheat malt has been successfully extruded to prepare snacks. The coarsely ground well modified wheat is mixed with small amount of sugar. The contents are brought to desirable moisture content and then extruded at 120°C. This work is still in progress in the Department of Food Science and Technology, PAU, Ludhiana.

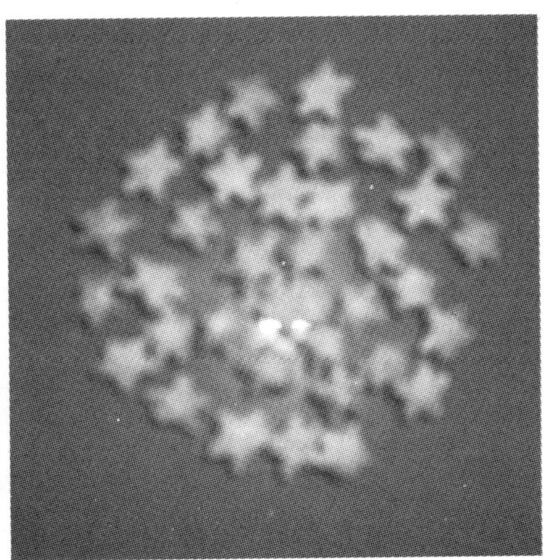

Extruded Malt Snacks

Extract concentrate – Use of whey for mashing

For food purposes, whey and buttermilk can be used to replace water by 50 and 100% for mashing of malt prepared from hulled barley, hull-less barley and wheat. The yield of extract, increased by 3 to 109% over those of the controls. Soluble nitrogen of the extract is more than the control. Reducing sugars of 100% whey-malt extracts were 3-6 fold higher than those of the corresponding buttermilk-malt extracts. At 20-50°C, flow behaviour index of malt concentrate varied from 0.40-0.97, indicating its mild to moderate pseudo-plasticity. Total replacement of water for mashing of malts with whey/ buttermilk produce concentrate with higher consistency index and yield stress values than those of the controls or with 50% whey/buttermilk. Buttermilk-mashed malt concentrates showed higher values for activation of flow than those of the whey. The concentrate might be used to prepare malt based milk products replacing expensive milk /milk solids.

Night drinks

Taking malt concentrate as base material, other ingredients such as milk concentrate, cocoa powder, sugar etc. are added to it. The blend is subject to puff drying under reduced pressure and controlled temperature to prevent browning. The dried material is scrapped from the trays and mildly ground to give granular product such as Bournvita, Boost etc.

Malted milk

Malt extract (wort) is mixed with fresh milk in the proportion of 50:50. Fat content of the beverage is adjusted to 3% (w/v). The contents are heated to about 60°C and then homogenized at 2500psi. Sugar (3%) is added to the blends, heated to 80°C and then hot filled in 200ml glass bottles. The filled bottles are processed at 121°C for 15 min. Cooling is done under running tap water. The beverage has long shelf life, excellent malty taste and smooth mouth feel.

Effect of whey and buttermilk on characteristics of barley and wheat malt extracts and concentrates

		Malt extract						Concentrate (20°C)			
		Yield		TA		RS		FBI		Consistency	
Malt	W/BM %	W %	BM %	W %	BM %	W %	BM %	W	BM	W Pa.s^n	BM Pa.s^n
Barley											
PL172	0	79	79	1.71	1.71	2.6	2.6	0.84	0.84	4.02	4.02
	50	102	87	2.70	1.80	10.1	4.0	0.90	0.75	5.31	6.72
	100	155	99	4.14	2.16	22.0	5.0	0.91	0.72	6.61	7.69
HLB223	0	80	80	2.34	2.34	1.1	1.1	0.79	0.79	4.89	4.89
	50	100	84	3.60	2.88	10.0	1.6	0.94	0.82	3.70	6:62
	100	159	97	4.77	3.24	18.0	3.3	0.88	0.88	8.80	43.69
Wheat											
PBW373	0	84	84	1.53	1.53	2.5	2.5	0.97	0.97	4.05	4.05
	50	118	88	2.43	1.76	12.8	4.5	0.96	0.93	4.82	9.88
	100	167	101	4.14	1.92	22.0	6.3	0.92	0.89	5.32	12.34
LSD (0.05)		1	1	0.15	0.15	0.5	0.5	0.02	0.02	0.28	0.28

Effect of malt extract on beverage characteristics

Malt	Extract (%)	TA (%) Cow	TA (%) Buffalo	Nitrogen (%) Cow	Nitrogen (%) Buffalo	Ash (%) Cow	Ash (%) Buffalo	NFS (%) Cow	NFS (%) Buffalo	Viscosity Cow	Viscosity Buffalo
Control	--	0.17	0.19	0.50	0.57	0.76	0.79	10.6	11.6	2.71	2.77
Barley	30	0.15	0.18	0.39	0.40	0.56	0.53	10.9	11.4	2.61	2.60
	50	0.14	0.15	0.30	0.33	0.42	0.42	10.0	10.3	2.32	2.34
Wheat	30	0.15	0.18	0.38	0.42	0.56	0.61	11.2	11.6	2.59	2.74
	50	0.15	0.16	0.29	0.33	0.33	0.37	9.0	9.3	2.38	2.38
Corn	30	0.17	0.20	0.33	0.46	0.42	0.65	10.3	12.1	2.24	2.82
	50	0.15	0.20	0.27	0.34	0.40	0.46	9.1	11.2	2.03	2.40
LSD 0.05		0.02	0.02	0.05	0.05	0.11	0.11	1.1	1.1	0.21	0.21

Baking

Malt is used either for taste and flavour or for supplementing enzymes in bread making. For taste and flavour, malt extract concentrate is used. Malt powder in small quantity is used for supplementing enzymes. Both are mixed with flour before making dough. Bread has excellent flavour but the crust becomes dark in colour when malt extract is used. Dough supplemented with malt flour has very good handling properties. Volume of the dough increases and the bread becomes soft and tasty.

Effect of wheat malt supplement on farinograph curve characteristics and loaf volume

Sample	α- Amylase SKB/100g	DDT min	Stability min	MTI min	Softening BU	Loaf volume, ml 75min	150min
Malt							
Control	–	3.3	4.3	90	105	530	480
Wheat	10	3.3	4.0	100	135	580	520
	20	3.2	3.2	110	160	–	–

Confectionery

Malt extract concentrate is used in several confectionery products for providing desirable colour and flavour.

Breakfast cereals

The breakfast cereals are coated with malt extract concentrate to impart flavour. The concentrate is either sprayed or the product is rotated along with the concentrate in the coating machine.

Pharmaceuticals

Malt extract concentrate is generally used as a base for many tonics and other medical syrups to impart flavour and mask the off flavour of the medicinal ingredients.

□□□

12

Nutritionally Balanced Formulated Functional Foods

MRIDULA D.

India is developing very fast in all sphere of activities viz. economic, technical, scientific, industrial and electronics but we are still placed at a cross road as far as the nutritional status of the population is concerned. India, being a country in developmental transition, faces the dual burden of pre-transition diseases like under-nutrition and infectious diseases as well as post-transition, lifestyle-related degenerative diseases such as obesity, diabetes, hypertension, cardiovascular diseases and cancers. According to recent National Family Health Survey and UNICEF Reports, 46% of preschool children and 30% of adults in India suffer from moderate and severe grades of protein-calorie malnutrition PCM as judged by anthropometric indicators. It is not only the PCM but other deficiency diseases like vitamin A deficiency in children, iron deficiency aneamia in all age groups particularly amongst women, children and pregnant women, B-complex deficiency and iodine deficiency are also prevent among the vulnerable group of India and other developing countries. According to most recent estimate released in October 2010 by FAO, it was reported that 925 million people are undernourished all over the world. As the figure below shows, the number of hungry people has increased since 1995-97, though the number is down from last year. The increase has been due to three factors: 1) neglect of agriculture relevant to very poor people by governments and international agencies; 2) the current worldwide economic crisis, and 3) the significant increase of food prices in the last several years which have been devastating to those with only a few dollars a day to spend. Nine hundred twenty five million people is 13.6 percent of the estimated world population of 6.8 billion. Nearly all of the undernourished are in developing countries.

The world produces enough food to feed everyone. World agriculture produces 17% more calories per person today than it did 30 years ago, despite a 70% population increase. This is enough to provide everyone in the world with at least 2,720 kilocalories (kcal) per person per day (FAO, 2002). The principal problem is that many people in the world do not have sufficient land to grow, or income to purchase, enough food. The world's undernourished face an average daily food deficit of 200–300 kcal per person per day in a diet dominated by cereals and starchy roots and tubers, in other words low in diversity (FAO, 2000). In some Asian countries, for example, 80% of calories per caput come from cereals and starchy roots and tubers. In addition to nutrient deficiencies, high carbohydrate diets raise insulin levels with consequences for increasing susceptibility to diabetes, myopia and other chronic diseases (Cordain *et al.*, 2002).

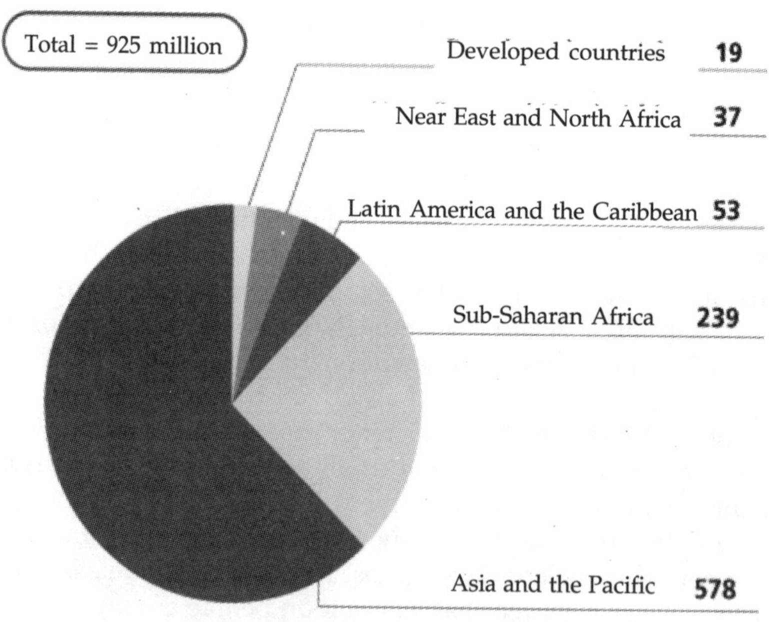

925 million hungry people in 2010
Source: FAO, 2000

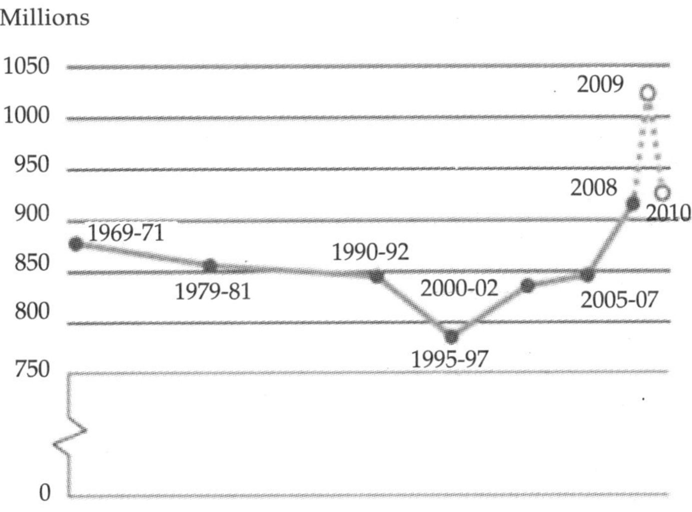

Number of hungry people, 1969-2010
Source: FAO, 2000

The implication of nutritious diet to health is well understood and supported by much scientific research. In a joint review by the World Health Organization (WHO, 2003) and the Food and Agricultural Organization of the United Nations, the role of diet and nutrition as determinants of non-communicable disease is well established: "Nutrition is coming to the fore as a major modifiable determinant of chronic disease, with specific evidence increasingly supporting the view that alterations in diet have string effects, both positive and negative, on health throughout life. Most importantly, dietary adjustment may not only influence present health, but may determine whether or not an individual will develop such diseases as cancer, cardiovascular disease and diabetes much later in life. However, these concepts have not led to a change in policies or in practice".

Consumers interest in formulated functional foods

For good health and normal life, our diet should also contain functional foods/ ingredients. Now days, consumers decide to purchase food products not only in terms of taste and nutritional needs, but also in terms of the ability to improve their health and well being. Functional foods/ ingredients exert a beneficial influence on body functions to improve the well being and health and/or reduce the risk of chronic diseases. Functional food can be produced by addition of health promoting ingredient/ food constituent(s), by reducing/ removing harmful components/ food constituent(s), and/or by modifying the nature or the bioavailability of specific food constituents. On the road to optimize nutrition that will be one of the major challenges of nutrition in the twenty first century, functional foods have their own role to play. But the development of claims for already existing food products, as well as the development of new products and their own claims, should remain first a scientific challenge and not only a marketing challenge. This is the condition for success to the benefit of both human health and the food industry (Verschuren, 2002).

Now days, consumers particularly in urban areas, generally prefer convenient and ready to eat processed foods, which are generally prepared from refined cereals and contains lot of calories. Hence, there is a huge scope for nutritionally balanced formulated foods, which are high in growth promoting nutrients and rich in dietary fibre and low in saturated fat, cholesterol and refined carbohydrates suitable for adults. Low glycemic and low fat and cholesterol foods are in demands particularly amongst obese, diabetic and individuals with cardio-vascular disease because of their association with the reduced risk for chronic disease, which are high in growth promoting nutrients and rich in dietary fibre and low in saturated fat, cholesterol and refined carbohydrates suitable for adults. Because of the recent research development and scientific data demonstrating the relationship between food consumption and incidence of diseases, consumers are accepting that **"Health is a Controllable Gift"**. Consumer interest in health-promoting foods will almost assuredly increase, because of the following reasons.

- *Clinical evidence*- The functionality of many familiar and new food components is being established by clinical and epidemiological research. For example, studies related to effect of dietary supplementation with food ingredients naturally rich in cancer-preventing substances, i.e. aged garlic extract, licorice extract, citrus juice, a mixed-vegetable beverage, and soybean meal, *etc.*
- *Age wave*- Elderly people are more interested in functional foods than the young adults and children to improve health and extend life expectancy.
- *Health care cost* - Due to expensive medical treatment, people are now believing and following healthy dietary practices and wants to consume nutritious and functional foods rich in antioxidants and health promoting constituents.
- *Nutritional labelling*- As consumers are becoming more health conscious, the promotion of certain nutritional attributes can be very effective for selling certain products.

- *Mission of medicine*- Medical school are beginning to place more emphasis on the prevention of chronic diseases. Physicians and the progressive health care clinics will make emphasise on nutrition education and healthy diet, a more important part of their services.
- *Food technology*- The development of formulated functional food products is convincing consumers that they can have tasty foods that are also healthy.
- *Brand differentiation*- Most food products represent mature categories where brand differentiation is a difficult challenge. Firms with strong Research and Development capabilities are using the interest of consumers in health and nutrition, an opportunistic way to differentiate and add value to their brand.

Functional foods

The fast growing knowledge of the role of physiologically active food components, from both plant (phytochemicals) and animal (zoochemicals) sources, has notably changed the role of diet in health. This led to the development of a new generation of foods termed functional foods. These foods provide health benefits beyond the basic nutrition needs (National Academy of Sciences, 1998).

The term Functional Foods was first introduced in Japan during the eights of the last Century with the development of a special seal to denote Foods for Specified Health Use (FOSHU) which refers to processed foods containing ingredients that enhance specific body functions, in addition to being nutritious. According to Huggett and Schliter, 1996, functional foods includes foods or food ingredients that exert a beneficial effect on human health and/ or reduce the risk of chronic disease beyond basic nutritional functions. The International Food Information Council (IFIC, 1998) defines functional foods as foods that provide health benefits beyond basic nutrition. The nutritional science evolved from identifying and correcting nutritional deficiencies to design foods that promote optimal health and reduce the risk of chronic disease. Focus group research showed that functional foods was recognized readily and was preferred by consumers over other terms such as nutraceuticals or designer foods (IFIC, 2002). The American Dietetic Association (ADA) takes one of the most inclusive views to define functional foods. The American Dietetic Association (1999) described functional foods as any potentially healthful food or food ingredient that may provide a health benefit beyond the traditional nutrients it contains and also made the following very important points:

- Functional foods may be whole, fortified, enriched, or enhanced foods
- It's likely that all foods are functional at some physiological level
- To have a beneficial effect on health, a functional food should be consumed as part of a varied diet on a regular basis, at effective levels

Table 1: Some functional foods available in the market

Functional food	Active food component	Target function
Yogurts, sugar	*Probiotics:* Foods with beneficial live cultures as a result of fermentation or that have been added to improve intestinal microbial balance, such as Lactobacillus sp., Bifidobacteria sp *Prebiotics:* A non-digestible component that has beneficial affects by stimulating the growth of bacteria in the colon eg. inulin and oligofructose.	Optimal intestinal function and intestinal microbial balance.
Margarines	Added plant sterols and stanols esters	Decreased LDL-cholesterol Decreased risk of coronary heart disease (CHD).
Omega-3 fatty acids enriched eggs	Omega-3 fatty acids	Control of hypertension, lipids metabolism.

Functional foods include foods that contain specific minerals, vitamins, fatty acids or dietary fibre, foods with added biologically active substances such as phytochemicals or other antioxidants and probiotics that have beneficial effects on life. According to this definition, unmodified whole foods such as fruits and vegetables may be considered as the simplest form of a functional food. For example, broccoli, carrots, or tomatoes would be considered functional foods because they are rich in physiologically active components such as sulforaphane, β-carotene and lycopene, respectively. Also, modified foods, including those that have been fortified with nutrients or enhanced with phytochemicals or botanicals fall within the realm of functional foods.

Table 2 : Functional components in foods grains

Food source	Food components	Potential health benefit(s)
Wheat bran	Insoluble fibre	Believed to reduce the risk of breast and/or colon cancer. Good for digestive health
Whole cereal grains	Insoluble fibre	* May reduce risk of coronary heart disease and cancer; may contribute to reduced risk of diabetes * May also affect body-weight regulation through effects on hormonal factors (Koh-Banerjee & Rimm 2003). * WHO/FAO (2003) expert committee report on nutrition and chronic diseases suggested that a high intake of non-starch polysaccharide (NSP) may be a protective factor against overweight and obesity. * Products that contain 51% or more of whole grain ingredients by weight can also make the following FDA approved health claim: "Diets rich in whole grain foods and other plant foods, and low in total fat, saturated fat and cholesterol, may reduce the risk of heart disease and certain cancers.
Oats (oat bran, whole oat flour and rolled oats)	Soluble fibre 'β-glucans' (indigestible polysaccharides)	*Believed to lower Low Density Lipo-protein (LDL-"bad" cholesterol), and possibly to reduce the risk of heart disease.
Whole grain barley and dry milled barley	Soluble fibre 'β-glucans'	* Associated with reduced risk of coronary heart disease
Corn, soy, wheat	Stanol ester	* Lowers blood cholesterol levels
Millets (Pearl millet, Finger millet, Foxtail millet, Proso millet)	Phytochemicals: Phytic acid and Phytate,	believed to lower cholesterol, associated with reduced risk of cancer.
Fenugreek	Saponins	* Decreases glucose absorption in the digestive tract,
	trigonelline, nicotinic acid and coumarin	* Shown hypoglycemic activity
	4-hydroxyisoleucine	* Shown to increase insulin secretion in studies with animals

Contd...

	Galactomannan	* May decrease glucose absorption from the gastrointestinal tract
	Galactomannan fiber and saponins	* May decrease absorption of cholesterol and lipids in the digestive tract
Guar (gum)	galactomannan	* Reduce postprandial glycaemia, (Jenkins *et al.* 1978)
Flaxseed	Lignans,	* Possess chemo-protective properties. Act as anti-carcinogenic agent and helps to avoid prostrate, breast and endometrial cancers.
	oemga-3 fatty acid,	* Reduce the risk of cardiovascular disease (CVD) osteoporosis, rheumatoid arthritis and cancer, * Increase blood HDL.
	Phytic acid,	* Recent researches have shown that phytic acid has antioxidant, anticancer, hypocholesterolemic and hypolipidemic effects (Mazza, 2008).
	Dietary fibre	* Digestive health.
Psyllium (seed husks)	Dietary fibre	* Used to relieveconstipation, irritable bowel syndrome, diverticular disease, and diarrhea. May lower cholesterol and control diabetes.

Source: (IFIC, 2008; Slavin. J, 2000)

Table 3 : Wholegrain cereal consumption and risk status: prospective cohort studies

Condition	Study subjects	Decreased Risk	Reference
Stroke	75,521 women	36% (0.64)	Liu *et al.* 2000
CHD	75,521 women	25% (0.75)	Liu *et al.* 1999
Fatal CHD	34,492 women	30% (0.70)	Jacobs *et al.* 1998
Non-fatal heart attack	31,208 men & women	44% (0.56)	Fraser *et al.* 1992
CHD	31,208 men & women	11% (0.89)	Fraser *et al.* 1992

Source: Hu, (2003)

Nutritionally rich formulations for product development

Food is essential for our existence. Most natural foods contain more than one nutrient, which can be divided into two major categories–macronutrients and micronutrients. Macronutrients are proteins, fats and carbohydrates which are often called 'proximate principles' because they form the main bulk of diet. In the Indian dietary pattern, proteins contribute 7-15 per cent, fat 10-30 per cent and carbohydrates 65-80 percent to the total energy intake. Vitamins and minerals are called micronutrients because they are required in small amounts which may vary from a fraction of a milligram to several grams. We derive these nutrients through the food we take in our daily diets including snacks and meals. The type and quality of foods we take vary widely from community to community and much depends upon the socio-economic status of the family. Each food in our diet, although contain a wide range of nutrients but serves as a major source of one or two main nutrients. Since, each food has a different nutritional profile; a judicious mix/proper formulation of

Table 4: Nutrients content in protein rich formulated and functional foods developed for school children (age group 7-9 years) using extrusion cooking (per 100g)

Formulations	Protein	Fat	Minerals	Crude Fibre	Carbohy-drates	Energy	β-carotene	Iron
	g	g	g	g	g	kcal	µg	mg
Corn: Rice: SPI (60:30:10)	17.7	2.3	1.1	1.7	63.2	308.7	54.0	1.6
Corn Wheat: SPI (60:30:10)	19.3	2.7	1.7	2.2	60.5	307.5	62.7	1.4
Corn Wheat: DFSF: SPI (55:30:10.5)	18.9	2.6	1.6	2.5	61.1	323.3	58.2	2.2
Corn Wheat: FFSF: SPI (56:30:10:4)	17.8	4.5	2.1	2.5	60.0	337.0	101.7	2.3
Corn Wheat: FFSF: SPI (55:30:10:5)	18.6	4.4	2.1	2.4	59.3	333.0	100.8	2.3
Corn Wheat: DFSF:DW:SPI (51:30:10:6:3)	17.9	3.6	1.6	2.4	61.6	337.0	546.0	2.1
Corn Wheat: SPI Carrot (50:25:10:15)	17.8	2.3	1.7	2.2	53.3	269.2	562.6	1.4
Corn Wheat: SPI Carrot (50:20:10:20)	17.3	2.2	1.7	2.2	50.8	256.5	731.2	1.5
Corn Wheat: SPI Papaya (57:25:10:8)	18.9	2.6	2.0	2.0	55.1	280.2	637.7	1.7
Corn Wheat: SPI Papaya (55:25:10:10)	18.8	2.5	2.1	2.0	53.8	273.4	780.7	1.8
Corn Wheat: SPI Spinach (56:30:10:4)	19.0	2.6	1.8	2.1	58.1	295.8	483.2	1.3
Corn Wheat: SPI Spinach (52:30:10:8)	18.7	2.5	1.9	2.1	55.7	284.1	403.7	1.2

Source: Bargale and Kulkarni, 2010; nutrients content is based on NIN. 2000; SPI-soy protein isolate; DFSF-defatted soy flour; FFSF-full fat soy flour

different types of food are desired to maintain good health. The proper formulation of ingredients to specific flour and foods based on the dietary habits and nutritional status of the target group is one of the most popular nutritional interventions for improving the nutritional health of the vulnerable population, exploring the possibility of using modern food technological practices.

Defatted soy flour fortified biscuits were prepared using different levels of beetroot (0-10%) powder, other ingredients and additive, and evaluated for different quality attricutes. Expansion in diameter and thickness of biscuits decreased with increased levels of beetroot powder. Hardness and breaking strength was found increased with increasing levels of beetroot powder in flour blends but addition of egg or ascorbic acid did not show any significant difference in the both of these properties. Redness (a* values) of biscuits increased while lightness (L* values) decreased with increasing levels of beetroot powder in flour blends. Although all biscuits samples were accepted on sensory evaluation but the mean sensory scores for overall acceptability for biscuits prepared with different levels of beetroot powder and ascorbic acid were more than other two types of biscuits, indicated the commercial scope for manufacturing of good quality vegetarian biscuits with beetroot powder up to 8% level in flour blends (Mridula, 2009). Extrudates prepared from rice, 10% defatted soya flour or chickpea *dhal* and dried beetroot (0 to 8%) at 14% feed moisture using twin screw extruder at 10.5 kg/h feed rate and screw speed 250 rpm/min (Table 5). Expansion ratio and bulk density of both types of extrudates were in the range of 3.207-2.833 and 3.244-3.029; and 0.192-0.26 and 0.162-0.24 g/cm^3, respectively. Whiteness (L* values) decreased while redness (a* values) increased with increased level of dried beetroot in both types of extrudates. Protein, fat, ash content, crude fiber, carbohydrates and calories in both types of extrudates were in the range of 9.48-13.05%, 0.57-1.08%, 0.83-1.73%, 0.79-1.47%, 76.53-81.45%, and 364.37-373.06 kcal/100g, respectively. Levels of dried beetroot affected the sensory quality significantly. Mean sensory score more than 7 (at nine point hedonic scale) for extrudates having DB up to 6% in feed material, indicated the commercial scope for beetroot supplemented extruded products from rice and deffated soy flour or chickpea *dhal*.

β-carotene rich defatted soy flour fortified biscuits were developed following the ingredients in the proportions given in Table 6 and evaluated for different quality attributes. Although spread ratio decreased, hardness increased, brightness decreased with increasing levels of carrot powder in flour blends but all biscuits samples were accepted on sensory evaluation (Mridula, 2011). The mean overall sensory acceptability scores more than 8 for biscuits samples up to 7.5% carrot powder indicated the commercial scope for manufacturing of good quality vegetarian biscuits with carrot powder, which will also be helpful in providing daily dietary requirement of β-carotene to the consumers at 100 g consumption level of these carotene rich biscuits. β-carotene rich extrudates were prepared from rice, wheat, 8% chickpea *dhal* or defatted soya flour and dried carrot (0 to 12.5%), using single screw (collet) extruder at 14% feed moisture, 25 kg/h feed rate and 500 rpm/min screw speed. Expansion in diameter and bulk density of both type of extrudates with dried carrot were in the range of 2.86-4.17 and 2.89-3.89, and 0.132-0.319 and 0.14-0.328 g/cm^3, respectively, which were significantly affected with the level of dried carrot and type of legume in feed material. Increased level of dried carrot in feed material resulted in increase of water absorption index while water solubility index was unaffected but it was affected with the type of legume in feed material. Colour quality was found affected and resulted in darker extrudates at increased level of dried carrot in feed material, which ultimately affected the sensory acceptability for appearance and colour. Protein, fat, ash, crude fibre and β-carotene content (wet basis) in both types of carrot incorporated extrudates were in the range of 8.74-10.96%, 0.67-1.21%, 1.02-2.06%, 0.48-1.61 and 35.96-1105.84 µg/100g, respectively.

Table 5: Cereal, legumes, vegetables (dried beetroot) based formulation for making acceptable quality extruded and baked product and their nutrients content (g/100g)

Proportion of ingredients in formulation	Moisture	Protein	Fat	Ash	Crude fibre	Total carbo-hydrates#	Total calories (kcal)
Biscuits							
WF: DSF: DB (89:5:6)	3.55	8.41	21.12	1.01	0.92	64.99	483.84
WF: DSF: DB (87:5:8)	3.58	8.33	21.12	1.07	1.07	64.82	482.82
Extrudates							
Rice:chickpea:DB (86:10:04)	6.42	9.76	1.07	1.13	1.05	80.55	370.94
Rice:chickpea:DB (84:10:06)	6.50	9.77	1.08	1.26	1.12	80.26	369.85
Rice: DSF: DB (86:10:04)	6.42	12.86	0.61	1.42	1.31	77.38	366.43
Rice:DSF:DB (84:10:06)	6.52	12.71	0.67	1.54	1.46	77.09	365.24

WF: refined wheat flour; DSF: defatted soy flour; DB: Dried beetroot

Although addition of dried carrot affected the overall sensory quality of extruded snack food samples but type of legumes (chickpea *dhal*/defatted soy flour) did not bring significant difference in the overall sensory quality. Sensory acceptability score more than 7 and β-carotene content for extrudates up to 7.5% dried carrot indicated the commercial scope of its incorporation in extruded products based on cereal and legume.

Table 6: Cereal, legumes, vegetables (dried carrot) based formulation for making acceptable quality extruded and baked product and their nutrients content (g/100g)

Proportion of ingredients in formulation	Moisture	Protein	Fat	Ash	Crude fibre	Total carbohy-drates	Total calories (kcal)	β-Carotene (mg)
Biscuits								
WF: DSF: DC (90:05:05)	3.53	7.55	21.34	1.05	0.85	65.69	484.99	1.68
WF: DSF: DC (87.5:05:7.5)	3.51	7.50	21.42	1.05	0.94	65.55	484.98	3.10
Extrudates								
Rice: wheat: DSF: DC (77:10:8:5)	5.67	10.92	0.74	1.38	1.06	80.23	371.28	0.311
Rice: wheat: DSF:DC (74.5:10:8:7.5)	5.63	10.93	0.76	1.68	1.22	79.78	369.65	0.598
Rice: wheat: chickpea:DC (77:10:8:5)	5.69	8.78	1.11	1.37	0.83	82.22	373.99	0.434
Rice: wheat: chickpea:DC (74.5:10:8:7.5)	5.65	8.75	1.17	1.58	0.99	81.86	373.02	0.422

WF: refined wheat flour; DSF: defatted soy flour; DC: Dried carrot

According to FAO/WHO recommendations, vegetable oils are included in foods meant for infants and children, which will not only increase the energy density, but also be a transport vehicle for fat soluble vitamins. The fat can also provide essential fatty acids like omega-3 and omega-6 polyunsaturated fatty acids (PUFA) needed to ensure proper neural development of children. In general, ready-to-eat food products formulated from locally available food commodities can meet the recommended daily allowances for macro nutrients, if intended for infants and children but do not meet the recommended micronutrient (minerals) requirements. Therefore, fortification with appropriate micronutrients or micronutrient-dense food ingredients and healthful constituents is important. Hence there is huge potential for development of proper formulations and fortification of these formulations using locally available food ingredients, which can provide cost effective health foods. This is the practical food-based approach and will certainly be helpful in improving the nutritional status, preventing deficiencies and degenerative diseases among the population.

References

American Dietetic Association (ADA), (1999). Functional foods-position of the ADA. J. Am. Diet. Assoc., 99: 1278-1285.

Bargale PC, Murugkar DA and Nema A (2010). Formulated Functional Foods-Concept, Design and Development, In: Extruded Functional Foods using Plant and Dairy Ingredients. Editors Bargale PC and SD Kulkarni, Soybena Processing and Utilization Centre, Central Institute of Agricultural Enginnerign, Bhopal. p31-44.

Cordain, L, Eaton, SB, Miller, JB, Lindeberg, S, *et al.* (2002). An evolutionary analysis of the National Academy of Sciences, 1998. Institute of Medicine. Dietary Reference Intakes, Thiamin, Riboflavin, Niacin, Vitamin B_6, Folate, Vitamin B_{12}, Pantothenic Acid, Biotin and Choline. National Academy Press, Washington, DC, USA.

FAO, (2000). Food Insecurity: When People Live with Hunger and Fear Starvation. FAO, Rome.

Food and Agriculture Organization, International Fund for Agricultural Development, World Food Program (2002). "Reducing Poverty and Hunger, the Critical Role of Financing for Food, Agriculture, and Rural Development."

Fraser GE, Sabate J, Beeson WL and Strahan TM (1992). A possible protective effect of nut consumption on risk of coronary heart disease. The Adventist Health Study. Archives of Internal Medicine 152: 2249-2258.

Hu FB (2003). Plant-based foods and prevention of cardiovascular disease: an overview. American Journal of Clinical Nutrition 78: 544S-551S.

Huggett, AC, Schliter, B (1996). Research needs for establishing the safety of functional foods. Nutr. Rev. 54, S143–S148.

IFIC. (International Food Information Council), (1998). Backgrounder: Functional Foods. Food Insight Media Guide. International Food Information Council Foundation, Washington, DC, USA.

IFIC. (International Food Information Council), (2002). Functional foods attitudinal research. International Food Information Council, Washington, DC. http://ific.org/research/funcfoodsres02.cfm.

International Food Information Council Foundtaion (IFIC), (2008). Functional Foods 1100 Connecticut Avenue, NW Suite 430 Washington, DC 20036 http://ific.org/nutrition/functional.

Jacobs DR, Meyer KA, Kushi LH, Folsom AR (1998). Whole-grain intake may reduce the risk of ischemic heart disease death in postmenopausal women: the Iowa Women's Health Study. The American Journal of Clinical Medicine 68: 218-219.

Jenkins, DJA, Wolever, TMS, Leeds, AR, Gassull, MA, Haisman, P, Dilawari, JB, Goff, DV, Metz, GL & Alberti, KGMM (1978). Br. Med. J. i : 1392.

Koh-Banerjee P & Rimm EB (2003). Wholegrain consumption and weight gain: a review of the epidemiological evidence, potential mechanisms and opportunities for future research. Proceedings of the Nutrition Society (2003) 62: 25-29.

Liu S, Manson JA, Stampfer MJ, Rexrode KM, Hu F, Rimm E & Willett WC (2000). Whole Grain Consumption and Risk of Ischemic Stroke. Journal of the American Medical Association 284: 1534-1540.

Liu S, Stampfer MJ, Hu FB, Giovannucci E, Rimm E, Manson JE, Hennekens CH & Willett WC (1999). Whole-grain consumption and risk of coronary heart disease: results from the Nurses' Health Study. American Journal of Clinical Nutrition 70:412-419.

Mazza, G (2008). Production, Processing and uses of Canadian Flax. First CGNA International Workshop, Temuco, Chile, August 3-6.

Mridula D (2011). Physico-chemical and sensory characteristics of beta-carotene rich defatted soy fortified biscuits. African Journal of Food Science Vol 5(5): 305-312.

Mridula D, Naveen Kumar M, Dilip Jain, SK Tyagi and DR Rai (2009). Effect of Addition of Beetroot Powder on Physical Properties and Acceptability of Defatted Soya Flour Fortified Biscuits, 2009, 46(2): 59-69.

National Academy of Sciences, (1998). Institute of Medicine. Dietary Reference Intakes, Thiamin, Riboflavin, Niacin, Vitamin B6, Folate, Vitamin B12, Pantothenic Acid, Biotin and Choline. National Academy Press, Washington, DC, USA. Huggett, A.C. and B. Schliter, 1996. Research needs for establishing the safety of functional foods. Nutr. Rev., 54: S143-S148.

National Family and Health survey (NFHS-3), IIPS (2005-2006).

Slavin J (2000). Dietary Fibre and the prevention of cancer. Department of Food Science and Nutrition, University of Minnesota, 1334 Eckles Avenue, St. Paul, MN 55108, USA, Woodhead Publishing Limited Ltd.

UNICEF: The State of the World's Children (2008). United Nations Children's Fund, United Nations Plaza, New York, NY 10017, USA, December 2007. Website: www.unicef.org

Verschuren, PM, (2002). Functional foods: Scientific and global perspectives. British J. Nutr., 88: 125-S130.

WHO and FAO. Diet, Nutrition and the Prevention of Chronic Disease. WHO Technical Report Series, 2003, Geneva; 916.

□□□

13

Frozen Bakery Foods

DEEPIKA GOSWAMI

Freezing is one of the several food preservation methods in which the food is preserved by reducing its temperature to the point at which a transformation in the state of the food occurs. Quick freezing is the process of lowering the temperature of foods much below their freezing point in a short period of time as to -20°C in 3-30 minutes. The process involves direct immersion or indirect contact of foods with the refrigerant and air blasts over the foods being frozen. It is the process whereby, virtually all the properties of most of the parent foodstuffs can be preserved (Pruthi, 1999). The bakery industry in India occupies a significant place in the industrial sector and is the largest of the food industries with an annual turnover of about Rs. 3000 crores. The production of bakery products has increased from 5.19 lakh tonnes in 1975 to 18.95 lakh tonnes in 1990 recording four-fold increase in 15 years (http://business.mapsofindia.com). There are around 60,000 bakeries, several pasta food units and 20,000 traditional food units in India. Products of bakery include bread, biscuits, pastries, cakes, buns, rusk etc. The quantities of bread and biscuits produced are more or less the same. Bakery products are increasingly becoming popular in India as indicated by over 2.5 fold increase in their production during the last 2 decades (Puranik 2003). The two major bakery industries, *viz.*, bread and biscuit account for about 80% (MOFPI 2008-09) of the total bakery products with annual production of around 4.0 million tonnes (MOFPI 2007) and individual shares of over 50 and 30 % of total bakery products, respectively (Shukla *et al* 2000). The bread market has a business volume of 1.5 million tonnes and the biscuit market has a total production volume of 1.1 million tonnes. The bread industry is represented by both the organized and unorganized sectors with 55% and 45%contribution to production (http://www. naukrihub. com/india/fmcg). MOFPI (2007) reported 40% and 60% share of organized and unorganized sectors for bread production, respectively. The reported share of unorganized sector in the production of biscuits is over 70% (MOFPI 2008-09). Bakery products are an important source of nutrients viz. energy, protein, iron, calcium and several vitamins. Commercial bread and biscuits contain around 7.5% to 7.8% protein respectively. These are manufactured from wheat-flour, sugar, baking powder, condensed milk, ghee (fat), salt, jelly, dry fruits, various essences and flavorings etc. Different type of bakery products can be classified as dry bakery products and moist bakery products.

Dry bakery products

Biscuits: Soft biscuits, hard biscuits, cookies, crackers, fancy biscuits, cream wafer biscuits.

Moist bakery products

Bread: Sweet bread, Milk bread, Masala bread, Garlic bread, Fruit bread *etc.*

Buns: Fruit buns, hamburger buns, dinner rolls, crisp bread, pizza.

Others: Cakes, pastries, doughnuts, muffins *etc.*

The sectors that have recorded a high growth rate from 10 per cent to 20 per cent during 2004-05 in bakery segment include bread, cakes and pastry (http://www.fnbnews.com). During the last decades, the social and scientific modernization, the boom in customers' needs and demands, along with the major changes in the way food products are manufactured, distributed and retailed, led to the development of alternative or novel methods for the production and preservation of bakery products such as freezing or MAP technology (Bhattacharya *et al*, 2003; Laaksonen and Roo, 2000). Freezing has been claimed as an almost ideal method of preserving nearly all baked products. It reduces the cost of production and distribution. Bread and rolls were the first frozen baked products to be marketed successfully: frozen cakes, pies, pastries and pizzas appeared in the market soon after (Pruthi, 1999). The practice of freezing yeast-raised bread dates from the 1930s. However, the commercial production of frozen bakery products mainly began during the 1950s, and started to gain popularity ever since. In the case of baked goods, the effect of freezing and frozen storage of bread was only studied in the fifties (Pence and Standridge, 1955); conversely the frozen doughs have been extensively studied during the last decades (Havet *et al* ,2000; Bail *et al*, 1999; Ribotta *et al* 2003a; Wolt and D'Appolonia, 1984). Nowadays, the frozen bakery products occupy an increasingly great portion of the food market covering sectors such as supermarkets, restaurants, institutional and catering businesses (Kennedy, 2000; Rouillé *et al*, 2000) In India the frozen bakery products are broadly divided into two classes: (a) freshly baked products, quick frozen to retain freshness for long periods of time, which need to be thawed prior to consumption, and (b) freshly prepared dough, frozen quickly to retain its physical and biological properties till baked afresh for consumption. The various bakery products manufactured by freezing process are bread, cakes, cookies, muffins, pies, pizzas, rolls, short cakes, and several other specialty bakery products like frozen waffles, pancakes, noodles, etc (Pruthi, 1999).

Freezing of bakery foods

The basic principle followed is to freeze bakery products so that the temperature in the centre of the product reaches -18°C as fast as possible. The process takes 3 to 4 hours in case of bread and 1 to 1.5 hours in case of rolls and other bakery products, using a circulating air temperature of -25°C to -30°C. The product so frozen may be stored at -18°C to -20°C for nearly 6 months without significant quality changes. Among the frozen bakery products, that are of both kinds- baked and unbaked, the unbaked frozen products are favoured for commercial freezing operations because of ease of handling and superior freshness at serving time (Pruthi, 1999).

Frozen batter

Tressler *et al* (1968) reported that to obtain a quality frozen batter, there should be slight modifications in formulation and mixing procedure. Temperature maintenance is one of the most important factors to consider while freezing the cake batters. To retain good quality the product must be stored at -23°C or below. Storage of batter above -12°C tends to lose air inclusion (Harder, 1938). Storage temperature has a profound effect on the quality of cakes made from frozen batters. Frozen batters stored at -18°C gave cakes with poor aroma, flavor, colour and volume. Batters stored at -34°C retained good volume up to six months (Pruthi, 1999).

Frozen doughs

Preparation of frozen, unbaked yeast dough presents many problems. Tressler *et al.* (1968) suggested one method of achieving stability in frozen bread dough by using double the concentration of yeast normally used in the recipe and by using the straight dough method to minimize the fermentation time. According to Ribotta *et al* (2003b) freezing and frozen storage of dough made with fresh yeast had more negative effects on baking quality than the addition of frozen yeast to dough. Studying the correlation of fermentation process and grain structure of baked breads from frozen dough using freeze tolerant yeast, Takano *et al* (2002) reported that prefermentation before freezing, punching and remolding, or resheeting and molding treatments increased loaf volume by 10 to 110% for the baked breads using freeze-tolerant yeast, while these treatments decreased loaf volume by 70% using freeze-sensitive yeast. The high freeze tolerance of Baker's yeast (*Sachharomyces cerevisiae*) is found to be correlated with higher intracellular glycerol content of yeast (Izawa *et al*, 2004) and higher level of L- proline in yeast cells (Terao *et al*, 2003). Multiple freeze-thaw procedure is also suitable for the selection of cryoresistant yeast strains that may be useful for frozen dough manufacture (Piasecka *et al*, 2003). The first fermentation before freezing and the second fermentation with punching or resheeting after thawing are useful for obtaining good quality breads from frozen dough using freeze-tolerant yeast. Low temperature during mixing inhibits fermentation. A very rapid freeze is ultimate goal in order to reduce large ice crystal formations (Harder, 1938). A study conducted by Hady *et al* (2002) reported that the emulsifier DATE (Diacetyl Tartaric Acid Ester) was more effective than lecithin in improving the frozen dough stability, resistance to extension and dough strength over that of control samples (no emulsifier). Presence of emulsifier also shortened final proof time, increased loaf volume and improved final baking score of bread made from the frozen dough. A study conducted by Aibara *et al.* (2001) revealed that a shortening with short chain fatty acids maintained the loaf volume of frozen dough even after two months of frozen storage. Reason of this may be due to strong interaction of the shortening with gluten proteins. Sweet and white bread doughs prepared using freeze tolerant yeast and stored at -20°C when subjected to multiple freeze-thaw cycles then it was observed that increase in number of freeze-thaw cycles increased both the coarseness of the gluten fibrils and the size of the knots (Naito *et al.*, 2004). Some of the most important problems concerning frozen dough such as decreased yeast viability and gas production ability or diminished gluten network ability to retain CO_2 and progressive loss of dough cohesiveness and stability, have not been fully resolved yet (Perron *et al.*, 1999; Ribotta *et al.*, 2001).

Baked breads and rolls

To meet the high demand for bread, bread dough can be pre-baked then cooled and either frozen or placed under vacuum; when needed it only requires baking for a further 10-15 minutes. The delicate nature of baked goods has to be considered while freezing. To prevent staling, the baked products have to be cooled, packaged and frozen quickly after baking. The important factors contributing to staling are time and temperature of freezing. Freezing process should be very fast to control these factors (Harder, 1938). Condition and rate of freezing have profound effect on quality of bread. Fast freezing of bread to below -30°C retard the staling process and enhance water stabilization. Storage at -18°C or lower causes immobilization of starch granules in the frozen crumb that leads to the stoppage of virtual change relating to firmness in bread (Pruthi, 1999). Studying the effect of hydrocolloids on frozen stored and microwave reheated breads; Mandala (2005) observed that bread quality was poorer with addition of guar gum, while properties of bread made with xanthan gum were more desirable when using in lower concentration. Barcenas *et al.* (2003) reported staling retardation of pre-baked frozen bread by use of bread improvers (alpha-amylase, kappa-carrageenan, hydroxypropylmethylcellulose). Hydroxypropylmethylcellulose (HPMC) reduces water activity (aw), decreases bread crumb hardness, increases moisture retention and specific vol. of the bread. Addition of HPMC also inhibited the effect of frozen storage on staling (Barcenas *et al.*, 2004).

Faster freezing of dough (in dry ice, -80°C) produces a superior loaf than that obtained from dough frozen more slowly (in the freezer, -20°C). But bread with a larger loaf volume, softer texture and fewer fish-eye like blisters can be made by freezing of dough at much higher temperature (-15°C) after addition of ice nucleation-active (INA) protein (Funaki *et al.*, 1996). The quality of frozen foods is closely related to the size and amount of ice crystals formed during freezing. Songming *et al* (2005) compared the Pressure Shift Freezing (PSF) method with liquid immersion freezing (LIF) and conventional air freezing (CAF) at -20°C and observed that PSF of gelatin gels encourages the formation of ice crystals with smaller size and thus may improve the texture of the frozen product.

Pies and pastries

Freezing of unbaked fruit pies requires prior control of enzymatic changes in fruits and use of thickening agents such as tapioca, waxy maize starch etc to produce good quality product. To avoid large ice crystal formation, rapid freezing is recommended. Fast freezing also prevents separation of the liquid from the solid portion. Formation of moist soggy bottom crust is the most common problem in baking frozen, unbaked pies. Frozen pastries retain high quality up to a year without deleterious changes in form, flavor, appearance and texture (Harder, 1938). Pastries are not materially changed by freezing and thawing. However, in case of presence of unsaturated fats, these products may become rancid if stored for longer than a few months at -18°C. For longer storage, a storage temperature of -23°C is recommended (Pruthi, 1999). Moder *et al.* (2003) has got a patent for manufacturing frozen, unproofed, laminated dough that can be baked to a desired volume without being thawed or proofed. The product consists of alternate layers of dough and shortening. The dough layer contains flour, a water-binding agent, a leavening agent, a fat and water. The frozen dough when baked can be used to prepare Danish pastries, croissants etc. Reuter and Guo (2003) have patented a frozen, rolled, raw pastry composition that can be cooked in a microwave oven to give a bakery product with fresh baked quality. It comprises dough, a cinnamon coating on a portion of the dough and a colourant on the surface of the pastry which, upon heating, gives a brown fresh baked colour to the finished bakery product. Zubatyi *et al.* (1992) described the manufacture of quick-frozen pies of layered dough with fruit filling in their patent. The product is prepared by mixing the dough, cooling it, forming dough envelopes, introducing a filling, processing the pie pieces and freezing. To increase storage life, the pie pieces are treated with air at 240-260°C for 7.0-7.5 min before freezing.

Cakes, cookies, muffins, shortcakes and waffles

The quality of frozen cakes depends on the baking time and temperature. Under baking or over baking creates problems. Freezing is the only method to preserve muffins easily. The storage life is about six months at -18°C and considerably longer at still lower temperatures (Pruthi, 1999). Packaging is one of the most important factors in preserving the quality of frozen cakes as it protects cakes from dehydration, hardening of the icing and condensation of moisture on product surfaces during thawing. Frozen cookies have an extremely long freezer shelf life and hence these are the foolproof desserts to prepare and freeze (Harder, 1938). Gupta *et al.* (2010) studied the effect of freeze thaw cycle of dough on the textural characteristic of cookies incorporated with barley flour. He reported a crispier texture of cookies as compared to the normal one. Muffins, waffles and pan cakes are not difficult to prepare for freezing. The only change necessary in the preparation of these is that a very hot griddle is used to brown the product quickly and at the same time, to retain desirable moisture (Harder, 1938).

Effect of freezing and frozen storage on product quality

Hady *et al.* (2002) reported that frozen storage and repeated freeze-thaw cycles decreased the polar lipids to non detectable levels. Storage of bread dough in frozen condition for 6 to 8 weeks slightly increased the protein solubility (Aminlari and Majzoobi, 2002). One of the reasons for reduction in quality of bakery products produced from frozen dough with increased frozen storage time may be a result of moisture transfer from hydrated gluten to ice phase especially at temperature close to the glass transition temperature of the dough. The freezing process mainly damages the protein matrix (Berglund *et al.*, 1991, Ribotta *et al.*, 2004) and also causes an internal damage in the starch granules (Berglund *et al.*, 1991). Changes in rheological properties of frozen doughs have been reported to be due to the physical damage of the gluten network caused by ice crystallization and recrystallization. For maximum storage stability, a product must be stored at temperatures below its glass transition temperature where the viscosity of the unfrozen water fraction is very great.

Packaging and storage

The key functions of a food package can be summarized as protection, preservation and presentation. Each of these, in turn, requires other factors to be considered, namely temperature stability, barrier properties, insulation properties, compatibility with packaging machinery and consumer appeal. The packaging material should be both physically and chemically stable over a wide range of temperature. This should encompass the range of temperatures likely to be experienced by the product during each stage of production, distribution, storage and consumer handling. Generally this should extend from freezer temperature (-40 °C and above) to normal ambient temperatures. However, if the food product is also likely to be reheated or cooked in the package then the packaging material should also be able to withstand temperatures up to 300 °C and above (George, 2004). To maintain frozen foods in perfect conditions packaging material should provide protection against dehydration caused by moisture evasion through the wall or seals of the package to avoid freezer burn and to inhibit oxidation caused by either enzymes or light and heat as in case of high fat foods (Lee, 2006). Most frozen foods are cooked in a microwave oven. To make it easier for the consumer, manufacturers have developed packaging that can go straight from freezer to microwave. Today there are multiple options for packaging frozen foods-boxes, cartons, bags, pouches, heat-in-bag pouches, lidded trays and pans, crystallized PET trays, and composite and plastic cans (Russel *et al.*, 2003). The plastic materials used for packaging of frozen foods are LDPE, HDPE, Ethylene vinyl acetate copolymer polypropylene, PVC, PET, Polystyrene and nylon (Lee, 2006). Paper and card are commonly used for packaging frozen foods. Paper may also be used as a surface coating that provides a smooth surface for high quality printing. Card or paperboard is used to produce both folding and rigid cartons, often not in direct contact with food product. The board often consists of plies made from different materials (Tanner and Amos, 2006).

Frozen bakery foods have been analyzed by many researchers and the result showed that storage at -18°C or lower is necessary for effective retention of the high quality of these products for at least 2 months (Pruthi, 1999). The frozen dough samples presented a rapid degrading behavior during the first months of storage under freezing conditions. They became thenceforth stable and remained stable even after nine months and all of the quality parameters which were examined remained in fairly acceptable levels (Giannou and Tzia, 2007). Increasing the frozen storage time decreased the yeast count and gas formation capacity of the dough and increased the proofing time (Koca *et al.*, 1997).

Defrosting or thawing

Thawing of frozen products requires the similar attention as the process of freezing because an improper defrosting can easily destroy any gain in quality obtained by good freezing and good storage practices. The gain or loss of moisture by the frozen product during defrosting, which is influenced by duration of thawing, temperature and relative humidity can be substantial and serious for the product. For example, Defrosting bread at 50°C, 60°C, or 70°C with air of 60 to 70% R.H. and an air velocity of about 25 m/s till the centre of the loaves reaches 0°C, yield products with crisp and short crusts resembling closely to those of freshly baked products. Humidities in the range of 50 to 60% are generally recommended for defrosting frozen bakery foods (Pruthi, 1999).

Advantages of frozen bakery foods

Frozen bakery foods have the following advantages over conventional bakery foods:

- Require less time for preparation for consumption
- Require less space for preparation at the outlets (restaurants, retail chain etc.)
- Greater convenience in handling and preparation
- Less fuel required at outlets
- Greater value for money, notably during off-season
- Ready-to-heat and serve frozen bakery foods possible
- All year around availability of a great variety of bakery foods

Limitations

- Higher investment in deep freezer/freezing equipments/frozen stores
- Requirement of cold chain throughout the length and breadth of the country
- Requirement of refrigerated display cabinets for retail marketing

- Small fluctuation in storage temperature can have detrimental effect on product quality
- Variable performance and loss of stability of frozen bread doughs over long term frozen storage and problems due to improper thawing.

Constraints to frozen bakery industry in India

- Lack of a Frozen Food Association vis-a-vis lack of publicity
- Lack of proper infrastructural facilities
- Lack of cold chain retail outlets
- Higher freight and power cost
- Raw material quality

Among the traditional methods used for extending the shelf life of the food products during long periods of time, the frozen storage is the unique method that allows obtaining a product with similar characteristics to those of the fresh one in a wide variety of food products. The growing demand of bakery products has aroused need of their instant availability at a number of bakery outlets. Furthermore the overall process of manufacturing bakery products is time consuming and requires a lot of space. Hence the frozen bakery products have emerged as a solution which is manufactured in separate units, stored frozen in the bakery outlets and can be prepared and served immediately whenever demanded by the consumers. Availability of household refrigerators and microwave oven has also made it convenient to prepare and serve frozen bakery foods even at home. The scope of frozen bakery foods is increasing day by day along with recent advances and researches in this field. There is required more awareness among consumers, more infrastructural facilities like cold chain storage, better refrigerated transport facilities and above all an authority/ law that will govern the frozen bakery foods' sector in Indian market.

References

Aibara S, Nishimura K, Esaki K (2001). Effects of shortening on the loaf volume of frozen dough bread. Food Science and Biotechnology 10: 521-528.

Aminlari M, Majzoobi M (2002). Effect of Chemical Modification, pH Change, and Freezing on the Rheological, Solubility, and Electrophoretic Pattern of Wheat Flour Proteins. Journal of Food Science 67: 2502–2506.

Bail AL, Grinand C, Cleach SL, Martínez S and Quilin E (1999). Influence of storage conditions on frozen French bread dough. Journal of Food Engineering 39:289-291.

Barcenas ME, Haros M, Rosell M (2003). An approach to studying the effect of different bread improvers on the staling of prebaked frozen bread. European Food Research and Technology 218: 56-61.

Barcenas ME, Benedito C, Rosell CM (2004). Use of hydrocolloids as bread improvers in interrupted baking process with frozen storage. Food-Hydrocolloids 18: 769-774.

Berglund PT, Shelton DR, Freeman T (1991). Frozen bread dough ultrastructure as affected by duration of frozen storage and freeze-thaw cycles. Cereal Chemistry 68:105-107.

Bhattacharya M, Langstaff TM, Berzonsky WA (2003). Effect of frozen storage and freeze–thaw cycles on the rheological and baking properties of frozen doughs. Food Research International 36: 365–372.

Funaki J, Michigami Y, Yano M, Shiraishi A, Abe K, Arai S (1996). Applying a bacterial ice-nucleation active protein to preparation of frozen bread dough. Journal of Japanese Society of Food Science and Technology 43:835-838.

Giannou V, Tzia C (2007). Frozen dough bread: quality and textural behavior during prolonged storage-prediction of final product characteristics. Journal of Food Engineering 79: 929–934.

George M (2004). Selecting packaging for frozen food products: In Managing frozen foods. pp: 195-211, Ed. Christopher J Kennedy, Woodhead Publishing Limited, Cambridge.

Gupta M, Bawa AS, Ghannam NA (2010). Effect of barley flour and freeze thaw cycles on textural nutritional and functional properties of cookies. Food and Bioproducts Processing Article in Press, Accepted Manuscript.

Hady AE, Shouk AA, Yaseen AAE (2002). Effects of lipids and emulsifiers on frozen bread dough quality. Egyptian Journal of Food Science, 30:77-94.

Harder EL (1938). Preparing baked goods for blast freezing. In: Blast freezing, pp. 111-117, CBI Publishing Company, Boston.

Havet M, Mankai M, Bail AL (2000). Influence of the freezing condition on the baking performances of French frozen dough. Journal of Food Engineering 45:139–145.

http://business.mapsofindia.com/india-industry/biscuits.html India Biscuits Industry, India Business directory as on 30/07/10 12:00 pm.

http://www.fnbnews.com/article/detnews.asp?articleid=18005§ionid=32. Phani Raj, 10 June 2006.

http://www.naukrihub.com/india/fmcg/overview/bakery-products/ Bakery products, Overview of FMCG market. As on 30 /07/10 12:34 pm

Izawa S, Ikeda K, Maeta K, Inoue Y (2004). Deficiency in the glycerol channel Fps1p confers increased freeze tolerance to yeast cells: application of the fps1 DELTA mutant to frozen dough technology. Applied Microbiology and biotechnology 66: 303-305.

Kennedy CJ (2000). Managing frozen foods (pp. 143–144), Woodhead Publishing Limited and CRC Press LLC, Cambridge, UK.

Koca AF, Anil M, Evren M (1997). The effects of bulk fermentation time and storage period on frozen dough stability. Gida, 22:129-135.

Laaksonen TJ and Roos YH (2000). Thermal, dynamic-mechanical, and dielectric analysis of phase and state transitions of frozen wheat doughs. Journal of Cereal Science 32: 281–292.

Lee KH (2006). Plastic packaging of frozen foods: In Handbook of frozen food processing and packaging. Ed: Da Wen Sun. pp: 641-651.

Mandala IG (2005). Physical properties of freash and frozen store, microwave-reheated breads, containing hydrocolloids. Journal of Food Engineering 66:291-300.

Moder G, Cammarota C, Hajovy M (2003). Freezer-to-oven, laminated, unproofed dough and products resulting therefrom. PN - US 6 579 554 B2; (US6579554B2), PA: The Pillsbury Co. AN: 2003-10-Mv1469.

MOFPI (2007). Ministry of Food Processing Industry, India. Present Status and Future Prospects of Indian Food Processing Industries. http://mofpi.nic.in/Content Page.aspx?CategoryId=122.

MOFPI (2008-09). Annual report (2008-09) Ministry of Food Processing Industry, India.

Naito S, Fukami S, Mizokami M, Ishida N, Takano H and Koizumi M (2004). Effect of freeze-thaw cycles on the gluten fibrils and crumb grain structures of breads made from frozen doughs. Cereal Chemistry 81:80–86.

Pence JW and Standridge NN (1955). Effects of storage temperature and freezing on the firming of a commercial bread. Cereal Chemistry 32:519.

Perron CE, Lukow OM, Bushuk W and F. Townley-Smith (1999). The blending potential of diverse wheat cultivars in a frozen dough system. Cereal Foods World 44: 667–672.

Piasecka K, Badocha E, Chlabowska B, Stecka K, Grzybowski R (2003). Cryoresistance of chosen baker's yeast strains, Prace Instytutow-i-Laboratoriow Badawczych Przemyslu Spozywczego, 58:32-49.

Pruthi JS (1999). Quick freezing preservation of bakery foods. In: Quick freezing preservation of foods. pp. 146-169, Allied publishers Limited, Bombay.

Puranik DB (2003). Milk byproducts as functional ingredients in bakery products. Beverage and Food World 30: 21-22.

Reuter LW, Guo A (2003). Microwave prepared foodstuff, PN: US 2003/0190398 A1; (US2003/0190398A1), AN: 2004-02-Mv0184.

Ribotta PD, León AE and Añón MC (2001). Effect of freezing and frozen storage of doughs on bread quality. Journal of Agricultural and Food Chemistry 49: 913–918.

Ribotta PD, León AE and Añón MC (2003a). Effect of freezing and frozen storage on the gelatinization and retrogradation of amylopectin in dough baked in a differential scanning calorimeter. Food Research International 36: 357–363.

Ribotta PD, León AE and Añón MC (2003b). Effects of yeast freezing in frozen dough, Cereal Chemistry 80: 454-458.

Ribotta PD, Pérez GT, León AE and Añón MC (2004). Effect of emulsifier and guar gum on micro structural, rheological and baking performance of frozen bread dough. Food Hydrocolloids 18: 305–313.

Rouillé J, Bail AL and Courcoux P (2000). Influence of formulation and mixing conditions on breadmaking qualities of French frozen dough. Journal of Food Engineering 43:197–203.

Russell, Nicholas J, and Grahame W Gould (2003). Food Preservatives. 2nd ed. New York, NY: Kluwer Academic/Plenum Publishers, New York.

Shukla FC, Shilpa M, Thind SS (2000). Bakery industry in India – present quality control and future scenario – A review. Beverage and Food World 27: 11-15

Songming Zhu, Ramaswamy HS, Bail AL (2005). Ice-crystal formation in gelatin gel during pressure shift versus conventional freezing. Journal-of-Food-Engineering 66:69-76.

Takano H, Naito S, Ishida N, Koizumi M and Kano H (2002). Fermentation Process and Grain Structure of Baked Breads from Frozen Dough Using Freeze-Tolerant Yeasts. Journal of Food Science 67: 2725-2733.

Tanner D and Amos N (2006). Paper and card packaging of frozen foods: In Handbook of frozen food processing and packaging. Ed: Da Wen Sun. pp: 653-663.

Terao Y, Nakamori S, Takagi H (2003). Gene dosage effect of L-proline biosynthetic enzymes on L-proline accumulation and freeze tolerance in Sachharomyces cerevisiae, Applied and Environmental Microbiology 69: 6527-6532.

Tressler, Donald K, Van Arsdel, Wallace B, Copley, Michael J (1968). The Frrezing Preservation of Foods, The AVI Publishing Company, Inc., Westport, Connecticut.

Wolt MJ and D'Appolonia BL (1984). Factors involved in the stability of frozen dough. II. The effects of yeast type, flour type and dough additives on frozen stability. Cereal Chemistry 61:213–221.

Zubatyi AL, Kovaleva RI, Kalmysh VI (1992). Manufacture of quick-frozen pies of layered dough with fruit filling, PN: SU 1 722 369; (SU1722369), PA: Union of Soviet Socialist Republics, AN: 1994-01-M0121.

□□□

14

Tempeh : A Nutritional and Functional Food

S.N. BHOWMIK

Tempeh, or *tempeh kedelee,* is a fermented product usually made from soybeans that have been soaked and cooked to soften them[1]. Like sour dough bread, tempeh requires a starter substance, which is added to the cooked beans. This mixture is left, for 24 hours and the result is a firm textured product with a somewhat nutty flavour and a texture similar to a chewy mushroom. Because tempeh is firm and it can be formed into a patty, it is used as a substitute for animal products in what in the West is typically called as "mock" burgers or sandwiches[2]. Unlike many indigenous fermented products such as *miso,* soy sauce, *natto, hamanatto, ang-kak,* and *sufu,* which are mainly used as flavouring agents, colouring agents, side dishes, breakfast items, and condiments, tempeh is used as a main dish in Indonesia. It has the necessary characteristics of a dietary staple that is high in protein and fibre and is rich in other nutrients. It also has the advantage of containing Vitamin B_{12}, which is a by-product of the fermentation process[3]. The purpose of this chapter is to present the properties, microbiological aspects, nutritional qualities, and functionality of legume tempeh in general and manufacturing practices of rice-soy tempeh in particular[4].

Role of microorganisms in tempeh production

Tempeh is the result of mixed culture fermentation by a diverse group of microorganisms including moulds, yeasts, lactic acid bacteria and different gram-negative bacteria[5]. *Rhizopus oligosporus* is the dominant tempeh fungus[6] although some other fungi, such as *R. oryzae* and *Mucor* spp, may also contribute to the flavour, texture or nutritive value[7]. Heseltine *et al.*[8] isolated many fungi from different lots of tempeh made in Indonesia and found that only *Rhizophus* could make tempeh in pure culture fermentation. They also found that the 40 strains of *Rhizophus* studied, 25 of them are *R. oligosporus* others are *R. stolonifier, R.arrhizus, R. oryzae,* and *R. formosaensis.*

Mulyowidarso *et al.*[9] also reported variable growth of bacteria during fermentation of soybeans into tempeh with *R.oligosporus*. Liem *et al.*[10] first reported that *Klebsiella pneumonia* was responsible for the production of vitamin B_{12} in commercial tempeh. Yeasts are frequently detected in tempeh, but their role is still unknown[11]. *Rhizopus* moulds produce natural, heat-stable antibiotic agents against some disease-causing organisms. Indonesians who eat tempeh as a regular part of their diet recognize it as medicine for their dysentery and rarely fall victim to the intestinal diseases to which they are constantly exposed[12].

Manufacturing process of legume tempeh

The process involved in tempeh production differs from each other and everyone reported their own process based on various factors. The overall production method is followed by flowchart representation as depicted (Fig. 1)[13, 14].

General tempeh preparation

Although the principal step in making tempeh is the fermentation of soybeans by *Rhizopus* mould, many factors affect the success in production and the quality of the product. In the method employed by Steinkraus *et al.*[15] raw, dried soybeans are hydrated by soaking in cold water overnight or for a shorter period of time if warm water is used. Lactic acid or acetic acid is added to the water to lower the pH of the solution and soybeans to pH 5 or below. This step is performed to discourage growth of undesired microorganisms. The *Rhizopus* mould is largely unaffected by the acidic environment. It was found that mould growth rate remained stable as long as the pH is at or above 3.5 and was slightly slower when the beans were more acidic. After hydration, the skins on each bean are removed to encourage better growth of the mould. This step can be done by rubbing the beans together with hands or by other mechanical means. The skinned beans are then partially cooked by boiling them in the acidic soaking water at 100°C for

90 minutes. When the beans have cooled to 37-38°C, the ideal temperature to begin the inoculation and the beans have been dried, the *Rhizopus* starter culture is mixed to the beans. Sufficient amounts of the culture, or inoculums, needs to be present. In the 1960 study by Steinkraus *et al.*[15], the inoculums were added at 1 gram per kg of cooked soybeans. The inoculated beans are then placed on covered trays, in a layer no thicker than 2 inches, so that sufficient oxygen is available on the bean surfaces for mould growth. Small holes in the trays are also beneficial in ensuring sufficient oxygen supply.

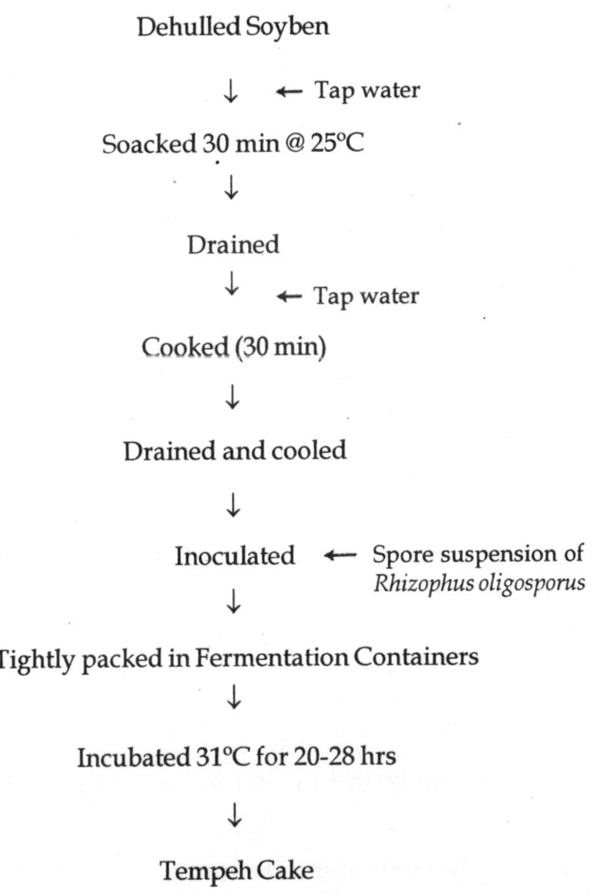

Fig. 1 : Production process of soybean tempeh

In addition to oxygen, the mould grows better in high humidity levels (75-78%). However, no liquid water should be in contact with the beans[16]. The ideal temperature for *Rhizopus* growth is 37°C, so the trays of beans are best kept at this temperature. During the phase of active mould growth, however, the temperature of the beans will naturally rise above the ambient temperature. Temperatures of up to 49°C were measured, where further mould growth was inhibited by this high temperature. It was found that keeping the temperature below 45°C and above 25°C was sufficient for satisfactory mould growth, although at 25°C, fermentation takes 5 days, which is 5 times as long as the time it takes to ferment at 37°C[14].

Stages of fermentation

Steinkraus *et al.*[5] reported that the initial few hours after inoculation, germination of spores takes place. Slow mould growth takes place for several hours after the germination phase. Later, rapid mould growth and temperature increase continues for 4 to 5 hours. As mould growth declines, the beans should be bound into a solid mass by the mycelium. After mould growth declines, sporulation and ammonia production due to protein breakdown appears. Beuchat[17] reported that the pH level on the beans also changes as mould growth occurs. From an initial pH of 5.0, the pH level reaches 6.0 to 6.7 during the peak of mould growth and continues to increase to 7.6 as time passes. It was found that the tempeh is most palatable when the pH is between 6.3 and 6.5. In short, two characteristic events from mould growth need to take place for the fermented soybeans to become tempeh. Firstly, the individual beans will be bound into a solid cake by the mould mycelium. Secondly, the soybeans become partially digested by the mould enzymes.

Homemade rice-soy tempeh production

Homemade rice-soy tempeh as developed by Bhowmik *et al.*[4] is prepared by:

1. Weighing whole rice (preferably parmal variety) and dehulled soybean splits at ratio 40:60 (w/w).
2. Separately washing the rice in clean water and soaking the soybean for 1 hr and giving subsequent wash.
3. Placing the rice and soybean in water just brought to boil for 10 and 30 min respectively.
4. Mixing of the blanched ingredients thoroughly, and draining and cooling (20 to 30 min).
5. Inoculating with tempeh starter @ 5 gm culture/kg (*R. oryzae* ITCC 7382-09 spores).
6. Packing in cylindrical stainless steel container (5 x 9.5 cm) with 50-60 perforations @ 150 g (wb).
7. Incubating at 31°C for 30-35 h.

The result is fresh tempeh cakes (Figure 2).

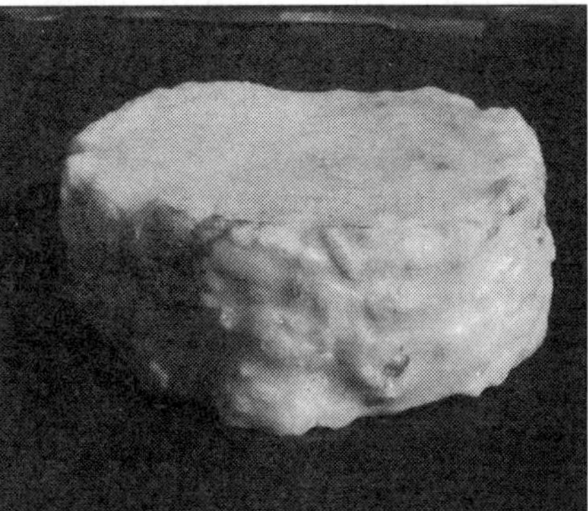

Fig. 2 : Rice-soy 40:60 (w/w) tempeh, a) production in stainless steel container (5 x 9.5 cm) with 50-60 perforations, and b) freshly harvested cake

Tempeh Based Products

Traditionally, tempeh was made from soybean. Yellow-seeded soybeans are usually preferred as raw material[6], but many different substrates can be used to produce tempeh (Table 1). Some substrates for example cereal-soy

Table 1 : Use of alternative ingredients for development of tempeh based products

Raw materials	Organisms	References
Legumes		
Soybeans (*Glycine max*)	*R. oligosporus*	Nout *et al.*,[18] Varzakas[19]
Black gram (*Phaseolus mungo*)	*R. oligosporus*	Jha and Verma[20]
Cereals		
Barley (*Hordeum vulgare*)	*R. oligosporus*	Nout *et al.*, [21] Hachmeister and Fung,[22] Berg *et al*.[23]
Wheat (*Triticum vulgare*)	*R. oligosporus*	Hesseltine and Wang,[24] Nout *et al.*, [21] Hachmeister and Fung[22]
Sprouted broken rice	*R. oligosporus*	Dinesh[25]
Mixture of legumes with nonlegumes		
Sesame and soybean	*R. oligosporus*	Shurtleff and Aoyagi[12]
Maize and soybean	*R. oligosporus*	Nout and Kiers[14]
Press cake (by-products)		
Okara (soy pulp residue)	*R. oligosporus*	Ko Swan and Hesseltine,[26] Matsuo[27]
Peanut presscake	*R. oligosporus*	Gandjar[28]

can be processed to obtain tempeh of high protein quality upon reducing the aminoacid imbalance in cereal based food per se.

Preservation of tempeh

Freshly made, raw Tempeh remains edible for a few days at room temperature. The tempeh is neither acidic nor does it contain large amounts of alcohol. Tempeh does, however, possess stronger resistance to lipid oxidation than unfermented soybeans, due to its antioxidant contents[16]. Traditionally, fresh tempeh is preserved by sun-drying, but modern methods are faster and reduce the risk of contamination. In the study by Steinkraus et al.[15] the tempeh were preserved by quick freezing (-15°C) and lyophilisation, or by drying in a circulating hot air oven at 69°C. Deep fat frying is another possible preservation method for tempeh. In our study, several ways of drying and storing of rice-soy tempeh has been explored (Table 2). In general 5% final moisture in the dried rice-soy tempeh was found to be suitable for storage while freeze drying was more effective method of preservation. Sun drying to 10% moisture was not only time taking but also presents poor keeping quality

Table 2 : Effect of drying process and storage on the shelf life of tempeh

S. No.	Isolate	Strain no.	Rice-soy ratio (w:w)	Drying process	Moisture (%)	Store	Shelf-life (Months)			
							1	2	4	6
1.	*Rhizopus*	ITCC 7382-09	40:60	Sun drying	5	Room	M*	A**	-	-
2.	*Rhizopus*	ITCC 7382-09	40:60	Sun drying	10	Room	A	-	-	-
3.	*Rhizopus*	ITCC 7382-09	40:60	Sun drying	5	Fridge	M	M	A	-
4.	*Rhizopus*	ITCC 7382-09	40:60	Sun drying	10	Fridge	M	M	A	-
5.	*Rhizopus*	ITCC 7382-09	40:60	Recirculatory air drying	5	Room	M	M	A	-
6.	*Rhizopus*	ITCC 7382-09	40:60	Recirculatory air drying	10	Room	M	A	-	-
7.	*Rhizopus*	ITCC 7382-09	40:60	Recirculatory air drying	5	Fridge	M	M	A	-
8.	*Rhizopus*	ITCC 7382-09	40:60	Recirculatory air drying	10	Fridge	M	M	A	-
9.	*Rhizopus*	ITCC 7382-09	40:60	Freeze drying	5	Room	M	M	M	A
10.	*Rhizopus*	ITCC 7382-09	40:60	Freeze drying	10	Room	M	M	A	-
11.	*Rhizopus*	ITCC 7382-09	40:60	Freeze drying	5	Fridge	M	M	M	M
12.	*Rhizopus*	ITCC 7382-09	40:60	Freeze drying	10	Fridge	M	M	M	A

* Mushroom smell **Alcoholic smell

Source: Bhowmik *et al.*[4]

as compared to recirculatory air drying and freeze drying. The basic reason is that these modern techniques are quick and also reduces the risk of contamination by lactic acid bacteria, alcohol fermenting microorganism and other pathogens. Mushroom smell is usually a characteristic aroma of a fresh tempeh while alcoholic aroma is a feature of an aged or deteriorating tempeh usually implicated by contamination[15]. Freeze drying and subsequent storing in fridge can extend the shelf life to six months which was the best among all the treatments. Biotechnology improved the inoculum, which makes a tempeh variant possible that may be protected by an international patent[29]. Modern processing methods mainly entail improving equipment needed, maintaining a sanitary and hygienic environment, creating an inoculum by selected microorganisms and modernizing the methods of packaging[30].

Uses and preparation of tempeh

Tempeh is a very versatile product and can be used in combination with many different recipes and dishes. Tempeh can be served with grains and eggs for a breakfast item or in salads, sandwiches, burgers, sauces, or soups for a lunch or dinner. When using tempeh as an ingredient in recipes such as for salads, soups, sauces, or casseroles, frying is a recommended step to ensure a crisp texture[31]. Deep-fat frying and pan frying tempeh in vegetable oil or margarine yields a crisp, golden brown coloured product. Before frying, sliced tempeh can be dipped in soy or fish sauce, and a batter made from corn or rice flour and coconut milk[5]. According to Bates *et al.*[32] American vegetarians consume tempeh burgers of about 1.5 cm thickness. Furthermore, the fried tempeh can be served on a bun with tomato, onion slices, and lettuce.

Organoleptic properties of legume tempeh

Good quality tempeh should appear as a white, compact cake with a clean, fresh odour, yeasty odour and without a beany flavor[33]. Tempeh can be served in numerous ways such as fried in oil, baked, or used as a soup. Before eating tempeh, the bean cake is usually sliced and deep-fat fried until the surface is crisp and has a golden brown colour; the flavour becomes nut-like and peppery due in part to the presence of free fatty acids[33,34]. Tempeh scored best at the end of the first phase of fermentation (30 h at 32°C), kept its food quality during the second phase (one additional day at 32°C), and deteriorates rapidly during the third phase[35]. The reasons given for tempeh's improved organoleptic qualities over unfermented soybeans have centred primarily on changes to the food due to the manufacturing steps (including fermentation)[36]. Histological observations and the degree of penetration of hyphae of *R. oligosporus* into the soybean cotyledons are partially responsible for the rapid physical changes in soybeans during tempeh fermentation[15,36].

Nutritional quality of tempeh

Soy tempeh has a good high fibre, cholesterol free protein, which can be easily used as a meat alternative as depicted in Table 3.

Table 3 : Nutritional comparison chart (100 g serving)

Product	Protein*	Saturated Fat	Total Fat	Cholesterol	Fibre
Soy tempeh	13 g	0.6 g	3 g	0 mg	7 g
Ground beef	15 g	9.0 g	18 g	58 mg	0 g
Lean ground beef	18 g	4.5 g	9 g	56 mg	0 g
Chicken breast	14 g	1.0 g	4 g	45 mg	0 g
Chicken drumstick	10 g	2.0 g	6 g	45 mg	0 g

*protein comparison to egg albumin
Source: Nutrition Almanac MC Graw-Hill and Turtle Island Foods Reports[38]

Tempeh is:

- An excellent source of protein, contains all the essential amino acids; same quality protein as meat or poultry
- Excellent source of calcium
- Low in saturated fat
- High in essential fatty acids & B vitamins
- Cholesterol free
- High in soluble dietary fiber
- Fermentation neutralizes the phytic acid present in the soybeans; therefore tempeh does not restrict the body's absorption of minerals
- Easy to digest because the fermentation process breaks down the complex proteins found in soybeans, making tempeh more easily digested than non-fermented soy foods or whole soybeans
- High in isoflavones
- A good source of folic acid
- Low in sodium

Rice-soy tempeh has the necessary characteristics of a dietary staple in that it has high protein and fibre and is rich in calcium as compared to white rice (Fig. 3).

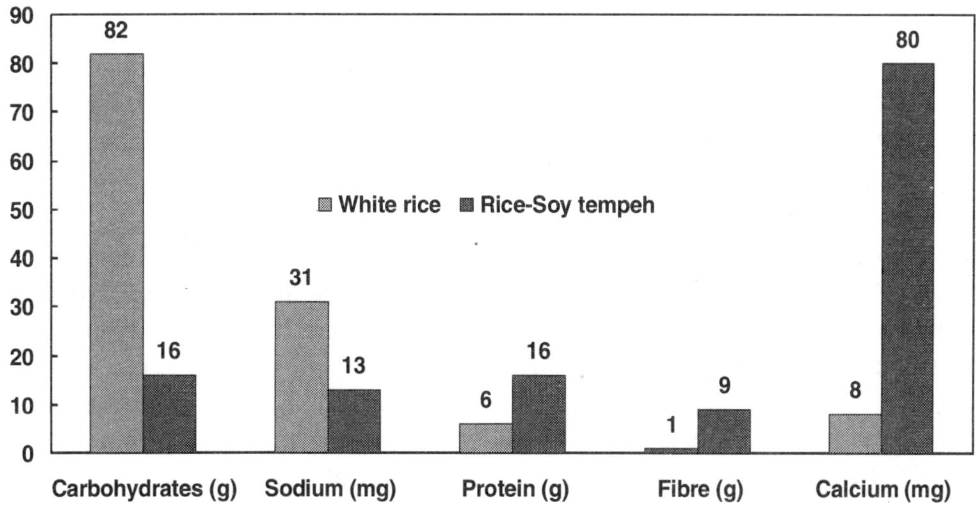

Fig. 3 : Comparison of nutritional profile of rice-soy tempeh with white rice (per 100 g)[4]

Functionality and safety of soybean tempeh

Digestion and absorption

During World War II, prisoners of war suffering from dysentery could not tolerate soybeans but were able to subsist on tempeh; an experience that underscores the easy digestibility of tempeh. This is related with enzymatic degradation of soybean polymeric substances resulting in soluble solids, such as soluble nitrogenous compounds. Macromolecules are degraded into oligomeric and smaller units improving the tempeh digestion[39]. Digestibility of cereals and legumes increases during cooking and fermentation[40]. Digestibility was influenced by fungal strain and fermentation time.

Diarrhoea

As early as the 1960s, tempeh was reported to contain an antibacterial substance, confirmed as demonstrated antimicrobial activity against selected species of Gram-positive bacteria[41, 42, 43]. Recent work showed that several tempeh extracts were able to inhibit adhesion of *E. coli* to piglet small intestinal brush border membranes. The protective effect of tempeh might be related to the production of a high molecular weight fraction (>5 kDa) by the specific mould. Severity of diarrhoea was significantly less on the diet with tempeh compared with the control diet containing toasted soybeans. An immune modulating effect of tempeh was suggested, but further evidence will have to be sought[44]. Tempeh can also be very useful as a nutritional supplement in oral rehydration therapy. Because of its protective effects during infection and its improved digestibility/nutrient availability, tempeh may be beneficial in case of (postweaning) diarrhoea and accelerating the recovery of young animals and young children, who are most 'at risk' for enterotoxic diarrhoea and malnutrition.

Intestinal growth and proliferation

Weaning is often associated with marked histological and biochemical changes of the small intestine (e.g. villous atrophy and crypt hyperplasia), causing decreased digestive and absorptive capacity and contributing to postweaning diarrhoea. Biopsies from the small intestinal mucosa showed improved repair after intestinal inflammation as a result of tempeh supplementation[44]. Tempeh, compared with cooked soybeans, contains high levels of easily accessible compounds such as peptides and free amino acids and possibly other factors which could affect intestinal growth and cell proliferation.

Anti-oxidative properties of fermented soybeans

Soybeans contain natural antioxidants. Tocopherols and phosphatides can be found in soybean oil, while the non-oil compound contains many isoflavones. Of the isoflavones, 99% occur as 7-0-monoglucosides. Of these, three isoflavones predominate: genistin, daidzin and glycitin[45]. Their aglycones are genistein, daidzein and glycitein respectively. Fermentation of soyabean foods causes increased antioxidative capacity. During fermentation at least a partial cleavage or change in the glucosides takes place associated with increased glucosidase and glucuronidase activities[46], releasing potent anti-oxidant substances by transformation of flavonoids[47]. Besides the above mentioned, factor 2 (6, 7, 4'-trihydroxyisoflavone) is also found in tempeh[48]. The level and type of isoflavones generated during fermentation depends on the composition of the inoculum; for instance, it was reported that not only *Rhizopus* sp. but also bacterial species, e.g. *M. lotus* and *Bacillus* epidermis determine the formation of specific isoflavones[49, 50]. Another anti-oxidative substance formed in tempeh was identified as HAA[51]. Of several soyabean foods[52] tempeh had somewhat lower isoflavone content than tofu but contained elevated levels of the aglycones formed by enzymatic hydrolysis during fermentation[53]. Fermentation of soya increased the bioavailability of isoflavones in humans.

Chronic degenerative diseases

In addition to the role of antioxidants in protecting the food against oxidative spoilage, antioxidants in soyabean (tempeh) are of medical interest because of their protective role against oxidative stress which is related to the pathogenesis of various chronic degenerative diseases. Soy (protein) has a potential role in the prevention and treatment of chronic diseases, most notably cancer[54] and heart disease[55], but also osteoporosis and menopausal symptoms[56]. Isoflavones in soybeans also have favourable effects on the postmenopausal reproductive system and bone health. Soyabean protein has been known for many years to have a hypocholesterolaemic effect. It is therefore not surprising that tempeh has also been found to lower blood cholesterol levels[57] and that it may thus be of benefit as a protective agent against cardiovascular disease. Copper induced human blood LDL oxidation in the presence of tempeh isoflavonoids showed that factor 2 was the most active inhibitor[50]. Tempeh, similar to other soyfoods and related products, has beneficial effects on metabolic outcome insofar as lipid metabolism is concerned, and probably inhibits atherogenesis as well[58, 59]. Hypocholesteraemic effects may be due to the protein, fibre, carotenoids, sitosterol, isoflavones, calcium, lecithin, niacin, and/or unsaturated fatty acids in tempeh[44]. It was demonstrated that tempeh, especially its glucolipids, inhibits the proliferation of tumour cells in mice[60]. Gamma-aminobutyric acid (GABA) has been overproduced in soyabean tempeh by means of a short anaerobic fermentation stage following the aerobic main fermentation. The GABA-enriched tempeh was reported to have an antihypertensive effect on rats[61].

Safety

Most traditional fermented foods are generally regarded as safe. Many traditional Asian fermented foods rely on growth of moulds of which strains are known that can produce potent mycotoxins. However, certain strains used in fermented foods production appear to have been domesticated over thousands of years. *Rhizopus oligosporus* is considered to be a potential human pathogen because of its ability to grow at 37°C. Inhalation of its sporangiospores might potentially cause pathogenic reactions, but this has not been studied in vivo; neither have pathological cases been reported. There have been no reports of poisonings, other than from a tempeh-like product made of coconut presscake (tempeh bongkrek). No reports of food poisoning or aflatoxin contamination in soyabean tempeh exist, although some species of *Rhizopus* were found to be toxic when tested in animals. Levels of biogenic amines were found to be low, whereas ethyl carbamate could not be detected[21]. *Rhizopus* spp. has even been shown to detoxify toxic substances in seeds[63]. Nevertheless, for modernized industrial-scale processing, toxicological screening of *Rhizopus* strains grown on various substrates is recommended[13]. The long use of tempeh at all stages of life, without recognized adverse effects, suggests that it poses no undue safety risk at the levels of intake known in Central Java.

During the past decade, significant scientific discoveries and technological innovations relating to the fermentation of soybean tempeh have been achieved. There is an increased interest in nutritional and health effects of food and feed. An important question is what fermentation has to offer in this field. A variety of indigenous fermented foods exist, many of which date back hundreds or even thousands of years. An important function of the micro-organisms involved in the fermentation process is the synthesis of enzymes that hydrolyse the food ingredients and contribute to the development of a desirable texture, flavour and aroma of the product. In the development of cereal-soy tempeh it is observed that the ratio of rice soy mix influences the textural and nutritional quality of tempeh. The ratio 40:60 (w/w) registered the best firmness with higher amount of protein and calcium. Fermentation also decreases antinutritional constituents, and the nutritional quality and digestibility of the fermented products is improved. Fermented foods could enhance the barrier function of the gastrointestinal tract, possibly resulting in reduced diarrhoeal disease risks. The potential impact of diarrhoeal episodes on nutritional status seems obvious, and the synergistic interactions of diarrhoea and malnutrition are well recognized. Fermented soybean foods being nutritious, easily digested and absorbed, culturally acceptable, palatable and possibly protective against diarrhoea might break the vicious cycle of diarrhoea and malnutrition. Furthermore, the present or potential role of fermented soybean foods in prevention and treatment of chronic diseases merits increasing attention.

References

1. Astuti, M, Meliala, A, Dalais, FS and Wahlqvist, ML (2000). Tempe, a nutritious and healthy food from Indonesia. Asia Pacific Journal of Clinical Nutrition, 9: 322-325.
2. Nout, MJR and Rombouts, FM (1990). Recent developments in Tempe research. Journal of Applied Bacteriology, 69: 609-633.
3. Irene, T, Liem, H, Steinkraus, KH and Cronk, TC (1977). Production of Vitamin B-12 in Tempeh, a Fermented Soybean Food. Applied and Environmental Microbiology, pp: 773-776.
4. Bhowmik, SN, Yadav, DN and Balasubramanian, S (2010). Development of a technology for home production of cereal-soy tempeh. RPF III, 1(147), CIPHET, Ludhiana.
5. Steinkraus, KH, Cullen, RE, Pederson, CS, Nellis, LF and Gavitt, BK (1983). Indonesian tempeh and related fermentations. In Handbook of indigenous fermented foods. Edited by.KH. Steinkraus, R.E. Cullen, C.S. Pederson, L.F. Nellis & B.K. Gavitt. New York, Marcel Dekker. pp: 1-94.
6. Sharma, R and Sarbhoy, AK (1984). Tempeh-a fermented food from soybean. Current Science, 53: 325-326.

7. Wiesel, I, Rehm and Bisping, B (1997). Improvement of tempe fermentations by application of mixed cultures consisting of Rhizopus sp. And bacterial strains. Applied Microbiology and Biotechnology, 47: 218-225.

8. Hesseltine, CW, Smith, M, Bradle, B and Djien, KS (1963). Investigations of tempeh, an Indonesian food. Developments in Industrial Microbiology, 4: 275-287.

9. Mulyowidarso, RK, Fleet, GH and Buckle, KA (1990). Association of bacteria with the fungal fermentation of soy bean Tempe. J. Appl. Bacteriol., 68: 43-47.

10. Liem, ITH, Steinkraus, KH and Cronk, TC (1977). Production of vitamin B-12 in tempeh fermented soybean food. Applied and Environmental Microbiology, 34: 773-776.

11. Samson, RA, Kooij, JAV and Boer, ED (1987). Microbiological quality of commercial tempeh in the Netherlands. Journal of Food Protection, 50: 92-94.

12. Shurtleff, W and Aoyagi, A (2001). The Book of Tempeh, 2nd edition. Pgs 48-50: 103-113.

13. Hachmeister, KA and Fung, DYC (1993). Tempeh: a mold-modified indigenous fermented food made from soybeans and/or cereal-grains. Critical Reviews in Microbiology, 19: 137-188.

14. Nout, MJR and Kiers, JL (2005). Tempe fermentation, innovation and functionality: update into the third millennium. Journal of Applied Microbiology, 98: 789-805.

15. Steinkraus, KH, Yap, BH, Van Buren, JP, Provvidenti, MI and Hand, DB (1960). Studies on tempeh an Indonesian fermented soybean food. Food Res., 25: 777-788.

16. Farnsworth, ER (2008). Handbook of Fermented Functional Foods. CRC Press.

17. Beuchat, LR (2001). Traditional fermented foods. In Food Microbiology. Edited by L.R.B. Michael P. Doyle, Thomas J. Montville. ASM press, American Society for Microbiology. Washington, DC. pp: 701-719.

18. Nout, MJR, Dreu, MAD, Zuurbier, AM, and Bonants van Laarhoven, TMG (1987). Ecology of controlled soybean acidification for Tempe of manufactures. Food Microbiology, 4: 165-172.

19. Varzakas, T (1998). Rhizopus oligosporus mycelial penetration and enzyme diffusion in soya bean Tempe. Process Biochemistry, 33: 741-747.

20. Jha, K and Verma, J (1980). Removal of flatulence principles from legumes by mold fermentation. Indian Journal of Experimental Biology, 18: 658-9.

21. Nout, MJR, Ruikes, MMW, Bouwmeester, HM and Beljaars, PR (1993). Effect of processing conditions on the formation of biogenic amines and ethyl carbamate in soybean tempe. Journal of Food Safety 13: 293–303.

22. Hachmeister, KA and Fung, DYC (1993). Tempeh – a mold modified indigenous fermented food made from soybeans and/or cereal grains. Critical Reviews in Microbiology 19: 137–188.

23. Berg, S, Olsson, J, Swan Berg, Schnauzer, MJ and Eriksson, A (2001). Method for the production of fermented cereal food products and products thereof. In World Intellectual Property Organization. Sweden: Olligon AB.

24. Hesseltine, CW and Wang, HL (1980). The importance of traditional fermented foods. Bioscience, 30: 402-404.

25. Dinesh Babu, P (2009). Designing a low cost nutritious antioxidant food using Rhizophus oligosporous. Thesis submitted for the award of Bachelor of Technology in Biotechnology to Bharathidasan University.

26. Ko Swan, D and Hesseltine, CW (1979). Tempe and related foods. Economic Microbiology, 4: 115-140.

27. Matsuo, M (1990). Suitability of ,okara tempe' as foodstuff. Nippon Nogeikagaku Kaishi, 64: 1235-1236.

28. Gandjar, I (1981). Soybean fermentation in Indonesia. In The 6th international fermentation Symposium, Edited by M.A.R. Young, C. W. London, UK: Pergamon Press. pp: 531-534.

29. Adnan, M and Sudarmadji, (1997). Contribution of tempeh for the economy and health of Indonesian. Proceedings International Tempe Symposium, July 13-15, Bali, Indonesia.

30. Kuswanto, KR (2004). Industrialization of Tempe Fermentation, in Steinkraus, K.H. (Ed). Industrialization of Indigenous Fermented Foods; 2nd ed. pp: 587-635, Marcel Dekker, Inc., New York.

31. Shurtleff, W and Aoyagi, A (1979). The Book of Tempeh, Harper and Row, New York.

32. Bates, C, Lyon, A, Sorenson, S, Keller, B, and Jenkins, S (1977). Utilization of tempeh in North America, Symposium on Indigenous Fermented Foods, Bangkok, Thailand,.

33. Winarno, FG and Reddy, NR (1986). Tempe, Legume-Based Fermented Foods, Reddy, N.R., Pierson, M.D., and D. K. Salunkhe, D.K., Eds., CRC Press, Boca Raton. FL, pp. 95–117.

34. Wang, HL and Hesseltine, CW (1965). Studies on the extracellular proteolytic enzymes of Rhizopus, Can. J. Microbiol., 11: 727–732.

35. Sudarmadji, S and Markakis, P (1978). Lipid and other changes occurring during the fermentation and frying of tempeh, Food Chem., 3: 165–170.

36. Hesseltine, CW (1976). A millennium of fungi, food and fermentation, Mycologia, 57: 149–197.

37. Jurus, AM and Sundberg, WJ, Penetration of Rhizopus oligosporus into soybeans in tempeh, Appl. Environ. Microbiol., 2: 284–287.

38. http://www.tofurkey.com/discover/what_is_tempeh.htm. Turtle Island Foods, Inc.

39. Matsuo, M (1996). Digestibility of Okara-tempe protein in rats. Journal of the Japanese Society for Food Science and Technology - Nippon Shokuhin Kagaku Kogaku Kaishi 43: 1059–1062.

40. Kiers, JL, Nout, MJR and Rombouts, FM (2000). In vitro digestibility of processed and fermented soya bean, cowpea and maize. Journal of the Science of Food and Agriculture 80: 1325–1331.

41. Hermana, H, Mahmud, MKMS, and Karyadi, D (ed.) (1990). Second Asian Symposium on Non-salted Soybean Fermentation. Bogor, Indonesia: Nutrition Research and Development Centre.

42. Kobayasi, SY, Okazaki, N and Koseki, T (1992). Purification and characterization of an antibiotic substance produced from Rhizopus oligosporus IFO 8631. Bioscience Biotechnology and Biochemistry 56: 94–98.

43. Kiers, JL, Nout, MJR, Rombouts, FM, Nabuurs, MJA and Van der Meulen, J (2002). Inhibition of adhesion of enterotoxic Escherichia coli K88 by soya bean tempe. Letters of Applied Microbiology 35: 311–315.

44. Agranoff, J (ed.) (1999). The Complete Handbook of Tempe. Singapore: American Soybean Association. Anderson, R.L. and Wolf, W.J. (1995). Compositional changes in trypsin inhibitors, phytic acid, saponins and isoflavones related to soybean processing. Journal of Nutrition 125: S581–S588.

45. Berghofer, E, Grzeskowiak, B, Mundigler, N, Sentall, WB and Walcak, J (1998). Antioxidative properties of faba bean-, soybean and oat tempeh. International Journal of Food Sciences and Nutrition 49: 45–54.

46. McCue, P and Shetty, K (2003). Role of carbohydrate-cleaving enzymes in phenolic antioxidant mobilization from whole soybean fermented with Rhizopus oligosporus. Food biotechnology 17: 27–37.

47. Stadler, P, and Kreysa, G (ed.) (1997). 15th Dechema Biotechnology Conference. Frankfurt am Main, Germany: DECHEMA.

48. Hoppe, MB, Jha, HC and Egge, H (1997). Structure of an antioxidant from fermented soybeans (tempeh). Journal of the American Oil Chemists Society 74: 477–479.

49. Klus, K and Barz, W (1995). Formation of polyhydroxylated isoflavones from the soybean seed isoflavones daidzein and glycitein by bacteria isolated from tempe. Archives of Microbiology 164: 428–434.

50. Sudarmadji, S, Suparmo, S, and Raharjo, S (ed.) (1997). Proceedings, International Tempe Symposium. Den Pasar, Bali, Indonesia: Indonesian Tempe Foundation, Jakarta, Indonesia.

51. Esaki, H, Onozaki, H, Kawakishi, S and Osawa, T (1996). New antioxidant isolated from tempeh. Journal of Agricultural and Food Chemistry 44: 696–700.

52. Anderson, RL and Wolf, WJ (1995). Compositional changes in trypsin inhibitors, phytic acid, saponins and isoflavones related to soybean processing. Journal of Nutrition 125: S581–S588.

53. Wang, HJ and Murphy, PA (1996). Mass balance study of isoflavones during soybean processing. Journal of Agricultural and Food Chemistry 44: 2377–2383.

54. Kennedy, AR (1995). The evidence for soybean products as cancer preventive agents. Journal of Nutrition 125: 733S–743S.

55. Messina, M (1995). Modern applications for an ancient bean: soybeans and the prevention and treatment of chronic disease. Journal of Nutrition 125: 567S–569S.

56. Liu, K (2000). Expanding soybean food utilization. Food Technology 54: 46–58.

57. Guermani, L, Villaume, C, Bau, HM, Nicolas, JP and Mejean, L (1993). Modification of soyprotein hypocholesterolemic effect after fermentation by Rhizopus oligosporus spT3. Sciences des Aliments 13: 317–324.

58. Karyadi, D and Lukito, W (1996). Beneficial effects of tempeh in disease prevention and treatment. Nutrition Reviews 54: S94–S98.

59. Karyadi, D and Lukito, W (2000). Functional food and contemporary nutrition-health paradigm: tempeh and its potential beneficial effects in disease prevention and treatment. Nutrition 16: 697.

60. Kiriakidis, S, Stathi, S, Jha, HC, Hartmann, R and Egge, H (1997). Fatty acid esters of sitosterol 3 beta glucoside from soybeans and tempe (fermented soybeans) as antiproliferative substances. Journal of Clinical Biochemistry and Nutrition 22: 139–147.

61. Aoki, H, Furuya, Y, Endo, Y and Fujimoto, K (2003). Effect of gamma-aminobutyric acid-enriched tempeh-like fermented soybean (GABA-tempeh) on the blood pressure of spontaneously hypertensive rats. Bioscience Biotechnology and Biochemistry 67: 1806–1808.

62. Kuo, YH, Bau, HM, Quemener, B, Khan, JK and Lambein, F (1995). Solid-state fermentation of Lathyrus sativus seeds using Aspergillus oryzae and Rhizopus oligosporus sp T-3 to eliminate the neurotoxin beta-ODAP without loss of nutritional value. Journal of the Science of Food and Agriculture 69: 81–89.

□□□

15

Probiotics : An Approach towards Health Foods

MONIKA SHARMA AND MRIDULA D.

Nutritional science has been expanding the knowledge on how foods influence consumers in relation to specific role of functional foods. Functional foods are defined as 'foods that contain some health-promoting component(s) beyond traditional nutrients'. Within the functional foods, probiotics are the small but rapidly expanding area. At the beginning of this century, Nobel Laureate Elie Metchnikoff, at the Pasteur Institute, linked health and longevity to ingestion of bacteria present in yoghurt (monograph reprinted in: Metchnikoff, 2004). Due to their perceived health benefits, probiotic bacteria have been increasingly included in yoghurts and fermented milks during the past two decades. Most commonly they have been lactobacilli such as *Lactobacillus acidophilus,* and *Bifidobacteria* often referred to as 'bifidus' (Daly and Davis, 1998).

A major development in functional foods pertains to foods containing probiotics and prebiotics which enhance health promoting microbial flora in the intestine. There is growing scientific evidence to support the concept that the maintenance of healthy gut microflora may provide protection against gastrointestinal disorders including gastrointestinal infections, inflammatory bowel diseases, and even cancer (Haenel and Bendig, 1975).

The word "Probiotic" was initially used as an antonym of the word "antibiotic". It is derived from Greek words pro and biotos and translated as "for life" (Hamilton-Miller *et al.*, 2003). It was probably Vergio (1954) who first introduced the term "probiotic", when he compared in his manuscript "Anti-und Probiotika" the detrimental effects of antibiotics and other antimicrobial substances on the gut microbial population, with factors ('Probiotika') favourable to the gut microflora.

In October, 2001, an Expert Consultation meeting convened by the Food & Agriculture Organization (FAO) and World Health Organization (WHO) defined probiotics as "live microorganisms which when administered in adequate amounts confer a health benefit on the host." When combined with the specifications outlined by the FAO/WHO Working Group for the Evaluation of probiotics in Food (2002), this definition has the following characteristics:

- Probiotics must deliver a measured physiological benefit, substantiated by studies conducted in the target host,
- Probiotics need not be restricted to food applications or oral delivery,
- Definition of probiotics should not limit the mechanism of action,
- Probiotics must be alive,
- Probiotic must have undergone controlled evaluation to document health benefits in the target host,
- Probiotic must be a taxonomically defined microbe or combination of microbes (genus, species and strain level) and,
- Probiotic must be safe for its intended use.

Although it was recognized that dead cells may have health benefits, the consultants suggested that a different term be used for dead cells to reflect to the consumer that probiotics are living microbes (Tannis, 2008).

History of probiotics

The work of one Nobel prize-winning scientist, Elie Metchnikoff, is often considered for the birth of probiotics (Mazza, 2008). In Metchnikoff's book, *The Prolongation of Life,* he postulated that the lactic acid bacteria (LAB) content of fermented milk offers health benefits. Metchnikoff further suggested that ingestion of LAB can lead to increased longevity of the host, as he observed that Bulgarian peasants who consumed large amounts of sour milk lived longer than the average human (Metchnikoff, 1907). He also bolstered his data with claims of his own 'feeling of general health and well-being' after ingesting the LAB. The bacteria in the sour milk were what would be later named *Lactobacillus bulgaricus*. Metchnikoff's subsequent work on cellular immunity won him the Nobel prize in 1908. Meanwhile, around 1900, Dr. Henry Tissier discovered *Bifidobacteria*. Tissier recommended the administration of *Bifidobacteria* to infants suffering from diarrhoea, claiming that *Bifidobacteria* gets rid of the pathogenic bacteria that cause the disease. He also showed that *Bifidobacteria* were predominant in the gut flora of breast-fed infants. The focus then shifted to other types of LAB that could survive in the human gut but also retained the beneficial features for health. The next strains examined fitting these criteria included *Lactobacillus acidophilus*, which was discovered by Moro in 1900 and promoted by Leo Rettger, a bacteriology professor at Yale University in the 1930s (Smolyansky, 2010). Rettger and his colleague Harry Cheplin claimed that the intestinal flora of the human gut is dependent almost entirely by the nature of the individual's diet, and that *L. acidophilus* may help in curing certain GI ailments (Cheplin, *et al.*, 1923 and, Rettger, 1935).

Criterion for consideration as probiotic

Probiotics must have good technological properties so that it can be manufactured and incorporated into food products without losing viability and functionality or creating unpleasant flavours or textures; it must survive passage through the upper gastrointestinal (GI) tract and arrive alive at its site of action; and it must be able to function in the gut environment. Following are some of the important criteria, need to be fulfilled for the organisms to be considered as probiotic:

- Be non pathogenic,
- Be able to survive transit through the gastrointestinal tract,
- Adherence and colonization of intestinal epithelium / tissue,
- Have a demonstrated beneficial effect on the host,
- Ability to metabolize prebiotics,
- Be isolated from the same species as its intended host,
- Hydrophobicity,
- Production of antimicrobial substances and antagonism to pathogens,
- Demonstrable efficacy and safety and,
- Be viable over storage periods.

Types of probiotic microbes

A number of genera of bacteria (and yeast) are used as probiotics, including *Lactobacillus, Leuconostoc, Pediococcus, Bifidobacterium*, and *Enterococcus*, but the main species believed to have probiotic characteristics are *L. acidophilus, Bifidobacterium* spp., and *L. casei*.

LAB are usually described as Gram-positive microorganisms, devoid of cytochromes and preferring anaerobic conditions but are aerotolerant, fastidious, acid-tolerant, and strictly fermentative, producing lactic acid as a main product (Stiles & Holzapfel, 1997). Members of the LAB are usually subdivided into two distinct groups based on their carbohydrate metabolism. The homofermentative group consisting of Lactococcus, Pediococcus, Enterococcus, Streptococcus and some lactobacilli utilize the Embden–Meyerhof–Parnas (glycolytic) pathway to transform a carbon source chiefly into lactic acid. As opposed to homofermentors, heterofermentative bacteria produce equimolar amounts of lactate, CO_2, ethanol or acetate from glucose exploiting phosphoketolase pathway. Members of this group include Leuconostoc, Weissella and some lactobacilli.

Bifidobacterium are normal inhabitants of the human gastrointestinal tract. Selected strains survive stomach and intestinal transit and reach the colon in abundant numbers. Newborns are colonized with Bifidobacterium within days after birth and the population appears to be relatively stable until advanced age, when a decline in their numbers occurs. However, diet, antibiotics, and stress are reported to influence the population of Bifidobacterium in the intestines (Shah, 2007).

Presently, there are 29 species in the genus Bifidobacterium, 14 of which are isolated from human sources (i.e., dental caries, faeces and vagina), 12 from animal intestinal tracts or rumen, and 3 from honeybees (Shah & Lankaputhra, 2002). The optimum pH for the growth of Bifidobacterium is 6.0–7.0, with virtually no growth at pH 4.5–5.0 and below or at pH 8.0–8.5 and above. Optimum growth occurs at a temperature of 37–41°C, the minimum and maximum growth temperatures are 25–28 and 43–45°C, respectively (Shah, 2007).

Nonpathogenic microorganisms that occupy important niches in the host gut or tissues, such as yeasts, enterococci and Enterobacteriaceae, are used as human and animal probiotics. *Saccharomyces boulardii* is the only probiotic fungus which has been successfully used for curing the intestinal infections, especially diarrhoea (Mombelli and Gismondo, 2000).

Spore probiotics: Spores being heat-stable have a number of advantages over other non-spore formers such as *Lactobacillus* spp., namely, that the product can be stored at room temperature in a desiccated form without any deleterious effect on viability. A second advantage is that the spore is capable of surviving the low pH of the gastric barrier (Barbosa *et al.*, 2005) which is not the case for all species of Lactobacillus (Tuohy *et al.*, 2007) so in principle a specified dose of spores can be stored indefinitely without refrigeration and the entire dose of ingested bacteria will reach the small intestine intact.

Bacillus species have been used as probiotics for at least 50 years with the Italian product known as Enterogermina® registered in 1958 in Italy as an OTC medicinal supplement. Of the species, *Bacillus subtilis,* *Bacillus clausii, Bacillus cereus, Bacillus coagulans* and *Bacillus licheniformis* have been most extensively examined. Spore probiotics are being used extensively in humans as dietary supplements, in animals as growth promoters and competitive exclusion agents and lastly in aquaculture for enhancing the growth and disease-resistance of cultured shrimps. The single and most important advantage of these products is that they can be produced easily and the stability of the finished product can be assured, further they can be incorporated into everyday foods (Cutting, 2010).

Health benefits of probiotics

Health benefits imparted by probiotic bacteria are very strain specific; therefore, there is no universal strain that would provide all proposed benefits, not even strains of the same species. Moreover, not all the strains of the same species are effective against defined health conditions. The strains L. *Rhamnosus* GG (Valio), *Saccharomyces cerevisiae Boulardii* (Biocodex), *L. Casei Shirota* (Yakult), and *B. Animalis* Bb-12 (Chr.Hansen) are certainly the most investigated probiotic cultures with the established human health efficacy data against management of lactose malabsorption, rotaviral diarrhoea, antibiotic associated diarrhoea, and *Clostridium difficile* diarrhoea.

According to reviews by various authors, probiotics can alleviate lactose intolerance, lower serum cholesterol, reduce diarrhoeal incidence, stimulate the immune system, control infections, act as antibiotics, suppress tumors and protect against colon/bladder cancer by maintaining a healthy intestinal microflora balance. Justification for health claims comes mainly from animal studies.

Enhance immune response

Modulation of host immunity is one of the most commonly purported benefits of the consumption of probiotics. Presumably, to modulate immunity, probiotic organisms must "talk" to immune cells that are endowed with recognition receptors or that are otherwise sensitive to probiotic-derived products (e.g., metabolites, cell wall components, DNA) (Fig. 1). Given the diversity of inflammatory or immune responses that can be mounted by the intestinal epithelium, association of probiotics with epithelial cells might be sufficient to trigger signaling cascades that ultimately activate underlying immune cells in the lamina propria (Menard, *et al.*, 2004). Alternatively, probiotics may also release soluble factors that themselves trigger signaling cascades at the level of the epithelium or associated immune system (Fig. 1). Research on immune system stimulation by probiotics has focused on:

i. Response of mammalian cell cultures to exposure to LAB or their cellular products;

ii. Response of mice to intraperitoneal injection of LAB or their cellular products;

iii. Response of mice to oral administration of LAB or fermented milks;

iv. Human feeding studies.

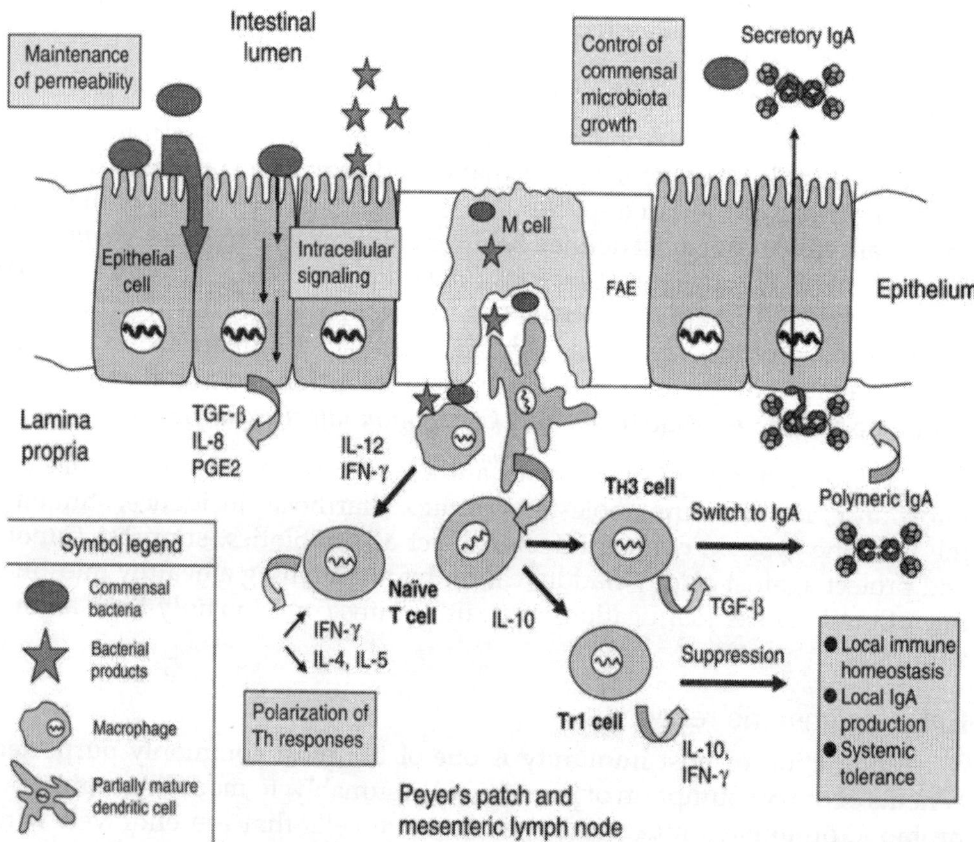

Fig. 1 : Schematic representation of the multiple consequences of the cross-talk between the probiotic bacteria and the intestinal mucosa. (*Source:* Corthesy *et al.*, 2007).

Studies involving (i) or (ii) have shown effects, but these do not involve oral consumption and their significance is unclear. Feeding lactobacilli or yogurt to mice stimulated macrophages and increased secretory IgA concentrations (Perdigon *et al.*, 1986, Scheinbach, 1998). Corthesy *et al.* (2007) tried to explain with the help of a simplified scheme obtained from various in vivo and in vitro experimental models; the specific effects of a particular probiotic on the development of local and systemic responses. At the intestinal epithelial level, probiotic bacteria may allow beneficial effects through transient colonization and/or release of bioactive compounds.

Probiotics in diarrhoea

Despite the multiple advances in the management of acute diarrhoea, it continues to cause some three million deaths in children. Furthermore, it is one of the main causes of deterioration of their nutritional status. Probiotics may have preventive or therapeutic effects on diarrhoea of various etiologies. However, not all probiotics are effective and physicians must select preparations with proven efficacy.

a) Antibiotic-associated diarrhoea

Antibiotic-associated diarrhoea is a common problem occurring in 5% to 25% patients receiving antibiotics, with rates varying depending upon the population studied and the antibiotic used (Bartlett, 2002). Diarrhoea related to antibiotic use can begin after a single dose of antibiotics or persist as long as 6 weeks after antibiotics have been stopped. The most commonly diagnosed and potentially severe form of antibiotic-associated diarrhoea is caused by *Clostridium difficile* accounting for 15% to 25% of antibiotic associated diarrhoea (Kelly *et al.*, 1994). Antibiotics alter the normal faecal flora and this disturbance of gastrointestinal microecology leads to a loss of colonization resistance predisposing the gut to colonization by potentially pathogenic microbes, such as C. *difficile* (Katz, 2006). The possible advantages of probiotics as biotherapeutic agents are that they can lower the dependence on antibiotic use, they are relatively inexpensive, they can have multiple mechanisms of action, and they are generally safe and well tolerated (Elmer and, McFarland, 2001).

b) Traveller's diarrhoea

Travel is a risk factor for infectious gastroenteritis. A recent meta-analysis revealed evidence of a protective effect by *S. boulardi* and by mixture of *Lactobacillus acidophilus* and *Bifidobacterium bifidum* (McFarland, 2007). However, evidence of the efficacy of probiotics in the prevention of traveller's diarrhoea is preliminary.

c) Probiotics for treating persistent diarrhoea

The World Health Organization (WHO) defines persistent diarrhoea as an illness of proven or presumed infectious aetiology that lasts 14 days or more (Anonymous 1988). Persistent diarrhoea accounts for 3% to 20% of all diarrhoeal episodes in children aged less than five years (IWGPD 1996).

Bernaola, *et al.* (2010), carried out a study to evaluate the efficacy and adverse effects of probiotics for the treatment of persistent diarrhoea in children. This study indicated that probiotics reduced the duration of persistent diarrhoea and stool frequency.

Use of probiotics in combating IBD

Inflammatory bowel disease (IBD) refers to a group of disorders of unknown aetiology that are characterized by chronic or recurrent mucosal inflammation. Ulcerative colitis (UC) and Crohn's disease (CD) are the two most frequent causes of this chronic inflammation of the gastrointestinal tract. Probiotic administration, either through regulation of the inflammatory response or modulation of gut microbiota composition and/or activity might bring about relief in IBD symptoms. The well-defined, nonpathogenic strain *E. coli* (Rembacken, *et al.*, 1999) has proven more effective in preventing relapse in Crohn's disease patients compared with a placebo. *S. boulardii* has shown some success in relieving the symptoms of active Crohn's disease and in reducing the risk of relapse (Guslandi, *et al.*, 2000).

Probiotics in relief of lactose intolerance

Most of the world's population (60-90% of non-Caucasians and 6-12% of Caucasians) become lactose intolerant after weaning (Bayless, *et al.*, 1971). This stems from a 90-95% decline in the production of lactase or *β-galactosidase* (Gilat, *et al.*, 1972).

Alleviation of lactose intolerance is probably the best established health claim for probiotics. Yogurt contains less lactose than milk and delays gastric emptying, which partly explains why lactose-intolerant individuals tolerate yogurt. However, yogurt tolerance is mainly due to the supply of lactase activity from the lactic acid bacteria present in the yogurt itself. Evidence shows that bacteria must be live and present in sufficient quantity to be of benefit; yogurts containing 10^8 bacteria/ ml are required (Pelletier *et al.*, 2001).

Probiotics in the management of allergies

A change in the proper functioning of the immune system can present itself as an allergy. Large-scale studies have indicated an alteration in the composition of the gut microflora, such as decrease in the numbers of lactobacilli, preceding the development of an allergy. Probiotics have been shown to reduce the incidence of childhood eczema by half, compared to placebo, when administered during pregnancy and up to 6 months postnatally. The incidence of asthma or rhinitis was not altered. A follow-up study demonstrated a two-fold increase in transforming growth factor β2, an anti-inflammatory cytokine, in the breast milk receiving probiotics compared to placebo (Kalliomaki, *et al.*, 2001).

Probiotics may exert a beneficial effect on allergic reaction by improving mucosal barrier function. In addition, probiotics consumption by young children may beneficially affect immune system development. Probiotics such as *Lactobacillus* GG may be helpful in alleviating some of the symptoms of food allergies such as those associated with milk protein (Majamaa and, Isolauri, 1997).

Food Allergies: It is estimated that food allergies affect 3.5% of adults and 8 to 10% of children; the food allergens most frequently involved are egg, peanuts, milk, fish, nuts, shellfish, wheat, kiwi and mustard. There are three ways in which probiotics are hypothesized to offer beneficial effects in the individuals with food allergies (Tannis, 2008):

i. Children with a food allergy have an increased intestinal permeability. Probiotics have been shown to reverse this, reducing the permeability of the intestines, preventing or treating food allergies.

ii. Research has proven that probiotics increase IgA in the intestines. IgA reduces the ability of microbes to adhere to the intestinal lining. This defense is frequently defective in children with food allergies.

iii. Changes in number and type of intestinal microbes are common in allergic individuals. Probiotics have the ability to promote a healthy, normal intestinal microfl ora, thus reducing the development of allergies.

Hypocholesterolemic effect of probiotics

Probiotic strains, especially lactic acid bacteria have a major role to play in the cholesterol lowering mechanism. One of the possible mechanisms for microbially-induced lowering of serum cholesterol have been suggested, that, since cholesterol is used in the production of bile acids, enhanced catabolism and excretion of bile acids might reduce serum cholesterol (Chikai *et al.*, 1987) by deconjugating the cholesterol to bile acids, thereby reducing the total body pool. Deconjugation by different lactic acid bacterial cultures was also tried using two forms of bile salts, viz. taurocholates and glycocholates. Most of the strains could deconjugate glycocholates.

The dietary cholesterol absorption is reduced by three ways - assimilating, binding or by degradation. Probiotic strains assimilate the cholesterol for their own metabolism. Probiotic strains can get bound to the cholesterol molecule, and they are capable of degrading cholesterol to its catabolic products.

According to a study, conducted by Gilliland *et al.* (1985) on pigs, fed with high cholesterol diet for ten days followed by feeding with probiotic strains of *L. acidophilus* P-47 and RP-32, found significant difference in treatments and also, *L. acidophilus* RP-32 performed better than *L. acidophilus* P-47.

Anticarcinogenic effect of probiotic microbes

Diet makes an important contribution to cancer, e.g., up to 75% of colorectal cancer cases are thought to be associated with diet, implying that risks of cancer are potentially reducible. Probiotics have been given credit for numerous health-promoting effects; one of which is their anticarcinogenic properties. Consumption of large quantities of dairy products such as yogurt and fermented milk containing *Lactobacillus* or *Bifidobacterium* may be related to a lower incidence of colon cancer (Shahani and, Ayebo, 1980). A number of mechanisms for the anti-tumor action of probiotics have been proposed as follows (Mclntosh, 1996):

i. Suppressing the carcinogen/procarcinogen by binding, blocking or removing it;

ii. Curbing the growth of bacteria with enzyme activities such as β-glucuronidase, azoreductase, nitroreductase, that directly or indirectly convert procarcinogens to carcinogens;

iii. Reducing the intestinal pH, thereby altering microflora activity and bile acid solubility;

iv. Altering colonic transit time to remove faecal mutagens more rapidly;

v. Stimulating the immune system.

vi. Production of short-chain fatty acids such as butyrate and propionate, is another mechanism by which probiotics may help in the treatment for colorectal cancer (Geier *et al.*, 2006).

vii. It has also been suggested that LAB or compounds produced by these microorganisms may directly interact with tumor cells in culture and inhibit their growth, supporting the idea that they can directly produce anti-tumorigenic or antimutagenic compounds (Reddy *et al.*, 1983).

Inhibition of helicobacter pylori and intestinal pathogens

The gastrointestinal environment contains a wide range of contents ranging from harmless beneficial dietary and microbial flora to harmful pathogenic bacteria. The mammalian organism fights against these pathogenic bacteria through various ways. A wide range of antibacterial compounds including organic acids (e.g., lactic acid and acetic acid), hydrogen peroxide, bacteriocins, various low-molecular mass peptides, and antifungal peptides/ proteins, fatty acids, phenyllactic acid, and OH-phenyllactic acid are being produced by probiotic cultures. Lactic and acetic acids are the main organic acids produced during the growth of probiotics and their pH lowering effect in the gastrointestinal tract has a bacteriocidal or bacteriostatic effect. Low-molecular mass compounds such as lactic acid have been reported to be inhibitory towards gram-negative pathogenic bacteria (Alakomi *et al.*, 2000). Fig. 2 explains the strategy through which probiotics can inhibit *Helicobacter pylori* based on the ability of probiotic bacteria (B) to bind pathogens (C) in intestinal epithelial tissue (A). Anti-pathogenic action of probiotics consists in production of lactic acid (D) which decreases the pH, interacts with the toxins produced by pathogens (E), with the production of hydrogen peroxide (F) and synthesis bacteriocine (G). Production of antimicrobial substances (bacteriocins), in situ in the intestine can be improved by increasing the ability of probiotic bacteria to adhere to the intestinal mucosa (Corcionivoschi *et al.*, 2010).

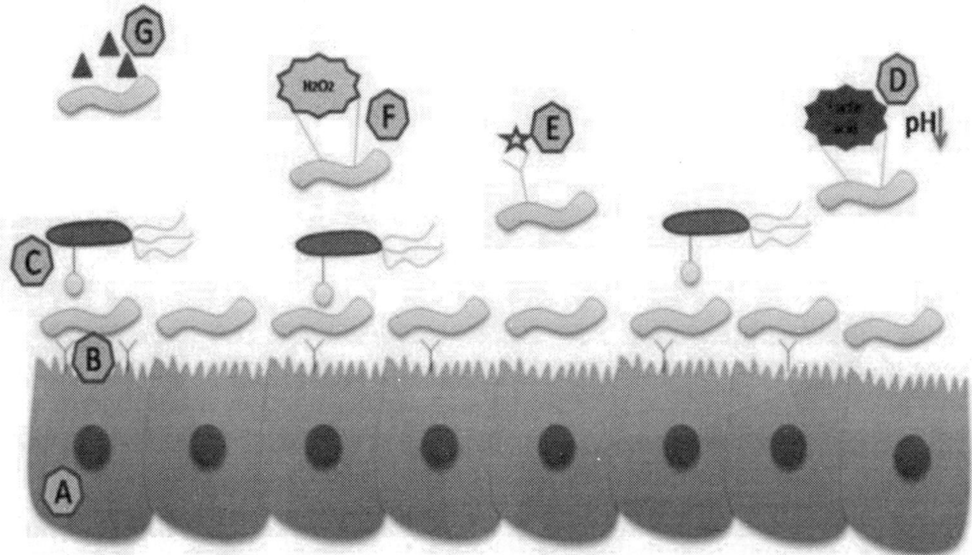

Fig. 2 : Schematic representation of the mode of action of probiotics in the intestines (Corcionivoschi *et al.*, 2009).

Helicobacter pylori is an intestinal pathogen, long-term infection by which leads to chronic gastritis, peptic ulcer and increases the risk of gastric malignancies. Currently *H. pylori* infection is treated by a combined therapy consisting of two antibiotics and a proton pump inhibitor, which, although in many cases appeared very effective, presents a very expensive treatment with many side effects including antibiotic-associated diarrhoea and likelihood of induction of the antibiotic resistance in intestinal pathogens. Clinical studies and experimental animal models have shown that *L. acidophillus* can affect growth and development of *Helicobacter pylori* both in vitro and in vivo (Felley *et al.*, 2003). However till date there is insufficient data to suggest the use of probiotics in the absence of antibiotics to prevent infection with *H. pylori*.

Probiotics and AIDS treatment

Even with the development in HIV medical treatment research, many social and clinical factors can affect functional status and quality of life for HIV-infected children. HIV is recognized to alter the structure and the function of human small intestines. Since intestinal dysfunction is common in HIV infection – that is, carbohydrate malabsorption, frequent diarrhoea, steatorrhea – the intestinal permeability is increased, and might cause enteric pathogens translocation, propitiated by bacterial gastrointestinal overgrowth (Miller, 2002). The probiotics can enhance mucosal immune defense (Trois, 2010) through enhancing macrophage activity, elevating the numbers of killer cells, T cells, and interferon; and their action against pathogenic microbial colonization and translocation.

Trois *et al.* (2008), studied 77 HIV-infected children (2–12 years old), divided into two groups. In one, 38 children received probiotics (formula containing *Bifidobacterium bifidum* with *Streptococcus thermophilus* – 2.5×10^{10} cfu/g) and in the other, 39 children received a standard formula (control group), for 2 months this study showed that probiotics might have immune stimulatory properties and can be helpful in the treatment of HIV-infected children, since the antiretroviral therapy had not changed during the study.

Urogenital health

Similar to the intestinal tract, the urogenital tract in women is highly colonized and susceptible to infection upon colonization disruption (Reid *et al.*, 1998). Bacteria originating from the colon are often linked to both urinary and genital tract infections. Several studies have correlated vaginal health (absence of infections) with the presence of lactobacilli, and specifically hydrogen peroxide-producing lactobacilli (Hillier *et al.*, 1992). Most of these studies were conducted with intravaginal instillations of probiotic bacteria. Oral consumption of certain probiotic-containing products, however, was found to mediate decreased recurrence of *Candida* infections and bacterial vaginosis (Shalev *et al.*, 1996).

Other health benefits

Probiotic influence on a variety of other clinical targets has been evaluated. Probiotic-mediated reduction in the severity of reaction to exposure to radioactive isotopes has been shown in mice (Dong *et al.*, 1987) and humans (Korschunov *et al.*, 1996). The effects of endotoxemia associated with alcoholic liver disease were reduced by probiotics (Nanji *et al.*, 1994). Probiotics have been used in enteral feeding to provide nutritional support of surgery patients (Bengmark and Gianotti, 1996). In addition to proposed direct effects on humans, probiotics may also have implications for human health through their use in animal agriculture.

Mechanism of action of probiotics

The mode of action of probiotic strains is likely to be multifactorial and, from existing evidence, appears to be strain specific. In the Table 1, an attempt has been made to summarize the mechanisms of probiotic action in the human and animal intestinal tract pertaining to various ailments and infections, discussed in the previous part of the chapter.

Table 1: Potential and established effects of probiotic bacteria

Target Health Benefit	Postulated Mechanism
Aid in lactose digestion	Bacterial lactase hydrolyses lactose
Resistance to enteric pathogens	• Secretory immune effect • Colonization resistance • Alteration of intestinal conditions to be less favorable for pathogenicity (pH, short chain fatty acids, bacteriocins) • Alteration of toxin binding sites • Influence on gut flora populations • Adherence to intestinal mucosa, interfering with pathogen adherence • Upregulation of intestinal mucin production, interfering with pathogen attachment to intestinal epithelial cells
Anti-colon cancer effect	• Mutagen binding • Carcinogen deactivation • Inhibition of carcinogen-producing enzymes of colonic microbes

Contd...

	• Immune response • Influence on secondary bile salt concentration
Small bowel bacterial overgrowth	• Influence on activity of overgrowth flora, decreasing toxic metabolite production • Alteration of intestinal conditions to be less favorable to overgrowth flora activities or populations
Immune system modulation	• Strengthening of non-specific defense against infection and tumors • Adjuvant effect in antigen-specific immune responses • Enhancement of secretory IgA production
Allergy	• Prevention of antigen translocation into blood stream
Blood lipids, heart disease	• Assimilation of cholesterol within bacterial cell • Increased excretion of bile salts due to deconjugation by bile salt hydrolase • Antioxidative effect
Antihypertensive effect	• Peptidase action on milk protein yields tripeptides which inhibit angiotensin 1 converting enzyme • Cell wall components act as angiotensin converting enzyme inhibitors
Hepatic encephalopathy	• Inhibition of urease producing gut flora
Urogenital infections	• Colonization resistance • Adhesion to urinary and vaginal tract cells
Helicobacter pylori infections	• Production of inhibitors of *H. pylori* (lactic acid and others)

Source: Sanders and Huis in't Veld, 1999

Probiotic bacteria in food safety applications

In addition to health benefits, researchers have demonstrated that probiotic bacteria produce bacteriocins and organic acids that act as antimicrobials to inhibit the growth of pathogens as well as can be used to extend the shelf life of food products (Goktepe, 2006).

Today's consumers expect and demand a safer food supply than ever before. Yet, despite the improvements in food safety, the number of food-borne illnesses continue to rise. Lactic acid bacteria produce bio-preservatives such as lactic acid, hydrogen peroxide, and bacteriocins that are used to retard both spoilage and the growth of pathogenic bacteria (B Ray, 1992). One of the most studied bacteriocins is nisin, which is used as a food additive in more than 48 countries to preserve the quality of processed cheese, dairy products, and canned foods. Up to now, a variety of bacteriocins from *Lactobacillus* species have been identified (Table 2).

Table 2: Application of *Lactobacillus* Bacteriocins as Food Biopreservatives

Bacteriocin	Example of Food Application	Active Against
Pediocin AcH	Muenster cheese, cooked sausage	*L. monocytogenes*
Sakacin A	Meat, pork	*L. monocytogenes*
Sakacin K	Dry fermented sausage	*L. monocytogenes*
Sakacin P	Cold smoked salmon, chicken cold-cuts	*L. monocytogenes*
Reuterin	Beef sausages	*L. monocytogenes*
Curvacin	Pork steaks and ground beef	*L. monocytogenes*
Plantaricin D	Ready-to-eat salad	*L. monocytogenes*

Source: Goklepe, 2006.

Selection of probiotic strains

Several aspects, including safety, functional and technological characteristics, have to be taken into consideration in the selection process of probiotic microorganisms. Careful screening of probiotic strains for their technological suitability can also allow selection of strains with the best manufacturing and food technology characteristics. The theoretical basis for the selection of probiotic micro-organisms including safety, functional and technological aspects is illustrated in Table 3.

Factors related to the technological and sensory aspects of probiotic food production are of utmost importance since only by satisfying the demands of consumers, the food industry can achieve success in promoting the consumption of functional probiotic products in the future. Good viability and activity of probiotics are considered prerequisites for optimal functionality. Probiotic foods should include specific probiotic strains at a suitable level throughout their shelf life. However, several studies have shown that non-viable probiotics can have beneficial effects such as immune modulation and carcinogen binding in the host. Thus, for certain probiotic strains it might be sufficient that they grow well during initial production steps (to obtain high enough cell numbers in the product) but they do not necessarily need to retain good viability during storage (Saarela, *et al.*, 2000).

Table 3: Selection criteria for probiotics

Particulars	Selection criteria
General aspects	Origin/ Definition/Characteristics
	Strain & genus safety properties
Safety and Stability	Activity and viability in products, adherence, invasive potential
	Resistance to low pH, gastric juice, bile juice, pancreatic juice, colonization/ survival in vivo
Functional and physiological aspects	Adherence to intestinal epithelium/ tissue/virulence and/or
	Antagonism to pathogens antimicrobial activity and/or
	Stimulation/suppression of beneficial bacteria and suppression of harmful bacteria and/or
	Clinical side effects
	Volunteers/patients

Source: Gasser, 1994; Donohue and Salminen, 1996.

Production of probiotic foods

Food substrate/diet is considered as one of the major factors in regulating colonization of micro-organisms in gastrointestinal tract. Food helps to buffer the bacteria through the stomach and may contain other functional ingredients that could interact with probiotics to alter their functionality. Colonic foods, which encourage the growth of favourable bacteria, are referred to as prebiotics. By increasing the amount of prebiotics in the diet, it is possible to increase and maintain healthy bacterial gut flora in the host (Gibson, *et al.*, 2003). Ingredients in certain food products may naturally contain prebiotics which help to improve the functional efficacy of probiotics. Foods can be fortified with prebiotics during manufacturing process to increase probiotic efficacy (Ranadheera, *et al.*, 2010). When both prebiotics and probiotics are present in a food then (Table 4), those functional foods are referred to as synbiotic.

Better growth and survivability during food manufacturing and storage as well as in gastrointestinal tract, protection against acid, bile and gastrointestinal enzymes, adhesion to intestinal epithelium, antimicrobial properties and antibiotic resistance could be considered as factors that might be important in maintaining probiotic efficacy.

Foods used for dissemination of probiotics are usually fermented foods even if probiotics also could be present in infant formula, fruit drinks, whey drinks and sweet milk. Food, particularly dairy products are considered as an ideal vehicle for delivering probiotic bacteria to the human gastrointestinal tract. Different products have demonstrated different viability levels of probiotics over shelf lives. Limitations of dairy products such as the presence of allergens and the requirement for cold storage facilities, as well as an increasing demand for new foods and tastes have initiated a trend in non-dairy probiotic product development (Lavermicocca, 2006). Furthermore, lactose intolerance and the cholesterol content are two major drawbacks related to the fermented dairy products. Nowadays there is an increasing consumer demand for non-dairy-based probiotic products and these organisms are being incorporated into drinks as well as marketed as supplements in the form of tablets, capsules and freeze dried preparations (*e.g.* Multibionta, Enterogermina, Reuterina, UltraLevure, Florastor) (Yadira and, Yoja, 2010).

Table 4: Development of Synbiotic Food

Probiotics	Synbiotics	Integrated synbiotics
*Exogenous strains *Known health properties *Immunomodulation potential *Specificity to individual hosts *Colonization ***Prebiotics*** *Proliferation of endogenous strains *Effects on the microbiota's metabolic activity as well as population dynamics *Specificity *Oligosaccharides- relatively rapid fermentation in the proximal colon	*Synergy between probiotic and prebiotic effects in the GI tract *Limited specificity of prebiotic for added probiotic strain	*Specific synergy between probiotic and prebiotic ingredients in the GI tract *Prebiotics that additionally protect probiotics during manufacture, storage, formulation and, intestinal transit (carriers/encapsulation) *Controlled, site specific release of probiotics in the GI tract *Larger, more slowly fermentable prebiotics with SCFA production more distally in the colon

Source: Mattila-Sandholm *et al.* 2002

Adding probiotics to the juices is more complex than formulating in the diary products because the bacteria need protection from the acidic conditions in the fruit juice. However, with microencapsulation technologies, the probiotics can become an important ingredient in the functional foods, expanding the probiotic application outside the pharmaceutical and supplement industries. Microencapsulation technologies have been developed and successfully applied using various matrices to protect the bacterial cells from the damage caused by the external environment (Del Piano *et al.*, 2006). There is a genuine interest in the development of fruit juice based functional beverages with probiotics because they have taste profiles that are appealing to all age groups and because they are perceived as healthy and refreshing foods (Sheehan *et al.*, 2007). Although LAB has been considered a difficult microorganism that demands various essential amino acids and vitamins for growing (Salminen and Von Wrigh, 1993), it has been found that some probiotic strains have the capability to grow in fruit matrices. Researchers have reported that cell viability depends of the strains used, the characteristics of the substrate, the oxygen content and the final acidity of the product (Shah, 2001).

In recent years, cereals have also been investigated regarding their potential use in developing functional foods. Lactic acid fermentation of cereals is a long-established processing method and is being used in Asia and Africa for the production of foods in various forms such as beverages, gruels, and porridge. Cereals contain water-soluble fiber (such as β-glucan and arabinoxylan), oligosaccharides (such as galacto-and fructo oligosaccharides) and resistant starch, and thus have been suggested to fulfill the prebiotic concept (Shah, 2001). Whole grains are also sources of many phytochemicals, including phytoestrogens, phenolic compounds, antioxidants, phytic acid and sterols. In Western countries, cereals, like wheat and rye, are used for sourdough production, which is traditionally prepared by adding a prefermented sourdough of good quality to the dough. The population of lactobacilli in fully fermented sourdoughs is more than 10^9cfu/g, while the lactic acid bacteria (LAB)/yeast ratio is generally 100:1 (Salovaara, 1998).

Probiotic products are usually standardised based on the presumption that culture viability is a reasonable measure of probiotic activity, thus the ability of the strain to attain high cell population is of primary importance. A concentration of approximately 10^7 cells/ ml at the time of consumption is considered functional (Shortt, 1999). Malt, wheat and barley extracts have a positive influence in increasing bile tolerance of *L. acidophilus, L. reuteri* and *L. plantarum*. However, addition of malt demonstrated a greater positive influence while barley and wheat extracts demonstrated similar levels of influence in retaining viability (Patel, *et al.*, 2004).

Lactic acid fermentation improves usually the nutritional value and digestibility of cereals (Charalampopoulos, *et al.*, 2002). Lactic acid fermentation of different cereals, such as maize, sorghum, finger millet, has been found effectively to reduce the amount of phytic acid, tannins and improve protein availability (Chavan *et al.*, 1988; Lorri and Svanberg, 1993). Increased amounts of riboflavin, thiamine, niacin, and lysine due to the action of LAB in fermented blends of cereals were also reported (Sanni *et al.*, 1999). Khetarpaul and Chauhan (1990) reported improved minerals availability of pearl millet fermented with pure cultures of lactobacilli and yeasts.

Safety considerations for probiotics

An enormous amount of money and energy can be spent in assessing the risk of each probiotic strain. A low risk may be accepted, but the risk to benefit ratio needs to be clearly established. This requires relevant information on the efficacy and safety of the products. Knowledge on survival of the probiotics within the GI-tract, their translocation and colonization properties, and the fate of probiotic-derived active components is important for the evaluation of possible positive and negative effects of probiotic consumption (*Saarela et al.*, 2000). The minimium requirements needed for probiotic status include the assessment of strain identity, in vitro tests to screen potential probiotics, assessment of safety above all, and in vivo studies for substantiation of effects. Good manufacturing practices must be applied in the manufacture of probiotic foods with quality assurance, and shelf-life conditions established. Assessment of lack of infectivity by a probiotic strain in immune-compromised animals would add a measure of confidence in the safety of the probiotic (Pineiro and Stanton, 2007).

The use of probiotic bacterial cultures stimulates the growth of preferred micro-organisms, crowds out potentially harmful bacteria, and reinforces the body's natural defence mechanisms. Today, plenty of evidence exists on the positive effects of probiotics on human health. However, to be effective, probiotics must be capable of being prepared in a viable manner and on large scale (e.g. for industrial purposes), whilst during use and under storage the probiotic should remain viable and stable. India's probiotics market has been highlighted by researcher Frost & Sullivan as a *"major growth market of the future"*, with annual growth of 22.6 per cent until 2015. With India's urban middle class growing rapidly it is easy to see why there is so much interest in a country with a population approaching 1.2bn (Starling, 2010). When considered together, the above mentioned evidences reinforce the fact that probiotics can influence human health in a positive manner. Thus, it can be concluded that, the field of probiotics is developing rapidly as evidenced by expansion of research and increased familiarity of probiotics to the general public.

References

Alakomi, HL, Skytta, E, Saarela, M, Mattila-Sandholm, T, Latva-Kala, K, Helander, IM, (2000). Lactic acid permeabilizes Gram-negative bacteria by disrupting the outer membrane. Applied and Environmental Microbiology, 66: 2001–2005.

Anonymous, (1988). Persistent diarrhoea in children in developing countries: Memorandum from a WHO Meeting. Bulletin of the World Health Organization, 66 (6): 709–17.

Barbosa, TM, Serra, CR, La Ragione, RM, Woodward, MJ, Henriques, AO (2005). Screening for Bacillus isolates in the broiler gastrointestinal tract. Appl. Environ. Microbiol., 71: 968-978.

Bartlett, JG (2002). Antibiotic-associated diarrhea. N Engl J Med., 346: 334–339.

Bayless, TM, Paige, M, Ferry, GD (1971). Lactose intolerance and milk drinking habit. Gastroenterology, 60: 605-608.

Bengmark, S, Gianotti, L (1996). Nutritional support to prevent and treat multiple organ failure. World J. Surg., 20: 474–481.

Bernaola Aponte, G, Bada Mancilla, CA, Carreazo Pariasca, NY, Rojas Galarza, RA (2010). Probiotics for treating persistent diarrhoea in children. The Cochrane Collaboration, 11. Published by JohnWiley & Sons, Ltd.

Charalampopoulos, D, Wang, R, Pandiella, SS, Webb, C (2002). Application of cereals and cereal components in functional foods: a review. International Journal of Food Microbiology, 79: 131–141.

Chavan, UD, Chavan, JK, Kadam, SS (1988). Effect of fermentation on soluble proteins and in vitro protein digestibility of sorghum, green gram and sorghum green gram blends. J. Food Sci., 53: 1574–1575.

Cheplin, HA, Post, CD, Wiseman, JR (1923). Bacillus acidophilus milk and its therapeutic effects. Boston Medical and Surgical Journal, 189: 405–411.

Chikai, T, Nakao, H and Uchida, K (1987). Deconjugation of bile acids by human intestinal bacteria implanted in germ-free rats. Lipids, 22: 669-671.

Corcionivoschi, N, Drinceanu, D (2009). Probioticele-la timpul prezent. Editura Mirton.

Corcionivoschi, N, Drinceanu, D, Stef, L, Luca, I, Julean, C, Mingyart, O (2010). Probiotics – identification and ways of action. Innovative Romanian Food Biotechnology, 6: 1-11.

Corthesy, B, Gaskins, HR, Mercenier, A (2007). Cross-talk between probiotic bacteria and the host immune system. J. Nutr., 137: 781–790.

Cutting, SM (2010). Bacillus probiotics, Food Microbiology, 10: 1016

Daly, C, Davis, R (1998). The biotechnology of lactic acid bacteria with emphasis on applications in food safety and human health. Agric. Food Sci. 7: 219–250.

Del Piano, M, Morelli, L, Strozzi, GP, Allesina, S, Barba, M, Deidda, F (2006). Probiotics: From research to consumer. Digestive and Liver Diseases, 38(2): 248–255.

Dong, MY, Chang, TW, Gorbach, SL (1987). Effects of feeding lactobacillus GG on lethal irradiation in mice. Diagn. Microbiol. Infect. Dis., 7: 1–7.

Donohue, DC, Salminen, S (1996). Safety assessment of probiotic bacteria. Asia Pac. J. Clin. Nutr., 5: 25–28.

Elmer, GW, McFarland, LV (2001). Biotherapeutic agents in the treatment of infectious diarrhea. Gastroenterology Clin NA., 30: 837–853.

Felley, C, Michetti, P (2003). Probiotics and Helicobacter pylori. Best Pract Res Clin Gastroenterol, 17: 785-791.

Gasser, F (1994). Safety of lactic acid bacteria and their occurrence in human clinical infections. Bull. Inst. Pasteur., 92: 45–67.

Geier, MS, Butler, RN, Howarth, GS (2006). Probiotics, prebiotics and synbiotics: A role in chemoprevention for colorectal cancer. Cancer Biology & Therapy, 5: 1265–1269.

Gibson, GR, Rastall, RA, Fuller, R (2003). The Health Benefits of Probiotics and Prebiotics. In R. Fuller & G. Perdigon (Eds.), Gut flora, nutrition, immunity and health (1st ed., pp. 52–76). Oxford: Blackwell Publishing.

Gilat, T, Russo, S, Gelman-Malachi, E, Aldor, TA (1972). Lactose in man: A non-adaptable enzyme. Gastroenterology, 102: 1269-1277.

Gilliland, SE, Nelson, CR, Maxwell, C (1985). Assimilation of cholesterol by Lactobacillus acidophilus. Appl. Environ. Microbiol., 49: 377–381.

Goktepe, I (2006). Probiotics as Biopreservatives for Enhancing Food Safety. In. Probiotics in Food Safety and Human Health. Edited by Ipek Goktepe Vijay K. Juneja and Mohamed Ahmedna. CRC Press. Boca Raton, FL USA. 285-307.

Guslandi, M, Mezzi, G, Sorghi, M (2000). Saccharomyces boulardii in maintenance treatment of Crohn's disease. Dig. Dis. Sci. 45: 1462–1464.

Haenel, H, Bendig, J (1975). Intestinal flora in health and disease. Progr. Food Nutr. Sci. 1: 21–64.

Hamilton-Miller, JMT, Gibson, GR, Bruck, W (2003). Some insight into the derivation and early uses of the word =probiotic'. British Journal of Nutrition, 90: 845.

Hillier, SL, Krohn, MA, Klebanoff, SJ, Eschenbach, DA (1992). The relationship of hydrogen peroxide-producing lactobacilli to bacterial vaginosis and genital microflora in pregnant women. Obstetrics Gynecol., 79: 369–373.

International Working Group on Persistent Diarrhoea (IWGPD), (1996). Evaluation of an algorithm for the treatment of persistent diarrhoea: a multicentre study. Bulletin of the World Health Organization, 74 (5): 479–489.

Kalliomaki, M, Salminen, S, Arvilommi, H, Kero, P, Koskinen, P, Isolauri, E (2001). Probiotics in primary prevention of atopic disease: A randomized placebo-controlled trial. Lancet, 357: 1076–1079.

Katz, J, A (2006). Probiotics for the Prevention of Antibiotic-associated Diarrhea and Clostridium difficile Diarrhea. J. Clin. Gastroenterol, 40: 249–255.

Kelly, CP, Pothoulakis, C, LaMont, JT (1994). Clostridium difficile colitis. N Engl J Med., 330: 257-262.

Khetarpaul, N, Chauhan, BM (1990). Effect of fermentation by pure cultures of yeasts and lactobacilli on the available carbohydrate content of pearl millet. Trop. Sci. 31: 131-139.

Korschunov, VM, Smeyanov, VV, Efimov, BA, Tarabrina, NP, Ivanov, AA, Baranov, AE (1996). Therapeutic use of an antibiotic-resistant Bifidobacterium preparation in men exposed to high-dose g-irradiation. J. Med. Microbiol., 44: 70-74.

Lavermicocca, P (2006). Highlights on new food research. Digestive and Liver Disease, 38(2): 295-299.

Lorri, W, Svanberg, U (1993). Lactic-fermented cereal gruels with improved in vitro protein digestibility. Int. J. Food Sci. Nutr., 44: 29- 36.

Majamaa, H, Isolauri, E (1997). Probiotics: A novel approach in the management of food allergy. J. Allergy Clin. Immunol. 99: 179-185.

Mattila-Sandholm, T, Myllarinena, P, Crittenden, R, Mogensen, G, Fonden, R, Saarela. M (2002). Technological challenges for future probiotic foods. International Dairy Journal, 12: 173-182.

Mazza G. (Ed.), (2008). Handbook of Fermented Foods. Boca Raton, FL: CRC Press.

McFarland, LV (2007). Meta-analysis of probiotics for the prevention of traveler's diarrhea. Travel Med Infect Dis., 5: 97-105.

McIntosh, GH (1996). Probiotics and colon cancer prevention. Asia Pac. J. Clin. Nutr., 5: 48-52.

Menard, S, Candalh, C, Bambou, JC, Terpend, K, Cerf-Bensussan, N, Heyman, M (2004). Lactic acid bacteria secrete metabolites retaining anti-inflammatory properties after intestinal transport. Gut. 53: 821-828.

Metchnikoff, E (1907). The Prolongation of Life: Optimistic Studies. London: Butterworth-Heinemann.

Metchnikoff, LE (2004). The prolongation of life: Optimistic studies (reprinted edition 1907). New York, USA, Springer Publishing Company.

Miller, TL (2002). Nutritional interventions in pediatric HIV: It's hard to hit a moving target. JPGN, 34(4): 353-356.

Mombelli B, Gismondo MR (2000). The use of probiotics in medical practice. Int J Ant Agents. 16: 531-536.

Nanji, AA, Khettry, U, Sadrzadeh, SMH (1994). Lactobacillus feeding reduces endotoxemia and severity of experimental alcoholic liver (disease). Proc. Soc. Exp. Biol. Med., 205: 243-247.

Patel, HM, Wang, R, Chandrashekar, O, Pandiella, SS, Webb, C (2004). Proliferation of Lactobacillus plantarum in solid-state fermentation of oats. Biotechnol. Progr., 20: 110-116.

Pelletier, X, Laure-Boussuge, S, Donazzolo, Y (2001). Hydrogen excretion upon ingestion of dairy products in lactose intolerant male subjects: Importance of the live flora. Eur. J. Clin. Nutr., 55: 509–512.

Perdigon, G, Nader de Macias, ME, Alvarez, S, Medici, M, Oliver, G, Pesce de Ruiz Holgado, A (1986). Effect of a mixture of Lactobacillus casei and Lactobacillus acidophilus administered orally on the immune system in mice. J. Food. Prot. 49: 986- 989.

Pineiro, M, Stanton, C (2007). Probiotic Bacteria: Legislative Framework- Requirements to Evidence Basis. J. Nutr., 137: 850–853.

Ranadheera, RDCS, Baines, SK, Adams. MC (2010). Importance of food in probiotic efficacy. Food Research International, 43: 1–7.

Ray, B (1992). Bacteriocins of starter culture bacteria as food biopreservatives: an overview. In: B Ray, M Daeschel, Eds., Food Biopreservatives of Microbial Origin. CRC Press, Boca Raton, FL, 125–134.

Reddy, GV, Friend, BA, Shahani, KM, Farmer, RE (1983). Antitumor activity of yogurt components. Journal of Food Protection, 46: 8–11.

Reid, G, Bruce, AW, Smeianov, V (1998). The role of lactobacilli in preventing urogenital and intestinal infections. Intl. Dairy J., 8: 555–562.

Rembacken, BJ (1999). Non-pathogenic Escherichia coli versus mesalizine for the treatment of ulcerative colitis: a randomised trial. Lancet., 354: 635–639.

Rettger, L (1935). Lactobacillus acidophilus and its therapeutic application. New Haven: Yale University Press.

Saarela, M, Mogensen, G, Fonden, R, Mattom J, Mattila-Sandholm, T (2000). Probiotic bacteria: safety, functional and technological properties. Journal of Biotechnology, 84: 197–215.

Salminen, S, Von Wrigh, A (Eds.), (1993). Lactic Acid Bacteria. Marcel Dekker Inc., New York.

Salovaara, H (1998). Lactic acid bacteria in cereal-based products. In Lactic Acid Bacteria: Microbiology and Functional Aspects. Marcel Dekker, New York, pp. 115–137.

Sanders, ME, Huis in't Veld, J (1999). Bringing a probiotic-containing functional food to the market: Microbiological, product, regulatory and labeling issues. Antonie van Leeuwenhoek, 76: 293–315.

Sanni, AI, Onilude, AA, Ibidabpo, OT (1999). Biochemical composition of infant weaning food fabricated from fermented blends of cereals and soybean. Food Chem. 65: 35-39.

Scheinbach, S (1998). Probiotics: Functionality And Commercial Status. Biotechnology Advances, 16(3): 581-608.

Shah, NP (2001). Functional foods, probiotics and prebiotics. Food Technology, 55: 46-53.

Shah, NP (2007). Functional cultures and health benefits. International Dairy Journal, 17: 1262–1277.

Shah, NP, and Lankaputhra, WEV (2002). Bifidobacterium spp.: Morphology and physiology. In H. Roginski, J. Fuquay, P. Fox, & Patrick (Eds.), Encyclopedia of dairy science London, UK: Academic Press, pp. 141–146.

Shahani, KM, Ayebo, AD (1980). Role of dietary lactobacilli in gastrointestinal microecology. Am. J. Clin. Nutr., 33: 2448–2457.

Shalev, E, Battino, S, Weiner, E, Colodner, R, Keness, Y (1996). Ingestion of yogurt containing Lactobacillus acidophilus compared with pasteurized yogurt as prophylaxis for recurrent candidal vaginitis and bacterial vaginosis. Arch. Fam. Med., 5: 593–596.

Sheehan, VM, Ross, P, Fitzgerald, GF (2007). Assessing the acid tolerance and the technological robustness of probiotic cultures for fortification in fruit juices. Innovative Food Sci. Emerg. Technol., 8: 279–284.

Shortt, C (1999). The probiotic century: historical and current perspectives. Trends Food Sci. Technol., 10: 411- 417.

Smolyansky, J (2010). Probiotics: A Historical Perspective. In. Bioactive Foods in Promoting Health: Probiotics and Prebiotics. Ed: Ronald Ross Watson and Victor R. Preedy. Academic Press. London/ USA.

Starling, S (2010). Indian probiotics market to quadruple by 2015. www.nutraingredients.com as on 21-Jan-2010.

Stiles, ME, Holzapfel, WH (1997). Lactic acid bacteria of foods and their current taxonomy. International Journal of Food Microbiology, 36: 1–29.

Tannis, A (2008). A History of Probiotics In Probiotic Rescue: How You Can Use Probiotics to Fight Cholesterol,Cancer, Superbugs, Digestive Complaints and More. John Wiley & Sons Canada, Ltd., pp: 13-20.

Trois, L (2010). Probiotics and AIDS Treatment. In. Bioactive Foods in Promoting Health: Probiotics and Prebiotics. Ed: Ronald Ross Watson and Victor R. Preedy. Academic Press. London/ USA. PO:229-233.

Trois, L, Cardoso, EM, Miura, E (2008). Use of probiotics in HIV-infected children: A randomized doubleblind controlled study (Epub 2007 Sep 17). Journal of Tropical Pediatrics, 54(1): 19–24.

Tuohy, KM, Pinart-Gilberga, M, Jones, M, Hoyles, L, McCartney, AL, Gibson, GR (2007). Survivability of a probiotic Lactobacillus casei in the gastrointestinal tract of healthy human volunteers and its impact on the faecal microflora. J. Appl. Microbiol., 102: 1026-1032.

Vergio, F (1954). Anti- und Probiotika. Hippokrates, 4: 116–119.

□□□

16

Plastics for Food Packaging Applications

DEEPAK RAJ RAI

Plastics are man made materials of great utility, commercial and economic importance. With the invention of bakelite during the early part of 20th century, a lot of different types of materials suitable for applications ranging from insulation to packaging have been developed. Now-a-days, it is really difficult to think of an application without a slight component of plastics involved in it. The research and development efforts in this direction have made it possible to replace the traditional metal components with that made from plastics. These materials have same or some times better properties than the materials they are replacing in terms of strength, flexibility etc.

Development of plastics

During 1940s, rigid metal containers were the lone choice for the food processors for canning of processed food material for the armies engaged in World War II. Around this time, a flexible packaging material known as cellophane was also invented and it was in great demand due to being attractive in nature. However, as the research and development in the production of different plastics continued, the invention of polymers marked the beginning of new plastic era. Polymers are basically the long chain of molecules of unsaturated hydrocarbons known as monomers. These days, many different types of polymers either made from single type of monomers or from different types of monomers are available in market and are being used for many food related applications such as components of food processing machinery, handling and transportation equipment, packaging of fresh and processed food material etc.

Plastics in food packaging

In food packaging, plastics are used in the form of either rigid containers or as flexible pouches. Rigid containers are mostly used for packaging of those processed food materials which requires to be protected from moisture, oxidation etc. e.g., now a days, baby foods or nutritional supplements are packaged in coloured rigid plastic containers to protect them from sunlight and moisture. Similarly, cookies and candies are packaged in the rigid containers for these very reasons. Flexible plastic pouches are used as re-fill packs, for ready to eat snack food, packaging of processed and fresh produce etc. Whatever is the application, the plastic containers (rigid or flexible) have certain advantages over the rigid metal or glass containers as detailed below:

a) They are lighter than the metal containers, which results in low handling and transportation costs. The lightness is attributed to the fact that depending upon the individual requirement, the thickness of plastic container can be adjusted accordingly which reduces the weight of container. Less weight of rigid containers allows transporting large volumes of packaged produce at low costs.

b) The plastic films can be sealed to prepare the film packages as per the individual requirement. They can even be sealed at home, without the need of any specialized equipment. It is a common scene in supermarkets and even in small shops that sealed plastic packages containing fresh, semi processed or processed food materials are available off the shelf.

c) Since the fabrication of plastic materials is not a complicated process, different types of containers can be easily prepared as plastics can be easily moulded. Similarly different types of single layer or multilayer flexible films can be produced to suit the requirements of food packaging.

d) Rigid plastic containers act as a barrier for the transmission of water vapour and various gases. Therefore, they are mostly used for packaging of crisp processed products and products which required to be protected from the effects of oxidation such as rancidity. A common example in this case can be cited as that of packaging of crispy snacks at household level. These snacks require an inert atmosphere or vacuum packaging for a longer shelf life, however, it may not possible to provide such conditions at individual household level. Therefore, rigid plastic containers are preferred for their storage as it is easy to keep them air tight.

Flexible film packages possess selective gas permeability and moisture permeability characteristics which makes them suitable for packaging of a variety of fresh produce having very low to very high respiration rates. This would result in modification of in-pack atmosphere in terms of O_2 and CO_2 which would ultimately lead to shelf life extension of the fresh produce. Further, if required, a single flexible film as a combination of different types of films such as multilayer films can also be prepared to exploit the different properties of individual layers. This is particularly helpful in specific applications where it is not possible to attain the desired results with the use of single layer film packages.

e) Most of the plastic film materials are transparent in nature. They increase the aesthetic value of the packaged produce by making it look attractive. e.g. in case of certain fresh fruits and vegetables , an additional surface gloss results by packaging them in polypropylene (PP) or oriented polypropylene (OPP) film, packages which increases their market value. Now a days, food packaging industry has reached to such a level that certain freshness indicators, antimicrobial agents, time-temperature indicator etc. are also being incorporated in to the plastic packaging materials to indicate the freshness of the packaged produce. In this case, certain specialized chemical are added in to the packaging material which change their colour on interaction with improper in-pack atmosphere which ultimately reflects the state of the packaged fresh fruit or vegetable.

f) Another important advantge of the plastics is their printability. Both rigid and flexible plastic materials such as bottles, tray , container, films etc. can be very easily printed making it easier to present information about package contents and other matter of commercial importance to the retailers and consumers.

Inspite of the numerous advantages offered by different types of plastics and their applications in food and other industries, there are certain disadvantages associated with their use as detailed below:

a) Plastics are less resistant to heat and temperature fluctuations. Further, their gas and water vapour permeability characterstics are temperature dependent. As respiration rate of fresh produce is comparatively more dependent on temperature, therefore under storage under fluctuating temperature conditions may lead to in-pack anaerobic conditions in no time.

b) They have lower dimensional stability. It is a common observation that after a prolonged storage period, common household container looses their dimensional strength. Similarly, in case of flexible film packages under modified atmosphere packaging (MAP), the collection of water inside these packages make them lose their dimensional stability.

c) It is very disheartening to note that plastics have limited gas permeability rates for O_2, CO_2, N_2, H_2O etc. This is particularly very important in case of highly respiring produce, where one has no choice to have lesser volume of packaged produce to have a properly modified in-pack modified atmosphere.

d) It is not environmental friendly, it is a well known fact that it does not degrades even after a substantial amount of time. Further, plastic bags and items made of plastic materials may lead to death of animals swallowing them accidently. Efforts have been made World over to have biodegradable plastics; however, they are not suitable for many commercial applications of packaging due to their low strength and gas permeability chracterstics.

Types of plastic materials

These days, a wide range of plastics is available. This is possible because monomers of different types can be combined to create a new polymer having altogether different characteristics. Through variations in manufacturing process, it is possible to produce different polymers from a single monomer. It is also possible to incorporate certain specific compounds such as anti microbial agents scavenging or indicating agents in the polymer during film manufacturing to obtain a polymer having special properties which are useful for active packaging of food materials. Also, different type of polymers can also be combined to produce multilayer films useful for specific applications of food packaging.

The broad classification of packaging films includes a large variety of films such as Polyethylene film, Polypropylene film, Polyester film, Polystyrene film, Polyvinyl Alcohol film, Ethylene-Vinyl Alcohol Copolymer, Nylon film, Vinylidene Chloride film, Polycarbonate film, Acrylonitrile Copolymer film, Shrink film, Multi-layer film, Co-extruded film, Perforated and Edible films. However, not all of them are suitable for MAP of fresh fruits and vegetables. Therefore, in industry only selected films find a useful place for packaging purposes. Films made from different materials having a range of thickness are available in the market. Different thickness of film imparts a specific range of gas-water permeability and mechanical characteristics to these films. Some of the common polymers used for food packaging purposes are discussed below:

a) Polyethylene

Polyethylene is one of the earliest and most frequently used polymer .The basic monomer used in this case is ethylene (C_2H_4). Various polymerized products of ethylene include Low Density Polyethylene (LDPE), High Density Polyethylene (HDPE), Linear Low Density Polyethylene (LLDPE), Mono-Oriented Polymer film, Blown films and Ethylene copolymers. They are available in different widths and thickness but they have low transmission rates for gas and water vapour. They have excellent flexibility and phenol resistance. In fresh produce handling, they are mainly suitable for bulk packaging of products with low respiration rates such as apples, potato etc. Further, they are also useful as lining materials for cardboard boxes since they have very less water vapour permeability characteristics.

b) Polypropylene

Polypropylene or PP is a polymer prepared from propylene. Some of the different polymers prepared from propylene include cast polypropylene film, biaxially oriented polypropylene film and mono oriented polypropylene foam/ film. Cast Polypropylene films have excellent stiffness, slippability, machinability, transparency and cold resistance. They are useful for packaging of various fresh food products and bread wrapping. Biaxially Oriented Polypropylene film is available in different variants such as uncoated, co-extruded and coated. Uncoated films have excellent transparency, stiffness and gloss. Their printability is excellent and they are particularly useful for bag making and packaging of fresh vegetables.

c) Polyester

They have two variants namely ordinary polyester film and biaxially oriented polyester film. Ordinary polyester films have high gas and water vapour resistance , good printability and transparency. Biaxially oriented polyester films have excellent mechanical strength, good electrical insulation, low gas and water vapour permeability characteristics. These films find their application in packaging of processed foodstuff, stickers and labels.

d) Polystyrene

Polystyrene films are available commercially in different thickness levels. They have good gas and water vapour permeability characterstics. They have excellent transparency, good resistance to temperature and are flexible and tough. They are also of two types namely ordinary polystyrene and biaxially oriented polystyrene films. Various applications of these films include fruit and vegetable packaging, carton box covering and lamination.

e) Polyvinyl alcohol

Among all type of plastic films, ordinary or biaxially oriented polyvinyl alcohol films have the best gas barrier property at all humidity levels. They have outstanding mechanical strength, low moisture permeability and provide excellent barrier to aroma and flavour transmission. They are useful for packaging of processed soups, pastes, marine and confectionary products.

f) Ethylene vinyl alcohol

These films have excellent gas and odour barrier characteristics, high resistance to oil, hot water and organic solvents, good printability, luster and transparency. They do not contain plasticizers or any harmful substances. They can both be used as surface layer or intermediate layer in copolymer formation. Various applications of these films include packaging of meat, coffee, tea, sauces and snacks.

g) Nylon film

Nylon films are also available as ordinary and biaxially films. They are transparent and glossy with high gas barrier properties, excellent printability, non-toxic and odourless. They have excellent burst strength, mechanical resistance and crack resistance. They are used for packaging of frozen food, dairy, marine products and meat.

h) Vinylidene chloride film

These films have high barrier against gas and water vapour. They have high resistance for heat and water absorption. They are microwaveable and can be easily thermoformed into rigid or semi-rigid containers. Different applications for these films include pouches, lid stock and containers for retorting, dessert, microwaveable foods, drinks, spices and medicinal products.

i) Polycarbonate film

Polycarbonate has excellent toughness and heat resistance. They can withstand temperatures in the range of –100° to 135 °C and have good flavour retention properties which make them an ideal material for retort, processed and frozen food packaging.

j) Acrylonitrile copolymer films

Rubber modified acrylonitrile based copolymer films provide excellent resistance to gas and chemicals. They are heat sealable and can retain the food aroma and flavour to satisfactory levels. Range of thickness of these films varies between 20-40 mm. They are mostly used for packaging of food and chemicals.

Blown films made from these copolymers have similar properties and are mostly used in packaging of pharmaceutical products.

k) Cellulose film

Vinyl copolymer coated cellulose film has good dimensional stability, heat sealable and moisture proof. Uncoated films have high transparency and gloss. These are suitable for carton overwrapping and decorative wrappings.

l) Shrink film

Shrink films are versatile, stretchable and have excellent merchandising appeal. They are abuse resistance, tough and have a broad spectrum of shrink properties.Their seal and tensile strength are high enough to withstand rough handling in distribution and at point of sale. They are ideal for packaging of irregular shaped bodies, window type boxes and heavy products.

m) Multi-layer films

Multilayer films comprise of various combinations of different plastic films. These films are stable at high temperatures and are suitable for microwaveable/retort packaging of processed foods. Multilayer films have OPP, PET or nylon films as the outer layers to withstand high temperature, whereas the middle layers are composed of materials such as EVOH, PVDC and Al which have low gas transmission rates.

Films for modified atmosphere

Creation of an in-pack atmosphere is not the only criterion for a successful MAP. The appropriate atmosphere requires to be sustained for a longer period to take full advantage of MA. It is possible to achieve this for the fresh produce having low and medium O_2 consumption and CO_2 evolution rates, using the ordinary LDPE, PP, OPP or PVC film packages. However, in case of certain fresh and fresh cut produce with high rates of respiration, it is not possible to sustain the appropriate atmosphere in these packages and the atmosphere becomes anoxic in a short period. This problem can be averted using low storage temperatures, but that will involve additional refrigeration costs. Geeson (1990) observed that sweetcorn cobs packaged in PVC overwrapped trays had less than 2% O_2 and high (2%) CO_2 levels after 1 day of storage at 20°C, which induced off-flavours and strong alcoholic smell in these packages.

Further, various researchers have tried to extend shelf-life of highly respiring fresh asparagus by MAP. Geeson (1990) simulated normal market conditions by packaging freshly harvested asparagus spears in PVC trays overwrapped with PVC films. This study showed that CO_2 concentrations inside the packs increased to 6-9% and O_2 levels fell to 0.5% during storage. Further, the spears lost their quality and became severely tainted. Tenario *et al.* (2004) observed that at 2°C, MA was effective in extending the shelf-life and in preserving the green colour of fresh asparagus packaged in ordinary PP packages. Varoquax *et al.* (1996) packaged fresh beansprouts in MA packages made of films with different permeabilities. During storage at various temperatures from 1 to 20°C, oxygen uptake rates and the development of aerobic microorganisms increased with time. The OPP film packages had anoxic conditions within 2 days, favoured anaerobic metabolism and necrosis of sprouts. In all packages, there was rapid development of aerobic microorganisms and lactic acid bacteria.

Therefore, improper atmospheric modification observed in case of highly respiring fresh produce led to alternative innovations. Burton *et al.* (1987) used a combination of highly permeable and less permeable films to pack fresh mushroom. However, the film area of highly permeable film was so small that accidental labelling could easily cover it, thus negating the effect of MA. Geeson (1990) tried limited number of microperforations (2-3) by piercing of PVC film overwrapped packs containing leeks, asparagus, parsnips and cherry tomatoes. In all these cases anaerobic conditions and off-flavours were avoided besides an extension of shelf-life. Although piercing of film packages were observed to be beneficial, but it was just an empirical approach. Further, over-piercing was another problem as in case of asparagus spears, toughness of spears was observed. Such types of conclusions led to the feeling that in order to have an appropriate MA for highly respiring produce, they should be packaged in such film packages, which provide higher gas permeability for O_2 and CO_2. Efforts to develop such films were carried out at various research laboratories. Horita *et al.* (1998) observed that such films known as micro-perforated films could be prepared by the use of press perforated rollers during the film manufacturing process. They developed several of such films using this process and observed that they had sufficient strength, could be mass produced and their O_2 permeability values did not vary much at different temperatures. They concluded that as compared to normal polymeric films, gas permeability ratio for CO_2 and O_2 varied between 1.1:1 and 1.5:1 at 15 °C. However, their gas permeability values decreased on covering the micro-perforations with water. Rai *et al.* (1999) evaluated the gas permeability characteristics of several types of micro-perforated films and obtained similar results (Table 1).

Table 1: Gas permeability and gas permeability ratio (GPR) for some oriented polypropylene micro-perforated films at 15°C

Film name	Gas permeability (ml/m2/day/atm)		GPR
	CO_2	O_2	CO_2:O_2
MA-B	8487	8182	1.03:1
MA-C	25141	23922	1.05:1
MA-D	69177	67958	1.01:1

Applications of micro-perforated films

Taking into account, the usefulness of microperforated films and packaging requirements of highly respiring and high value produce such as asparagus, broccoli, lettuce, mushroom etc., various researchers have tried to use them for shelf-life extension and quality retention purposes. Horita *et al.* (1998) packaged broccoli (*Brassica oleracea L. italica*) in micro-perforated film packages and observed that as compared to normal polymeric film packages, broccoli quality was preserved better in micro-perforated film packages.

Saito *et al.* (2000) packaged asparagus spears in micro-perforated OPP film packages and stored at 15°C for 10 days. Weight loss was less than 2% in micro-perforated film packages as compared to 15% in air-stored samples under similar conditions. O_2 concentrations (1-6%) in micro-perforated film packages were found to be largely responsible for higher glutathione levels, but ascorbic acid content was adversely affected. Results of the study indicated that to preserve ascorbic acid and glutathione simultaneously in stored asparagus through MAP under high CO_2 levels (up to 23%), O_2 concentration must be between 1 and 6% to get acceptable levels of ascorbic acid and glutathione. Rai *et al.* (2000a) studied the effect of MA on low molecular weight carbohydrates of oyster mushrooms (*Pleurotus ostreatus*), during post-harvest storage at 5 and 15°C, using highly permeable micro-perforated OPP film and ordinary OPP film packages. Among free sugars, *trehalose* content was observed to be varying between pileus and stipe portion of mushroom, in response to in-pack gaseous environment. Increase in pileus and stipe portion trehalose and total sugar content was observed for mushrooms packaged in micro-perforated film packages at storage temperatures of 5 and 15°C, while they reduced in ordinary OPP film packages and the reduction was observed to be more in stipes than pileus, indicating that low O_2 (0.4%) and high CO_2 (12-13%) observed in these packages had more adverse effect on stipe portion trehalose and total sugar content. These results showed that the consumption of glucose substrate and thus, the rates of respiration of mushroom were more pronounced in ordinary OPP film packages. Among sugar alcohols, only mannitol was found to be affected under low O_2 and high CO_2 conditions, prevailing in ordinary OPP film packages at both the storage temperatures.

Rai *et al* (2000b) studied the effect of MA on glutathione, cap development, disease suppression and mycelial growth of oyster mushroom (*Pleurotus ostreatus*) at 15°C using micro-perforated OPP film, PVC film overwrapped tray and ordinary OPP film packages. MA of 15.5-16.5% O_2 and 4.5-5.5% CO_2 inside micro-perforated OPP film was beneficial for preservation of oyster mushroom compared to control packages. However, among different packaging treatments, mushrooms packaged in micro-perforated film packages had highest amount of surface moisture due to covering of package microperforations by water droplets. Saito and Rai (2005) evaluated the gas concentrations, weight loss, ascorbic acid, chlorophyll, β-carotene and smell rating of radish (*Raphanus* spp.) sprouts, packaged in four different highly permeable micro-perforated film packages, stored for 6 days at 15 °C. Results indicated that a combination of 9-13% O_2 and 8-11% CO_2 generated inside two types of micro-perforated film packages used in the study was a suitable gaseous composition to maintain most of the qualitative parameters as well as simultaneously avoiding the off-flavours till 6 days of storage.

References

Burton K S, Frost CE and Nichols R (1987). A combination of plastic chlorophylls in fresh green asparagus (Asparagus officinalis L) stored under modified atmosphere packaging. J Sci Food Agric 84: 357-65.

Geeson J (1990). Micro-perforated films for fruit and vegetable packing. Professional Hort 4: 32-35.

Geeson JD, Browne KM, Maddison K, Shepherd J, Guaraldi F (1985). Modified atmosphere packaging to extend shelf-life of tomatoes. J Food Technol 20: 339-349.

Horita H, Nawa Y, Sato K, Ishitani T, Takeshita K, Takahashi A, Nakagawa Y (1998). Quality preservation of broccoli by modified atmosphere packaging with two type of film having a high gas permeability. J Pack Sci Technol 7: 185-195.

Rai DR, Saito M, Masuda R and Y Ishikawa (1999). Effect of MA packaging on glutathione content and quality of hiratake mushroom. Proc Spring Con. Japan Soc Hort Sci. pp 347. Tsukuba University, Tsukuba, Ibaraki, Japan.

Rai DR, Saito M and Masuda R (2000). Effect of modified atmosphere (MA) packaging on glutathione and some other qualitative parameters of hiratake mushroom. J Japan Soc Hort Sci 69: 435-439.

Rai DR, Masuda R, Satio M (2000a). Effect of modified atmosphere packaging on low molecular weight carbohydrates of oyster mushrooms. J Food Sci Technol 37: 384-387.

Rai DR, Satio M, Masuda R (2000b). Effect of modified atmosphere (MA) packaging on glutathione and some other qualitative parameters of hiratake mushroom. J Japan Soc Hort Sci 69: 435-439.

Rai DR, Saito M (2004). Effect of controlled atmospheres on antioxidants and pigments of chingensai (Brassica rapa var. chinensis). Bev Food World 31: 65-66.

Saito M, Rai DR and Masuda R (2000). Effect of modified atmosphere Packaging on glutathione and ascorbic acid content of asparagus spears. J Food Proc Preserv 24: 243-51.

Saito M, Rai DR (2005). Qualitative changes in Radish (Raphanus spp.) Sprouts under modified atmosphere packaging in micro-perforated films. J Food Sci Technol, 42(1): 70-72.

Tenario MJ, Villanueva D and Sagardoy M (2004). Changes in carotenoids and chlorophylls in fresh green asparagus (Asparagus officinalis L) stored under modified atmosphere packaging. J Sci Food Agric 84: 357-65.

Varoquax P, Mazollier J and Albaganae G (1996). The influence of raw material characteristics on the storage life of fresh-cut butterhead lettuce. Post-harvest Biol Technol 9: 127-39.

□□□

17

Co-Products for Disease Prevention

VIDYADHAR KAWALKAR

Chronic diseases account for 58.1 percent of all deaths globally. The number of people with cancer is globally projected to double by the year 2030. The coronary artery disease has assumed epidemic proportions the world over and more so in India. Heart disease is considered the number one killer as it is responsible for above 40% of the deaths all over the world. In India about 10 to 14% of the adult population suffer from CAD. The concept that natural components in plant foods could have a beneficial effect on human health is not new. In India it forms the basis of the practice of traditional system of medicine, while in A.D. 390 Hippocrates the most celebrated Greek physician said "Let food be your medicine and medicine be your food". Particular vegetables and fruits have long been believed to be useful in the prevention and cure of disease. Western medicine also involved the prescription of specific plants and foods. Cruciferous vegetables were used therapeutically against several acute and chronic disorders. Medical traditions of India and China have retained this therapeutic knowledge.

The evidence

The cohort studies have examined the relationship between vegetables and fruit consumption and different types of cancers indicating significant reductions in risk mostly due to their co-products components . The case control studies found evidence that allium compounds were protective for stomach cancer. The ecological and experimental studies also showed anticarcinogenic effects at various sites including stomach. Most of other studies have shown evidence for isoflavones, polyphenols, glucosilates and indoles, flavonoids and coumarines for their effects on cancers.

Seventeen studies (65%) out of 26 experimental studies that involved diets containing soybeans have reported protective effects (Messina *et al.*, 1994). Because soy co-products isoflavones and lignans action as phyto-oestrogens. The cohort and case control studies evidenced the potentially preventive action of bio-active compounds-isoflavones and lignans in breast cancer (Wcrf, 1997). The known risk exposure for breast cancer are mainly related to exposure to unopposed oestrogen which may have more of a promotional effect (Potter, 1997). In a controlled human feeding trial found an increase in the length of menstrual cycle, in women fed moderate amount of soy co-product isoflavone (Cassidy *et al.*, 1993). The oestrogen like properties of isoflavones created interest as an alternative to conventional hormone replacecement therapy (HRT). In addition to possessing weak estrogenic activity, genestein has been shown to inhibit protein–tyrosine kinase (Herman *et al.*, 1995) and to have antioxidant activity (Wei *et al.*, 1995). Natural edible antioxidant nutrients can effectively supplement the endogenous defenses, which add significantly to the anti- free radical forces (Kawalkar, 1996).

Atherosclerosis Studies show that the co-products bioflavonoids reduce over all levels of circulating lipids in the blood stream and inhibit the formation of dangerous LDL-ox. In the blood oxygen can damage low density lipoprotein (LDL). Once oxidized they are deposited in the inner layer of the blood vessels, and the atherosclerosis process begins. A diet high in co-products - antioxidants/ bioflavonoids/ phytochemicals appears to inhibit oxidation and subsequent atherosclerosis, preventing the coronary artery blockages. Publication of meta analysis in 1995 showed that co-product soy protein lowered serum cholesterol in hypercholesterolenic individuals. In 1999 US FDA approved the claim for reducing heart disease using 25 gm/day of soy protein. American Heart Association also endorsed the use for cholesterol reduction.

The co-products : The functional food

As nutritional science is moving from the concept of "adequate nutrition" to "optimal nutrition", the new co-products which have the potential to improve physical and mental well being and which may also reduce the risk and aid in prevention of diseases are being developed as medicinal – functional foods.

There are many biologically plausible reasons why consumption of vegetables and fruits, soybean and other foods might deter the occurrence of cancer, heart disease and some other. These include the presence of potentially anti-carcinogenic and heart friendly substances-co-products, such as carotenoids, vitamin C, vitamin E, selenium, dietary fibre, dithiolthiones, isothiocynates, indoles, phenols, protease inhibitors, allium compounds, plant sterols, phyto-estrogens and others. These are collectively being called phytochemicals, antioxidants and bio-active compounds - the chemicals of plant origin that play a crucial role in metabolism. However, the crucial issue is the complexity and variety of the mixtures and not the specific dose of one or two compounds. These byproducts are now utilized as value added co-products-functional foods. Functional foods are so called because they are believed or claimed to have special qualities, such as promoting wellbeing and/or protecting against diseases. What marks them out from 'normal' foods as being functional in specified ways, for which claims are made.

The term "functional food" hints specially developed foods will protect consumers from a variety of diseases and discomforts. British Nutrition Foundation and in China and Japan the functional foods are also called medicinal food that have health giving properties and benefits beyond that of basic nutrition and a claim about this benefit. Very often the' functional ingredients–co-products' are known or claimed to be bio-actives/ phytochemicals.

Phytochemicals : The frontiers : The phytochemical family and major food sources

1. Diallyl Sulphide : Onions, garlic, leeks and chives.
2. Carotenoides (Beta-carotene, lycopene, lutein, zeaxanthine) : Carrots, cooked tomatoes, leafy greens. Sweet popatoes, apricots.
3. Flavonoids : Tea, coffee, citrus fruits.
4. Indoles : Cruciferous vegetables (broccoli, cabbage, kale, cauliflower, brussels sprouts).
5. Isoflavones : Soyabeans (tofu, soymilk).
6. Isothiocynates : Cruciferous vegetables.
7. Phenolic Acids (Ellagic acid, Ferulic acid) : berries, citrus fruits, apples, whole grains, nuts.
8. Polyphenols : Green tea, grapes, wines.

9. Saponins : Beans and other legumes.
10. Terpens (Perillyl Alcohole, Limonene, Camosol) : Citrus Fruits peel, Rosemerry.

The disease and prevention

The disease : Source –The Free Radicals : The damaging roll of free radicals: All this free radical damage is based on a single mechanism - an electron hunger, caused by an unpaired electron. This causes free radicals to damage biological molecules. Japanese biochemist Yukie Niwa estimates that at least 85% of chronic and degenerative diseases are the result of oxidative stress and damage, caused by free radicals. Previously there has been a great deal of publicity associating atherosclerosis with high levels of cholesterol. No doubt it has also a roll. Now it is shown that atherosclerosis actually correlate with circulating LDL that has been attacked by oxygen or free radicals - known as LDL-ox. This rancid cholesterol is itself a free radical. They can attack cell membranes that line our blood vessels and can lead to hardening and thickening of the arteries and eventually heart attack and stroke. Free radicals can attack DNA, leading to dysfunction, mutation and cancer.

How free radicals are created : Oxygen based tiny molecules and their production in body is continuous and inescapable feature. These tiny molecules can be devastastingly destructive wreaking havoc through out the body. (a) The cell energy system inadvertently operates as the most fundamental free radical factory in the body. (b) Immune system creates oxyradicals and Reactive Oxygen Species (ROS) deliberately as weapons. (c) Pollution and other external substances: In modern life we are constantly exposed to external substances, which generates free radicals. (d) Stress : The fast pace of modern life is also a recipe for free radicals. When mind and body come under stress free radicals are mass produced. Oxidative stress can originate as a part of life giving processes that create energy in every cell where free radicals are created as toxic waste. A free radical is intrinsically unstable, highly reactive with other items and molecules and for these reasons short lived. They can attack enzymes and proteins disrupting normal cell activities. Excess free radical activity in the skin caused by radiation from the sun can reduce skin suppleness and increase wrinkling.

The cancer process : At the earliest stages in the cancer process the dietary carcinogens such as HCA, PAH and N-Nitroso compounds, may contribute directly to the body's carcinogenic load. The extent to which such dietary carcinogens may initiate the cancer process depends on the availability of microconstituents - the bioactive compounds, which induce detoxification enzymes, which in tern, reduce the exposure of DNA to carcinogen. The initiation of cancer process also depends on the exposure to procarcinogen and conversion of normal cell into cancer prone cell is carried out by the activation of procarcinogen. Abnormal cells become more malignant and able to spread.

These are stages that themselves involve further changes in the cellular DNA. At almost everyone of the above stages known phytochemicals/ bioactive compounds can alter the likelihood of carcinogenesis.

- Glucosinolates and indoles, Isothiocynates and thiocynates, Phenols and coumarins can induce a multiplicity of solubilizing and usually inactivating enzymes.
- Ascorbate and phenols block the formation of carcinogens such as nitrosamines.
- Flavonoids and Carotenoids act as antioxidants disabling carcinogenic potential.

The prevention : The co-products

Free radicals management : Every cell in the body creates its own antioxidants enzymes to diffuse oxyradicals and ROS. The defense enzyme super oxide dismutase takes hold of most destructive free radical super oxide. The body has its own free radical fighting system and we can effectively supplement these systems with edible nutrients - antioxidants, bioactives and phytochemicals - the known co-products.

Antioxidants : Oxygen is the basis of all plants and animal life. In normal biochemical reactions oxygen can be unstable and capable of oxidizing neighboring molecules leading to cellular damage, which are linked to chronic diseases and aging. These are known as free radicals - reactive oxygen species (ROS). The chemicals capable of disarming free radicals are called antioxidants. Antioxidants are microconstituents of diet that are involved in DNA and cell maintenance and repair. They specifically protect DNA and cell membranes against oxidative damage by free radicals including that induced by carcinogenic agents. These include Beta-carotene, other carotenoids, vitamin C and E and trace element selenium. Other bio-active microconstituents-phytochemicals of food also have antioxidant properties.

Oxidative damage to DNA occurs, most, but not all, damage is corrected by internal surveillance and repair systems. Each cell has an antioxidant defence system that includes various enzymes, antioxidants, and iron and copper binding proteins.

Carotenoids and Vitamin C and E scavenge reactive oxygen. Other Bioactive compounds that have antioxidative properties include some phytoestrogens, glutathione, some phenols and some flavonoids. Selenium is found in plant foods such as cereals and seeds.

In addition to a role in the initiation, promotion, and progression of cancer, long- term oxidative stress has been linked to cardiovascular disease, because oxidized low density lipoprotein appear to be involved in atherogenesis. Oxidative stress has further been suggested to contribute to diabetes mellitus, age related eye diseases, neurodegenerative diseases (such as Parkinson's disease) and the aging process in general. Because antioxidants combat oxidative stress in the body, these diseases may be prevented by antioxidants.

Bioactives – phytochemicals : These are plant chemicals. Vitamin is a food substance essential for health. In contrast, phytochemicals, are substances in plants that may possess health protective effects, even though they are not essential for life. These are complex chemicals which include thousands of compounds, pigments and natural antioxidants. Many are associated with protection from heart disease, hypertension, cancer and diabetes. These are allium compounds, dithiolthiones, D-limonene-terpenoids, phytoestrogens, lignans, flvonoids, phenolic compounds, Ellagic acid, protease inhibitors, phytic acid, glucosinolates and indoles, plant sterols, saponins, coumarins. Plants contain phytochemicals because these substances are of benefit to the plants itself. An orange has at least 170 distinct phytochemicals. Phytochemicals in citrus fruits help protect against cancer and also have additional benefits. Whole grains phytochemicals are important in prevention of cancer and heart disease. Terpenoids produces significant reduction in total and LDL cholestesterol levels, thus reducing the risk of heart disease. Phytochemicals work to prevent cancer with the mechanism of modification of hormone effects and neutralization of free radicals. Many appear to act synergistically, both as antioxidants and as modifier of hormonal influences. There is no doubt that consumption of plant foods – co-products containing multiple antioxidants is strongly associated with health benefits. The bio-active-phytochemicals are capable of attacking and neutralizing the free radicals and offsetting their many pathogenic manifestations in the body.

The micronutrients, phyto-chemicals/bioactive compounds can alter the likelihood of carcinogenesis. The bio-actives-glucosinolates and indoles, isothiocynates and thiocynates, phenols and coumarins can induce multiplicity of solubilizing and inactivating enzymes. The ascorbate and phenols block the formation of carcinogens such as nitrosamines. Flavonoids and carotenoids can act as antioxidants, essentially disabling carcinogenic potential (Potter, 1997).

Health synergy : The likelihood of carcinogenesis and Chronic degenerative diseases such as heart disease and Diabetes can be altered by the phytochemicals-bioactive compounds-phytonutrients which are considered beyond nutrition.

Increased intake of intact plant foods–co-products with the multiplicity and complexity of agents that they contain, it is less likely that any clone of malignant cells can survive the poly pharmacy of plant constituents because of their synergistic action.

References

Boyce Burge : nutrients and nourishment. Phytochemicals: Why you should eat more fruits, Vegetables, whole grains and legumes.

Cassidy A, Bingham S, Carlson J *et al.* (1993). Biological effects of plant oestrogen in premenopausal women. FASEB J, 7: A866.

Hari Sharma, Freedom from Disease (1993).

Herman C, Adlercreutz T, Goldin BR *et al.* (1995). Soybean phyto-estrogen intake and cancer risk. J.Nutr 125: 7575-7705.

Kawalkar Vidyadhar (1996). Free radical Hazard – Aging and nutrient Defences. Proc-abstract of scientific presentation at National conference of Indian Dietetic Association.

Kawalkar Vidyadhar (1998). Breast Cancer and Nutritional Intervention. Proc. Guest lecture presentation at National conference on Breast Cancer.

Messina MJ, Persky V, setchell KDR, Barnes S (1994). Soy intake and cancer risk. Nutr cancer 21: 113-131.

Potter John D (1997). Plant Foods and cancer risk – Science and tradition. Proc. Nutr Soc.India Vol. 44.

Wcrf/ Aicr (1997). Global perspective-prevention of cancer.

Wei H, Bowen R, Cai Q, Barnes S, Wang Y (1995). Antioxidant and antipromotional effects of the soybean isoflavon genistein Proc. Soc, Exp. Biol Med 208 (1): 124-30.

WHO (1997). The world health report (Geneva WHO).

Wu Anna (2000). Cited in science news. Wcrf 25 March, 2000.

WCRF/ AICR (2007). Global perspective- Prevention of cancer.

❑❑❑

18

Conversion of Milling Co-Products, Oilseeds Derivatives, Animal Co-Products and Agro-industrial Wastes into Animal Feed

DEVINDER DHINGRA

Animal feed manufacturing involves the use of a variety of raw materials to produce compound feeds. The feeds are defined according to certain specification with regard to nutritive composition based on specified description for nutritional, hygienic and physical quality. Together, these specifications require knowledge of a vast number of different properties of ingredients to optimize processing while maintaining and controlling nutritional quality for a given feed form. Animals which receive pelleted feeds generally have high performance in terms of average daily gain and lower feed conversion compared with mash feeds as reported for pigs (Vanschoubroek *et al.*, 1971; Pond and Maner, 1984) and poultry (Calet, 1965; Quemere *et al.*, 1988; Moran, 1989). Feed processing influences rate of degradation and rate of passage of feed components of ruminants (Van der poel *et al.*, 1995) for instance, pelleting reduces the resistance of starch against ruminal degradation by about 15% (Tamminga and Goelema, 1995). Nielsen (1994) showed that expander processing reduces the effective protein degradability (EPD) of a raw material by an average of 8%. These effects may lead to differences in milk production and composition. It can therefore be argued that the nutritional value is both influenced by the raw material used and by processing conditions during operation (Skoch *et al.*, 1981, Skoch *et al.* 1983). Different animal species require different physical properties for their respective feeds, for fish feeds, additional pellet characteristic such as flowability, sinking velocity, water absorption and water solubility are important. Furthermore, the hygienic quality of feeds is important. Hygienic quality involves the control of microbiological contamination of feeds based on levels of enterbacteriaceae and salmonella (McCapes *et al.*, 1989). Feed processing includes the treatment (physical, chemical, thermal) of a feed prior to consumption by animals (Maier and Baker- Arkema 1992). In routine raw materials are blended and ground to obtain mash feeds that are further subjected to some form of steam and / or water conditioning before *e.g.* granulation is applied. Granulation and its products (Granulates, pellets) offers a product form that has many advantages over meal and that is why application is wide spread for both raw material, feeds additives and animals feed. Some advantages of pellets over meal are given here under (Vanschoubroek *et al.*, 1971) :

- Pellets have better flow properties, necessary for foods transport in conveying equipment and discharging behavior from silos, than the meal they were prepared from.
- The bulk density of pellet is generally higher than that of meal, so that more tonnage can be carried by truck.
- The composition of pellet as obtained from carefully blending and mixing ingredients remain fixed, no segregation of additives occurs.

However the use of shaping equipment like pellet presses, requires additional costs in terms of energy demand and necessary additional equipment; like boiler to generate steam, conditioning equipment, a pellet press and cooler. Making of mash in the form of pellets originated from almost 60 years ago when Patton *et al.* (1937) defined the nutritional superiority of pelleted poultry feeds over mashes. Pellets are obtained by a pressing process in animal feed manufacturing. In former times the mash was simply pressed between two rollers or in a cake press to obtains pellets/cakes without any preliminary treatment. Relatively low pressure was used and feed did no heat up (Calet 1965). In modern feed mills the mash is pelleted in a so called roller and dies pellet press, both vertically and horizontally. Before entering the pellet press, the mash is subjected to some form of pretreatment before granulation such as mixing with molasses or fats conditioning with steam (Skoch *et al.*, 1981) or being subjected to the use of an expander (Veenendal

1990; Pickford, 1992) to increase temperature and moisture level Maier and Bakker- Arkema, 1992). When steam is used, the temperature of pellet after leaving the die is generally higher in comparison with that of the conditioning meal due to the frictional heat in the die. Finally, pellets are cooled with ambient air

Various hypothesis have been postulated in order to clarify the factor that determine the structural integrity in a pellet. In general causative factor that affect pellet quality are diet ingredient composition and its properties, process technology and specific pellet blinders.

Table 1 : Factors affecting the pellet quality

Factor	References
Diet ingredient composition	
Physical	
Particle size (distribution)	Stevens (1987), Payne (1978)
Specific density	
Bulk density	Macmohan and Payne (1991)
Angle of repose	Mohsenin (1986)
Surface area	
Chemical	
Moisture	Skoch *et al.* (1983)
Ether extract	Salmon (1985), Richardson and Day (1976)
Crude fibre	
Crude protein and ash	
Functional	
Viscosity	Keller (1983)
Protein solubility	Hermansson (1979), Kinsella (1979), Wood (1987)
Starch gelatinization	Wood (1987), Smith (1983)
Diet ingredients	MacMohan and Payne (1991), Israelsen *et al.* (1981), Payne (1978)
Process technology	
Conditioner	Leaver (1984)
Temperature	Skoch *et al.* (1983), Winowiski (1988), Stevens (1987)
Time	Israelsen *et al.* (1981)
Moisture	
Steam	Maier and Gardecki (1992), Stevens (1987)
Pellet press die specification	
Diameter to length ratio	Stevens (1987), Hanrahan (1984)

Cooler/drier	
Air speed	
Time	Maier and Bakker-Arkema(1992)
Air moisture level	
Pellet layer thickness	Maier and Bakker-Arkema(1992)
Pellet blinders	
Adhesion and cohesion between particles	Winowiski (1988), Payne (1978)

Processing steps

Feed processing includes the treatment (physical, chemical, thermal) of a feed prior to consumption by animals (Maier and Bakker-Arkema, 1992). In general, the variability in processing effects is associated with the choice of equipment, with processing conditions as well as with the processing system, e.g. the combination and sequence of processing equipment (Melcion and Vander Poel, 1993). The commonly used equipments for production of animal feed are:

1. Cleaner
2. Destoner
3. Hammer mill/grinder or pulveriser
4. Mixer/Blender
5. Conditioner
6. Vertical pelletiser/Horizontal pelletiser
7. Extruder with pre-conditioner
8. Drier
9. Weighing and bagging machines

In addition to the above bins for storage, elevators for conveying and handling and silos for storage of ingredients are required. The selection of the equipment depends on the type of feed, capacity and the ingredients to be used. The individual machines are discussed under various unit operations:

Cleaning: Most of the grain ingredients have 5-8% dockage and other impurities. These are cleaned by grain cleaners fitted with aspiration system and sieves. The aspiration system removes the lighter impurities, whereas scalping sieve rejects the bigger impurities. The grading screen helps in separating the finer impurities. The grain is also passed through a vibrating deck destoner to remove stones which have size and shape similar to that of grains. In modern units a magnetic separator is also used to separate ferrous particles from feed ingredients. Cleaners are available in the market in various capacities. The same cleaner can be used for different grains by changing the screens. Studies showed that differences in pellet quality of animal feed using different diet ingredients, e.g. grains or legume seeds, which can be attributed to differences in their physico-chemical properties affected by processing history, geographical and climatic origin and cultivar.

Grinding: The weighed amounts of ingredients are grinded in hammer mill or pulveriser. The ingredients are grinded in the mill until they are reduced to fine particle size. Particle size is an important influencer of pellet durability. Generally, finer the grinding, higher the durability. Fine particles usually accept more moisture then large particles and, therefore, undergo a higher degree of conditioning. Also, large particles fissure point that cause cracks and fractures in pellets. The recommended particle size for good pellet quality is 0.6 to 0.8 mm. A particle size of 0.5 to 0.7 mm is recommended to produce durable pellets. Feed mills that used a hammer mill screen size of 3.2mm or 4.0-3.2mm split screen produced high quality pellets. Feed mills that used a hammer mill screen size of 4.0 mm had the lowest pellet quality. This process generally takes 20-25 minutes per quintal of the ingredients.

Blending: The ground ingredients are mixed in ribbon blender for proper mixing and binding of the ingredients. Blending in pellets most probably is due to solubiliztion and subsequent crystallization of feed stuff components e.g. starch, sugars, fat or 'liquid necking'. Liquid necking is a binding mechanism which uses the surface tension of water, in the three phase system of air, water and particles to maintain structural integrity of the pellet. The best known binders used in animal feed production are bentonite (a clay mineral), carboxymethyl cellulose (CMC: modified cellulose) and lignosulphonates. The effect of binder materials depends on the presence of water, which is required for a binding agent to become active. The blending of ingredients in the form of mash is mostly done in 30 minutes.

Conditioning

Conditioning in animal feed manufacturing can be defined as the process of converting the mixed mash with the use of heat, water, pressure and time to physical state that facilitate compaction of the feed mash. Conditioning increases production capacity and simultaneously affects the physical, nutritional and hygienic quality of the produced feed (Skoch *et al.*, 1981).

Steam conditioning

The conditioner most widely used in animal feed manufacturing to incorporate water, steam and/other liquids is barrel type conditioner. The overall effect of steam addition is an increase in moisture content and heating of the feed-mash. In literature, five classes of feed are often distinguished each with own specific conditioning requirements. These classes includes feeds which are (a) high in starch (b) heat sensitive (diets include sugar or milk) (c) high in natural protein (d) high in fiber and (f) rations high in urea and/or molasses (Payne, 1978; Maier and Gardecki, 1992). In Table 2 some of the conditioning recommendations are summarized.

Table 2 : Steam conditioning recommendations for different feeds Maier and Gardecki, 1992)

Feed Type	Specific Component	Recommended steam pressure (kPa)	Temp of final meal (°C)
High starch feed	50-80% Starch from cereals or tapioca	102	80 - 85
Dairy rations	High fibre, 12-16% protein	442	60 - 65
High protein supplements and concentrates	25-45% protein	442	80
Heat sensitive	5-25% Dry milk powder, Sugar and / or whey	102	< 55
Urea containing rations	6-30% Urea	442	-

Water addition

Water is generally added by means of a conventional barrel type conditioner. Feed components exhibit different behavior when water is present. Gelatinization and solubilization process are facilitated by the presence of water. Subsequently the water in this respect changes the structure of other materials to such an extent that binding between the feed particles becomes possible. In order to enhance the binding properties and therefore increase the pellet quality characteristic, heat is often a necessary pre-requisite. Found an increase in pellet durability when water was exchanged for steam.

Residence time

Only a few reports deal with the subject of long time conditioning of feed mash with the aid of ripening vessels. In the technical literature the positive effects of ripening, with or without steam, points to better quality of feed pellets. More moisture or other liquids (molasses) can be added when ripening time of 20 min is incorporated in the process line, which is then followed by steam injection with no loss in productivity and feed quality. In case of ripening without steam injection before pelleting, pellet quality was lacking. These effects were attributed to the diffusion of liquid in the feed particles during holding in the vessel. The relative dry surface of the particle which appear will have negative effect on the physical quality of the pellets. From this it can be concluded that long time ripening followed by stream injection permits higher uptake of liquid and improves pellet quality.

However modern feed manufacturing involves the use of great number of different animals' diets that have to be produced in relatively short time spans. This means that equipment which incorporate some form of hold time will decrease the flexibility of the plant with respect to the number of feed changes per time unit. Interest is therefore, raise in equipment with short hold times (< 2 min) with the same favorable effect on pellet and nutritional quality.

Pelleting: Advantages and disadvantages of this key process in animal feed manufacturing have been described by Vanschoubroek *et al.* (1971). In the pelleting process, conditioned feed mash is pressed through a die. A majority of the pellet presses in feed manufacturing are of the ring die design. Different designs exist in which usually two or three rollers are used. In most designs the die revolves around the fixed rollers. A minority is designed as flat die presses in which the die is static and the horizontal rollers rotate around a vertical axis while forcing the meal through the die plate. Die sizes may differ in their length to diameter ratio, normally expressed as width of the bore times the length of the die hole. Increasing die thickness or decreasing bore width will increase the amount of shear which the feed mash receives. This is limited since a too high amount of shear (shear dies or small bore holes) will block the pellet press.

Formulation is said to be the single most important variable affecting the physical quality of the pellet feeds. Die speed of the pelletiser is also an important factor in pellet making as die plugging was observed at a die speed of 126 rpm for corn based swine diet. The speed of 152 -268 rpm did not affect pellet durability of corn based swine diet, and resulted in pellet durability of about 90%. Also, die speed of 126 to 268 rpm did not affect pellet durability of wheat based swine diet, and resulted in pellet durability of about 98% (Stevens, 1987). High die speed is suggested for small pellets (3 to 6 mm diameter), and low die speed (about 6 to 7 ms^{-1}) is suggested for large pellets and cubes recommended a low peripheral die speed of 4 to 5 ms^{-1} for low bulk density materials, where a large volume must be expelled during pelleting.

Pellet quality is usually measured and expressed by means of the Pellet Durability Index (PDI), and high PDI results are associated with pellets that are able to withstand handling including auguring, transportation and feeding pressure without substantial breakage that would then lead to a final product with an excessive amount of loss particles commonly referred to as fines. In place of pelletiser an extruder or expander can also be used for the production of cattle feed (Pipa and Frank, 1989). Horizontal pelletiser with a ring die and vertical pelletiser with a flat die are commonly used. These are presented in Fig. 1&2.

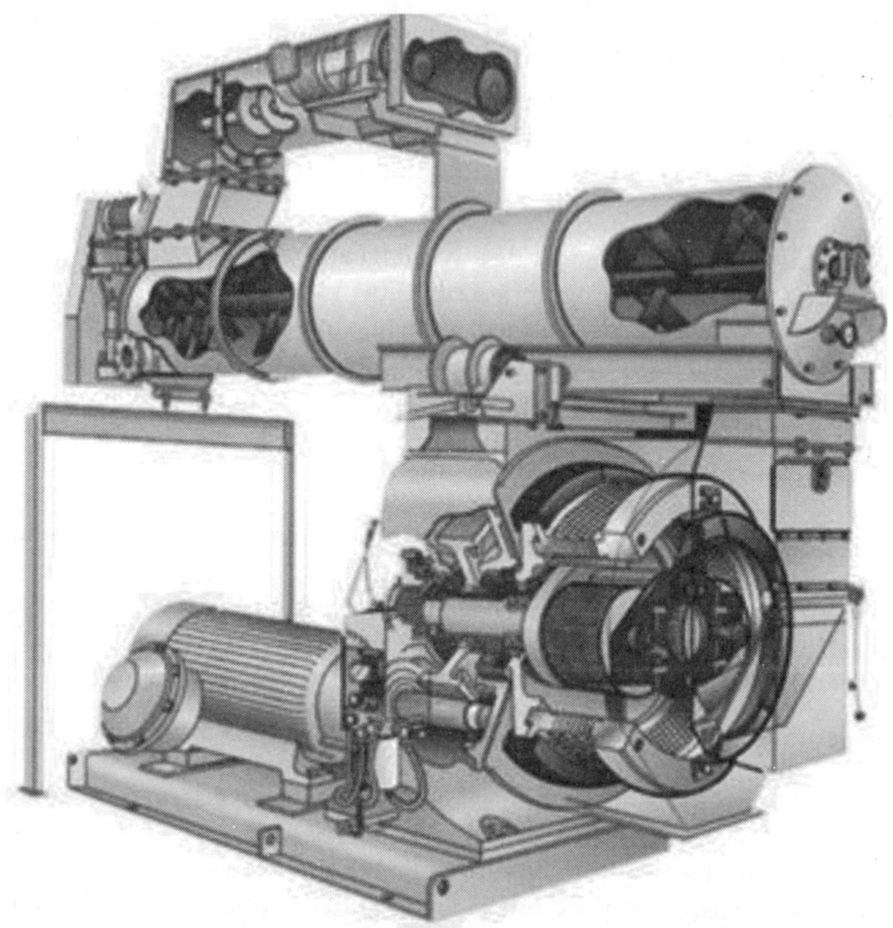

Fig.1 : Horizontal pelletiser with ring die

Extrusion: A process by which moistened, expansible, starchy and/or proteinaceous materials are plasticized and cooked in a tube by a combination of moisture, pressure, temperature and mechanical shear. This is a developing feed manufacturing process and is now considered important for both, animal and human foods in India. Extrusion technology is presently being used for human food, pet food and fish feed. This is owing to higher initial investment and higher operating costs in comparison to pelletising. The advantages of extrusion technology are listed as under:

- Good for processing ingredients & feed
- Provides formulation flexibility
- Versatile in giving physical shape/size
- Higher water stability
- Higher physical integrity
- Higher fat absorption capacity
- Ability to control product density
- Enhances feed efficiency
- Prolongs shelf life/sterilizes
- Good gelatinization of starches (90%)

Single screw extruders with pre-conditioners are in use for pet foods and fish feeds. In case of pet foods special mixing and filling heads are used to prepare pet treats.

Fig. 2 : Vertical pelletiser with flat die

Cooling / drying: In order to decrease moisture and latent heat, pellets need to be cooled after the conditioning and pelleting steps. Pellets generally leave the die at temperature ranging from 60-95°C with moisture content of 120-175 g kg^{-1}. The free water content of the feed is decreased in the cooling process, making it possible to store the feed for a sufficient time period. The amount of water and heat that is dissipated from the pellets is a function of air flow and air properties, raw material characteristics and pellet–size. (Maier and Bakker-Arkema, 1992).

Packaging and storage: Finally the prepared animal feed pellets are packed in HDPE (High density polyethylene) plastic bags. The combined effect of temperature (8-40°C) and relative humidity (60-90%) on the physical quality of alfalfa cubes after 66 or 90 days of storage period. Generally, durability decreased as temperature and relative humidity increased. In high humidity storage study, found that an increase in the moisture content of the poultry litter pellets from 6.0 to 10.4% (w.b.) increased the durability from 92 to 95%. Further increase in moisture content from 10.4 to 22% (w.b.) reduced the durability from 95 to 88%.

Effect of diet ingredients on physical quality of pellets

It has been shown that the physical quality of pellets (pellets durability and hardness) is influenced by the ingredient composition of the diet. In general, the effect of ingredient composition on pellet quality has been attributed to changes that occur if the ingredients are subjected to physical compression and shear during the pelleting process. The extent of these changes increases with conditioning of the material (Moran 1989) before pelleting and by the influence of the pelleting process itself. An overview is given on the influence of the diet ingredients on the physical quality of feed pellets.

Diet ingredients: The use of least cost formulation to formulate diets leads to large number of feedstuffs incorporated at different inclusion levels. This may lead to variation in physical quality of the feeds after pelleting, although the calculated nutritional requirements are met. In addition, some pelleting parameter such as pressure (distribution) in die hole and porosity which change during pelleting. Also the amount of energy required to overcome friction in the die hole can only be roughly estimated rather than exactly measured, and these parameters are strongly dependent on the physico-chemical properties of the diet ingredients themselves.

To predict pellet quality, a pragmatic approach often used, as, *e.g.*, proposed by MacMohan and Payne (1991). They tried to relate different raw material to pelleting criteria used in animal feed manufacturing. For certain raw material, a classification was made according to type of raw material, e.g., cereals, oilseeds, or by-products. They estimated the effect of inclusion of a specific raw material on the physical quality of pellets, pelleting capacity of the pellet press, and wear of the die. These three physical factors were scaled so that virtually all raw materials fell in a range between 0 and 10, exception being fat and binding agent (Table 3). Israelsen *et al.* (1981) used linear regression to relate pellet quality to different diet ingredients. They estimated the effect of inclusion of different diet ingredients on pellet durability (%) and energy consumption (kWh ton^{-1}) of the pelleted raw material (Table 4) by substituting barley or cotton seed meal by a number of other ingredients. From their investigation, it was concluded that some byproducts alter significantly pelleting properties of the feed. The authors concluded that inclusion of about 10% cane, or beet molasses in combination with solid by products reduced the formation of fines and improved specific press capacity.

Table 3 : Feedstuffs and their respective nutritional and pelleting properties (modified after MacMohan and Payne, 1991)

Raw material	Constituents					Physical factor*		
	Crude protein (g/kg)	Crude fat (g/kg)	Crude fibre (g/kg)	Starch (g/kg)	Bulk density (kg/m³)	Pellet quality	Press capacity	Die wear
Milling by product								
Barley meal	107	22	47	490	480	5	6	5
Maize meal	87	38	21	585	610	5	7	6
Milo meal	90	28	40	580	540	4	6	7
Oat meal	112	48	106	378	520	2	3	7
Rice	78	20	88	580	480	5	5	4
Wheat meal	119	17	23	555	540	8	6	3
Oilseed and derivatives								
Coconut cake	207	82	130	5	480	7	8	6
Cotton decorticated	403	308	28	0	640	7	8	6
Cotton meal extracted	436	30	118	8	610	8	6	7
Ground nut cake deo.	469	79	54	63	620	7	8	4
Ground nut meal extracted	503	5	125	20	670	8	6	5
Linseed meal extracted	334	31	94	16	560	7	6	5
Palm kernel cake expeller	146	91	188	4	480	6	7	4
Palm kernel meal extracted	152	21	189	3	700	6	5	5
Palm kernel (whole)	93	478	102	0	750	3	8	3
Rapeseed meal extracted	343	22	114	11	510	6	6	6

Sesame meal expeller	451	114	62	14	560	7	7	4
Soybean meal extracted	449	18	53	8	500	4	5	4
Soybeans full fat	356	189	53	9	480	4	8	3
Sunflower cake expeller	383	71	167	35	560	6	6	4
Soybean meal extracted	339	20	192	26	530	6	5	5
Animal by products								
Blood meal	878	7	0	0	560	3	5	3
Fat (added at mixer)	0	1000	0	0	900	< - 10	> 10	0
Feather meal	824	76	0	0	400	4	5	5
Fish meal	564	116	0	0	640	4	7	5
Meat meal	569	100	22	0	620	5	7	3
Meat and bone meal	498	91	16	0	690	4	7	4
Poultry by product meal	711	133	27	0	590	3	8	4
Legumes								
Field beans	254	13	73	376	690	7	5	5
Peas	206	11	5	410	720	6	5	5
Lentils	229	13	45	422	800	4	4	5
Locust beans	40	0	71	0	400	4	4	6
Others								
Brewers grains, dried	252	67	126	24	320	3	4	5
Citrus pulp	61	22	118	13	330	7	3	6
Maize germ meal	142	53	69	339	480	5	8	3
Maize gluten feed	185	38	70	188	540	3	4	6
Maize gluten meal	607	37	11	156	480	4	5	5
Tapioca	24	4	43	655	640	5	3	7
Minerals	0	0	0	0	1000	2	4	10
Beet molasses	110	0	0	0	1230	7	6	
Rice bran	129	130	110	200	320	2	3	6
Skim milk powder	349	23	0	0	640	9	2	9
Sugar beet pulp (molassed)	n.a.	n.a.	n.a.	n.a.	240	7	3	5
Lignosulphonate	n.a.	n.a.	n.a.	n.a.	500	>10	> 10	0

* Figures on a scale of 0 (poor contribution) to 10 (high contribution), based on pellet press operators, experience

Effect of raw constituents

Ingredients' constituents can be classified as starch, protein, sugar, and non-starch poly-sachharide (NSP), fat, fiber, inorganic matter and water. The matrix structure in which the different components are arranged is very complex, and this complexicity may prevent the expression of single constituents on pellet quality.

A well known example for instance, is the diet inclusion of fats. Free fats added in the mixer negatively affect pellet durability and hardness, but will improve press capacity in terms of tons h^{-1} produced. This effect is attributed to lubricating effect of added fat on the mash–die interface during pelleting. The same amount of naturally occurring fats enclosed in cell walls does not have such a marked effect on these parameters.

Table 4 : Effect of diet ingredients on specific power consumption and durability of pellets (after Israelsen *et al.*, 1981)

Diet ingredient (kWh ton^{-1})	Specific power consumption	Durability (%)
Control	6.5	97.2
Grain substitute		
Beet pulp, dried, pelleted	6	98.9
Barely malt culms, dried	4.2	98.3
Citrus pulp, dried, pelleted	4.9	97.6
Barely, ground	7.3	97.6
Coconut meal	7.1	97.4
Alfalfa meal, dried	6.8	97.2
Grass dried pelleted	7.6	97.2
Wheat bran pelleted	6.9	96.9
Palm kernel cake	8.1	96.8
Cotton seed substitute		
Sunflower seed meal	7.09	4.9
Soybean meal	5.6	94.5
Rape seed meal	5.8	91.2

Starch and sugar

In human food engineering, starches are used for various purposes. According to Smith (1983), starch may function, among others, as an adhesive or binding agent. Starches used for these purpose have undergone a heat or chemical treatment in which the properties of the natives starch are changed. Gelatinizing of starch in the presence of water and heat, in the presence or absence of shear, is the most common way to affect functional properties of starches (Smith, 1983). Chemical reagents are used to modify starches to specific needs; an example of the various processing technique using for example soda, borax, or urea to create different adhesives for the paper and board industry. Processing of starch leads to the cross linking which may give the starch its desired properties as for instance, changed rate and amount of swelling (Smith, 1983), which may be important in food applications. To obtain feeds with a certain specified quality standard in terms of hardness and durability, starch needs to be modified by either the feed manufacturing process functionally in terms of binding or adhesion to produce durable fed pellets. Wood (1987) showed that the amount of pre-gelatinized starch in feed is related to the physical quality of the feed pellets. Using a design in which a native starch was gradually replaced by pre-gelatinized starch (feed model system containing 40% starch), an increase was found in physical quality as determined by Kahl hardness test and Holmen durability test. Within the range tested, no optimum was found, indicating that pre-gelatinized starch included in the diets of percentages up to 40% might still lead to harder and more durable pellets (Wood, 1987).

Heffener and Pfost (1973) found a higher degree of gelatinization of feed starch with the use of a smaller diameter of the die hole. Gelatinization and durability of the pellets were also higher with the smaller die holes. However, the ratio of length to diameter was also slightly altered, making it difficult to attribute higher gelatinization percentage to the smaller die diameter. The underlying mechanism of the contribution of gelatinized starch in binding properties in pellets is still not fully elucidated. The presence of water, for instance, is a pre-requisite to initiate gelatinization. A ratio of 0.3:1 (water:starch) is needed. Various authors state that, to fully gelatinize starch, water is needed in a ratio of at least 1.5:1 (water:starch). This implies that during the manufacturing stage of compound animal feeds, water is a limiting factor to fully gelatinize starch. Depending on feed ration used, approximately 3% water in the form of steam is added. To act as a binder, bonds have to be formed at the contacting sites of particles, so it would be sufficient if the starch in the outer layers of the particle is gelatinized.

Protein

The properties of proteins in foods or food-systems have been classified: (1) sensory and kinaesthetic (*e.g.* flavour, odor, color, texture); (2) hydration, dispersibility, solubility and swelling; (3) surface – active properties (required for e.g. emulsion and/or foam formation); (4) rheological properties, including gelation and texturization; (5) other properties (e.g. Including the adhesions, cohesion, dough making, film and fibre formation during extrusion). In feed manufacturing especially the adhesive forces that protein may exert, are of interest. Protein can act as a binding agent between different feed particulates. Processing involves the combined effect of shear, heat, residence time and water resulting, among others in partial denaturation of the protein in the feed. As has been shown by Wood (1987) partial denaturation of the protein during processing may positively affect the hardness and durability of the feed pellets. Denaturation involves the breakdown of the (Spatial) three dimensional structure of the proteins (the secondary, tertiary and quaternary structures), thereby changing the bioactivity of the protein. Upon cooling protein reassociate and so bonds can be established between the different particles. No report could be found in which the effect of processing variable on feed manufacturing has been described on denaturation of protein and consecutive effect on physical quality.

Fibre

Fibre can be classified as water soluble part, which may cause a high viscosity, and a water insoluble part. The classification may be useful in explaining the effect of different fibre sources on pelleting characteristics.

Water-soluble fibre like glucans, arabinoxylans and pectin strongly raise the viscosity and may act as filler in the feed; the viscous material embeds the more coarse particles, thereby reducing the porosity in the feed. Subsequently, the structural integrity of the feed agglomerate is higher, resulting in higher durability and hardness of the pellets. During processing, process variables, such as residence time, moisture content, pressure and heat, alter the state of soluble fibre. To obtain the desired functional properties, the effect of time, pH and temperature are of major importance (Keller, 1983). Due to the complexity of the feed mash (blends of several raw materials that differ in composition and functional properties) and processing conditions (heat, pressure and time relationships), quantitative description of the effect of fibres, which forms with water a viscous solution / paste, on pellet quality are difficult to acquire.

The effect of insoluble plant fibres may be two fold. Plant fibres may prove beneficial in the pelleting process since they have the opportunity of entangling and folding between different particles or strands of fibre. On the other hand, due to their stiffness and elasticity, they may impair problems to the pellets-press operator since resilience characteristics of the material oppose good contact between particle fibres. Moreover, when large fibres are present within the pellet, they might induce a weak spot in the pellet. Large fibre particles may lead to homogeneity at which pellets are more likely to break. Increasing the residence time in the die of the pellet -press would diminish the effect due to the resilience the material. This is in line with result, who found that an increase in hold time within a wafering machine resulted in the further stress decay of the materials under investigation, and subsequently increased durability of the formed wafer.

Fat

Added fat in the compound animal feed is known for its adverse effect on pellets hardness and pellets durability. Since most binding of feed particles incorporates water or, when involved, solubilise starches, protein, fibres, fat with its hydrophobic nature may interfere with binding properties of water-soluble components in the feed. Moreover added fat (and to lesser extent, fat enclosed in the matrix of cell walls) acts as a lubricant between particles and feed mash and the die-wall resulting in a lowering pelleting pressure. This will already cause lower pellet quality in many cases. Also, as a result of fat addition various properties of a raw material constituent might be influenced. For instance, gelatinization of starch can be inhibited in the presence of lipids or delayed to higher temperatures. This may directly lead to weaker pellets, vulnerable to deterioration in transport systems. Salmon (1985) showed that more fat in the pellet (0, 30, 60, 90 g kg^{-1}) significantly increased the amount of fines returned; 2.7, 7, 10.3, 15.6% with no bentonite added. Richardson and Day (1976) showed that an increasing amount of fat added in mixer increased the amount of fines returned from the pellet mill and decreased the amount of mechanical energy (kWh $tonne^{-1}$) needed to convert 1 tonne of mash into pellets (Table 4).

Fat enhances the production rates in pellet mills, primarily because of the lubricating effects of fats between the meal and die surface (Richardson and Day, 1976). Because of low friction and pressure in the die decreased which in turn, would result in feed pellets with lower hardness and durability. However, some authors suggest that during blending, the feed mash with steam, natural oils and waxes would accumulate at the contacting sites of two particles and create a binding point either solid (waxes) or via liquid necking (oil or water) between particles upon cooling. This could have positive effect on pellet hardness and durability because of immiscibility of water and fats, there would be an optimum in the amount of bonds established by waxes and fats on one side and water on the other side.

Pellet blinders

When physical quality of pellet is not sufficient to obtain a saleable product or it does not meet the manufacturer's quality standards, binder material can be incorporated to increase the physical quality of the pellets. For this purpose, different types of binders are available. The best known binders used in the animal feed production are bentonite (a clay mineral), carboxymethyl cellulose (CMC; modified cellulose) and lignosulphonates. The effect of binder material depends on the presence of water, which is required for a binding agent to become active.

Lignosulphonates are by product form the paper industry. During the process of delignifying of fibres, $NaHSO_3$ and or Na_2SO_3 is added to which the alkaline components $(SO_3)^{-2}$/ $(HSO_3)^-$ react with chemical bonds of the lignin to form water soluble ligno- sulphonates. Common inclusion levels range form 0.5% to 3%.

Bentonite as binder is used in feed technology to improve physical pellet quality. Its main target is acting as filler, thereby decreasing porosity in pelleted feed. In addition bentonite works as a lubricant in the die hole it decreases pressure and subsequently energy requirement of the pellet press. Especially with voluminous feed mashes as for instance, fibre - rich diets, bentonite may prove advantageous.

Carboxy-methylcellulose (CMC) can also be used as binder in animal feed production. CMC gives a viscous solution/paste when mixed with water. Hydration occurs during conditioning. Subsequent compaction brings particulates close together and allows for the development of ionic attractions between the CMC and particulates.

The production of food grains is almost stagnant whereas the demand is on the rise. Under such a situation nothing can be wasted. In order to feed the livestock in a productive manner it is very important now to process / convert by-products, crop residues and agro-industrial wastes into animal feed. It needs a holistic approach involving animal nutritionists, engineers, bio-chemists and microbiologists to work together to process the so called waste into animal feed.

References

Calet, C (1965). The relative value of pellets versus mash and grain in poultry nutrition, World Poult., 21: 23-52.

Hanrahan,TJ (1984). Effect of pellet size and pellet quality on pig performance. Anim. Food Scie. Technol. 10: 277-283.

Heffner, LE and Pfost, HB (1973). Gelatinization during pelleting. Feedstuffs, 45(23): 32.

Hermansson, AM (1979). Methods of studying functional characteristics of vegetable proteins. J. Am. Oil Chem. Soc., 56: 272-279.

Israelsen, M, Busk, J and Jensen J (1981). Pelleting properties of dairy compounds with molasses, alkali-treated straw and other byproducts, Feedstuffs, 7 Dec.: 26-28.

Keller, J (1983). Food starches and their uses. In: D.L. Downing (Editor) Gum and Starch Technology Eighteenth Annual Symposium. New York State Agricultural Experiment station, Geneva, pp. 1-9.

Kinsella, JE (1979). Functional properties of soy proteins, J. Am. Oil Chem. Soc., 56: 242-258.

Leaver, R (1984). Feedstock conditioning. Food Int., Nov. 37-44.

MacMahon, MJ and Payne, JD (1991). The Pelleting Handbook. Borregaard Lignotech, Sarpsborg Norway, 65 pp.

Maier, DE and Bakker- Arkema, FW (1992). The counterflow cooling of feed pellets, Agric. Eng. Res., 53: 305-319.

Maier, DE and Gardecki, J (1992). Food mash conditioning field case studies, Paper no. 92-1541, ASAE International Winter Meeting, 15-18 December, Nashville.

McCapes, RH, Ekperigin, HE, Cameron, WJ, Ritchie, WL, Slagter, J, Stangeland, V and Nagaraja, KV (1989). Effect of a new pelleting process on the level of contamination of poultry mash by Escherichia coli and Salmonella, Avian Dis., 33: 103-111.

Melcion, JP and van der Poel, AFB (1993). Process technology and antinutritional factors: principles, adequacy and process optimization. In: A.F.B. van der Poel, J. Huisman and H.S.Saini (Editors), RecentAdvances of Research in Antinutritional Factors in Legume Seeds. Wageningen, Proceedings of the Second International Workshop on Antinutritional Factors (ANF's) in Legume Seeds, Wageningen, The Netherlands. pp.419-434.

Mohesnin, NN (1986). Physical Properties of Plant and Animal Materials. Gordon and Breach, New York, 891 pp.

Moran, ET (1989). Effect of pellet-quality on the performance of meat birds . In: W. Haresign and D.J.A. Cole (Editors), Recent Advances in Animal Nutrition. Butterworths, London, pp. 87-108.

Nielsen, I (1994). Effect of expanding on protein degradability in different raw materials and mixtures. Intercoop Feedstuffs Congress, 23-26 June 1994, Lofoten, Norway. 3 pp.

Patton, JW, Buskirk, HH and Rauls, LA (1937). A study of the relative merits of pellets and mash poultry feed, Vet. Med., 32: 423-427.

Payne, JD (1978). Improving quality of pellet feeds, Milling Feed Fertil., 162: 34-41.

Pickford, JR (1992). The potential and use of expanders and extruders in animal feed production. Animal Feed production for the 1990s. Feed Expo Ireland.

Pipa, F and Frank, G (1989). High-pressure conditioning with annular gap expander. A new way of feed processing, Adv. Feed Technol., 2: 22-30.

Pond, WG and Maner, JG (1984). Swine Production and Nutrition Aniaml Science. Textbook Series AVI Publ. Co., Inc., Westport, CT, USA, 646 pp.

Quemere, P, Castaing, J, Chastanel, JP, Grosjean, F, Latimier, P, Saulnier, J and Willequet, F (1988). Inflence of the physical form of the feed on the performance of bacon pigs, J. Rech. Porcine France, 20: 351-360.

Richardson, W and Day, EJ (1976). effect of varying levels of added fat in broiler diets on pellet quality Feedstuff, 48(20): 24.

Salmon, RE (1985). Effects of pelleting, added sodium bentonite and fat in a wheat-based diet on performance and carcass characteristics of small white turkeys, Anim. Feed Sci. technol., 12: 223-232.

Skoch, ER, Behbke, KC, Deyyoe, CW and Binder, SF (1981). The effect of steam-conditioning rate on the pelleting process, Anim. Feed Sci. Technol., 6: 83-90.

Skoch, ER, Binder, SF, Deyoe, CW, Allee, GL and Behnke, KC (1983). Effects of pelleting conditions on performance of pigs fed a corn- soybean meal diet, J. Anim. Sci., 57(4): 922-928.

Smith, PS (1983). Food starches and their uses. In:D.L. Downing (Editor), Gum and Starch Technology Eighteenth Annual Symposium. New York State Agricultural Experiment Station, Geneva, pp. 1-9.

Stevens, CA (1987). Starch gelatinization and the influence of particle size, steam pressure and die speed on the pelleting process. Ph.D. Thesis Kansas State University, USA, 86 pp.

Tamminga, S and Goelema, JO (1995). The significance of rate and site of starch digestion in ruminants. Carbohydrates in feeds for ruminants. Proc. SCI, 28 February 1995, London, UK.

Van der Poel, AFB, Verstegen, MWA and Tamminga, S (1995). Chemical physical and nutritional effects of feed processing technology. In: 16th Western Nutrition Conference, 13-14 September 1995, Saskatoon Sask., Canada. 266 pp.

Vanschoubroek. F, Coucke, L and van Spaendonck, R (1971). The quantitative effect of pelleting feed on the performance of piglets and fatterning pigs, Nutr. Abstr. Rev., 41: 1-9

Veenendaal, J (1990). Extrusion in the compound feed industry, Adv. Feed Technol., 3: 60-72.

Winowiski, T (1988). Problems in optimizing pellet quality. Feed Manage., Sep.: 36-42.

Wood, JF (1987). The functional properties of feed raw materials and their effect on the production and quality of feed pellets, Anim. Feed. Sci. Technol., 18: 1-17.

□□□

19

Utilization of Co-Products of Horticultural Crops and its Impact on Environment

D.B. SINGH

India is ranked third in the world in terms of horticultural produce, with an annual production of over 102 million tones of fruits and vegetables. However, as estimated 30% of this goes co-products in the absence of facilities for preservation, the perishable fruits/vegetables are unable to reach the consumers in time. This is compounded by meager utilization (less than 5%) by the fruit/vegetable processing industry. This may be compared to 30-70% utilization by processing industries in Southeast Asian countries. Other factors for spoilage of the produce are lack of efficient transportation and low prices offered to the growers. As a result, growers are forced to sell their horticultural produce at throwaway prices to middlemen and contractors. These distress sales are all the more debilitating for small and marginal farmers who produce a larger proportion of fruits and vegetables. Thus, not only is a valuable source of nutrition got co-productsd, but its producers face impoverishment and misery. On the other hand consumers are left at the mercy of avaricious wholesale dealers and the organized sectors producing high cost branded products. Though in recent years India has witnessed a shift towards the use of processed fruits and vegetables, there exists a vast and growing demand for such products, both at the rural and urban levels.

India being a home of wide variety of fruits and vegetables holds a unique position in production figures among other countries. The diverse agro climatic zones the country make it possible to grow almost all varieties of fruits and vegetables in India. Fruits and vegetables are more prone to spoilage than cereals due to their nature and composition, and this spoilage occurs at the time of harvesting, handling, transportation, storage, marketing and processing. Due to lack of postharvest management infrastructure, developing countries lose a big chunk (20-40%) of their total production of fruits and vegetables as shown in Table 1. Conditions in India are no better. Left unattended, these co-productss can be a source of atmospheric pollution and contamination. Efficient disposal and recycling of these co-productss can help in minimizing the pollution hazards, supply of vital nutritional components in our foods and feeds, and bringing down the cost of production of processed foods. They can also provide valuable raw material for gas reactors, organic manures and extraction of several types of useful by-products.

Table 1 : Estimated loss of fruits & vegetables in developing countries

Commodity	Estimated loss (%)	Commodity	Estimated loss (%)
Fruits		*Vegetables*	
Banana	20-80	Onions	16-35
Papaya	40-100	Tomato	5-50
Avocado	43	Plantain	35-100
Apricot	28	Cabbage	37
Citrus	20-95	Cauliflower	49
Grapes	27	Lettuce	62
Raisins	20-45	Roots/Tubers	
Apples	14	Carrots	44
		Potatoes	5-40
		Sweet potatoes	35-95

Co-products utilization

There are many options for the utilization of co-products especially from the processing factories. The covered area under the fresh fruits in India was 5510000 Ha with the production of 58740000 MT in 2005-06. In India utilization of fruits and vegetables for processing purposes is hardly 1.8% of total production. There are about 4000 processing units in India with a turn-over of about 8.0 lakh tonnes of fruit and vegetable products. About 90% of these units are in small scale sector with an annual production limit between 50-250 tonnes. These units have no organized method of co-products disposal. Most of the solid co-productss are dumped outside and are taken care of by the stray cattle. In certain large scale units, these co-productss are taken away by farmers for their disposal as *cattle feed* or *field manure*. With strict enforcement of the *Environmental Protection Act* (1986), discarding solid co-productss in outside dumps and liquid co-productss in municipal sewage system can attract a fine of Rs. 1.00 lakh and /or imprisonment of 5-7 years. Fruit and vegetable processing units located in congested areas with limited space and inadequate water supply are finding it difficult to lawfully manage these co-productss (Bhalerao *et al.*, 1989). Some of these fruit and vegetable co-productss are a rich source of vital constituents like carbohydrates, proteins, fats, minerals, fibres etc. Mango seed kernel is rich in carbohydrates, fats, proteins, and minerals; orange, melon and pumpkin seeds can provide fats and mineral matters. Most of these co-productss are nutritionally rich. Fruits and vegetables are grown all-over the country incurring co-productss everywhere. With smallholdings of the farmers, the quantum of these co-productss in individual farm are small except in large pockets growing individual type of fruits and vegetables. These co-productss have attracted the attention of scientists and a large volume of data has been generated on various fruits and vegetables defining their valuable constituents, mode of extraction of some of the byproducts and their end uses (Table 2). Estimated loss of fruits & vegetables in developing countries range from 14% in apple to 40-100% n papaya in case of fruits and as high as 35-100% in plantain and 35-95% in sweet potato in case of vegetables and The nature of some of these co-productss, value-added by-products which could be recovered or their use as animal feed are discussed here.

Possible co-products from solid co-productss in fruit processing units

Apple : 20-30% co-products (co-products), mostly pomace and possible by products are Juice, wine, vinegar, pectin, cattle feed Essential oil, pectin.

Mango : 40-60% co-products and mostly peel, pulp co-products, hull and kernels can be utilized for makingPectin, cattle feed, alcoholWine, vinegar, juiceFat, tannins, starch.

Table 2 : Composition of different fruit co-productss (per 100 g)

Co-products	Moisture (g)	Protein (g)	Fat (g)	Minerals (g)	Fibre (g)	Carbohydrate (g)
Apple pomace	-	2.99	1.71	1.65	16.16	17.35
Mango seed kernel	8.2	8.50	8.85	3.66	-	74.49
Jack fruit (inner and outer portion)	8.5	7.50	11.82	6.50	30.77	14.16
Jack fruit seeds	64.5	6.60	0.40	1.20	1.50	25.80
Jack s eed flour	77.0	2.64	0.28	0.71	1.02	18.12
Passion fruit peel	81.9	2.56	0.12	1.47	5.01	-
Banana peel	79.2	0.83	0.78	2.11	1.72	5.00
Sweet orange seeds	4.00	15.80	36.90	4.00	14.00	-
Watermelon seeds	4.3	34.10	52.60	3.70	0.80	4.50
Muskmelon seeds	6.8	21.00	33.00	4.00	30.00	-
Pumpkin seeds	6.0	29.50	35.40	4.55	12.00	12.53
Banana stem						
Central core	93.1	0.30	0.03	1.04	0.68	1.20
Outer hard fibrous sheath	91.9	0.12	0.06	0.98	1.81	2.44
Press juice from st em	98.6	0.05	-	0.63	-	0.41

Citrus : 50-60% co-products consisting Peels, seeds and pulp (Table 3 & 4). Mostly converted in to cattle feed, peel candy etc. Essential oil, pectin, cattle feed, peel candy etc.

Pineapple: 30-60% co-products consisting of can be utilized for Juice, wine, syrup, bromeline, cattle feed, biogas

Tomato: 20-30 co-products consisting core, peel and seeds can be utilized for making animal feed, seed oil and meal

Potato: Peel and coarse solids is converted in to animal feed, single cell protein

Banana: Major co-products consists of Pseudostems, leaves, suckers etc. Extraction of fibres and sap from banana pseudo stems sap— —generated while extracting fibres from pseudo stem is rich source of K(~4%) and can be recycled as liquid fertilizer and for beverages. - 450 lakh tonnes of pseudo stem biomass is generated annually. From this co-products biomass,~3390 lakh kg of fibre can be extracted (value of about 1500 crore) -Quantity of K available from the sap is worked out to be10-12 lakh tones (value of 900 cr-Banana pseudo stem fibre is extensively as blending material used in textile industries in Philippines, Malaysia, Japan and Korea. It can be blended easily with jute and mesta.raw material for manufacturing tissue paper,filter paper and decorative craft paper - for textile and paper industries.

Table 3 : Utilization of citrus co-productss

Total production	3 million tones
Orange co-products	3 million tonnes
Lime co-products	50% approx.
Mandarin oil extracted	60% approx.
Lime oil extracted	Bangalore, Nagpur, Abohar, Delhi
Dried orange peels, rags,	Uttarayan (MS), Jalgaon and Kodur
seed as livestock feed	Low in protein, fibre and fat but rich in carbohydrate, classed as carbohydrate concentrate
Liquid concentrate from above used for dairy and beef cattle	Molasses

Table 4 : Physico-chemical characteristics of Indian Mandarin and Lime oils

Physico-chemical Characteristics	Mandarin oil		Lime oil	
	Cold pressed	Steam distilled	Cold pressed	Steam distilled
Sp. gravity (25°C)	0.8470	0.8490	0.8620	0.8480
Refractive index	1.4748	1.4750	1.4860	1.4720
Optical rotation	92° 30'	92° 15'	48° 0'	52° 0'
Acid number	1.90	0.68	1.64	1.62
Ester number	3.88	2.52	3.50	1.86
Aldehyde	0.94	0.48	2.80	2.62
Evaporation residue (%)	3.60	1.15	3.40	2.08
Solubility in 90% alcohol	10.00	9.00	8.00	7.00
Colour (LovibondUnits)				
Red	28.00	-	23.00	1.50
Yellow	7.30	-	23.00	1.50
Blue	1.20	-	1.00	0.40

Table 5 : Utilization of banana, apple, and tomato co-products

Banana co-products	
Total production	8 million tonnes
Peels	22-30% (poultry and cattle feed)
Pseudostem yield	20-25 tonnes from 1000 plants
Starch in pseudostem	5% (in food products)
Each banana plant	Gives 20-25 leaves
Dried leaves	Cups and trays
Green leaves	To serve meals
Apple co-products	One million tonnes
Total production	2,00,000 tonnes
Cull apples	3 (Parwanoo, Sopore, Jarole)
No. of juice concentration unit	32,000 tonnes
Total crushing capacity	10,000 tonnes approx
Apple pomace	
Tomato co-products	
Total production	5 million tonnes
Tomato used for seeds (0.2%)	9000 tonnes
Total seed production	125-150 tonnes
Tomato juice co-productsd	7000-8000

Coconut: Production: ~15 billion nuts/year - one of the rarest plant whose all parts are utilized. Husk, Water, Leaves Stem, Coconut wood products, Shell, Carbon, Pith are the major left out material having potential for processing and value addition: fuel and coir for fibre : refreshing nutritional drink: roof covers in huts, making baskets and brooms: construction of houses as beams and columns floor tiles and panels-eco friendly :manufacture of activated —used as discoloring agent in chemicals, pharmaceutical, vegetable oil and allied industries : raising mushroom and for vermiculture are major products etc made out of coconut co-products.

Oil palm : Trunk/Fronds, Empty bunches, fibre and shell Palm kernel cake / pellet can be used in many ways. Palm oil mill effluent (POME) can be used for the industrial production of building materials (boards & panels),fuel and chemicals. Trunks contain carbohydrate that can be converted into gaseous or liquid fuels. viz;methane, methanol and ethane. Fronds, when left in the field, could supply nutrients to the soil, increasing its fertility as well as conserving moisture and reducing soil erosion. Bio polymer components such as cellulose, xylose and legnin can be fractionated by successive steam explosion treatment of trunk.

These can be used for mulching in the field/nursery as filler for biodegradable plastic mulch, medium density boards, pulp and papermaking, thermal insulating boards etc. Sterilized fruit bunches are used as substrate for mushroom cultivation. secondary/tertiary products, viz; phytin, phytic acid and inositol besides, edible protein can be derived from protein rich (12-15%) palm kernel/pellet. Need to develop pollution free processes to control dust,noise and other harmful gases and especially liquid effluents, POME. This is a liquid co-products generated by the mill having a BOD of 24,000 mg/litre more than 100 times as polluting as the domestic sewage. Appropriate technology has to be developed using anaerobic digestion which can also produce bio gas and generate electric power meeting the energy requirement of the whole processing plant. POME can also be used for bio gas production, fish rearing, animal feed and fertilizer.

Pomegranate : Rind powder, leaves and flowers can be processed/utilized in many use ful products. The rind powder is rich in beta-carotene, polyphenol, and minerals (Singh *et al.*, 2007). Pomegranate rind powder has potential for use in medicines, leather & drying industry and for making tooth powder. In India rind is used in diarrhea and chronic dysentery, often combined with opium. It is used as an injection in leucorrhoea, as a gargle in sore throat in its early stages & in powder in intermittent fever. Gargling with boiled water of rind of pomegranate helps in eliminating foul smell. Taking one-tablespoon ground dried rind of pomegranate with water twice a day stops excessive bleeding. Applying the paste of dried ground rind of pomegranate mixed with rose water and leaving it for sometime before wash, eliminate spots, freckles in the face & increases the glow of complexion thus making skin soft busters. Ground fresh leaves of pomegranate gives relief in heart palpitation. Applying paste of ground fresh leaves of pomegranate over the lower abdomen (in case of feeling pain) helps in controlling abortion or miscarriage. Taking 20-30 gm of ground fresh leaves of pomegranate with black pepper in one cup of water twice a day controls leucorrhoea.

Dry the flowers of Pomegranate in shade grind and strain. Massaging this powder on teeth helps in curing bleeding gums. By taking pomegranate flower (dried in shade) with honey twice a day cures too much urinating tendency.

Flowers also yield a red dry and with leaves and seeds were used by ancients as astringent medicines and to remove worms.

Keeping the nature of these co-productss in to consideration value added by products which could be recovered or their use as animal feed are discussed

I. Source of fat

The industrial processing of mangoes end up with a considerable proportion of *stones* and *peels* which are discarded and end up in *co-productss* (Narsimhachar *et al.,* 1977).

II. Essential oils

India is producing 3 million tonnes of citrus fruits like mandarins, lime, lemon, sweet orange accordingly generating large quantities of co-productss around the processing units. Citrus co-productss are a rich source of essential *oils, pectin* and a variety of by-products including *cattle feeds* which can be produced from such co-productss. *Mandarin essential oil* is extracted in small quantity at different processing units at Bangalore, Nagpur, Abohar (Punjab) and Sikkim. Mandarin oil is also extracted in small scale unit based on orange peels collected from fresh juice vending stands. *Lime oil* is extracted at Uttrayan, Jalgaon and Kodur.

Physico-chemical characteristics of a large number of Indian mandarins and lime oils, both cold pressed and steam distilled, have been analysed. *D-limonine* is a major by-product from citrus industry and at current level of production of citrus fruits in India; its potential availability is about 15,000 tonnes.

III. Source of animal feed

Orange co-products in the form of peels, rags and seeds are dried in Florida and California for use as a livestock feed. It is low in crude protein, fibre and fat but rich in carbohydrate. Often it is classed as a carbonate concentrate. The liquid concentrate made from crude citrus co-products is called *molasses.* Its use is similar to cane molasses and is widely used for feeding dairy and beef cattle. Growing yeast on it and using it for animal feed is further helpful. Ripe banana peels (constituting about 22-30%) are found littered all around the streets, causing pollution and skidding risks. In Vietnam, these peels are sun dried for use as poultry feed. There is always acute shortage of fodder in hilly areas during winter months. Apple drops and culls can be partly used to meet these shortages by the use of appropriate technology like slicing fruits, fumigating them and partially drying them and making them into compressed blocks. These blocks can be coated with molten wax to avoid *microbial* or *insect* attack. At the time of use, wax can be scrapped and apple block is used as animal feed. About 20,000 tonnes of apple pomace obtained in processing units is a source of great pollution in the factory surroundings. Method of conversion of pomace into ensilage has been developed in 20 tonnes batch. It can also be used to make edible products.

Pineapple bran has a great demand as a cattle feed in the Hawaii Island. *Cannary* co-productss are also being fed to the dairy cattle in Hawaii after ensiling. Proximate composition of the silage made out of pineapple tops and leaves with molasses and urea is almost comparable to any of the commonly used grass silage. Major co-products like mango seed kernel is a good source of carbohydrate and can replace around 20% of corn in growing and egg laying chicken rations. Drying of mango kernels and peel can make high-class energy food for animal feed.

IV. Extracting pectin

Majority of pectin produced in the world is extracted from cirrus peels. Pectin is used mainly for jam making, pharmaceutical and several other industries. Two citrus processing units in Uttrayan, and Kodur are producing lime pectin in a small quantity. India is still importing about 160 tonnes of pectin valued at about Rs.10 crores for fruit and vegetable processing industry alone. Manufacture of 'Genu' pectin in Denmark, almost commands a virtual monopoly by organizing year round production after procuring dried citrus peels of requisite quantity from different parts of the world. Mango peels which are also available in plenty from processing factories, have also been shown to be a good source of pectin (Table 6).

Table 6 : Quality of pectin obtained from different fruit source

Fruits	Yield (dry wt)	% Methoxyl content (%)	Anhydrouronic acid (%)	Jelly grade
Mango peels				
Dashehari	18-19	8.99	58.40	155
Langra	15-16	8.07	54.60	175
Alphonso	13.0	9.20	61.12	200
Totapari	15.7	8.31	-	200
Assam lemon	20.2	9.20	71.20	277
Mandarin	-	9.86	84.73	250
Pumello	-	9.31	79.89	200
Galgal	26.5	6.12	47.05	350
Apple pomace	15-18	8.40	59.40	200
Raw papaya	9.0	7.50	66.80	227
Guava	9.4	8.90	79.90	-
Cashew apple	-	3.50	45.10	75
Fig	-	9.28	78.36	150

Source : Srivas and Pruthi, (1976).

V. Source of starch

Banana pseudostem having a yield of 20-25 tonnes from about 1000 banana plants, when extracted, gives about 5% edible starch. The method for starch extraction from pseudostem has been standardized and physico-chemical properties of starch have been studied. Likewise, 1,40,000 tonnes of starch may also be available from mango seed kernels.

VI. Source of natural colours

Due to their carcinogenic effects, coaltar dyes are being replaced rapidly by natural colours or plant pigments for use in food products. Natural colours of blue grapes skin, kokum *(Garcinia indica)*, phalsa *(Grewia subinaequalis)* and Jamun *(Syzygium cumini)* have been thoroughly investigated for their nature, concentration, extractability, stability and suitability as food colours. Methods for making colour concentrates and powders from them have also been developed. A new fat soluble yellow pigment, garcinol was isolated from rind of kokum. Central committee on Food Standards is actively considering permitting the use of kokum and safflower pigments in the food industry.

Utilization of some less known fruit and vegetable co-products

Amla, Harad and Behra : India is growing a large variety of wild type of fruits and vegetables which find application in several preparations, some of which have a herapeutic role (Table 7). Fruits like Amla *(Phyllanthus emblicd)*, Harad *(Myrobalan)*, Behra are included in the list of vital 'Ayurvedic' drugs and widely used in several preparations like 'Chavanprash' 'Triphala' 'Arogya Wardhani' and 'Ashokarishta' (Kalara, 1988).

Cashew apple

After removal of cashew nuts, 'cashew apple' available to the tune of 2,50,000 tonnes finds a little application in conversion to different types of products like juice, syrup, candy etc. It is grossly under-utilized although methods are now available to remove its *acridness* through processing. A part of this fruit is used in preparing cashew brandy called 'Feni'. Cashew shell contains 20% oil and resin used in break lining, electrical insulators, foundry moulds and chores. In Brazil, fruit is finding better use than the cashew nut.

Wild Apricots Wild apricots are grown extensively in Ladakh (J&K) and Himachal Pradesh where these are locally called as 'Chulli'. The fruits are used extensively for the extraction of both sweet and bitter oils besides preparing sun dried apricots (Bhatia, 1976)

Raw papaya (Papain extracted) and watermelon rind

Papain extracted fruits, with cut marks on their surface; papayas are difficult to dispose-of of as ripe fruits. Such fruits are now picked, cut into small cubes, dipped in thick sugar syrup with different colours and finally disposed-off of as ' *Tuti-Fruitf* for use in bakery products and ice creams. This industry is centred in Maharashtra. In UP, watermelon rind after removing the green portion is used as a raw material for *tuti-fruiti*. It is estimated that 4000 to 5000 tonnes of this product is manufactured and the demand for *Tuti-fruitf* is further increasing.

Table 7 : Utilization of some less known fruit and vegetable co-products

Fruit/Vegetable	Location	Products
Amla, Harad and Behra (Wild types) Cashew apple (2,50,000 tonnes) Cashew shell 20% Oil) Apricot (wild) 'chulli' Pomegranate seeds (Wild type) Tamarind seeds Cucumber seed kernels, pumpkin seed kernels Musk- melon seed kernels Watermelon rind Wild beans and fruits	UP,MP,MS MS, AP, Orissa Jammu & H.P. Jammu&HP AP, MS and Karnataka Hathras Distt." UP, Lucknow, Gujarat, Hathras Distt. ofUP Lucknow, UP _ Raj as than	Ayurvedic drugs, chavanprash, Triphala, Arogya Vardhani and Ashikarishta Juice, syrup, candy, feni, resin used in breaklining, electrical insulators, foundary molds and chores Oil extraction Curry powders Starch for textile industry Four kernels collectively called four 'Maghaz' used in beverages and bakery products For tuti-fruiti manufacture, Sun dried products used as vegetables

Source : Bhatia, (1976)

Cucumber, pumpkin, watermelon, muskmelon, seed kernels *(Magaz)*

Production of cucumbers, pumpkins, watermelons and muskmelons in India is around 106, 533, 206 and 625 thousand tonnes, respectively. Kernels from their seeds are extracted on a cottage scale in UP and from watermelon in Gujarat. Watermelon kernels sell at Rs. 40/Kg while others at Rs. 100/Kg. These kernels are extensively used in sweets, bakery and ice creams. Beverages are prepared after grinding these kernels alongwith other herbs and are supposed to have cooling effect in scorching summers.

Sundried wild beans and fruits

People in desert areas of Rajasthan with chronic shortage of common vegetable adopted little used beans and fruits locally; known as *Kumatya* (kikar phalli), *Khejri*, Cluster beans, *Kachri*, *Lasura*, *Kair* and growing wild, in their vegetable menus. Some of these vegetables are soaked in whey to introduce *lactic fermentation* before *drying*. Dried vegetables are mixed thoroughly with ash to remove the last traces of moisture. These dried vegetables are relished by people of Rajasthan on festive occasions and, the turnover of dried vegetables is 4-5 crores of rupees.

Miscellaneous uses

Large quantities of green mangoes are gathered in fields after dust storm and rains, together with bird damaged fruits. Due to its extreme versatililty for use at every stage of maturity, most of the above mentioned mango co-productss are used for pickles, dried mango slices and powder (Amchur). Mango pickle is a national pickle and widely prepared, and used in every home. If each family prepares one kg pickle, total production of home made pickle will amount to 1,75,000 tonnes against the total organised production of about 8,00,000 tonnes for all the fruit and vegetable products.

Large proportion of raw mangoes are collected, peeled, sliced and dried in the sun. Dried raw mango slice powder is called *Amchur* and is widely used in the North India as *curry powders.* The central core of banana pseudostem constitute 10-15% of total stem and can be used as a vegetable. This can be easily dried for subsequent use. A great contribution has been made by CFTRI, Mysore to develop a machine to convert dried banana leaves into easily dispensable beautiful cups and trays. A banana plant contains 20-25 green leaves which find extensive use to serve meals in Indian homes. By ensuring their hygiene, we can introduce their use in high class hotels to serve typical Indian dishes.

India produces about 4.6 million tonnes tomatoes. It is estimated that 0.2 per cent of total tomato production i.e. 9,200 tonnes of fruits are used for seed extraction to the tune of 125-150 tonnes. According to the present practice, except for seeds, the entire quantity of 7000-8000 tonnes of tomato juice is allowed to go co-products. Appropriate method should be thought of to salvage this huge quantity of juice by using appropriate type of disintegrator-cum-pulper and collecting the juice in appropriate sized jars containing desired proportion of KMS and sodium benzoate. We may reduce the time of extraction of the seeds and also retrieve the juice.

Use of apple pomace has also been advocated in preparing mushroom compost. Pineapple leaves have been employed to extract high quality fibre for textile industry.

Issues and future thrust

1. To develop efficient processes and machinery/gadgets for production of densified particle boards, charcoal, power generation through gasification.

2. Suitable technologies/processes and hardware for production of fibre, yarn, mattresses, geotextiles, soil conservation structures etc. from empty fruit bunches; shell charcoal, power production through gasification from shell co-products and particle board, soil less culture, composting and conservation structures etc. from mesocarp fibers.

It is clear from the on-going discussion that co-productss from processing factories are rich source of several useful materials. In the developed countries, many value added components are recovered from such co-productss, but the scenario with developing countries is different where this co-products is not utilized; resulting in loss of precious resources besides a great pollution problem. The future should see greater utilization of this co-products for recovery of various by-products, which could also stabilize the economy of the processing.

References

Annual Report Solvent extractors Association of India (1985). p. 5.

Beerh, OP, Raghuramaiah, B and Krishnamurthy, GV (1976). Utilization of mango co-products peel as a source of petcin. J. FoodSci. Technol., 13:96.

Bhalerao, SD, Mulsmuley, GV, Ananthakrishna, SM and Potty, VH (1989). Effluent management in food industry. Indian Food Packer, 43(2): 5.

Bhatia, AK (1976). Apricots of Ladakh, status and problems. Indian Food Packer, 30(2): 35-37.

Gill, HS and Tomar, BS (1991). Vegetable statistics at a Glance. Project Directorate on Vegetable Research, IARI, New Delhi, Tech Bull. No. 4.

Joshi, C and Joshi, VK (1990). Food Processing Management Technology: Need for integrated approach. Indian Food Packer, 44(5): 56.

Kalara, CL (1988). The chemistry and technology of amla. Indian Food Packer, 42(2): 67.

Narsimhachar, BL, Reddy, BR and Thirmula, Rao, SD (1977). Processing of mango stones for fat. J. Am. Oil Chem. Soc., 54: 494.

Singh, DB, ARP, Kingsly, HR. Meena and Nirmaljit Kaur (2007). Studies on physico-chemical and nutritional properties of pomegranate rind powder. Indian J. Hort. 64(2): 155-158.

Srivas, SR and Pruthi, JS (1976). Studies on the quantitative estimation of chemical components of cashew apple, citrus and other fruit pectins. J. Inst. Chemists (India), 48: 95.

Swaminathan, MS (1981). Report of the group on perishable agricultural commodities. Ministry of Agriculture, Govt of India, New Delhi.

□□□

20

Co-Product Utilization of Kinnow Processing for Food and Industrical Use

UPPAL SINGH AND NEHA BABBAR

India is unique in having a long fruit season and varieties to suit the need of all of us. Among fruits, India produces about 6.3 million tones of citrus fruits every year. Among the different Citrus fruits grown in India, Kinnow mandarin is cultivated commercially in North India, particularly in Punjab, Haryana, Rajasthan, some parts of Himachal Pradesh and in south Kodaikanal region of Tamil Nadu and Coorg in Karnataka (Chundawat *et al.*, 1978) and are exported to different parts of the world from these regions. Kinnow has emerged as number one fruit of the Punjab state, both from area and production point of view. Kinnow occupies a major share of 31,788 hectares of total area under fruit in Punjab with an annual production of 5,91,319 MT (Anonymous, 2007). The districts of Hoshiarpur, Ferozepur and Faridkot occupy over half of the area under Kinnow in the state.

Kinnow, a hybrid between King Mandarin and Willow leaf (Mediterranean mandarin) [*Citrus nobilis* × *Citrus deliciosa*] was introduced in India in the early 1945 (Nijjar, 1979) and since then it has become the most favoured winter season fruit. Kinnow is known for its superior characteristics such as heavy bearing, wide adaptability, fruit quality and high juice content.

Nutritionally the fruit is a good source of vitamin C (31 mg/100 g) and minerals like calcium (40 mg /100 g), phosphorous (18 mg/100g), potassium (126 mg/ 100g) and antioxidants (Watt and Merrill, 1975).

The best time of picking of Kinnow fruit has been found to be from mid-January to mid-February with 12:1 to 14:1 solid to acid ratio. According to Nagar (1993) delaying harvesting period beyond second week of January increases loss of fresh weight. Kinnow fruit has high preference for table purpose but it also finds acceptability for processing in organised and unorganised sectors which yield large quantity of processing co-products-which are considered as wastes. However these wastes are very rich source of bioactive compounds. These co-products could be utilized in an organized manner to produce high value by-products. Kinnow peel and pulp alone account for about 55-60% of the total fruit weight and has good nutritional value (Kalra *et al.*, 1989). The processing of Kinnow Co-products constitute about 45-50% (Kalra *et al.*, 1989), respectively which is a substantial quantity. Presently more than 90,000 tonnes kinnow are processed in India in organized and unorganized sectors. The demand of processed products is increasing and the production and processing of these commodities is expected to increase manifolds in near future due to plantation of Kinnow orchards on large scale under Horticultural Mission Board, which will also generate large quantities of wastes. Their disposal in open spaces or in municipal bins compounds the environmental pollution problems. The production of nutraceuticals using bioactive compounds from fruit residues will help inefficient, inexpensive, and environmentally friendly use of these fruit residues. One of the most effective options is the recovery of the bioactive constituents, which could be used in food, cosmetic, and pharmaceutical industry (Makris *et al.*, 2007). This will not only help setting up commercial units for extraction of bioactive compounds in the villages or in rural catchments, but also help mitigate environmental pollution problems caused because of the poor disposal of such residues.

Major bioactive compounds of kinnow co-processing products

Pectin : It is a polygalacturonide linked to α-galacturonic acid with α-1,4 linkage. These are deposited in cell wall and middle lamella. Pectin has protopectin, petinic acid and pectic acid. These are capable of forming gels (jellies) with sugars and acids. Kinnow peel and pulp are good source of high quality of pectin, which is used for making 200 grade pectin for its use in manufacturing of food products. Organic acids and mineral acids are used to extract pectin efficiently and economically from Kinnow peel and pulp. Pectin content of about 20% from pulp and 35% from dried peel can be obtained with hydrochloric acid (pH 1.0-1.6). Pectin is used as gelling agent in jams, jellies and desserts, filling & topping in bakery preparations, dairy application, Nutritional and health products, modified Citrus Pectin for pharmaceutical industry. Pruthi *et al.*, 1984 found 2.0% pectin content on fresh weight basis in kinnow peel.

Biocolours : Kinnow peel has deep orange colour and contains large amount of carotenoids. Carotenoids are major source of colour production. It is a large family of C_{40} isoprenoids compounds, which accumulate in the chromoplasts of ripened fruit and they are responsible for bright redish colour of Kinnow fruits. The concentration of carotenoids is highly correlated with peel colour. Flavedo of fruit is one of the richest source of carotenoids. In citrus peel, main carotenoid is lutin and with fruit ripening specific xanthophylls also start accumulating in peel. To obtain maximum yield from fresh peel, different extraction solvents can be used. The most promising is acetone which gives maximum yield from fresh peel. Kinnow peel is a good source of biocolour (yield about 115- 123 mg carotenoids/kg) of peel. In the peel, as chlorophyll disappears, the total carotenoids increase several fold. According to Godwin and Goad (1970) rind or peel of citrus contained 50.75% of the total carotenoids of oranges. The more intense (red) peel colour of tangerines and mandarins is due to pigments cryptoxanthin, beta-apo-8-carotenal, Beta-citravin, voilaxanthin, antheraxanthin and Zeaxanthin. Beta-citravin being the main pigment contributing to the red tint of these fruits (Gross, 1977).

Carotenoids thus obtained have antioxidant and anticancer properties. β-cryptoxanthine has five times more activity to control cancer than β-carotene. Kinnow peel is the richest source of carotenoids and extraction and purification will give a natural pigments suitable for organic procducts. After extraction of carotenoids, dried peel cake can be used for pectin extraction. Biocolours are used for colouring and flavouring citrus drinks. The product is of deep yellow colour having good antioxidant capacity and is easily dispersible in water at room temperature.

Peel oil (d-Limonene) : Kinnow peel contains good amount of essential oils Citrus peel contains about 1.0-4.5 g oil/Kg which can be obtained by cold pressing and distillation. D-Limonene is generally used as general purpose cleaner, solvent replacer, cleaner for concrete in food industry, perfume industry, cosmetic industry, textile industry and leather industry. Kinnow peel was found to have 0.75-1.0% essential oil content (Pruthi *et al.*, 1984).

Vitamin C : Ascorbic acid is the major vitamin present in citrus fruits. Only 25% of Vitamin C is present in juice, rest is in peel (in flavedo). Kinnow peel has 69 mg/100 g of vitamin C.

Antioxidants : Antioxidants are specific organic compounds that are active in inhibiting or delaying the oxidation of other compounds in foods and humans. Natural antioxidants in human diet are the primary preventive ingredients against the human diseases and spoilage of foods. The most widely used synthetic antioxidants, BHA and BHT, have toxic effects and are not used freely as they influence liver enzymes. Phenolic and polyphenolic compounds are natural antioxidants found in plants. Natural phenolics are attracting more and more attention because they are anticarcinogenic and anti-inflamatory agents. Natural antioxidants also have antimicrobial effects. They may act as reducing agents, free radicals scavengers complexes of pro-oxidant and quenchers of formation of singlet oxygen.

Phenols : Phenolic compounds are secondary metabolites and found in plant derived food. Phenolic compounds can be extracted from Kinnow peel and seed by refluxing with 50 ml of 80% methanol. Kinnow peel and seed has been reported to contain 17.5 mg/g and 3.68 mg/g phenols, respectively (Neha *et al.*, 2011). Phenolics have potential health benefits as antioxidants, antibacterial, antiviral, anti-carcinogenic, anti-inflammatory, vasodilatory.

Flavonoids : Flavonoids are secondary plant metabolites. The basic flavonoid structure is the flavan nucleus consisting 15 carbon (C_6-C_3-C_6) rings. Flavonoids is a group of flavonols, flavanols, flavones, flavanones and anthocyanidins. Kinnow peel is rich in flavonoids which can be easily extracted. Flavonoids present in peel are hesperidin, naringin, isonaringin, nobiletin. They are water soluble and bitter in taste. Flavonoids have a large capacity of oxidizing the free radicals. Premi *et al.* (1994) found 4.69 mg/g limonin, 0.420 mg/g naringin content in Kinnow peel. The potential health benefits of flavonoids are: antioxidant, therapeutic agent, muscle relaxant, raw material for medicine production.

Limonoids : Kinnow fruits contain highly oxygenated triterpenoid compounds known as limonoids. Limonoids are involved in the development of bitterness in kinnow juices. Kinnow juice, peel, pulp and seeds have high concentration of limonoids. Some studies reveal that the kinnow grown in Punjab is the richest source of limonoids (218 ppm), among the world's commercial cultivars (www.expressindia.com). The important limonoids present in Kinnow are: Limonin, Nomilin, Obacunone, Ichangin and Nomilinic acid.

In the intact fresh fruit, most of the limonin is in the A-ring monolactone form, naturally occurring precursor of limonin which is tasteless and that is gradually converted to intensely bitter limonin in the juice after extraction from the fruit. This conversion proceeds under acidic condition and has been accelerated by the action of limonin D-ring lactone hydrolase which has been shown to be present in citrus (Maier *et al.*, 1969). Heat accelerated conversion has also been reported (Premi *et al.*, 1994).

The limonin content varies in different parts of kinnow. The seed contains the highest quantity of limonin (9.5 mg/g), followed by peel (4.6 mg/g) and juice (0.218 mg/g) (Premi *et al.*, 1994). Limonin is reported to have efficacy against a variety of cancers and cholesterol lowering properties. Limonoids offer various health benefits viz. protective against a wide range of cancers in humans, anti tumour activity, antioxidant activity to intervene in free radical reactions, hypocholesterolemic activity for lowering LDL cholestrol level and Insecticidal activity at larval stage.

Dietary fiber or sometimes *roughage* is the indigestible portion of plant foods having two main components: *Soluble* (prebiotic, viscous) fiber that is readily fermented in the colon into gases and physiologically active byproducts. *Insoluble* fiber that is metabolically inert, absorbing water as it moves through the digestive system, easing defecation. It acts by changing the nature of the contents of the gastrointestinal tract, and by changing how other nutrients and chemicals are absorbed. Soluble fiber absorbs water to become a gelatinous, viscous substance and is fermented by bacteria in the digestive tract. Insoluble fiber has bulking action and is not fermented. Although a major dietary insoluble fiber source, lignin may alter the fate and metabolism of soluble fibers. Chemically, dietary fiber consists of non-starch polysaccharides such as arabinoxylans, cellulose and many other plant components such as resistant dextrins, inulin, lignin, waxes, chitins, pectins, beta-glucans and oligosaccharides.

Kinnow peel is rich in dietary fibre. After extraction of pectin from kinnow peel, the residual material is dried and pressed in pellets for extraction of dietary fibre. This residue has high energy and nutritive value. Citrus pulp has been considered as good source of insoluble dietary fibre that ranks equal to cereal bran. Similarly Kinnow pulp can be exploited as dietary fibre for human use. The indigestible part of Kinnow peel can be consumed as dietary fibre which is beneficial for human health. Candied peel from kinnow peel can be eaten as such. Other uses are as a feed for animal and in cosmetic.

References

Anonymous (2007). Director of Horticulture, Punjab, Chandigarh. Personal Communication.

Chundawat BS, Gupta AK and Singh BP (1978). Storage behaviour of different grades of kinnow fruits. Pb Hort J., 18: 156-60.

Godwin TW and Goad LJ (1970). Carotenoids and triterpenoids. In: Hulme A C (ed) The biochemistry of fruit and their products, Vol 1. Academic Press, New York.

Gross J (1977). Carotenoid pigments in citrus. In: Nagy S, Shaw P E and Veldhuis M K (ed) Citrus Sci and Technol. Vol.1, pp 302-54. AVI Publishing Co.Inc.West port, CT.

Kalra, KL, Grewal, HS, and Kahlon, SS (1989). Bioconversion of kinnow-mandarin waste into single-cell protein. World Journal of Microbiology & Biotechnology, 5: 321-326.

Maier VP, Hasegawa S and Hera E (1969). Limonin D ring lactone hydrolase: A new enzyme of citrus seeds. Phytochemistry, 8: 405.

Makris, DP, Boskou, G, and Andrikopoulos, NK (2007). Recovery of antioxidant phenolics from white vinification solid by-products employing water/ethanol mixtures. Bioresource Technology, 98: 2963-2967.

Nagar PK (1993). Effect of different harvesting periods on shelf life and quality of kinnow fruits. J Food Sci Technol, 30: 44-45

Neha Babbar, Harinder Singh Oberoi, Dewinder Singh Uppal, Ramabhau Tumadu Patil (2011). Total phenolic content and antioxidant capacity of extracts obtained from six important fruit residues. Food Research International, 44: 391-396.

Nijjar GS (1979). Kinnow – A promising citrus fruit. Prog Farming, 15: 11-13.

Premi BR, Lal BB and Joshi VK (1994). Distribution pattern of bittering principles in kinnow fruit. J Food Sci Technol, 31: 1404-41.

Pruthi JS, Mawan JK and Teotia MS (1984). Studies on the utilization of kinnow and malta orange. J Food Sci Technol, 21: 123-27.

Watt BK and Merrill AL (1975). Composition of Foods. Vol 8. U.S. Department of Agriculture Handbook.

□□□

21

Novel Techniques in Extrusion Processing

MANDEEP KAUR, H.K. SHARMA AND B.C. SARKAR

Extrusion is a technique used to form shapes by forcing a food material through a region of high-temperature and/or pressure, and then through a die to form the desired shape. Food, commonly a cereal product, is heated under pressure, then extruded through fine pores, when the super-heated water evaporates rapidly, leaving a textured product. Extrusion process basically includes several unit operations including mixing, cooking, kneading, shearing, shaping and forming. Various food items such as cookie *dough*, *sevai*, pastas, breakfast cereals, *jalebi*, fig newtons, french fries, baby food, ready-to-eat snacks and dry pet food are most commonly manufactured using the extrusion process. Food extruders can be classified according to the method of operation (cold extruders or extruder-cookers) and the method of construction (single or twin-screw extruders).

The principles of operation are similar in all types: Food is compressed into a semi-solid mass, in the initial step and then it is forced through a small aperture in the next step to enhance the varieties of texture, color, and shape that can be acquired from basic food ingredients. Extrusion can form & even cook the raw ingredients into the finished products. If the food is heated above 100°C the process is known as extrusion cooking (or hot extrusion). Here, frictional heat and any additional heating that is used to cause the temperature to rise rapidly. The food is passed to the section of the barrel having the smallest flights, where pressure and shearing is further increased. Finally, it is forced through one or more restricted openings (dies) at the discharge end of the barrel as the food emerges under pressure from the die, it expands to the final shape and cools rapidly as moisture is flashed off as steam.

Different shapes, including spheres, tubes, doughnuts, rods, squirls, strips or shells can be formed. Typical products include a wide variety of low density, expanded snackfoods and ready-to-eat (RTE) puffed cereals (Table1).

In case of cold extrusion, temperature of the food remains at ambient. It is used to mix and shape foods such as pasta and meat products. Low pressure extrusion, at temperatures below 100°C, is used to produce, for example, liquorice, fish pastes, surimi and pet foods (Table 1). Extrusion cooking is known as a high-temperature short-time (HTST) process. It reduces microbial contamination and inactivates enzymes. However, the main method of preservation of both hot- and cold-extruded foods is by the low water activity of the product (0.1–0.4), and for semi-moist products in particular, by the packaging materials that are used.

Table 1 : Examples of extruded foods

Types of product	Examples
Cereal-based products	Pasta products
	Ready To Eat and puffed breakfast cereals
	Cookie dough
	Instant drinks, soup and beverage bases,
	Weaning foods
	Sevai
	Pre-gelatinised and modified starches, dextrins
	Crispbread and croutons
	Expanded snack foods
	Pre-cooked composite flours
Sugar-based products	Candy bars
	Chewing gum
	Chewy candies
	Fruit gums
	Toffee, caramel, peanut brittle
	Liquorice

Protein-based products	Hot dogs Texturised vegetable protein (TVP) Semi-moist and expanded petfoods Sausage products, frankfurters Animal feeds and protein supplements Surimi Caseinates Processed cheese

Adapted from Jones (1990), Best (1994), Harper (1979), Harper (1987), Heldman and Hartel (1997).

Extrusion has following advantages

Homogeneous and consistent cooking process: Extrusion is a homogeneous and consistent cooking process, which leads to a final product of higher quality with minimum waste.

Low costs: Extrusion has lower processing costs and higher productivity than other cooking or forming processes. Some traditional processes, including manufacture of cornflakes and frankfurters, are more efficient and cheaper when replaced by extrusion.

Versatility: A very wide variety of products are possible by changing the ingredients, the operating conditions of the extruder and the shape of the dies. Many extruded foods cannot be easily produced by other methods.

High capacity and automated control: Extruders operate continuously and have high throughputs. For example, production rates of up to 315 kg h^{-1} for snack foods, 1200 kg h^{-1} for low-density cereals and 9000 kg h^{-1} for dry expanded pet foods are possible (Mans, 1982).

No effluent is generated: Extrusion is a low-moisture process that does not produce process effluents. This eliminates water treatment costs and does not create problems of environmental pollution.

High product quality: Extrusion cooking involves high temperatures applied for a short time and the limited heat treatment therefore retains many heat sensitive components.

Novel techniques in extrusion

Although 'extrusion' itself is a novel technique used in food processing industry and various processed foods such as all snacks, breakfast cereals, confectionery and some pet food are now also manufactured by using this technique. But new techniques and methods are invented everyday to increase the quality of the processed food. These new techniques delay and sometimes completely stop the rotting process and also increase the variety and the flavor of the available food products. The novel techniques give the opportunity to increase the variety of designs, shapes and texture of the same products again and again. Novel techniques in extrusion processing can be categorized into 3 categories, which are:

- Machinery advancement
- Statistical
- Waste utilization

Machinery advancement

Extrusion process is done through an extrusion machine called 'extruder'. Normally, an extruder contains a power supply to run the main screw, a feeder to push all the raw ingredients in and a barrel to surround the main part i.e. screw. This screw passes on the raw material through and towards a shaped hole and the die from which the shape of the product forms. Advancements are made continuously to increase the variety and quality of the available food products.

Small variations in processing conditions affect process variables as well as product quality (Desrumaux *et al.*, 1999). Product quality can vary considerably depending on the extruder type, screw configuration, feed moisture, temperature profile in the barrel, screw speed, feed rate and die profile.

Basic components of a food extruder

While each manufacturer uses its own names for extruder parts, the basic components of a food extrusion line include:

i. A feeding unit or storage bin above the extruder that meters the raw ingredients into the extruder.

ii. Pre-conditioner: an assembly sometimes used to adjust the moisture content and temperature of ingredients (may partially or completely cook them before entering the extruder).

iii. Extruder barrel: a pipe-like retainer in which the screw turns. The barrel may be built in sections that contain their own heating or cooling sleeve.

iv. Screw: the core of any extruder, the screw conveys the product through the extruder. The flight of the screw pushes product forward. The screw determines not only the quality of the product but also the output of the extruder. Extruders can use either a single screw or twin screws.

v. Vent: an opening in the barrel before the die plate that allows pressure and steam to be removed.

vi. Die plate: final assembly for shaping the product as it leaves the extruder.

vii. Cutting station: ensures precise and consistent shape and size of product.

viii. Gear box, motor, and controls.

There are three major types of extruders used in the food industry: piston extruders, roller-type extruders and screw extruders.

Piston extruders

The piston extruder (Fig. 1) can consist of a single piston or multiple sets of pistons that deposit a precise amount of product onto conveyers or trays. Piston extruders are primarily used for forming product shape and are used in confectionery as well as bakery production facilities.

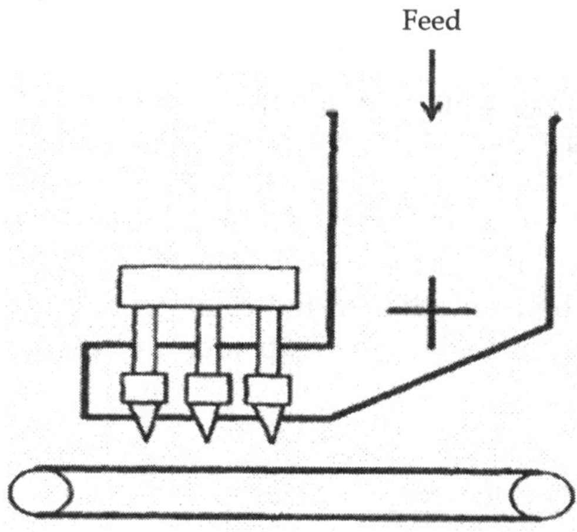

Fig. 1 : Piston extruder

Roller extruders

Roller extruders (Fig. 2) are used to form the shape of a product. A roller extruder consists of two counter-rotating rollers that either turns at similar or differential speeds. This process is also referred to as calendering in the dough industry. The roller surfaces can be smooth to create a long thin strip or can be perforated to form the dough into shaped products. The roller extruder can be altered to control the width of the layer of product moving in between them. Products such as crackers and hard cookies can be formed by creating the desired shape within the rollers and conveying the dough in between the rollers.

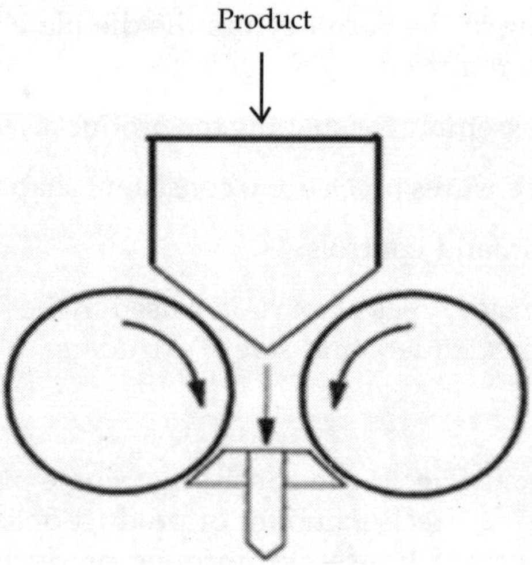

Fig. 2 : Roller extruder

Screw extruders

Screw extruder (Fig. 3) utilizes single, twin or multiple screws rotating within a metal cabinet called the barrel. The screws convey the material forward and through a small orifice called a die which can take on many shapes and sizes. Several external parameters such a screw speed and configuration, temperature of the barrel, die size and shape, and the length of the barrel effect the properties of the final product. Screw extruders are further categorized into single screw extruder and twin screw extruder.

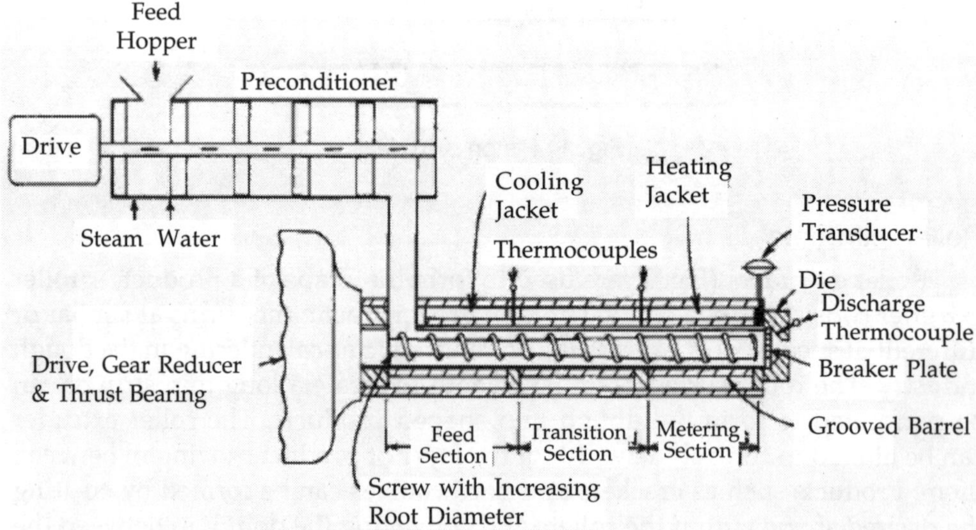

Fig. 3 : Screw extruder

Single-screw extruders

The equipment (Fig. 4) consists of a cylindrical screw that rotates in a grooved cylindrical barrel, made from hard alloys or hardened stainless steel to withstand the frictional wear. The length to diameter ratio of the barrel is between 2:1 and 25:1 (Hauck, 1993). The pitch and diameter of the screw, the number of flights and the clearance between the flights and the barrel can each be adjusted to change the performance of the extruder.

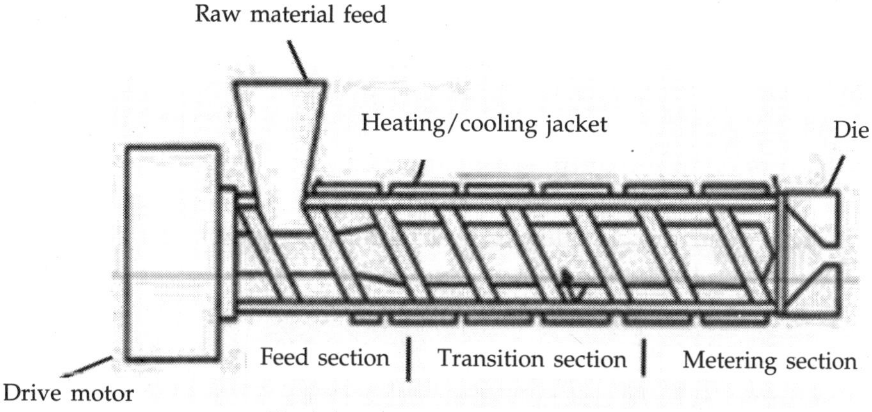

Fig. 4 : A single-screw extruder

The screw is driven by a variable speed electric motor that is sufficiently powerful to pump the food against the pressure generated in the barrel. The screw speed is one of the main factors that influence the performance of the extruder: it affects the residence time of the product, the amount of frictional heat generated, heat transfer rates and the shearing forces on the product. Typical screw speeds are 150–600 rpm, depending on the application. Compression is achieved in the extruder barrel by back pressure, created by the die and by:

- Increasing the diameter of the screw and decreasing the screw pitch.
- Using a tapered barrel with a constant or decreasing screw pitch.
- Placing restrictions in the screw flights.

Single-screw extruders can be classified according to the extent of shearing action on the food into:

High shear: These extruders are used for highly expanded products with low moisture and bulk density levels. They feature the shortest length-to-diameter ratio and extruder speeds and mechanical energy inputs are high. High speeds and shallow flights create high pressures and temperatures that are needed to make breakfast cereals and expanded snackfoods.

Medium shear: This type of extruders handle materials with a lower moisture level and mechanical energy inputs are higher. Typical products processed are pet food, breadings, aquatic feeds, semi-moist petfoods and textured vegetable proteins.

Low shear: Low shear stress extruders (forming extruders) are used to densify material that is generally high in moisture, such as pasta. They operate at a slow speed and feature a long length-to-diameter ratio, imparting low levels of mechanical energy per unit of throughput. Deep flights and low speeds create low pressures for forming pasta, meat products and gums.

In extrusion cooking, much of the energy from the extruder motor is lost as friction and rapidly heats the food (between 50 and 100% of total energy input (Harper, 1987). Additional heating can be achieved using a steam-jacketed barrel and/or by a steam heated screw (in some applications the jacket is also used to cool the product using cold water). In other designs, electric induction heating elements are used to heat the barrel directly. Some products also require the extruder die to be heated to maintain the viscosity and degree of expansion, whereas others require the die to be cooled to reduce the amount of expansion. The temperature and pressure profiles in different sections of a high-shear cooking extruder are shown in Fig. 5.

Single screw extruders can be further categorized as pasta extruder, high pressure forming extruder, low shear cooking extruder, collet extruder, and the high shear cooking extruder.

Pasta extruder

The pasta extruder is one example of a single screw extruder and is used for forming various types of pasta products from a dough. The pasta extruder utilizes low temperatures and creates relatively low shear. It somewhat resembles a meat grinder in configuration in that the product is conveyed forward with a single screw to a flat, smooth die with several openings. The shape of the die determines the final shape of the product.

High pressure forming extruder

The high pressure forming extruder is used for compressing and shaping pre-gelatinized dough into "half products" or products that will require another step in processing. An example of this is a snack product that after extrusion would need to be fried, microwaved or processed by some other method before consumption. Also ready-to-eat (RTE) breakfast cereal pellets that need to go to a flaking roll, for example, can be formed using this type of extruder.

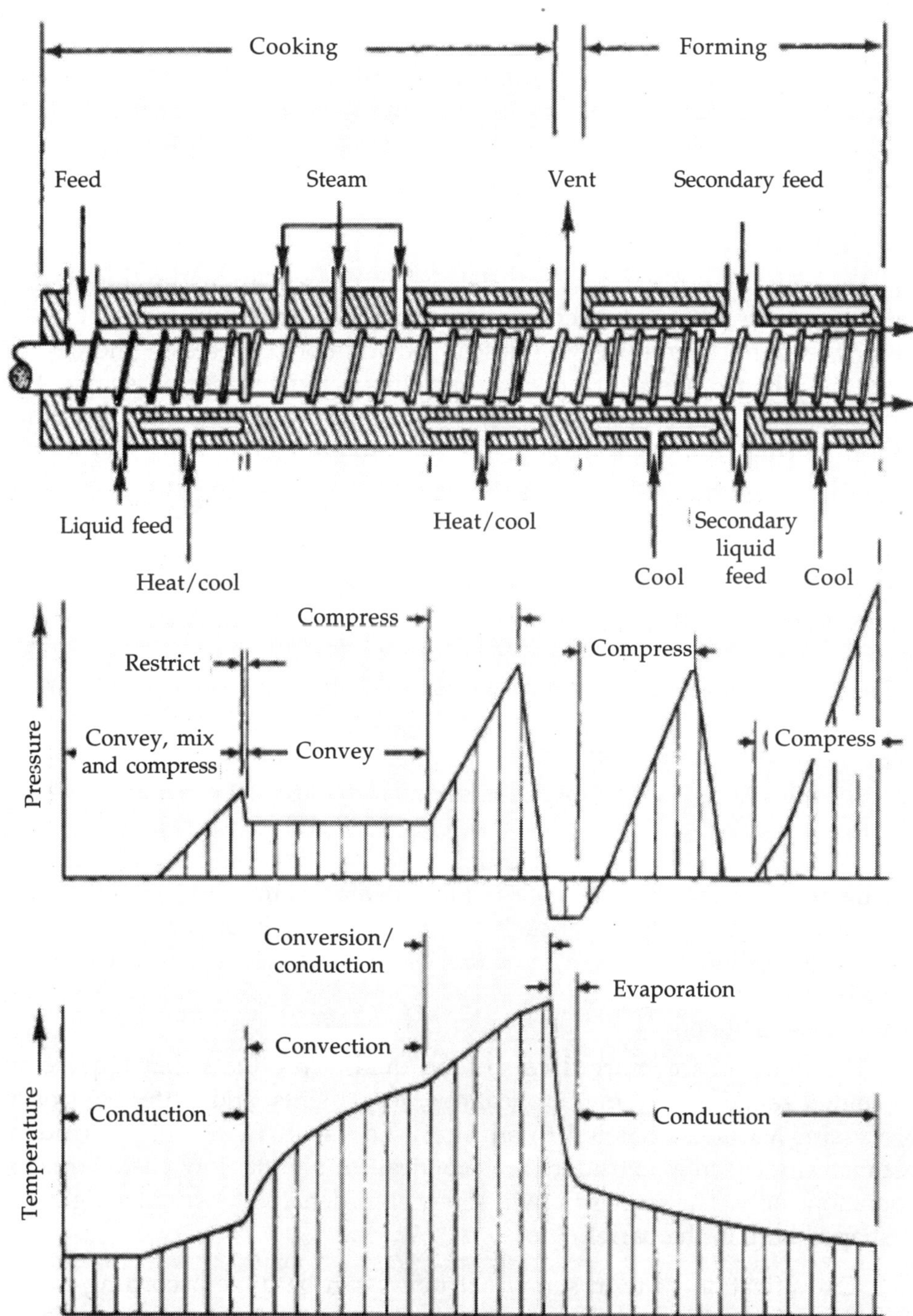

Fig. 5 Changes in temperature and pressure in a high shear cooking extruder for expanded food products. (From Fellows,2000; Miller, 1990).

Low shear cooking extruder

The low shear cooking extruder is a continuous cooker for high moisture dough. The cooked product can then be processed further by forming or drying. Soft moisture pet foods and gelatinized starches are examples of the products made with this type of extruder. The collet extruder is a high shear, short barreled extruder used to make highly expanded snack food products.

Collet extruder

The collet extruder has no external heating source. Its energy comes from the viscous dissipation of mechanical energy. The low moisture input creates a highly expanded product with little or no additional processing necessary. A special die arrangement gives the product a twisted puffed shape. The product is then fried in vegetable oil, and coated with cheese and some other flavor. During frying the moisture level reduces from 8% to 1-2% in this product. The most common material used for fried collet is corn meal. Some other cereal grains can also be used for this type of product. Baked collets are example of the expanded extruded snacks. This includes products such as baked corn curls, onion rings and potato sticks. Baked collets can be made with different cereal grains and tuber flours. Protein, fibers, cellulose, and bran can be blended with cereal grain up to 20% to make healthy snacks. Potato sticks usually made by mixing potato flour with corn or rice flour.

High shear cooking extruder

The high shear cooking extruder is similar to the collet extruder, however a longer barrel is used to increase residence time and excess heat is removed by cooling. The final product is usually less expanded and requires a final drying process. A wide variety of dry ingredients and moisture contents leads to a variety of uses. Products such as texturized vegetable protein, dry pet foods and modified starches can be processed using this type of extruder.

Twin-screw extruders

The demand for more diverse food quality products that require an expanded range of cooking applications, ingredients, and better control of processing variables has led to an increased use of twin screw extruders because single screw extruders are practical for simple, low cost extrusion operations only. The screws in twin-screw extruders rotate within a 'figure of 8' shaped bore in the barrel.

Classification of twin screw extruders can be done according to the direction of rotation and the way in which the screws intermesh. In food-processing applications, co-rotating intermeshing screws are most commonly found (Fig. 6). These are self-wiping (the flights of one screw sweep food from the adjacent screw). In such type of extruders, spacing between the flights can be adjusted so that large spaces initially convey the material to the cooking section and then smaller spaces compress the plasticized mass before extrusion through an inter-changeable die.

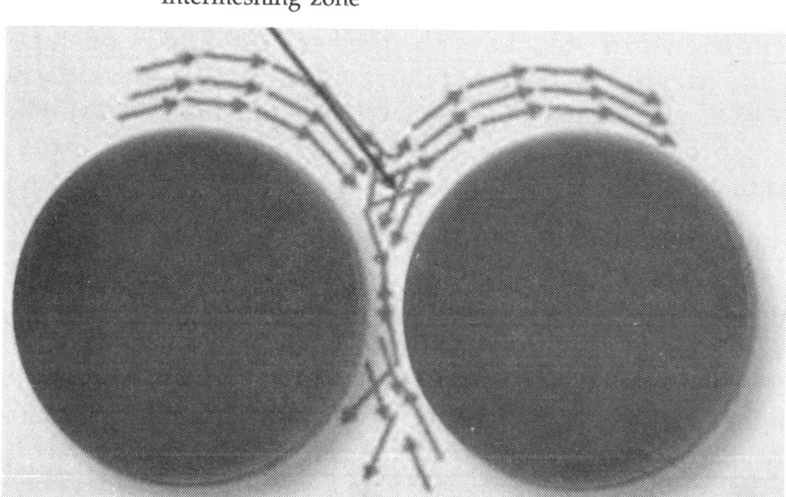

Fig. 6 : Mixing zone of a corotating twin screw extruder

Twin screw extruders can also be classified by the path of flow in the extruder channel. The flow can move either across or lengthwise in the channel or both depending upon the screw configuration. In a lengthwise open extruder, the material is conveyed from inlet to outlet moving from the channel of one screw to the channel of another. In a crosswise open extruder, there is an area common to both screws consisting of a path across the flights and material may be exchanged from one flight to the other flight in a direction perpendicular to the screw channel.

Screw configuration, whether it is lengthwise or crosswise opened or closed, affects the conveying conditions, mixing action, and the pressure build-up capacity of the screw system. The co-rotating intermeshing screws can not be lengthwise and crosswise closed, and partially intermeshing co-rotating screws can not be lengthwise open and crosswise closed. In contrast, a counter-rotating intermeshing lengthwise open and crosswise closed system does not exist either. The action of the intermeshing counter-rotating screw and the intermeshing co-rotating screw differ significantly. In the counter rotating system, the screws roll off each other in the wedge area, whereas the co-rotating screws are always in contact with each other creating natural wiping action. Counter-rotating fully intermeshing twin screw extruders have a positive displacement characteristic, but they require reduced screw speeds to lower the separating forces that cause wear on the screw by the calendaring effect at the nip between the screws. Counter-rotating fully intermeshing twin screw extruders are not widely used in the food industry, but are utilized in the production of gum, jelly and licorice confectionery products which have plastic-like characteristics.

Twin-screw extruders have the greater flexibility of operation that is possible by changing the degree of intermeshing of the screws, the number of flights or the angle of pitch of the screw. 'Kneading discs' can also be fitted to the screws so that the product passes between and through the discs to increase the kneading action.

Twin screw extruders are further categorized as:

Nonintermeshing twin screw extruders

The most complex design after the single-screw extruder is the counterrotating nonintermeshing twin-screw type (Fig. 7). Formation of nip at the flight tips, and a large feed throat causes improved performance in the feed zone. Because the screws do not intermesh, this extruder may be seen as two parallel single-screw devices. Figure 8 shows an end view of the twin-screw extruder. Several flows control performance: first, a drag flow that is closely related to that occurring in a single-screw extruder; second, a pressure flow similar to that in a single-screw extruder; third, a leakage flow between the screw flights and the barrel walls; and fourth, a leakage flow in the apex formed by the extruder shell and the two screws.

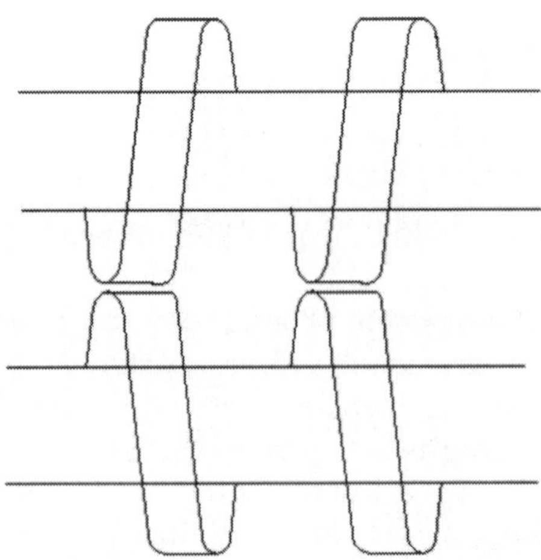

Fig. 7 : Nonintermeshing counterrotating, twin screw extruder (nonstaggered configuration). (*Source :* Levine and Miller, 2007).

A key parameter needed to describe this situation is the fraction of the barrel surface in proximity to the screw tips, where the drag mechanism operates. This is given by:

$$f = \frac{\alpha_a}{\pi}$$

Where, $\alpha_{a=}$ Angle of uncovered screw; f = fraction of screw perimeter covered by barrel

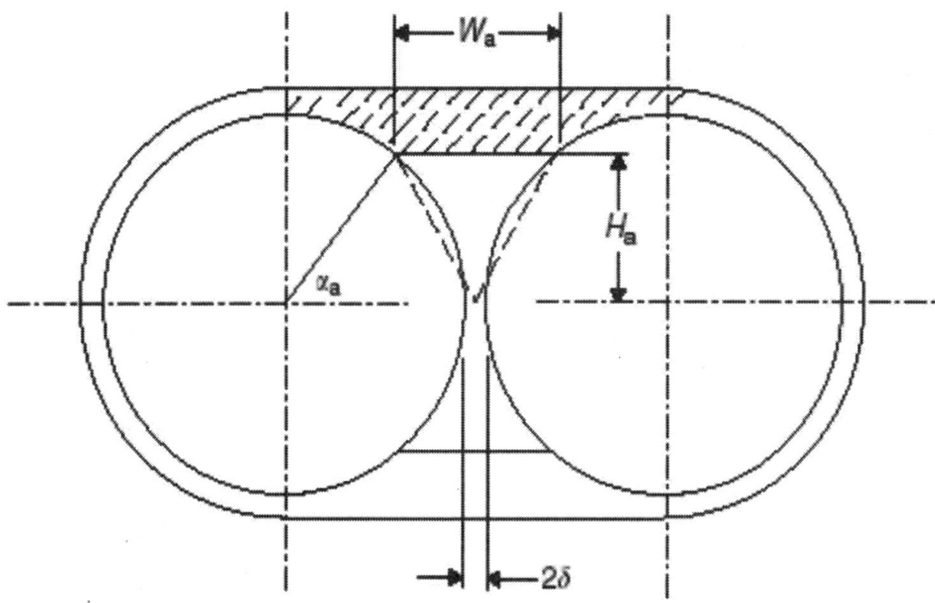

Fig. 8 : Apex region of nonintermeshing twin screw extruder (*Source :* Levine and miller, 2007) Wa= Width of apex between nonintermeshing screws; $\alpha_{a=}$ Angle of uncovered screw; Ha= Height of apex between nonintermeshing screws; δ = Screw clearance

Intermeshing counter rotating twin-screw extruders

The intermeshing counter rotating twin-screw configuration (Fig. 9) is the closest approximation of a positive-displacement pump that can be obtained with a conventional extruder design. During each rotation of the screws a C-shaped space, as illustrated in Fig. 10, is displaced in the forward direction by the screw. Without leaks, the output of this screw is:

$$Q = 2NV \quad \text{... (1)}$$

Where, Q= Volumetric flow rate; N= Rotational rate of screw; V=Volume of C-shaped chamber

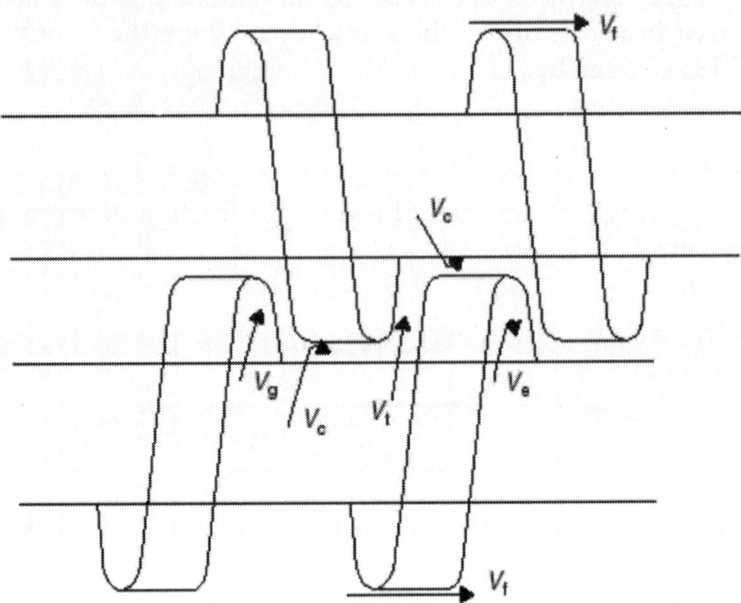

Fig. 9 : Intermeshing, counterrotating twin-screw extruder (*Source :* Levine & miller, 2007)

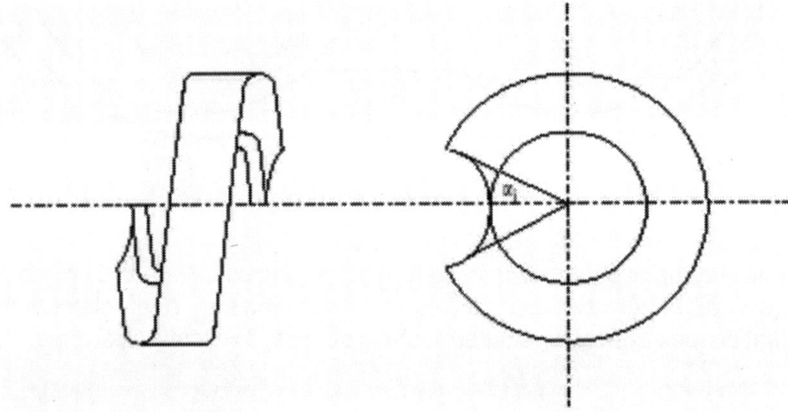

Fig. 10 : C-shaped section in an intermeshing, counterrotating twin-screw extruder (*Source :* Levine and Miller, 2007)

The volume of the C-shaped chamber is given by:

$$V = \frac{\pi DHW_m}{\cos\theta} - \frac{w_m D^2 \left(2\alpha_i - \sin 2\alpha_i\right)}{4\cos\theta} \quad \text{... (2)}$$

Where, D=Screw diameter, *H*= Channel depth, *ai* =Angle of twin-screw intermesh, θ = Screw helix angle, w = Channel width, φ = Flight flank angle

$$w_m = w + (D - H)\tan\varphi \qquad \text{... (3)}$$

$$W_m = \pi(D - H)\sin\theta - w_m \qquad \text{... (4)}$$

In this type of extruder there is no pressure flow such as that found in single-screw extruders. There are, however, several leakage flows that reduce the output of the extruder below that predicted by Equation (1). These leakages may be described as follows:

1. A leakage (*Vf*) between the flight tips and the barrel, analogous to the leakage flows in single-screw devices.
2. A calender leakage (*Vc*) between the root of one screw and the flight tip of the other. This leakage is the result of pumping by the nip formed by these components and is analogous to pumping in a two-roll mill.
3. Interscrew leakage (*Vt*) through the gap between the flights flanks in the radial direction. This leakage occurs in the tetrahedral gap.
4. A tangential leakage (*Vs*) through the side gap.

Net output is the displacement volume flow less the sum of these four leakage flows:

$$Q = 2NV - 2(V_c + V_s) - V_f - V_t$$

Intermeshing co-rotating twin-screw extruders

Co-rotating, intermeshing extruders, illustrated in Fig 11, are difficult to analyze. The screws exhibit the properties of single-screw extruders in that, unlike intermeshing counter rotating twin screw extruders, they provide a continuous flow path that allows pressure flow to occur.

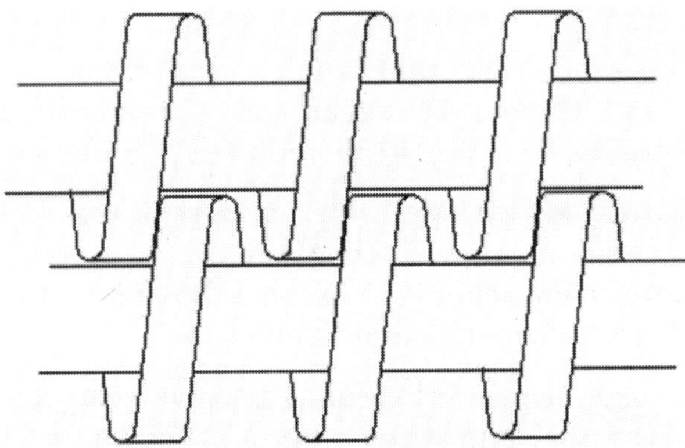

Fig. 11 : Intermeshing corotating twin-screw extruder (*Source :* Levine and Miller, 2007)

But analysis is complicated by the fact that the channel depth changes with axial and cross-channel positions. And there is a small degree of positive impetus to flow in the overlapping region between the two screws. Fig. 12 is a cutaway view of the screws. Region A is the completely intermeshing area. Pressure and drag flows occur in the regions marked B in Fig. 12.

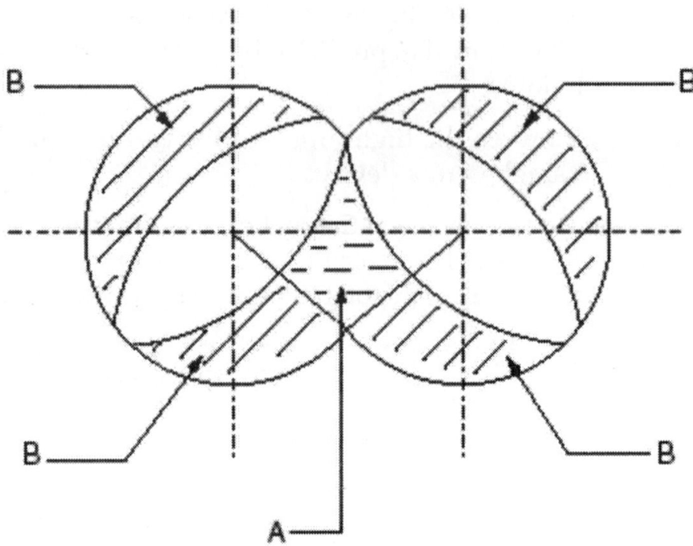

Fig. 12 : End view of intermeshing co-rotating twin-screw extruder. (*Source :* Levine and Miller, 2007).

Advantages of twin screw extruders over single screw extruder

Twin-screw extruders have the following advantages:

i. Sticky, oily, very wet materials or some other products can slip in a single screw but twin-screw machines can handle such type of materials easily.

ii. The limitations for single- and twin-screw machines, respectively are 4% and 20% fat, 10% and 40% sugar, and 30% and 65% moisture. There is therefore greater flexibility in operation using different raw materials.

iii. In case of twin screw extruders, the throughput is independent of feed rate, and fluctuations in production rate can be accommodated by the positive displacement action of the screws. Whereas, a single screw must be full of material to operate effectively.

iv. The positive displacement in case of twin screw extruder produces higher rates of heat transfer and better control of heat transfer than a single screw does.

v. A short discharge section develops the pressure required for extrusion and thus subjects a smaller part of the machine to wear than in single-screw extruders.

vi. A mixture of particle sizes, from fine powders to grains, may be used, whereas a single screw is limited to a specific range of granular particle sizes.

vii. Forward or reverse conveying is used to control the pressure in the barrel. For example, in the production of liquorice and fruit gums, the food is heated and compressed by forward conveying, the pressure is released by reverse conveying, to vent excess moisture or to add additional flavour ingredients, and the food is then recompressed for extrusion.

Statistical techniques

Statistical techniques make the process easy and number of experimental runs is reduced. Modeling of extrusion processing involves consideration of process parameters, system parameters, and product properties. Thus, extrusion cooking modeling is a multiple input and multiple output process. Though mathematical modeling of food extrusion process has benefited from available information on plastic extrusion, modeling of quality changes during food extrusion is a difficult task. Because of large number of process variables involved and nonlinear relationship among them, several approximations have to be made, which limit the result of mathematical models. The response surface methodology (RSM) has been widely used approach for modeling of extrusion processing (Shihani *et al.*, 2006). It is a collection of mathematical and statistical techniques which emphasize the modeling and analysis of the problem in which response of interest is influenced by several variables and the objective is to optimize this response (Montgomery, 1997). The main advantage of RSM is to reduce number of experimental runs needed to provide sufficient information for statistically acceptable results.

Upadhyay *et al.* (2010) employed response surface methodology for optimization of carrot pomace powder incorporation on extruded product quality. Rice and gram flour mix with carrot pomace powder (CPP) was used for the study. Response surface methodology was used to analyze the effect of die temperature, feed rate, feed moisture and % CPP on extrudate moisture, expansion index, bulk density and sensory characteristics of extrudates. The study demonstrated that an acceptable extruded product can be prepared by CPP incorporation.

Yagc *et al.* (2008) used response surface methodology to investigate the effects of extrusion conditions including moisture content, temperature, screw speed, and change in feed composition, % durum clear flour, % partially defatted hazelnut flour (PDHF) and % fruit waste contents on the physical and functional characteristics of the extruded snack food based on rice grit in combination with fruit waste, durum clear flour and PDHF. Response variables were bulk density, porosity, water absorption and water solubility indices and sensory properties of the extruded snacks. The product responses were most affected by changes in PDHF content and to a lesser extent by fruit waste content. The sensory evaluation of extrudates suggested that extrusion of PDHF, fruit waste and durum clear flour in combination with rice grit can produce acceptable extruded snack.

Altan *et al.* (2008) used response surface methodology for evaluation of snack foods from barley-tomato pomace blends by extrusion processing. Blends of barley flour and tomato pomace were processed in a co-rotating twin-screw extruder. Experimental design with die temperature, screw speed and tomato pomace level as independent variables produced 20 different combinations that were studied using response surface methodology to investigate the effect of these variables on system parameters (SME, die melt temperature and die pressure) and product responses (expansion, bulk density, water absorption and solubility indices, texture and color). Extrudate from five experiments within 20 samples was selected for sensory evaluation in terms of color, texture, taste, off-odor and overall acceptability. The system parameters and product responses were most affected by changes in temperature, pomace level and to a lesser extent by screw speed. The results suggest that tomato pomace can be extruded with barley flour into an acceptable and nutritional snack.

Chakraborty and Banerjee (2009) also employed response surface methodology for optimization of extrusion process for production of expanded product from green gram and rice.

RSM: Designs and features

Once the important factors have been identified, the next step is to determine the settings for these factors that result in the optimum value of the response. The optimum value of the response may either be a maximum value or a minimum value, depending upon the product or process in question. Methodologies that help the experimenter reach the goal of optimum response are referred to as Response Surface Methods. These methods are exclusively used to examine the "surface" or the relationship between the response and the factors affecting the response. Regression models are used for the analysis of the response, as the focus now is on the nature of the relationship between the response and the factors, rather than identification of the important factors. Response surface methods usually involve the following steps:

1. The experimenter needs to move from the present operating conditions to the vicinity of the operating conditions where the response is optimum. This is done using the method of steepest ascent (In this method, the contour plot of the first order model is used to decide the settings for the next experiment, in order to move towards the optimum conditions) in the case of maximizing the response. The same method can be used to minimize the response and is then referred to as the method of steepest descent.

2. Once in the vicinity of the optimum response the experimenter needs to fit a more elaborate model between the response and the factors. Special experiment designs, referred to as RSM designs, and are used to accomplish this. The fitted model is used to arrive at the best operating conditions that result in either a maximum or minimum response.

3. It is possible that a number of responses may have to be optimized at the same time. For example, an experimenter may want to maximize strength, while keeping the number of defects to a minimum. The optimum settings for each of the responses in such cases may lead to conflicting settings for the factors. A balanced setting has to be found that gives the most appropriate values for all the responses.

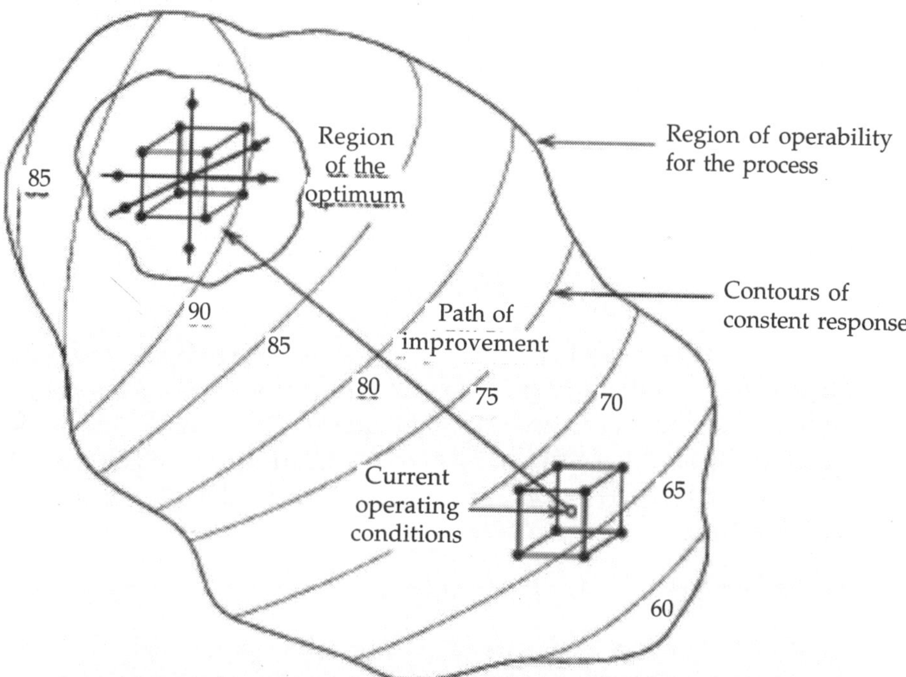

Fig.13 : Sequential nature of RSM (*Source :* Montgomery, 1997)

From Fig.13, we see that the analysis of a response surface can be thought of as "climbing a hill", where the top of the hill represents the point of maximum response if the true optimum is a point of minimum response then we may think of "descending into a valley" (Montgomery, 1997).

The eventual objective of RSM is to determine the optimum operating conditions for the system or to determine a region of the factor space in which operating requirements are satisfied.

Features of desirable design

i. Provides reasonable distribution of data points.
ii. Allows model adequacy, including lack of fit, to be investigated.
iii. Allows experiments to be performed in blocks.
iv. Allows designs of higher order to be built up sequentially.
v. Provides an internal estimate of error.
vi. Provides precise estimates of the model coefficients.
vii. Does not require a large number of runs.
viii. Ensures simplicity of calculation of the model parameters.

RSM Designs

This section can be divided into the following subsections:

- Central Composite Designs
- Rotatability
- Spherical Design
- Face-Centered Design
- Box-Behnken Designs

A second order model is generally used to approximate the response once it is realized that the experiment is close to the optimum response region where a first order model is no longer adequate. The second order model is usually sufficient for the optimum region, as third order and higher effects are seldom important. The second order regression model takes the following form for *k* factors:

$$y = \beta_0 + \beta_1 x_1 + \beta_2 x_2 + ... + \beta_k x_k + \beta_{11} x_1^2 + \beta_{22} x_2^2 + \beta_{kk} x_k^2 + \beta_{12} x_1 x_2 + ... + \beta_{k-1,k} x_{k-1} x_k + \varepsilon$$

The model contains $p=(k+1)(k+2)/2$ regression parameters that include coefficients for main effects ($\beta_1, \beta_2 ... \beta_k$)coefficients for quadratic main effects ($\beta_{11}, \beta_{22} ... \beta_{kk}$) and coefficients for two factor interaction effects ($\beta_{12}, \beta_{13} ... \beta_{k-1,k}$)

A full factorial design with all factors at three levels would provide estimation of all the required regression parameters. However, full factorial three level designs are expensive to use as the number of runs increases rapidly with the number of factors. For example, a three factor full factorial design with each factor at three levels would require 3^3=27 runs while a design with four factors would require 3^4=81 runs. Additionally, these designs will estimate a number of higher order effects which are usually not of much importance to the experimenter. Therefore, for the purpose of analysis of response surfaces, special designs are used that help the experimenter fit the second order model to the response with the use of a minimum number of runs. Examples of these designs are the central composite and Box-Behnken designs.

Central composite designs

Central composite design is the most popular of the many classes of RSM designs due to the following three properties:

- A CCD can be run sequentially. It can be naturally partitioned into two subsets of points; the first subset estimates linear and two-factor interaction effects while the second subset estimates curvature effects. The second subset need not be run when analysis of the data from the first subset points indicates the absence of significant curvature effects.
- CCDs are very efficient, providing much information on experiment variable effects and overall experimental error in a minimum number of required runs.
- CCDs are very flexible. The availability of several varieties of CCDs enables their use under different experimental regions of interest and operability.

Central composite designs are two level full factorial (2^k) or fractional factorial (2^{k-f}) designs augmented by a number of center points and other chosen runs. These designs are such that they allow the estimation of all the regression parameters required to fit a second order model to a given response.

The simplest of the central composite designs can be used to fit a second order model to a response with two factors. The design consists of a 2^2full factorial design augmented by a few runs at the center point (such a design is shown in Fig. 14 (a). A central composite design is obtained when runs at four other points - (-1, 0), (1, 0), (0, -1) and (-1, 0) are added to this design. These points are referred to as *axial points* or *star points* and represent runs where all but one of the factors are set at their mid-levels. The number of axial points in central composite designs having *k* factors is 2^k. The distance of the axial points from the center point is denoted by α and is always specified in terms of coded values. For example, the central composite design in Fig. 14 (b) has α=1, while for the design of Figure 14 (c) α =1.414. It can be noted that when α>1, each factor is run at five levels (-α, -1, 0, 1, +α) instead of the three levels of -1, 0 and 1. The reason for running central composite designs with α>1 is to have a rotatable design, which is explained next.

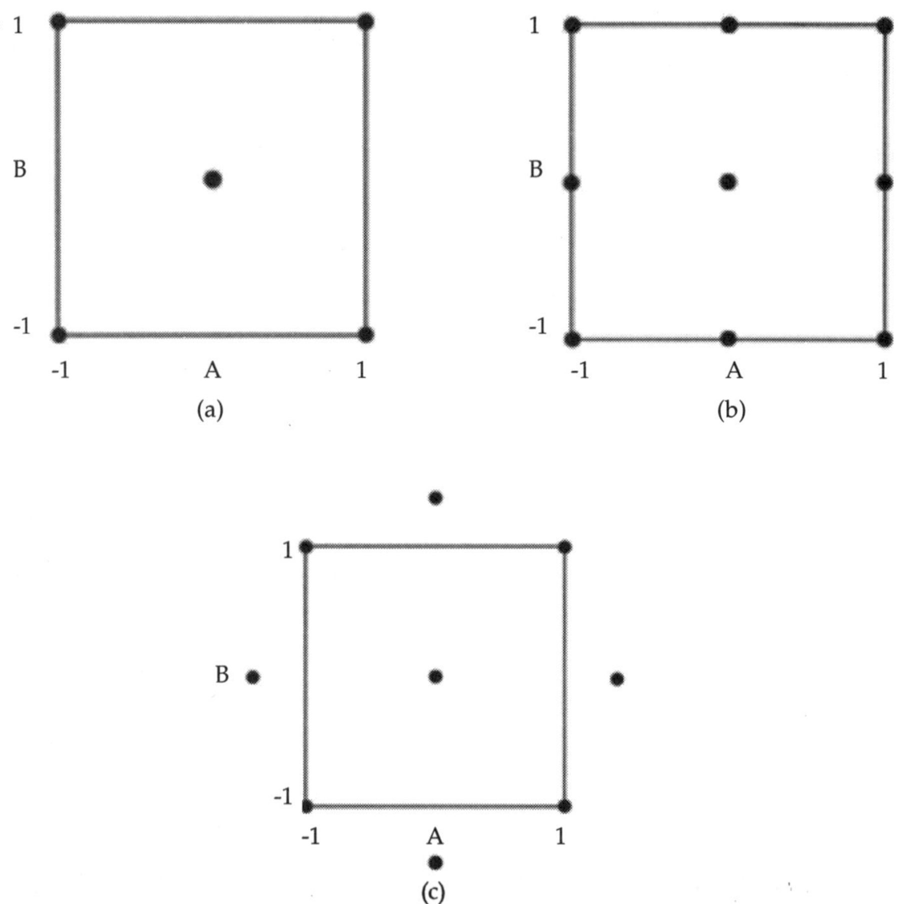

Fig. 14 : Central composite designs : (a) shows the 2^2 design with center point runs, (b) shows the two-factor central composite design with α =1 and (c) shows the two-factor central composite design with a=√2

Rotatability

A central composite design is said to be rotatable if the variance of any predicted value of the response, y_p for any level of the factors depends only on the distance of the point from the center of the design, regardless of the direction. In other words, a rotatable central composite design provides constant variance of the estimated response corresponding to all new observation points that are at the same distance from the center point of the design (in terms of the coded variables). The variance of the predicted response at any point x_p, is given as follows:

$$V[\hat{y}p] = \sigma^2 x_p' (X'X)^{-1} x_p$$

The contours of $V[\hat{y}_p]$ for the central composite design in Fig. 14 (c) are shown in Fig. 15. The contours are concentric circles indicating that the central composite design of Fig. 14 (c) is rotatable. Rotatability is a desirable property because the experimenter does not have any prior information about the location of the optimum. Therefore, a design that provides equal precision of estimation in all directions would be preferred. Such a design will assure the experimenter that no matter what direction is taken to search for the optimum, he/she will be able to estimate the response value with equal precision. A central composite design is rotatable if the value of α for the design satisfies the following equation:

$$\alpha = \left[\frac{2^{k-f} n_f}{n_s}\right]^{1/4}$$

where n_f is the number of replicates of the runs in the original factorial design and n_s is the number of replicates of the runs at the axial points. For example, a central composite design with two factors, having a single replicate of the original factorial design, and a single replicate of all the axial points, would be rotatable for the following α value:

$$\alpha = \left[\frac{2^{k-f} n_f}{n_s}\right]^{1/4} = \left[\frac{2^2(1)}{1}\right]^{1/4} = \sqrt{2} = 1.414$$

Thus, a central composite design in two factors, having a single replicate of the original 2^2 design and axial points, and with α=1.414, is a rotatable design. This design is shown in Fig. 14 (c).

Example : Kumar *et al.* (2010) used response surface methodology for development and characterization of extruded product using carrot pomace and rice flour. The experimental combinations were decided based on central composite rotatable design for four variables at five levels of each variable. The lateral expansion, bulk density, water absorption index, water solubility index, hardness and sensory characteristics were measured as responses.

The independent variables selected for the 1st set of experiments were: Feed proportion (Rice Flour: Pomace Powder), X_1, 70:30, 75: 25, 80:20, 85:15 and 90:10, Moisture Content, X_2, 17, 18, 19, 20 and 21%, Screw Speed, X_3,

270, 280, 290, 300 and 310 rpm and Barrel Temperature, X_4, 110, 115, 120, 125 and 130°C and the variables for 2nd set of experiments were same as that of the 1st set except in Feed proportion (Rice Flour: Pulse + Pomace Powder), x_1, 70:30, 75: 25, 80:20, 85:15 and 90:10. The variables and their levels were chosen based on the limited literature available on cauliflower by-products as a new source of dietary fiber, antioxidants and proteins in cereal based ready-to-eat expanded snacks (Stojceska *et al.*, 2008). The actual value of each level is detailed in Table 2. The five levels of the process variables were coded as -2, -1, 0, 1 and +2 (Montgomery, 1997) and design in coded (x) form and at the actual levels (X) is given in Table 3 & 4.

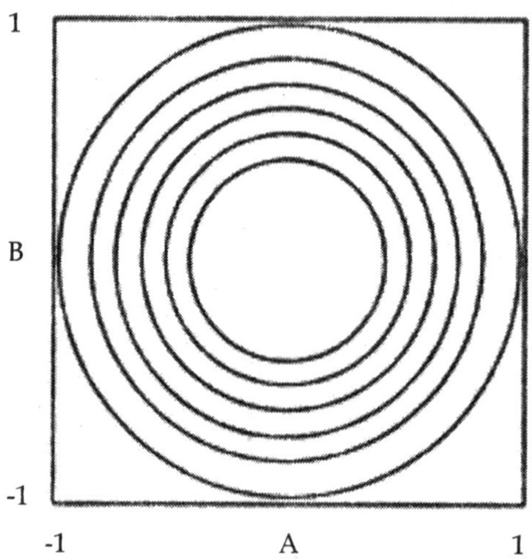

Fig. 15 : The contours of $V[\hat{y}_p]$ for the rotatable two-factor central composite design.

Table 2 : Values of independent variables at five levels of the CCRD design

Independent Variables		Levels in coded form				
	Code	-2	-1	0	+1	+2
Feed composition (ratio)	X_1	90:10	85:15	80:20	75:25	70:30
Feed moisture (%)	X_2	17	18	19	20	21
Screw speed (rpm)	X_3	270	280	290	300	310
Die temperature (°C)	X_4	115	120	125	130	135

Table 3 : Central composite rotatable experimental design (in coded and uncoded levels of four variables and five levels) employed for development of rice flour and carrot pomace based extrudates

S.No	Coded variables				Uncoded variables			
	x_1	x_2	x_3	x_4	X_1 Feed Proportion (Rice: Pomace)	X_2 Moisture content(%)	X_3 Screw speed(rpm)	X_4 Temperature (°C)
1	-1	-1	-1	-1	85:15	18	280	120
2	1	-1	-1	-1	75:25	18	280	120
3	-1	1	-1	-1	85:15	20	280	120
4	1	1	-1	-1	75:25	20	280	120
5	-1	-1	1	-1	85:15	18	300	120
6	1	-1	1	-1	75:25	18	300	120
7	-1	1	1	-1	85:15	20	300	120
8	1	1	1	-1	75:25	20	300	120
9	-1	-1	-1	1	85:15	18	280	130
10	1	-1	-1	1	75:25	18	280	130
11	-1	1	-1	1	85:15	20	280	130
12	1	1	-1	1	75:25	20	280	130
13	-1	-1	1	1	85:15	18	300	130
14	1	-1	1	1	75:25	18	300	130
15	-1	1	1	1	85:15	20	300	130
16	1	1	1	1	75:25	20	300	130
17	-2	0	0	0	90:10	19	290	125
18	2	0	0	0	70:30	19	290	125
19	0	-2	0	0	80:20	17	290	125
20	0	2	0	0	80:20	21	290	125
21	0	0	-2	0	80:20	19	270	125
22	0	0	2	0	80:20	19	310	125
23	0	0	0	-2	80:20	19	290	115
24	0	0	0	2	80:20	19	290	135
25	0	0	0	0	80:20	19	290	125
26	0	0	0	0	80:20	19	290	125
27	0	0	0	0	80:20	19	290	125
28	0	0	0	0	80:20	19	290	125
29	0	0	0	0	80:20	19	290	125
30	0	0	0	0	80:20	19	290	125

Table 4 : Central composite rotatable experimental design (in coded and uncoded levels of four variables and five levels) employed for development of rice flour, pulse powder and carrot pomace based extrudates

S.No	Coded variables				Uncoded variables			
	x_1	x_2	x_3	x_4	X_1 Feed Proportion (Rice: Pomace)	X_2 Moisture content(%)	X_3 Screw speed(rpm)	X_4 Temperature (°C)
1	-1	-1	-1	-1	85:15	18	280	120
2	1	-1	-1	-1	75:25	18	280	120
3	-1	1	-1	-1	85:15	20	280	120
4	1	1	-1	-1	75:25	20	280	120
5	-1	-1	1	-1	85:15	18	300	120
6	1	-1	1	-1	75:25	18	300	120
7	-1	1	1	-1	85:15	20	300	120
8	1	1	1	-1	75:25	20	300	120
9	-1	-1	-1	1	85:15	18	280	130
10	1	-1	-1	1	75:25	18	280	130
11	-1	1	-1	1	85:15	20	280	130
12	1	1	-1	1	75:25	20	280	130
13	-1	-1	1	1	85:15	18	300	130
14	1	-1	1	1	75:25	18	300	130
15	-1	1	1	1	85:15	20	300	130
16	1	1	1	1	75:25	20	300	130
17	-2	0	0	0	90:10	19	290	125
18	2	0	0	0	70:30	19	290	125
19	0	-2	0	0	80:20	17	290	125
20	0	2	0	0	80:20	21	290	125
21	0	0	-2	0	80:20	19	270	125
22	0	0	2	0	80:20	19	310	125
23	0	0	0	-2	80:20	19	290	115
24	0	0	0	2	80:20	19	290	135
25	0	0	0	0	80:20	19	290	125
26	0	0	0	0	80:20	19	290	125
27	0	0	0	0	80:20	19	290	125
28	0	0	0	0	80:20	19	290	125
29	0	0	0	0	80:20	19	290	125
30	0	0	0	0	80:20	19	290	125

Spherical design

A central composite design is said to be spherical if all factorial and axial points are at same distance from the center of the design. Spherical central composite designs are obtained by setting $\alpha=\sqrt{k}$. For example, the rotatable design in Fig. 14 (c) is also a spherical design because for this design $\alpha = \sqrt{k} = \sqrt{2} = 1.414$ ·

Face-centered design

Central composite designs in which the axial points represent the mid levels for all but one of the factors are also referred to as face-centered central composite designs. For these designs, $\alpha=1$ and all factors are run at three levels, which are -1, 0 and 1 in terms of the coded values (see Fig. 16).

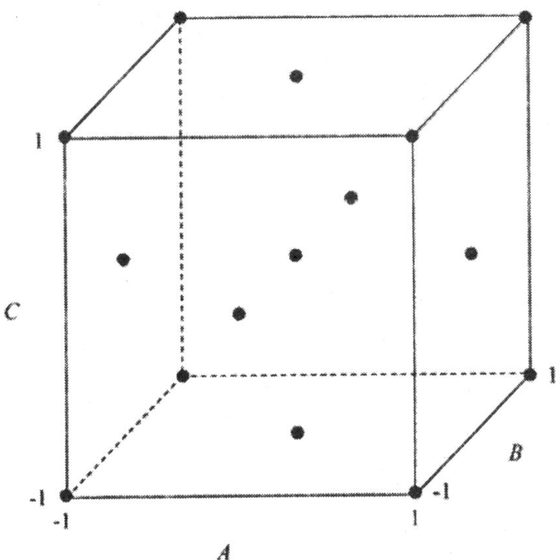

Fig 16 : Face-centered central composite design for three factors

Box-behnken designs

Box Behnken (1960) introduced designs for three level factors that are widely used in response surface methods to fit second-order models to the response. The designs are referred to as Box-Behnken designs. The designs were developed by the combination of two level factorial designs with incomplete block designs. For example, Fig. 17 shows the Box-Behnken design for three factors. The design is obtained by the combination of 2^2 design with a balanced incomplete block design having three treatments and three blocks.

The advantages of Box-Behnken designs include the fact that they are all spherical designs and require factors to be run at only three levels. The designs are also rotatable or nearly rotatable. Some of these designs also provide orthogonal blocking. Thus, if there is a need to separate runs into blocks for the Box-Behnken design, then designs are available that allow blocks to be used in such a way that the estimation of the regression parameters for the factor effects are not affected by the blocks. In other words, in these designs the block effects are orthogonal to the other factor effects. Yet another advantage of these designs is that there are no runs where all factors are at either the +1 or -1 levels. For example, in Fig. 17 the representation of the Box-Behnken design for three factors clearly shows that there are no runs at the corner points. This could be advantageous when the corner points represent runs that are expensive or inconvenient because they lie at the end of the range of the factor levels.

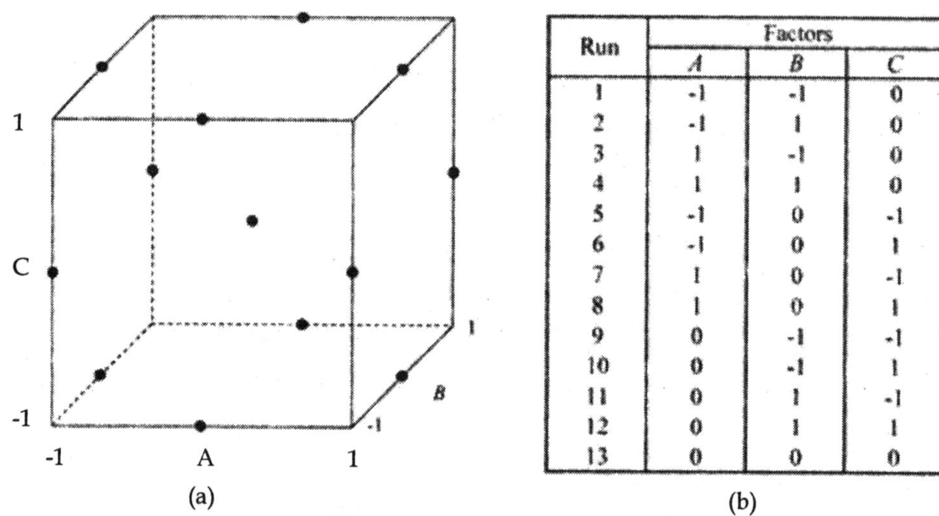

Run	Factors		
	A	B	C
1	-1	-1	0
2	-1	1	0
3	1	-1	0
4	1	1	0
5	-1	0	-1
6	-1	0	1
7	1	0	-1
8	1	0	1
9	0	-1	-1
10	0	-1	1
11	0	1	-1
12	0	1	1
13	0	0	0

Fig 17 : Box-Behnken design for three factors - (a) shows the geometric representation and (b) shows the design.

Co-products (Waste) utilization

The food processing industry produces large quantities of waste products. Food processing wastes are promising sources of valuable compounds such as dietary fiber, antioxidants, essential fatty acids, antimicrobials, minerals which may be used because of their favorable technological, nutritional and functional properties (Laufenberg *et al.*, 2003). Food processing wastes are inexpensive, available in large quantities, characterised by a high dietary fibre content resulting with high water binding capacity and relatively low enzyme digestible organic matter (Serena and Bach-Knudsen, 2007). Due to the high dietary fibre content and contrasting dietary fibre properties, the co-products could be used to change physicochemical properties of diets. Many fruit and vegetable residues i.e. apple pomace, orange peel are rich in dietary fiber which is useful for preventing and treating many diseases such as intestinal problems, cancer and diet related health problems (Larrea *et al.*, 2005; Wang & Thomas, 1989). Various sources of dietary fiber, such as apple pomace (Wang & Thomas, 1989) and orange pulp (Larrea *et al.*, 2005) incorporated into bakery products for their nutritional, functional and technological properties.

A number of researchers have used fruits and vegetable by-products such as apple, pear, orange, peach, blackcurrant, cherry, artichoke, asparagus, onion, carrot pomace (Grigelmo-Miguel and Martin-Belloso, 1999; Ng *et al.*,1999; Nawirska and Kwasnievska, 2005) as sources of dietary fibre supplements in refined food. Dietary fibre concentrates from vegetables showed a high total dietary fibre content and better insoluble/soluble dietary fibre ratios than cereal brans (Grigelmo-Miguel and Martin-Belloso, 1999).

Grape seeds are rich sources of polyphenolics such as catechin and epicatechin which have been shown to act as strong antioxidants and exert health promoting effects (Jayaprakasha *et al.*, 2001). Citrus seeds and peels were found to possess high antioxidant activity (Schieber *et al.*, 2001).

Waste utilization in extrusion processing

Upadhyay *et al.* (2010) developed an exrtuded product incorporating carrot pomace powder (CPP) in pre-extrusion mix and optimizated the carrot pomace powder incorporation on extruded product quality by response surface methodology. Innovative approach in this study involves utilization of a by-product through incorporation of CPP in pre-extrusion mix. Furthermore, a polynomial model is established for each process variable, and CPP incorporation is optimized for the desired quality of extrudates.

A lot of research work has been reported using rice flour for extrusion for its good extrusion characteristics. Kumar *et al.*(2010) worked on development and characterization of extruded product using carrot pomace and rice flour. Sensory evaluation revealed that carrot pomace could be incorporated into ready-to-eat expanded products upto the level of 11.75%.

Cauliflower has a very high waste index (Kulkarni *et al.*, 2001) and is an excellent source of protein (16.1%), cellulose (16%) and hemicelluloses (8%) (Wadhwa *et al.*, 2006) and it is considered as a rich source of dietary fibre and it possess both antioxidant and anticarcinogenic properties. Phenolic compounds and vitamin C are the major antioxidants of brassica vegetables, due to their high content and high antioxidant activity (Podsedek, 2007). Stojceska *et al.* (2008) studied the incorporation of cauliflower trimmings into ready-to-eat expanded products and their effect on the textural and functional properties of extrudates. Dried and milled caulifower at levels of 5–20% was added to the formulation mix. Extrusion cooking significantly ($P < 0.0001$) increased the level of phenolic compounds and antioxidants. The expansion indices, total cell area of the products, wall thickness showed negative correlation to the level of cauliflower. Sensory test panel indicated that cauliflower could be incorporated into ready-to-eat expanded products up to the level of 10%.

Durum clear flour is by-product which is obtained during the semolina production by collecting the line fractions of certain streams and the flour which is extracted during grinding and composed approximately 13–16% of the milled durum wheat. Durum clear flour is relatively high in bran, high in protein (14–16%), ash (1.5–2%) and starch (about 65%) and slightly grey in color (Kýlýc, 1999). Defatted hazelnut flour is residual of hazelnut after extraction of oil parts from the main fruit. Yagc *et al.* (2008) worked on evaluation of physical and functional characteristics of the extruded snack food based on rice grit in combination with fruit waste, durum clear flour and partially defatted hazelnut flour (PDHF). The sensory evaluation of extrudates suggested that extrusion of PDHF, fruit waste and durum clear flour in combination with rice grit can produce acceptable extruded snack.

Tomato seed from tomato cannery waste has been reported to have appreciable amounts of digestible amino acids, methionine, cystine and lysine (Persia *et al.*, 2003). During tomato processing a by-product, known as tomato pomace, is generated. Moreover, most of the lycopene in tomato is associated with the tomato pomace (Schieber *et al.*, 2001). This by-product represents, at most, 4% of the fruit weight (Del Valle, Camara, & Torija, 2006). Tomato pomace consists of the dried and crushed skins and seeds of the fruit (Tadeu-Pontes *et al.*, 1996). The skin, important component of pomace, is source of lycopene. Lycopene is an excellent natural food color and also serves as a functional ingredient with important health benefits beyond basic nutrition (Kaur *et al.*, 2005). A diet rich in lycopene is related to a decreased risk of certain cancers, particularly cancers of the digestive tract, prostate cancer and pancreatic cancer due to protective effect of lycopene against oxidative damage (Johnson, 2000). It was also found that tomato pomace significantly reduced cholesterol level in liver and heart by 15% and 18%, respectively (Bobek *et al.*, 1998). The use of tomato processing by-products could provide gaining valuable substances and at the same time reduce the waste disposal problem. Altan *et al.* (2008) worked on evaluation of snack foods from barley-tomato pomace blends by extrusion processing. Blends of barley flour and tomato pomace were processed in a co-rotating twin-screw extruder. Extrudate from five experiments within 20 samples was selected for sensory evaluation in terms of color, texture, taste, off-odor and overall acceptability. Extrudates with 2% and 10% tomato pomace levels extruded at 160^{0}C and 200 rpm had higher preference levels for parameters of color, texture, taste and overall acceptability. The results suggested that tomato pomace can be extruded with barley flour into an acceptable and nutritional snack.

Chakraborty and Banerjee (2009) optimized extrusion process for production of expanded product from green gram and rice using response surface methodology.

Ding *et al.* (2006) studied the effect of extrusion conditions on the functional and physical properties of wheat based expanded snacks. So there is an ample scope for inclusion of vegetable and fruits waste (co-products) in the snack food as a source of mineral, vitamin and dietary fibre.

Effect of extrusion processing on foods

Effects on sensory characteristics

- Extrusion conditions exert a considerable impact on microstructure and the size of air cells in the extrudates (Yaseen and Shouk, 2006). Production of characteristic textures is one of the main features of extrusion technology which depends on the microstructural changes of extrudates taking place during extrusion process. The extent of changes to starch, determined by the operating conditions and feed materials, produce the wide range of product textures that can be achieved.

- Chen *et al.* (1991) studied the effects of extrusion conditions on sensory properties of corn meal extrudates. Temperature was found to be the most significant factor affecting Munsell value, airiness, toasted corn aroma, and flavor, denseness, crispiness, chewiness and hardness of extrudates. Temperature and feed moisture had significant effects on Munsell value, surface texture, raw flour aroma, toasted corn aroma and flavor, chewiness and hardness. Interaction between temperature and screw speed had significant effects on airiness and denseness of extrudates.

- In case of cold extrusion, usually temperature remains at ambient temperature. So added flavours can be mixed with ingredients before extrusion, but this method is unsuccessful in extrusion cooking as the flavours are volatilised when the food emerges from the die. Micro-encapsulated flavours are more suitable for extrusion cooking, but these are highly expensive. Flavours are therefore more often applied to the surface of extruded foods in the form of sprayed emulsions or viscous slurries. However, this may sometimes cause stickiness in some products and hence require additional drying.

- During extrusion cooking temperature rises above 100°C. HTST conditions do not affect the natural colour and flavour of foods to a greater extent. However, in many foods the colour of the product is determined by the synthetic pigments added to the feed material as water- or oil-soluble powders, emulsions or lakes. Fading of colour due to product expansion, excessive heat or reactions with proteins, reducing sugars or metal ions may be a problem in some extruded foods.

Effects on nutritional value

- Beneficial effects of extrusion include destruction of antinutritional factors, gelatinisation of starch, increased soluble dietary fibre and reduction of lipid oxidation (Singh *et al.*, 2007).
- Maillard reactions between protein and sugars reduce the nutritional value of the protein, depending on the raw material types, their composition and process conditions. Heat-labile vitamins may be lost to varying extents. Changes in proteins and amino acid profile, carbohydrates, dietary fibre, vitamins, mineral content and some non-nutrient healthful components of food may be either beneficial or deleterious (Singh *et al.*,2007).
- Ascorbic acid and β-carotene losses are up to 50%, depending on the time that the food is held at the elevated temperatures (Harper, 1979), and loss of lysine, cystine and methionine in rice products varies between 50 and 90% depending on processing conditions (Seiler, 1984).
- Mild extrusion conditions (high moisture content, low residence time, low temperature) improve the nutritional quality, while high extrusion temperatures (200 °C), low moisture contents (<15%) and/or improper formulation (e.g. presence of high-reactive sugars) can impair nutritional quality adversely. To obtain a nutritionally balanced extruded product, careful control of process parameters is essential (Singh *et al.*, 2007).
- Type of food, the moisture content, the temperature of processing and the holding time has significant effect on Vitamin losses in extruded foods. Generally, losses are minimal in cold extrusion. The HTST conditions in extrusion cooking, and the rapid cooling as the product emerges from the die, cause relatively small losses of most vitamins and essential amino acids. For example at an extruder temperature of 154°C there is 95% retention of thiamin and little loss of riboflavin, pyridoxine, niacin or folic acid in cereals.
- Extrusion of steam-conditioned soyabean material inactivated the antinutritional factors making extruded soyabean an ideal ingredient for pet food and its stability (Purushotam *et al.*, 2007).

- Processing conditions and formulation in soy flour affect the changes in proteins. High temperatures and the presence of sugars cause millard browning and a reduction in protein quality. Lower temperatures and low concentrations of sugars result in an increase in protein digestibility, owing to rearrangement of the protein structure. Destruction of anti-nutritional components in soya products improves the nutritive value of texturised vegetable proteins.

Other developments in extrusion processing

A number of more recent developments in extrusion include:

- Bailey *et al.* (1995) described a computerised system of process control that is able to control the extrusion process, the start-up and shut-down sequences, alarm recognition, storage of formulations and presentation of colour displays of the process to the operator. The control system continuously monitors over one hundred process alarm conditions, including the product formulation in relation to the operating conditions. It visually alerts the operator if non-specified conditions exist, and in extreme cases, controls an orderly shutdown or an emergency stop.
- Modified starches can be produced by using extruders as enzymatic reactors with thermostable α-amylase. Caseinates are also subjected to partial hydrolysis in an extruder with selected proteases and the products are reported to have very good bacteriological quality, improved colour, flavour and water absorption properties (Jones, 1990).
- Jones (1990) worked on the production of oligosaccharide mixtures from potato starch, without the use of enzymes that are suitable for dietetic and infant food. It was found that oil can be extracted more efficiently from soybean flour by first using dry extrusion (10–14% moisture) at 135°C followed immediately by pressing in a continuous screw press. The process is said to produce very high quality oil, similar to a refined, deodorized product, together with an oil cake, containing 50% protein and 90% inactivation of trypsin inhibitors, that is suitable for human consumption.
- Decontamination and sterilization of spices prior to roasting can be done in extruders. It is reported that extrusion cooking results in a 1000-fold reduction in microorganisms and removal of off-flavours that eliminates the need for a time consuming and expensive conching stage (Jones, 1990).
- Computer modeling of fluid flow behaviour and heat transfer inside the extruder barrel has more recently led to a greater understanding of the operation of extruders.

- Artificial neural networks (ANN) are a powerful tool for modeling of extrusion processing of food materials. ANN models performed better than the RSM models in describing the extrusion process and characteristics of the extruded product from wheat flour as well as the blend of wheat- black soybean in terms of specific mechanical energy requirement, expansion ratio, water absorption index, water solubility index as well the sensory characteristics (Shihani *et al.*, 2006). Artificial neural networks (ANN) are being used in the field of food processing due to their ability for solving non-linear problems (Ganjyal and Hanna, 2002; Popescu *et al.*, 2001). Neural networks are useful when exact mathematical information is not available. Another advantage of a neural network model over a rule based model is that, if the process under analysis changes, new data can be added and the neural network can be trained again. This is much easier than determining new models or rules. Artificial neural networks are mathematical models of biological neural systems. It learns from examples through iteration, without requiring a prior knowledge of relationships between variables under investigation. Each example includes both inputs (information used to make a decision) and patterns (prediction or responses). ANN tries each example in turn using the inputs to calculate answers which it compares to provide pattern. If it is wrong, ANN corrects the network by making changes to internal connections (weights). The trial and error process continues until the network outputs are in good agreement with patterns to a certain specified level of accuracy (Shihani *et al.*, 2006).

Applications

Cold extrusion

Non-texturized whey protein isolates (WPI) addition to some snacks or baked products sometimes reduces the textural quality of the food. Texturization is the protein-modifying effect of extrusion processing and it produces textured proteins using moderate shear under near and/or sub-ambient extrusion conditions. Texturization prevents denaturing of the nutritious WPI, and converts it to a stretched or texturized dough-like gel that is in the form required for its addition to foods. Temperature of extrusion affects the solid/liquid behavior of the gels, both initially and during storage. Cold-textured WPI forms gels and may be used by food processors in meat analogues, high-protein snacks, and other products to provide health-enhancing benefits (Onwulata *et al.*, 2007).

Cold extrusion of moistened waxy corn starch mixed with methionine or with cysteine produced methional (potato chip flavor component) or hydrogen sulfide (meat flavor component) as key flavors among several other compounds. Cold extrusion process produced those flavors more efficiently than thermal cooking in a heating bomb or hot extrusion process. Natural flavors from cold extrudates were more pleasant than those generated from thermal treatment for these reaction systems (Cho *et al.*, 1995).

In cold extrusion process, the product is extruded without cooking or distortion of the food. The extruder has a deep-flighted screw, which operates at a low speed in a smooth barrel, to knead and extrude the material with little friction. Cold extrusion is used to produce pasta, hot dogs, pastry doughs and some types of confectionery.

Pastas are generally made of flour and water, although eggs are sometimes added. Durum semolina is the best material for making flour for pasta. Durum wheat is a hard wheat and gives pasta its yellow color and is generally different from common wheat. Generally, durums have poor wheat gluten and do not make good bread products. The semolina is mixed with water until 31% moisture is reached. The product will form into small dough balls. Small air bubbles in the product will weaken the final dried pasta product. The incorporation of air also leads to the activation of lipoxygenase enzymes that bleach the dough white. Hence, elimination of air in the system will limit the activation of these enzymes. A single screw extruder with deep flight channels to eliminate shear is used for the extrusion of pasta products. It acts as a mixer, kneads the dough and exerts pressure. The dough moves down the barrel to the extruder die. This process produces a lot of heat which needs to be eliminated. Generally, the pasta extruder is equipped with a jacket that is usually filled with cold water, keeping the temperature below 45°C. The low temperature combined with the low moisture content leads to little or no expansion which is desired in pasta production (Hoseney, 1986).

Hot extrusion (Extrusion cooking)

Extrusion cooking is a technique that makes it possible to create prepared foods within a very short period of time. The food product is heated under a high degree of pressure, then slowly forced through a series of pores into another cooking chamber. As this process takes place, the moisture content of the food is reduced significantly, leaving behind a product that is thoroughly cooked and dried low-density product. Hot extrusion is a HTST (high temperature and short time) process, which minimizes the loss of nutrients and reduces microbial contamination. The moisture content of some products, for example crispbread, snackfoods and breakfast cereals, is further reduced after extrusion by drying.

Confectionery products

Twin screw extruders are more common in the processing of confectionery products. This is due to the twin screw's ability to convey material, renew material at the heating surfaces, control temperatures of heat-sensitive materials, and incorporate fat, milk solids, nuts, color, and flavor portends. HTST extrusion cooking is used to produce a gelatinised, chewy product such as fruit gums and liquorice, from a mixture of sugar, glucose and starch. The heat gelatinises the starch, dissolves the sugar and vaporises excess water which is vented from the machine. Twin screw extruders used for confectionery products generally have a longer length to diameter ratio (25:1) to increase the heat transfer area that is required to produce products such as licorice, peanut brittle, caramel and toffee. For example, licorice is produced by adding flour, starch, corn syrup, sugar, and a gum such as carboxymethyl cellulose. Colourings and flavours are added to the plasticised material and, after mixing, it is cooled and extruded. The product texture can be adjusted from soft to elastic by control over the formulation and processing conditions, the shape can be changed by changing the die, and a variety of flavours and colours may be added. These different combinations permit a very large range of potential products, including licorice, toffee, fudge, boiled sweets, creams, and chocolate, each produced by the same equipment.

Cereal products

Ready to eat cereals (RTE)

Extrusion cooking of ready-to-eat cereals (RTE) provides several advantages over conventional processing methods. The twin screw extruder can be adapted to perform several different tasks including heating, cooling, conveying the product, feeding, compressing, reacting, homogenizing, melting or rendering amorphous, cooking, texturizing and shaping of the product. However, considering the most RTE cereals are of low moisture and high expansion formed under high shear and pressure, single screw extruders can also be utilized for this type of processing. The initial moisture content would be higher which in turn requires more energy for drying, but the same end product can be produced.

Crisp bread

Crisp bread is commonly made from wheat flour, milk powder, corn starch, sugar and water. Ingredients are mixed and crisp bread is prepared from dough under pressure, in an extruder. The sudden drop in pressure causes water to evaporate, creating bubbles in the dough. The product is extruded at a high temperature and pressure. The crisp bread is then toasted to reduce the moisture content further and to brown the surface.

Expanded snack foods

Snack products are generally extruded at a lower moisture content, higher shear and higher temperature to cause significant starch modification as opposed to the RTE cereals. This leads to a more expanded product that melts in the mouth. Collet extruder is primarily used for forming snack products with high expansion. The internal pressure, as well as the size and shape of the die orifice lead to the final shape of the product. Extruded snack products are produced in a variety of shapes and sizes which include rings, stars, curls, balls, lattices and squiggles.

Protein-based foods

Texturised vegetable protein (TVP)

TVP is a high-fiber, high-protein meat substitute made from soy flour and available in a variety of flavored and unflavored varieties, as well as different sizes, from large chunks to small flakes. Texturization of vegetable protein is the restructuring of protein molecules (usually soy protein) into a layered, cross linked mass which is resistant to disruption upon further heating and/or processing. Extrusion cooking destroys the enzymes present in soybeans, including a urease which reduces the shelf life, a lipoxidase which causes off flavours by oxidation of soya oil and also a trypsin inhibitor which reduces protein digestibility. This improves the acceptability, digestibility and shelf life of the product. Defatted soya flour, soya concentrate or isolate are moistened and the pH is adjusted. A lower pH (5.5) increases chewiness in the final product, whereas a higher pH (8.5) produces a tender product and more rapid rehydration. Colours, flavours and calcium chloride firming agent are added, and the material is plasticised in an extruder at 60–104°C. Either single or twin screw extruders can be used for texturizing vegetable protein. In both cases, the texture is affected by screw speed, barrel, temperature, moisture, raw material quality and die orifice. The die for processing TVP emits a smooth streamlined flow that eliminates shearing so not to disturb the newly cross-linked protein which is cooled and dried to 6–8% moisture content.

Processed cheese

Processed cheese is also an important example of protein based extruded foods. Adhikari *et al* (2009) studied the physical and sensory characteristics of processed cheeses manufactured by extrusion technology. In this study Two extrusion temperatures (80° and 90°C), two moisture levels (440 and 480 g kg^{-1}), and two emulsifying salt levels (10 and 15 g kg^{-1}) were studied. Instrumental measures (moisture and texture profile analysis), descriptive analysis, and consumer acceptability for the processed cheese samples were carried out. The instrumental measures and descriptive analysis showed differences among the samples. The major difference was a result of the moisture content. The lower-moisture cheeses were firmer/chewier in texture, and had more Cheddar flavor note. Consumers found the lower-moisture cheeses extruded at 80°C to be more acceptable.

Meat and fish products

The meat grinder is similar to the pasta extruder in that there is one solid screw with deep channels used for conveying the meat forward, and a smooth die is used to save on wear and forces the product through the desired die size. Application of extruders to meat and fish products has mostly focused on production of extruded snacks or shelf-stable starch pellets that incorporate previously under-utilized by-products from meat, fish or prawns. Additionally, the manufacture of shiozuri surimi from ground, minced fish has been investigated using extruders operating at a die temperature of 6°–27°C.

References

Adhikari Kaushik, Cole Andrea, Grun Ingolf, heymann Hildegarde, Hsieh Fu-Hung, Huff Harold (2009). Physical and sensory characteristics of processed cheeses manufactured by extrusion Technology. Journal of the Science of Food and Agriculture, 89(8): 1428-1433.

Altan, Aylin, L.Kathryn, Mccarthy, Maskan, Medeni (2008). Evaluation of snack foods from barley-tomato pomace blends by extrusion processing. Journal of Food Engg., 84: 231-242.

Bailey, LN, Hauck, BW, Sevatson, ES and Singer, RE (1995). Ready-to-eat breakfast cereal production. In: A. Turner (ed.) Food Technology International Europe, Sterling Publications International, London, pp. 127–132.

Best, ET (1994). Confectionery extrusion. In: N.D. Frame (ed.) The Technology of Extrusion Cooking. Blackie Academic and Professional, Glasgow, pp. 190–236.

Bobek, P, Ozdin, L, & Hromadova, M (1998). The effect of dried tomato grape and apple pomace on the cholesterol metabolism and antioxidative enzymatic system in rats with hypercholesterolemia. Nahrung, 42: 317–320.

Chakraborty Pratap and Banerjee Sournitra (2009). Optimization of extrusion process for production of expanded product from green gram and rice by response surface methodology. Journal of Scientific and Industrial Research, 68: 140-148.

Chen, Jing., Serafin Florence L, Pandya Rajesh N, Henryk Daun (1991). Effects of extrusion conditions on sensory properties of corn meal extrudates. Journal of Food Science, 56(1): 84-89.

Cho, MH, Zheng, X, Wang, SS, Kim,Y, Ho Chi-Tang (1995). Production of natural flavors using a cold extrusion process In: Flavour Technology, Vol. 610, Chapter 9, pp 120–126, ACS Symposium Series, ISBN 13: 9780841233263.

Del Valle, M, Camara, M, & Torija, ME (2006). Chemical characterization of tomato pomace. Journal of the Science of Food and Agriculture, 86: 1232–1236.

Desrumaux, A, Bouvier, JM, & Burri, J (1999). Effect of free fatty acids addition on corn grits extrusion cooking. Cereal Chemistry, 76: 699–704.

Ding, Qing-Bo, Ainsworth Paul, Plunkett Andrew, Tucker Gregory, Marson, Hayley (2006). The effect of extrusion conditions on the functional and physical properties of wheat based expanded snacks. Journal of Food Engg., 73: 142-148.

Fellows. P (2000) Extrusion In: Food Processing Technology: Principles and Practice, Woodhead Publishing Limited and CRC Press LLC, pp 294-308.

Ganjyal, GM and Hanna, MA (2002). Review on residence time distribution (RTD) in food extrudates and study on the potential of neural networks. Journal of Food Science, 68(4): 1996-2002.

George Box, Donald Behnken (1960). Some new three level designs for the study of quantitative variables, Technometrics, 2: 455–475.

Grigelmo-Miguel, N, Martin-Belloso, O (1999). Comparison of dietary fibre from by-products of processing fruits and greens and from cereals. LWT – Food Science and Technology, 503–508.

Harper, JM (1979). Food extrusion. CRC Crit. Rev. Food. Sci. Nur. pp.155–215.

Harper, JM (1987). High-temperature short-time extrusion cooking. In: A. Turner (ed.) Food Technology International Europe. Sterling Publications International, London, pp. 51–55.

Hauck, BW (1993). Choosing an extruder. In: A. Turner (ed.) Food Technology International Europe. Sterling Publications International, London, pp. 81–82.

Heldman, DR and Hartel RW (1997). Principles of Food Processing. Chapman and Hall, New York, pp. 253–283.

Hoseney, RC, ed. (1986). Principles of Cereal Science and Technology.Am. Assoc. Cereal Chem.: St. Paul, MN.

Jayaprakasha, GK, Singh, RP, & Sakariah, KK (2001). Antioxidant activity of grape seed (Vitis vinifera) extracts on peroxidation models in vitro. Food Chemistry, 73: 285–290.

Johnson, EJ (2000). The role of lycopene in health and disease. Nutrition in Clinical Care, 3: 35–43.

Jones, SA (1990). New developments in extrusion cooking. In: A. Turner (ed.) Food Technology International Europe, Sterling Publications International, London, pp. 54–58.

Kaur, D, Sogi, DS, Gary, SK, & Bawa, AS (2005). Flotation-cumsedimentation system for skin and seed separation from tomato pomace. Journal of Food Engineering, 71: 341–344.

Kulkarni, M, Mootey, R, Lele, SS (2001). Biotechnology in agriculture, industry and environment. In: Proceedings of the International Conference of SAARC Countries, Organized by Microbiologists Society at Karad, India during December 28–30, pp. 24-31.

Kumar, Navneet, Sarkar, B.C., Sharma, H.K. (2010). Development and characterization of extruded product using carrot pomace and rice flour, International Journal of Food Engineering, Vol. 6 : Iss. 3, Article 7.

Kýlýc, G (1999). Evaluation of Durum Clear Flour. A MSc Thesis. Gaziantep, Turkey: University of Gaziantep, Food Engineering Department.

Larrea, MA, Chang, YK, & Martinez-Bustos, F (2005). Some functional properties of extruded orange pulp and its effect on the quality of cookies. Lebensm.-Wiss. U.-Technology, 38: 213–220.

Laufenberg, G, Kunz, B, & Nystroem, M (2003). Transformation of vegetable waste into value added products: (A) the upgrading concept; (B) practical implementations. Bioresource Technology, 87: 167–198.

Levine Leon and Miller, Robert C (2007). Extrusion Processes In Handbook of Food Engineering, Edtd by Heldman, D.R., Lund, Daryl B., CRC press, New York, pp 799-846.

Mans, J (1982). Extruders, Prep. Foods, 11: 60–63.

Miller, RC (1990). Unit operations and equipment IV: extrusion and extruders. In: R.B. Fast and E.F. Caldwell (eds) Breakfast Cereals and How They are Made. American Association of Cereal Chemists, Inc., St Paul, MN, pp. 135–196.

Montgomery, Douglas C (1997). Design and Analysis of Experiments, 4th Edition, John Wiley and Sons. New York.

Nawirska, A, Kwasnievska, M (2005). Dietary fibre fractions from fruit and vegetable processing waste. Food Chemistry, 91: 221–225.

Ng, A, Lecain, S, Parker, ML, Smith, AC, Waldron, KW (1999). Modification of cell wall polymers of onion waste. III. E.ect of extrusion-cooking on cell wall material of outer fleshy tissues. Carbohydrate Polymers, 39: 341–349.

Onwulata, Charles, Tunick, Michael, Tomasula, Peggy (2007). Cold extrusion texturization of whey proteins, AOCS Book of Abstracts: p. 143.

Persia, ME, Parsons, CM, Schang, M, & Azcona, J (2003).Nutritional evaluation of dried tomato seeds. Poultry Science, 82: 141–146.

Podsedek, A (2007). Natural antioxidant capacity of brassica vegetables: a review. LWT - Food Science and Technology, 40: 1–11.

Popescu, O, Popescu, C and Wilder, J (2001). A new approach to modeling and control of a food extrusion process using ANN and an expert system. Journal of Food Process Engineering, 24: 17-36.

Purushotham, B, Radhakrishna, PM and Sherigara, BS (2007). Effects of steam conditioning and extrusion temperature on some anti-nutritional factors of soyabean (Glycine max) for pet food applications. American Journal of Animal and Veterinary Sciences 2 (1): 1-5.

Schieber, A, Stintzing, FC, & Carle, R (2001). By-products of plant food processing as a source of functional compounds-recent developments. Trends in Food Science and Technology, 12: 401–413.

Seiler, K (1984). Extrusion cooking and food processing. Food Trade Rev. March 124–125: 127.

Serena, A, Bach-Knudsen, KE (2007). Chemical and physicochemical characterisation of co-products from vegetable food and agro industries. Animal Feed Science and Technology, 139: 109–124.

Shihani, Neelam, Kumbhar, B.K., Kulshreshtha, Manoj (2006). Modeling of extrusion process using response surface methodology and artificial neural networks, Journal of Engineering Science and Technology, 1(1): 31-40.

Singh Shivendra, Gamlath Shirani, Wakeling Lara (2007). Nutritional aspects of extrusion: A review, International Journal of Food Science and Technology, 42(8): 916-929.

Stojceska, Valentina, Ainsworth, Paul, Plunkett Andrew, Ibanoglu Esra, Ibanoglu, Senol (2008). Cauliflower by-products as a new source of dietary fibre antioxidants and proteins in cereal based ready to eat expanded snacks, Journal of Food Engg., 87: 554-563.

Tadeu-Pontes, MA, Carvalheiro, F, Roseiro, JC, & Amaral-Colloc O, MT (1996). Evaluation of product composition profile during an extrusion based process of tomato pomace transformation. Agro-Food-Industry Hi-Tech, May/June, 39–40.

Upadhyay, A, Sharma, HK, Sarkar, BC (2010). Optimization of carrot pomace powder incorporation on extruded product quality by response surface methodology, Journal of Food Quality, 33(3): 350-369.

Wadhwa, M, Kaushal, S, Bakshi, MPS (2006). Nutritive evaluation of vegetable wastes as complete feed for goat bucks. Small Ruminant Research, 64: 279–284.

Wang, HJ, & Thomas, RL (1989). Direct use of apple pomace in bakery products. Journal of Food Science, 54: 618–620, 639.

Yagc Sibel and Gogus Fahrettin (2008). Response surface methodology for evaluation of physical and functional properties of extruded snack foods developed from food by-products. Journal of Food Engg., 86: 122-132.

Yaseen, AA, and Shouk, AA (2006). Effect of extrusion variables on physical, structure and sensory properties of wheat germ-corn grits extrudates. Egyptian Journal of Food Science 33(1): 57–71.

□□□

22

Application of Microwave Heating in Food Grains Processing

D.N. YADAV

Microwaves are invisible waves of energy that travel at the speed of light, $3x10^8$ m/s. In the electromagnetic spectrum, it lies between radio frequencies and infrared radiation. From the broad range of microwave frequencies available, a few are designated for industrial, scientific and medical applications (ISM). For all practical purposes, industrial applications are carried out at 915 MHz in the USA, 896 MHz in the UK, and 2450 MHz worldwide (Mullin 1995). A higher frequency of 5800 MHz is available for industrial purposes (Linn and Moller 2003). Microwaves are reflected by metals, transmitted through electrically neutral materials such as glass, most plastics, ceramics and paper, and absorbed by electrically charged materials (Mullin 1995). The use of microwaves in food processing was extended for killing insects (Wang *et al.*, 2003), blanching of fruits and vegetables (Boyes *et al.*, 1997), drying of fruits, vegetables and dairy products (Funebo and Ohlsson 1998, Mullin 1995), stabilization of rice bran and wheat flour (Yadav *et al.*, 2009, 2010), prevention of dough disclouration (Yadav *et al.*, 2008) and pre-treatment of oilseeds for efficient oil extraction (Irfan and Pawelzik 1999).

Cereals and cereal products like wheat, rice, various pulses, semolina, refined wheat flour, whole wheat flour commonly stored in pre-packed form in various storage devices like gunny bags, tin containers etc in tropical countries like India, adverse climatic conditions of high temperature and humidity results in insect proliferation. Insect causes significant losses in the stored grains particularly in the developing countries. Insect attacks the embryo, bore the grain or feed on the surface. They consume nutritious components, encourage moisture uptake by the infested grains and promote the development of microorganism. On many occasions insect fragments, frass and uric acid get mixed with the stored grain and beyond certain limit such grains are rejected by the consumers or declared unfit for human consumption. Molds are responsible for spoilage of cereals and pulses under high temperature and humid conditions. Several types of molds grow on the surface of grains and produce toxins. Insect infestations results in losses of viability, nutritional quality, processing properties and qualities such as luster and colour of the stored grains. Heat generated by the metabolic activities of microorganism causes the microbial infection, leading to complete loss of grains. When insect damage grain is used it is difficult to achieve clear separation of stock in preliminary operation as a result a high proportion is thrown out as offal. The larvae make holes in grains and thus cause damage. There is a reduction in the extraction of higher grade flour in proportion of degree of infestation. Insect infestation encourages fungus growth by increasing moisture content of the grain and losses in bread making quality (Subramanyan *et al.*, 1954). Gluten obtained from stored wheat flour infested by the rust red flour beetle was found brittle. The flour exhibited off flavour, bitter taste and finally become unacceptable. Girish *et al.* (1975) observed decrease in gluten content by 8-10% and sedimentation value by 12% in infested wheat. Thus infestations adversely affected baking quality. El-Dessouki and El.Kifl (1976) reported that protein content of wheat decrease by 48% and minerals and vitamins content by 25% due to the infestation. Besides quantitative losses in proteins, stored grain paste causes denaturation of proteins and there by decreases their solubility (Pingale *et al.*, 1954).

Basic components of microwave oven (Fig. 1)

- *Oven cavity* : oven cavity or cooking space is a rectangular metal box. Made of stainless steel, aluminum or acrylic coated cold rolled steel. Microwaves will be reflected from all sides to give a pattern of nodes and anti-nodes (or hot and cold spots from heating point of view). There is usually a shelf, a ceramic floor or a tempered glass tray which serves in position of the food load above the metal floor so that microwave energy can be reflected from the floor up into the food being heated as well as from all other directions.

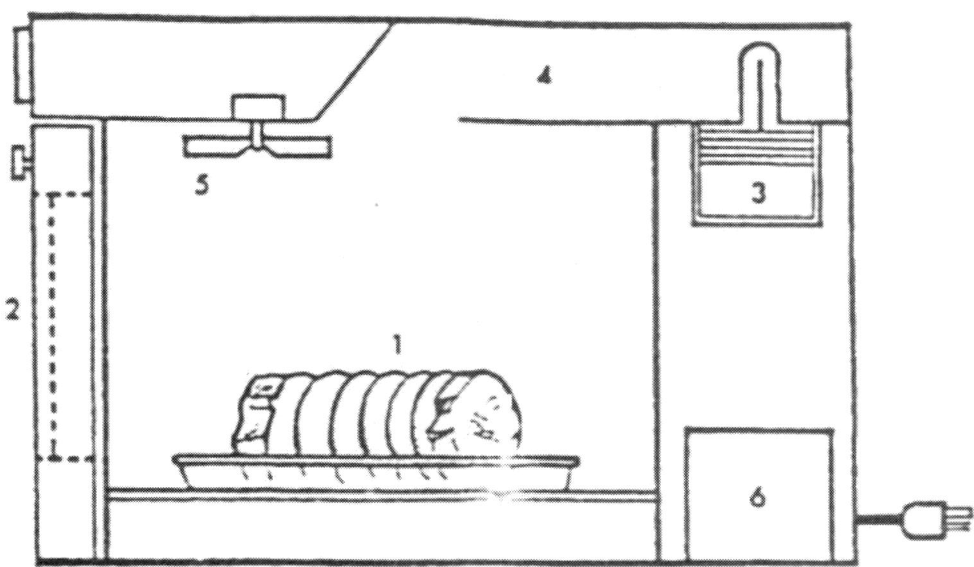

Fig. 1 : Basic components of the microwave oven: 1. oven cavity, 2. door that accesses cooking space, 3. magnetron (microwave generator), 4. waveguide, 5. mode stirrer and 6. power

- *Door that access cooking space:* it provides access to the cooking space and prevents emission of microwave energy from the oven during the operation.
- *Magnetron (microwave generator):* It (Fig 2a,b) generates the energy for cooking. This is a vacuum tube, that uses a built-in magnetic field to affect the flow of electrons from the cathode of the tube to the multi cavity anode in such a manner that the electric field is made to oscillate millions of times each second. The microwave and the energy they carry are then broadcast into the oven cavity.
- *Waveguide:* this is a rectangular duct that carries the microwave energy from the magnetron to the oven cavity
- *Mode stirrer:* it is a fan-like device that acts as a reflector of microwaves to distribute the microwaves more uniformly throughout the oven cavity.
- *Power supply:* this converts AC from the wall socket to the proper DC voltage needed to operate the magnetron and controls that permit adjustment of microwave power and cooking time.

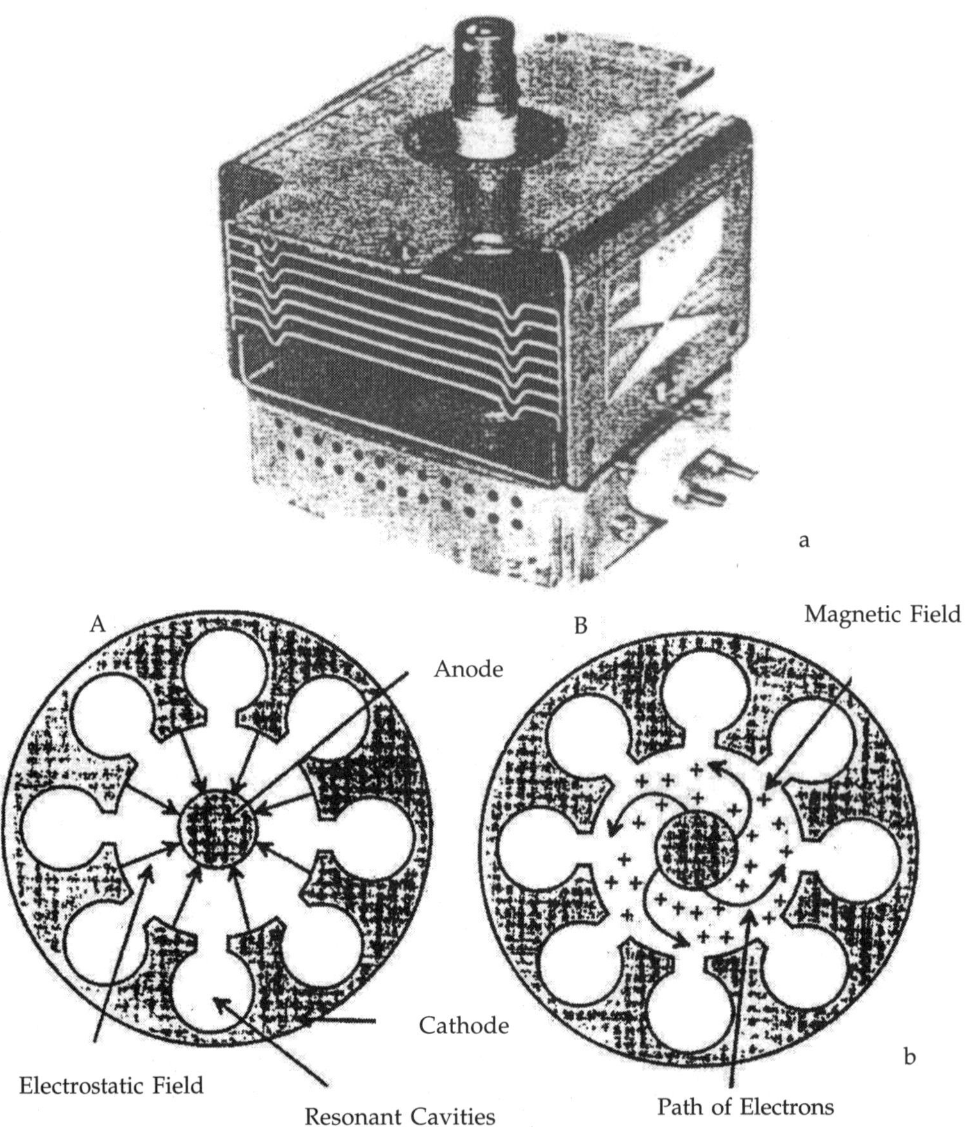

Fig. 2 : (a) and (b) Magnetron

Mechanism of heating

In microwave heating or drying, the microwave-emitted radiation is confined within the cavity and there is hardly heat loss by conduction or convection so that the energy is mainly absorbed by a wet material placed in the cavity. Furthermore, this energy is principally absorbed by the water in the material, causing the temperature to raise, some water to be evaporated, and the moisture level to be reduced. A domestic microwave oven works by passing microwave radiation, usually at a frequency of 2450 MHz (a wavelength of 12.24 cm), through the food. Water, fat, and sugar molecules in the food absorb energy from the microwave beam in a process called dielectric heating. Many molecules (such as those of water) are electric dipoles, meaning that they have a positive charge at one end and a negative charge at the other, and therefore rotate as they try to align themselves with the alternating electric field induced by the microwave beam. This molecular movement creates heat by friction as the rotating molecules hit other molecules and put them into motion.

Microwave heating is most efficient on liquid water, and much less on fats and sugars (which have less molecular dipole moment), and frozen water (where the molecules are not free to rotate). Large industrial/commercial microwave ovens operating in the 900 MHz range also heat water and food perfectly well. The power generated in a material is proportional to the frequency of the source, the dielectric loss of the material, and the square of the field strength within it. The microwave heating rates and potential non-uniformity are functions of oven factors and load characteristics (size, shape, dielectric properties, etc.).

In conventional heating, heat is transferred to the surface of the material to be heated by conduction, convection, and/or radiation, and into the interior by thermal conduction. In contrast, in dielectric heating, heat is generated directly inside the material, making possible higher heat fluxes and thus a much faster temperature rise than in conventional heating. However, heat conduction still plays an important role when heating thick samples by dielectric heating and for equilibrating temperatures when heat generation is uneven.Depending on water content the depth of initial heat deposition may be several centimeters or more with microwave ovens, in contrast to grilling, which relies on infrared radiation, or the thermal convection of a convection oven, which deposit heat shallowly at the food surface. Depth of penetration of microwaves is dependent on food composition and the frequency, with lower microwave frequencies being more penetrating.

The heat generated per unit volume of material (Q) represents the conversion of electromagnetic energy. Its relationship to the average electric field intensity (E_{rms}) at that location can be derived from Maxwell's equations of electromagnetic waves as shown by:

$$Q = 2\pi f \varepsilon_0 \, k' \, E^2_{rms} \qquad \ldots 1$$

Where, f is the frequency of microwaves, ε_0 is the dielectric constant of the free space ($8.854 \times 10^{-12} A^2 s^4/kg\ m^3$) and k' is the loss factor of the food being heated. At a given frequency, the dielectric loss factor is a function of the composition of the food materials and its temperature.

Penetration depth is another important factor in microwave heating. As electromagnetic energy is converted to heat energy as microwave comes in contact with the food material, its power decreases. Penetration depth (D_p) is defined as the depth below the surface of the material where the power density of a plane electromagnetic wave decays by 1/e (37%) from the original value at the surface.

The D_p is calculated as follows:

$$D_p = \frac{\lambda_0}{\sqrt{2}\pi\left[k'\left(\sqrt{1+\tan^2\delta}-1\right)\right]^{1/2}}$$

Where λ_0 is incident wavelength (12.24 cm), k' is dielectric constant, δ is loss angle (0.29 rad).

Industrial Microwave Systems

Typical industrial microwave heating or drying equipment is shown in Fig. 3. Basically there are three major components. The first component is the power supply unit where microwaves are generated at the required frequency band. The second component forms the applicator, where the material is subjected to intense microwave fields, and to which any additional ancillary process equipment such as pumps for operation under moderate vacuum conditions, steam or hot air injection, must be connected. Often the applicator forms the last part of a conventional processor.

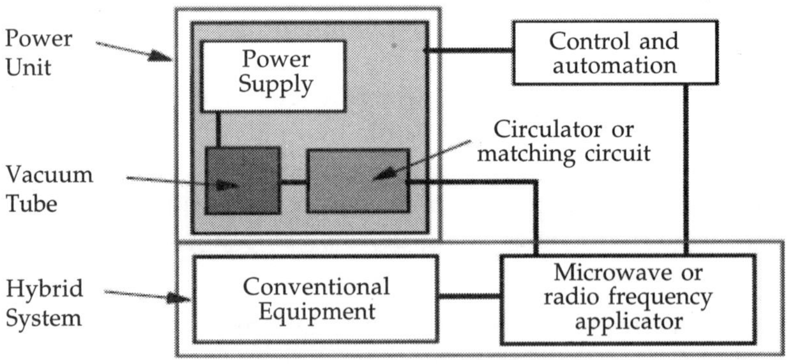

Fig. 3 : Typical microwave hearting system (Metaxas, 1991)

The other major component is the control circuit to optimize and regulate the overall performance of the microwave heater. Magnetron tubes are used primarily to generate the microwave power. It is by now usual and prudent practice to incorporate a ferrite iso-circulator protection between the magnetron source and the applicator (Metaxas, 1991).

Advantages of microwave heat

- Microwave drying of pasta is shown to require 25% less energy than the conventional drying.
- Microwave drying of macaroni is shown to enhance colour and better texture; infestations are also reduced (15 times lower) than the conventional drying.
- Time and space utilization are less.
- Potato dice, when dried by microwave has better rehydration property. Microwave drying minimizes case hardening and open porous structure.
- Quietness- no noise is created as in conventional dryers that require high velocity air-flow.
- In preparation of potato chips, partially fried slices are dried in conveyorised microwave oven, the finished product shows lighter colour and lower oil content.
- Microwave vacuum drying of fruit juices ensures minimal loss vitamin-C and better retention of volatile flavour compounds.

Microwave disinfestations of food grains

Insects that infest grain, cereal products, seed and other stored products, can be controlled through dielectric heating by microwave or lower radio frequency energy. Raising the temperature of infested materials by any means can be used to control insects if the infested product can tolerate the temperature levels that are necessary to kill the insects (Hurlock *et al.* 1979). The use of microwaves for killing insects is based on the dielectric heating of insects present in grain, which is a relatively poor conductor of electricity. Since dielectric heating depends upon the electrical properties of the material, there is a possibility of advantageous selective heating in mixtures of different substances (Antic and Hill 2003). In a mixture of dry food stuffs and insects, it is possible to heat the insects to a lethal temperature because they have high moisture content while leaving the drier foodstuff unaffected or slightly warm (Wang and Tang 2001). Hirose *et al.* (1975) conducted experiments to study the use of microwave heating to control insects in tobacco shreds used in cigarette manufacture. Control of all stages of tobacco moth and cigarette beetle in tobacco shreds were studied with a domestic microwave oven operating at 2450 MHz. The tobacco leaves at 18% moisture content (MC) are dried to 12% MC in a large rotary drier. When the tobacco shreds were heated by microwave, immediately following the rotary drier, sufficient temperature to kill the insects were achieved in a short exposure time without any detrimental effects on aroma and taste of tobacco. Wang and Tang (2001) reported a review on the use of radio frequency and microwave treatments as alternatives for control of insects in nuts. They suggested that radio frequency and microwave treatment seem attractive as a quarantine treatment because they are quick, safe and do not damage the product quality. Ikediala *et al.* (1999) studied the mortality of codling moth larvae in infested cherries using microwave heating at 915 MHz. They also compared the quality parameters of the microwave treated berries with those subjected to methyl bromide fumigation. The mortality was more than doubled when the microwave treatment was combined with cold storage. With a 2 min holding and 5 min hydro-cooling after microwave treatments, mortalities ranged from 5 to 62% and 39 to 98% without and with 1-2 days of cold storage, respectively. Quality parameters such as firmness, percentage soluble solids content, acidity, fruit weight and objective fruit color of microwave treated fruit were comparable with those of methyl bromide fumigated cherries. Shayesteh and Barthakur (1996) studied the mortality of life stages of *T. confusum* in wheat flour and *Plodia interpunctella* in wheat by exposing to continuous or intermittent microwave radiation at 2450 MHz. The effect of microwave radiation on *T. confusum* eggs was higher followed by pupae, adults and larvae. Their results showed that use of intermittent power supply was more effective in killing insects and the operational cost of microwave generator could also be minimized. The survival of insects decreased at 6 or 9% MC compared to 12% MC. Higher mortality was observed in *P. interpunctella* than *T. confusum.* The larger size of *P. interpunctella* would favor a higher probability of direct microwave absorption and heat transfer from the medium than the smaller insects. Tilton and Vardell (1982) studied the control of stored-product insects using combination of microwave and partial vacuum in rye, corn, and wheat. The four insects selected were *Rhyzopertha dominica, Sitophilus oryzae, Sitophilus zeamais motschulsky* (Maize weevil) and *Sitotroga cerealella* (Olivier). The infested grains were treated at three rates, high rate of 0.238 power density unit (PDU) for 10 min at 4.66 kPa, 0.083 PDU for 30 min at 4.66 kPa, and low rate of 0.028 PDU for 90 min at 4.66 kPa. They concluded that low rate of treatment was ineffective and produced only small reductions in the emerging adults whereas high rate of treatment was the most effective method and complete control of insects could be achieved. The major advantage of using microwave energy is that no chemical residues are left in the food and hence no adverse effects on human beings (Wang and Tang 2001; Wang *et al.*, 2003). Microwave energy has no adverse effect on the environment as chemical methods do (Wang and Tang, 2001). High frequency radiation may not only kill insects by the dielectric heat induced within them but may also affect the reproduction of the survivors.

Microwave drying

In recent years, microwave drying has gained popularity as an alternative drying method for a wide variety of food products such as fruits, vegetables, snack foods and dairy products. Several food products have been successfully dried by the microwave vacuum application or by a combined microwave assisted-convection process: cranberries (Yongsawatdigul and Gunasekaran 1996), carrot slices (Lin *et al.,* 1998), model fruit gels (Drouzas *et al.,* 1999), potato slices (Bouraout *et al.,* 1994), carrots (Prabhanjan *et al.,* 1995), grapes (Tulasidas *et al.,* 1996), apple and mushroom (Funebo and Ohlsson 1998), and banana (Maskan, 2000). Several experiments have reported microwave-assisted hot-air drying, where considerable improvements in the drying process have been evident: apple and potato (Huxsoll and Morgan, 1968), banana (Garcia *et al.,* 1988), carrot (Torringa *et al.,* 1993). Microwave drying results in a high thermal efficiency, shorter drying time and improved product quality compared to conventional hot air drying. Microwave drying helps to remove the moisture from the food products without the problem of case hardening (Prabhanjan *et al.,* 1995). Compared with hot air drying, combined microwave hot air could greatly reduce the drying time of biological materials without damaging the quality attributes of the finished products (Ren and Chen, 1998). Microwave drying requires a smaller floor space compared to conventional driers because the increase in processing rate makes it possible to design more compact equipment and hence plant capacity can be increased without additional building space. In microwave drying, operational cost is lower because energy is not consumed in heating the walls of the apparatus or the environment (Mullin, 1995). Heat generated by microwave energy occurs principally in the product, not in the oven walls or atmosphere. Therefore, heat losses from the oven to the surroundings are much lower, making for more comfortable working temperatures. Fast start-up and shut-down and precise process control are possible in microwave heating (Mullin, 1995). Microwave drying has been reported to improve product quality such as better aroma, faster and better rehydration, considerable savings in energy and much shorter drying times compared with hot air drying alone (Maskan, 2001). Nijhuis *et al.* (1998) compared the advantages and disadvantages of various technologies like freeze drying, microwave and radio frequency drying. Microwave drying has positive ratings for drying rate, flexibility, color, flavor, nutritional value, microbial stability, enzyme inactivation, rehydration capacity, crispiness and fresh-like appearance. However microwave drying is known to result in a poor quality product if not properly applied (Adu and Otten, 1996). It has also been suggested that microwave energy should be applied in the falling rate period or at low moisture content for finish drying (Funebo and Ohlsson, 1998). The reason for this is essentially economic. Due to high cost, microwave cannot compete with conventional air drying. However, microwaves may be advantageous in the latter stages of air drying.

Convective air-drying became very effective in the drying of many foods and is therefore commonly used in industries. In conventional cooking, heat is transferred to food by conduction, convection and radiation. In the microwave heating of foods, on the other hand, energy in the form of electromagnetic radiation at microwave frequencies is coupled directly to the food so that energy absorption and, hence, rapid heating take place throughout the food. However, the geometry of the product is an important consideration in designing microwaveable foods. Maximum thickness for samples should be limited to two to three times the penetration depth for uniform cooking at higher power. In other cases, low power must be used to heat foods more slowly and avoid overcooking of the surfaces. However, a number of disadvantages are associated with hot-air drying, such as case hardening, shrinkage, crusting, browning of food surfaces, moisture diffusion to the surface of food and product discolouration etc., which results long drying times and low energy efficiency and promotes research into alternative drying methods. These problems can be overcome through the selection of proper microwave package such as microwave passive, microwave active or microwave reflective. Microwave drying is relatively a cheap method, in which dipole heating causes water to move towards the product surface. Industrial applications of microwave energy include dehydration of pasta and post-baking of biscuits. However, the major problem of this method is uneven heating. Combination of microwave energy with convective air drying has been studied as a means to overcome the disadvantages associated with the application of each method alone. Microwave drying, in which air-drying is followed by the application of microwave drying, is useful for removing moisture trapped in the interior of foods at the end of the air-drying process. This method has been shown to decrease the drying time for both banana and kiwi. An alternative method involves applying microwaves simultaneously during the air-drying process. Microwave energy promotes the movement of water from the interior of the product, in the form of liquid or vapour, towards the surface of the product where it is then removed by convective air currents. Simultaneously combining microwave with convective hot-air drying has been shown to reduce the drying time of garlic cloves by 80–90%, in comparison to convective drying alone, while maintaining high quality product. Increasing microwave power level during the drying of apple resulted in higher weight gain and volume recovery, but a lower water holding capacity. Dried pulses are well known for their nutritional and functional value, however they are underutilized in Europe due to long soaking and cooking times. Drying of cooked chickpeas using a combination of microwave and air convection could produce highly rehydratable products in short drying times.

Microwave application in oil extraction

Irfan and Pawelzik (1999) studied the effect of microwave treatment on rapeseed (*Brassica napus L.*) oil extraction and oil quality. The seeds of rapeseed were heated in a laboratory type microwave oven operating at 2450 MHz and 1200 W power and the maximum temperature on the seed surface was limited to 100°C. The seeds were pressed for oil extraction and the oil quality was analyzed. There were no effects on the fatty acid composition and on the iodine value of the rapeseed oil, whereas acid and peroxide values were reduced. The reduction in the acid value was assumed to be as a result of successful inactivation of lipid esterase in the seeds. The peroxide value increased at 60 and 70°C but, as temperatures reached 80°C, there was a reduction in peroxide value which allows other lipid splitting enzymes to inactivate, which is very important in the storability and the further processing properties of oils. Oberndorfer and Lucke (1999) studied the effect of rapeseed treatment by microwave on the influence of mechanical oil extraction and concluded that microwave pretreatment resulted in an increase of the extraction rate in microwave treated samples compared to an untreated control. Light microscopy and electron microscopy showed clear differences in the microstructure of the oil bearing cells between the microwave treated and control samples. Velentova *et al.* (2000) studied the microwave and irradiation treatment to improve the oil extraction process in rapeseed. Thermal pretreatment is one of the important steps in the processing, which damages the oil bodies, improves the fluidity and adjusts the optimal moisture content of the seeds for further processing. Twenty five gram samples of rapeseed were spread to 20 mm depth and treated in a microwave oven (2450 MHz) for treatment times of 1 to 7 min. The irradiation was given at a dose rate of 8 kGy/h using cobalt-60 irradiator. The results of this study showed that oil extracted from microwave-treated seeds was comparable with the value obtained for flakes and was higher than those for untreated and irradiated seeds. No changes in enzyme activity were observed for untreated and irradiated seeds whereas activities of both the enzymes (phospholipase D and peroxidase) dropped to approximately 10% of the original value in microwave treated rapeseed. This study suggested that microwave treatment of rapeseed is recommended for processing operations due to improvement in the oil extraction process and quality.

Problems associated with microwave heating

One of the major problems associated with microwave heating is the non-uniform temperature distribution. The non-uniform temperature distribution has been studied by several researchers (Lee *et al.*, 2002; Sakai and Wang 2004; Manickavasagan *et al.*, 2006; Gunasekaran and Yang 2007;

Geedipalli *et al.,* 2007). Researchers have tried to develop a model for the microwave heating to predict temperature distribution in the microwave heated food (Yang and Gunasekaran 2004; Campanone and Zaritsky 2005). Microwave heating or drying sometimes results in poor quality of the end product (Krokida *et al.,* 2001; Blaszczak *et al.,* 2002). One of the serious concerns in heating food in a microwave is the incomplete kill of microbes due to uneven temperature distribution (Aleixo *et al.,* 1985; Rosenberg and Bogl 1987). Moisture accumulation at the surface of food during microwave heating was studied by Datta and Ni (2002). Ohlsson and Thorsell (1984) observed a couple of problems in microwave reheating of chilled foods such as uneven heat distribution between different meal components, dehydration of thin meat and fish slices, and skin formation on boiled white potatoes. Another issue with the microwave heating is the large number of factors that affect the microwave heat transfer behavior such as the thickness, the geometry, and the dielectric properties of the food. The heat capacity and the dielectric properties (dielectric constant "loss factor.") change with the moisture content and temperature which also complicates the microwave drying process (Funebo and Ohlsson 1998). In addition, several factors influence the uniformity of electromagnetic field. These factors can be divided into two groups: cavity effects; and workload or product interaction. Cavity effects are due to design limitation, location of the microwave inlet point, shape of the cavity, hanging parts such as mixer which are used for stirring of the product to assure more uniform electromagnetic field distribution. Workload interactions include loss factor, penetration depth, thickness, shape and size of the product that are different from product to product (Kelen *et al.,* 2006).

Non-uniform heating is prevalent in microwave heating, irrespective of the food product. Wide temperature variations were observed within heated samples, during microwave heating of different kinds of meat, ready-to-eat meals, grains, vegetables and model foods. Researchers have suggested some ways to reduce the intensity of uneven heating. Fung and Cunningham (1980) suggested that microwave heating in combination with conventional heating would result in more uniform heating of foods and destruction of bacteria. Ohlsson and Thorsell (1984) recommended that large food components of above 25-30 mm thickness should not be placed on top of each other but should be placed side by side whereas thin slices should be stacked edge to edge and uniform thickness should be maintained whenever possible. By controlling the food geometry, heating uniformity could be improved substantially. Ho and Yam (1992) studied the effectiveness of using metal bands to improve the microwave heating uniformity and concluded that shielding using metal bands was an effective way to improve the heating uniformity of the cylindrical samples under restricted conditions. They concluded that more experimental works are needed to describe the effect of metal shielding under conditions of food materials with various dielectric properties, sizes and geometries. Ryynanen and Ohlsson (1996), and Vilayannur *et al.* (1998) suggested some means for controlling the uneven heating, such as design of the microwave oven, manipulation of the heat cycle, ingredient formulation, design of the package and a combination of the above. Fakhouri and Ramaswamy (1993) suggested that uniform temperatures could be achieved by a combination of heating at high power for a short time and subsequent holding of the product or heating at low power for a longer period of time. Boyes *et al.* (1997) studied microwave and water blanching of corn kernels and suggested that non-uniformity could be overcome by over-blanching, but the product viability may be commercially lowered compared to a water or steam blanched product. Goksoy *et al.* (1999) suggested that heating on reduced power for longer times and shielding the overheated portion with aluminum foil resulted in improved surface temperature distributions. Funawatashi and Suzuki (2003) suggested that in the case of uneven heating due to standing waves, the non- uniformity could be reduced by moving with time the nodes of the standing wave which are done by metallic stirrers and turntables in domestic ovens. A device and method for uniform heating of food in microwaves was designed by Zhang *et al.* (2004) (US patent No: 6,777,655 B2) to reduce the problems of cold spot, uneven heating, and splattering of food. Datta *et al.* (2005) suggested that microwave heating in combination with air-jet impingement or infrared heating decreases the non-uniformity of temperature distribution. Gunasekaran and Yang (2007) concluded that pulsed microwave heating resulted in more uniform temperature distribution in the samples than the continuous microwave heating. Although researchers have given some solutions based on their studies, the results are confined to specific conditions and cannot be generalized.

Microwave processes in food industry

There are following major unit operations which can be carried out using microwave processing systems, such as,

(a) Dehydration for reducing moisture content: This can also be carried out in conveyorized microwave oven at 2450 MHz, vacuum drying assisted with vacuum for fruit juice concentrate at 48 KW, 2450MHz. Microwave assisted freeze drying can be applied for beef stick at 60-70 μm pressure and 2450MHz.

(b) Blanching for inactivating spoilage enzymes for fruit juice, in-package blanching

(c) Pasteurization for inactivating vegetative microbes

(d) Sterilization for inactivating microbial spores

(e) Cooking for modifying flavour and texture

(f) Heating at controlled temperature

(g) Thawing for frozen fish of meat

(h) Baking for biscuits, cookies, crackers

Future research

However, most of the reported studies on technologies of microwave-related combination drying were based on laboratory scale systems. There is a need for further studies to bridge the gap between laboratory research and industrial applications. Future research on microwave (MW)-related drying should focus on the following areas:

- Determination of combination order, type and conversion point of MW-related combination drying
- Process modeling and optimization of MW-related combination drying
- Improving hardware and automatic control of MW-related combination dryers
- For two- or three-stage combinations, a study of continuous and automatic control is very important for further potential applications
- Study of special phenomena (such as plasma discharge) taking place in MW-related combination drying

Although microwave heat treatment has many advantages compared to conventional methods, it is still not used widely for commercial purposes, which is due to both technical and cost factors. The quality of microwave-treated products is better or equal to that of conventional drying. However, there are certain drawbacks in microwave heating such as the non-uniform heat distribution and higher equipment costs. Equipment costs can change with time and developing technology. A major improvement in the efficiency of the treatment could change the economics of the microwave process. Thus, microwave heat treatment does appear to have a high potential for the processing of agricultural products in the near future. However, there are needs and opportunities for new technologies in various food sectors. The market place is also undergoing vast changes in all area from globalization, to changing consumer needs to technology impacting to processing and distribution of our foods. Quality, convenience and good taste still drive the ultimate consumers' decisions in buying. However, consumers continue to drive the boundaries for more and more convenience, at the same time quality, homemade style, prepared, quick-to-serve, fresh-like and with a long shelf-life requirement. Undoubtedly the food industry with its diverse operations such as tempering, blanching, sterilizing, cooking, puffing and vacuum drying offers the biggest opportunity for microwave processing, but the formidable challenge of other competitive techniques must be seriously addressed. Recent developments in the food industries point to major applications which may come on stream involving large microwave power in the near future.

References

Adu, B and L Otten (1996). Microwave heating and mass transfer characteristics of white beans. Journal of Agricultural Engineering Research, 64(1): 71-78.

Aleixo, JAG, B Swaminathan, KS Jamesen and DE Pratt (1985). Destruction of pathogenic bacteria in turkeys roasted in microwave ovens. Journal of Food Science, 50(4): 873-875.

Antic, A and JM Hill (2003). The double diffusivity heat transfer model for grain stores incorporating microwave heating. Applied Mathematical Modelling, 27(8): 629-647.

Blaszczak, W, J Gralik, E Klockiewicz-kaminska, J Fornal and JR Warchalewski (2002). Effect of radiation and microwave heating on endosperm microstructure in relation to some technological properties of wheat grain. Nahrung/Food, 46(2): 122-129.

Bouraout, M, P Richard and T Durance (1994). Microwave and convective drying of potato slices. Journal of Food Process Engineering, 17(3): 353-363.

Boyes, S, P Chevis, J Holden and C Perera (1997). Microwave and water blanching of corn kernels: control of uniformity of heating during microwave heating. Journal of Food Processing and Preservation, 21(6): 461-484.

Campanone, LA and NE Zaritzky (2005). Mathematical analysis of microwave heating process. Journal of Food Engineering, 69(3): 359-368.

Datta, AK and H Ni (2002). Infrared and hot-air assisted microwave heating of foods for control of surface moisture. Journal of Food Engineering, 51(4): 355-364.

Datta, AK, SSR Geedipalli and MF Almeida (2005). Microwave combination heating. Food Technology, 59(1): 36-40.

Drouzas, AE, E Tsami and GD Saravacos (1999). Microwave/vacuum drying of modelfruit gels. Journal of Food Engineering, 39(2): 117-122.

EL Dessouki SA and EL Kifl AH (1976). Sitotroge cerealella infestation and its influence on certain chemical and physical properties of stored wheat in Egypt. Zangewamdte Ent. 80: 83.

Fakhouri, MO and HS Ramaswamy (1993). Temperature uniformity of microwave heated foods as influenced by product type and composition. Food Research International, 26(2): 89-95.

Funawatashi, Y and T Suzuki (2003). Numerical analysis of microwave heating of a dielectric. Heat Transfer-Asian Research, 32(3): 227-236.

Funebo, T & Ohlsson, T (1998). Microwave-assisted air dehydration of apple and mushroom. Journal of Food Engineering, 38: 252-267.

Fung, DYC and FE Cunningham (1980). Effect of microwaves on microorganisms in foods. Journal of Food Protection, 43(8): 641-650.

Garcia, R, F Leal and C Rolz (1988). Drying of bananas using microwave and air ovens.International Journal of Food Science and Technology, 23(1): 73-80.

Geedipalli, SSR, V Rakesh and AUK Datta (2007). Modeling the heating uniformity contributed by a rotating turntable in microwave ovens. Journal of Food Engineering, 82(3): 359-368.

Girish GK, Jain SK, Ashok Kumar and Agrawal SN Part V (1975). Assessment of storage losses, quality and pesticidal contamination in wheat available in the markets of western Uttar Pradesh, Punjab and Haryana. Bulletin Grain Technology, 13: 8.

Goksoy, EO, C James and SJ James (1999). Non-uniformity of surface temperatures after microwave heating of poultry meat. Journal of Microwave Power and Electromagnetic Energy, 34(3): 149-160.

Gunasekaran, S and H Yang (2007). Effect of experimental parameters on temperature distribution during continuous and pulsed microwave heating. Journal of Food Engineering, 78(4): 1452-1456.

Hirose, T, I Abe, M Kohno, T Suzuki, K Oshima and T Okakura (1975). The use of microwave heating to control insects in cigarette manufacture. Journal of Microwave Power, 10(2): 181-189.

Ho, YC and Yam, KL (1992). Effect of metal shielding on microwave heating uniformity of a cylindrical food model. Journal of Food Processing and Preservation, 16(5): 337-359.

Hurlock, ET, BE Llewelling and LM Stables (1979). Microwaves can kill insect pests. Food Manufacture, 54(1): 37-39.

Huxsoll, CC and AI Morgan (1968). Microwave dehydration of potatoes and apples. Food Technology, 22(1): 47-51.

Ikediala, JN, J Tang, LG Neven and SR Drake (1999). Quarantine treatment of cherries using 915 MHz microwaves: Temperature mapping, codling moth mortality, and fruit quality. Postharvest Biology and Technology, 16(2): 127-137.

Irfan, I and E Pawelzik (1999). The effect of rapeseed treatment by microwave and radiofrequency application on oil extraction and oil quality. Part II: Influence on oil quality. Lipid, 101(5): 168-171.

Kelen, A, S Ress, T Nagy, E Pallai and K Pintye-Hodi (2006). Mapping of temperature distribution in pharmaceutical microwave vacuum drying. Powder Technology, 162 (2): 133-137.

Krokida, MK, ZB Maroulis and GD Saravacos (2001). The effect of the method of drying on the color of dehydrated products. International Journal of Food Science and Technology, 36(1): 53-59.

Lee. DS, D Shin and KL Yam (2002). Improvement of temperature uniformity in microwave-reheated rice by optimizing heat/cold cycle. Food Service Technology, 2(2): 87-93.

Lin TM, TD Durance and CH Scaman (1998). Characterization of vacuum microwave, air and freeze dried carrot slices. Food Research International, 31(2): 111-117.

Linn, H and M Moller (2003). Microwave heating. In Thermprocess Symposium, Dusseldorf, Germany, June 16-21.

Manickavasagan, A, DS Jayas and NDG. White (2006). Non-uniformity of surface temperatures of grain after microwave treatment in an industrial microwave dryer. Drying Technology, 24(12): 1559-1567.

Maskan, M (2000). Microwave/air and microwave finish drying of banana. Journal of Food Engineering, 44: 71-78.

Maskan, M (2001). Drying, shrinkage and rehydration characteristics of kiwifruits during hot air and microwave drying. Journal of Food Engineering, 48: 177-182.

Metaxas AC (1991). Microwave heating. IEE Power Engineering Journal, 5(5): 237-247.

Mullin, J (1995). Microwave processing. In New Methods of Food Preservation, ed. G.W. Gould, 112-134. Bishopbriggs, Glasgow: Blackie Academic and Professional.

Nijhuis, HH, HM Torringa, S Muresan, D Yuksel, C Leguijt and W Kloek (1998). Approaches to improving the quality of dried fruit and vegetables. Trends in Food Science and Technology, 9(1): 13-20.

Oberndorfer, C and W Lucke (1999). The effect of rapeseed treatment by microwave and radio-frequency application on oil extraction and oil quality. Part I: Influence on mechanical oil extraction. Lipid, 101(5): 164-167.

Ohlsson, T and U Thorsell (1984). Problems in microwave reheating of chilled foods. Food Research International, 3(1): 9-16.

Pingale SV, Rao N and Swaminathan M (1954). Effects of insect's infestation on stored grains, 2-Studies on soft wheat. Journal of Science and Food Agriculture, 5: 51.

Prabhanjan, DG, HS Ramaswamy and GSV Raghavan (1995). Microwave assisted convective air drying of thin layer carrots. Journal of Food Engineering, 25(2): 283-293.

Ren, G and F Chen (1998). Drying of American gingseng (Panax quinquefolium) roots by microwave-hot air combination. Journal of Food Engineering, 35(4): 433-445.

Rosenberg, U and W Bogl (1987). Microwave pasteurization, sterilization, blanching, and pest control in the food industry. Food Technology, 41(6): 92-97.

Ryynanen, S and T Ohlsson (1996). Microwave heating uniformity of ready meals as affected by placement, composition, and geometry. Journal of Food Science, 61(3): 620-624.

Sakai, N and C Wang (2004). An analysis of temperature distribution in microwave heating of foods with non-uniform dielectric properties. Journal of Chemical Engineering of Japan, 37(7): 858-862.

Shayesteh, N and NN Barthakur (1996). Mortality and behavior of two stored-product insect species during microwave irradiation. Journal of Stored Products Research, 32(3): 239-246.

Subramanyan V., M Swaminathan, SV Pingale and SB Kadkol (1954). Uric acid as an index of insect filth in cereals and milled cereal products. Bulletin, CFTRI, Mysore, India, 4: 86.

Tilton, EW and HH Vardell (1982). Combination of microwaves and partial vacuum for control of four stored-product insects in stored grain. Journal of Georgia Entomological Society, 17: 96-106.

Torringa, EM, HH Nijhius, HHJ Remmen and PV Bartles (1993). Electromagnetic drying of vegetable produce. In Proceedings of the Fourth International Congress on Food Industry, 331-344. Cesme, Turkey. April 19-23.

Tulasidas, TN, GSV Raghavan and ER Norris (1996). Effects of dipping and washing pre-treatments on microwave drying of grapes. Journal of Food Process Engineering, 19: 15-25.

Valentova, O, Z Novotna, Z Svoboda, W Schwarz and J Kaš (2000). Microwave heating and irradiation treatment of rapeseed (Brassica napus). Journal of Food Lipids, 7(4): 237-245.

Vilayannur, RS, VM Puri and RC Anantheswaran (1998). Size and shape effect on non uniformity of temperature and moisture distributions in microwave heated food materials: Part 1 Simulation. Journal of Food Process Engineering, 21(3): 209-233.

Wang SJ Tang, JA Johnson, E Mitcham, JD Hansen, G Hallman, SR Drake and Y Wang (2003). Dielectric properties of fruits and insect pests as related to radio frequency and microwave treatments. Biosystems Engineering, 85(2): 201-212.

Wang, S and J Tang (2001). Radio frequency and microwave alternative treatments for nut insect control: a review. International Agricultural Engineering Journal, 10: 105-120.

Yadav DN, Patki PE, Pandit Srihari, Sharma GK, Bawa AS (2010). Studies on polyphenol oxidase activity of heat stabilized whole wheat flour and its chapati making quality. International Journal of Food Properties, 13: 142-154.

Yadav DN, PE Patki, Pandit Srihari, GK Sharma and AS Bawa (2009). Process for the stabilization of whole wheat flour. Patent filed No. 479/DEL/2009.

Yadav, DN, PE Patki, GK Sharma and AS Bawa (2008). Effect of microwave heating of wheat grains on the browning of dough and quality of chapattis. International Journal of Food Science and Technology, 43(7): 1217-1225.

Yang, HW and S Gunasekaran (2004). Comparison of temperature distribution in model food cylinders based on Maxwell's equations and Lambert's law during pulsed microwave heating. Journal of Food Engineering, 64(4): 445-453.

Yongsawatdigul, J and S Gunasekaran (1996). Microwave vacuum drying of cranberries: Part I: Energy use and efficiency. Journal of Food Processing and Preservation 20(1): 121-143.

Zhang, H, LH Bonneveau, W Yout, GC. Helstern and G Loizeau (2004). Uniform microwave heating of food in a container. US Patent No: 6: 777,655.

□□□

23

Infrared Processing of Cereals and Pulses

D.N. YADAV

Infrared (IR) energy is radiant energy which passes through space in the form of electromagnetic waves (Figure 1). Like light, it can be reflected and focused. Infrared energy does not depend on air for transmission and is converted to heat upon absorption by the work piece. In fact, air and gases absorb very little infrared. As a result, infrared energy provides for efficient

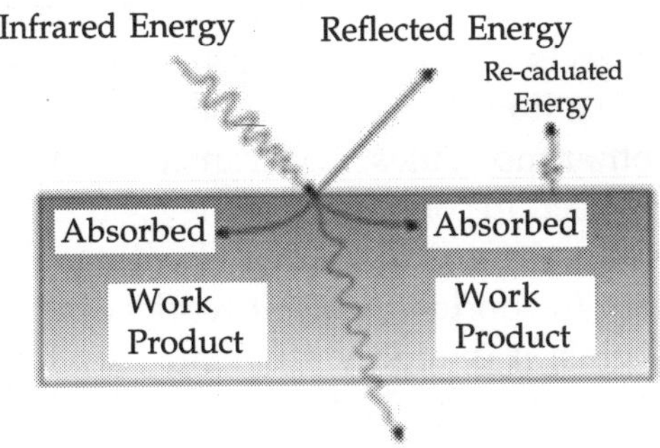

Fig. 1 : Extinction of radiation (absorption, transmission, and reflection).

heat transfer without contact between the heat source and the work piece. Infrared heating is frequently misapplied and capacity requirements underestimated due to a lack of understanding of the basic principles of radiant heat transfer. When infrared energy from a source falls upon an object or work piece, not all the energy is absorbed. Some of the infrared energy may be reflected or transmitted. Energy that is reflected or transmitted does not directly heat the work piece and may be lost completely from the process

Another important factor to consider in evaluating infrared applications is that the amount of energy that is absorbed, reflected or transmitted varies with the wave length of the infrared energy and with different materials and surfaces. These and other important variables have a significant impact on heat energy requirements and performance. In the simplest terms, infrared energy is light that we cannot see, but our bodies can detect as heat. It is part of the electromagnetic spectrum that includes radio waves, X-rays and visible light. All of these forms of energy have a specific frequency, as represented in the chart below.

Infrared energy is comprised of those frequencies that exist just below the red end of the visible spectrum, and for cooking properties they have a very unique benefit - when they strike organic molecules (such as any type of food), they cause the molecules to vibrate, thereby creating heat. Although almost any type of electromagnetic energy can cause heating, for the purpose of cooking, infrared energy is the perfect choice. Heat transfer can occur one of three ways - conduction, convection or radiation. Conduction requires direct contact between two objects and is used frequently in cooking. Convection relies on heat transfer via a liquid or gas. The heated molecules of air or water physically bump into other molecules, and transfer some of their energy. This process is also regularly used in cooking, whether you are boiling, deep frying or cooking on a conventional gas grill. Radiant heating is different. It does not rely on any intermediary to conduct heat - it works directly on the molecules in its path. Because of this, it is highly efficient. Any heat transfer involves some loss, so when you have a gas flame, heating air, then heating a radiant, then heating air some more, then finally heating your food (as in a traditional grill), you have much more loss than if that heat energy goes directly from the source to the food being cooked.

Basic laws of infrared radiation

The amount of the IR radiation that is incident on any surface has a spectral dependence because energy coming out of an emitter is composed of different wavelengths and the fraction of the radiation in each band, dependent upon the temperature and emissivity of the emitter. The wavelength at which the maximum radiation occurs is determined by the temperature of the IR heating elements. This relationship is described by the basic laws for blackbody radiation such as Planck's law, Wien's displacement law, and Stefan-Boltzman's law.

Sources of IR heating

Two conventional types of infrared radiator used in heating process are electric and gas-fired heaters. These two types of IR heaters generally fit into three temperature ranges (Hung *et al.*, 1995): 343 to 1100°C for gas and electric IR, and 1100 to 2200°C for electric IR only. IR temperatures are typically used in the range of 650 to 1200°C to prevent charring of products. The capital cost of gas heaters is higher, while the operating cost is cheaper than that of electric infrared systems. Electrical infrared heaters are popular because of installation controllability, ability to produce prompt heating rate, and cleaner form of heat. Electric infrared emitters also provide flexibility in producing the desired wavelength for a particular application. In general, the operating efficiency of an electric IR heater ranges from 40% to 70%, while that of gas-fired IR heaters ranges from 30% to 50% (Hung *et al.*, 1995). The spectral region suitable for industrial process heating ranges from 1.17 to 5.4 μm, which corresponds to 260 to 2200°C (Sheridan and Shilton 1999). Infrared radiation is transmitted through water at short wavelength, whereas at longer wavelengths it is absorbed at the surface (Sakai and Hanzawa 1994). Hence, drying of thin layers seems to be more efficient at the far infrared (FIR) region, while drying of thicker bodies should give better results at the near infrared (NIR) region. Studies to investigate the superiority of FIR to NIR radiation have also been found in the literature. Sakai and Hanzawa (1994) have discussed the effects of the radiant characteristics of heaters on the crust formation and color development at the surfaces of foods such as white bread and wheat flour. Radiant heating with an NIR heater led to a greater heat sink into food samples, resulting in formation of relatively wet crust layers, compared to dry layers formed by FIR heaters. However, the rate of color development by FIR heaters was greater than NIR heaters, primarily due to a more rapid heating rate on the surface. Sheridan and Shilton (1999) evaluated the efficacy of cooking hamburger patties using infrared sources at λ max of 2.7 μm (MIR) and at λ max of 4.0 μm (FIR). With a higher energy source (MIR), change in core temperature followed closely the change in surface temperature with a shorter cooking time. Fat content of the food was found to be independent of core temperature. However, with the lower energy source (FIR), the increasing rate of core temperature was dependent on the fat content, showing that targeted core temperature was achieved more quickly as fat content increased. FIR energy penetration into the food has gained ceaseless concern. Hashimoto *et al.* (1990, 1994) studied the penetration of FIR energy into sweet potato and found that FIR radiation absorbed by the vegetable model was damped to 1% of the initial values at a depth of 0.26 to 0.36 mm below the surface, whereas NIR showed a similar reduction at a depth of 0.38 to 2.54 mm. Sakai and Hanzawa (1994) reported the penetration depth of the FIR energy did not affect the temperature distribution inside the food. Further, they indicated that FIR energy penetrates very little, almost all the energy being converted to heat at the surface of the food, which was consistent with the study of Hashimoto *et al.* (1993) evaluating FIR heating technique as a surface heating method. Table 1 shows the penetration depth of NIR energy into food products (Ginzburg 1969).

Table 1 : Penetration depth of NIR (0.75 to 1.4 μm) into food products

Product	Spectral peak (μm)	Depth of penetration (mm)
Dough, wheat	1.0	4 to 6
Bread, wheat	1.0	11 to12
Bread, biscuit dried	1.0	4
Grain, wheat	1.0	2
Carrots	1.0	1.5
Tomato paste, 70% to 85% water	1.0	1
Raw potatoes	1.0	6
Dry potatoes	0.88	15 to 18
Raw apples	1.16	4.1
	1.65	5.9
	2.36	7.4

Types of infrared heaters

Infrared heaters are commonly used in infrared modules (or emitter banks) combining several heaters to achieve larger heated areas. Infrared heaters are usually classified by the wavelength they emit. Near infrared (NIR) or short-wave infrared heaters operate at high filament temperatures above 1800°C. Their peak wavelength is below the absorption spectrum for water, making them unsuitable for many drying applications. They are well suited for heating of silica where a deep penetration is needed. Medium-wave and carbon (CIR) infrared heaters operate at filament temperatures of around 1000°C. Far infrared heaters are typically used in low-temperature. An infrared heater is a body with a higher temperature which transfers energy to a body with a lower temperature through electromagnetic radiation. Depending on the temperature of the emitting body, the wavelength of the infrared radiation ranges from 780 nm to 1 mm. The relationship between temperature and wavelength is expressed by the Wien's displacement law. No contact or medium between the two bodies is needed for the energy transfer. Near infrared (NIR) for the range from 780 nm to 1400 nm, these emitters are also named bright because still some visible light with glare is emitted. Medium infrared (MIR) for the range between 1400 nm and 3000 nm. Far infrared (FIR) or dark emitters for everything above 3000 nm.

IR heating in food processing operations

The usefulness of IR heating has been demonstrated in various food processing applications such as roasting, frying, broiling, heating, and cooking meat and meat products, soy beans, cereal grains, cocoa beans, and nuts. With the growing interest in flame-broiling and rapid cooking methods, conveyorized IR broiling is a unique and innovative method. Khan and Vandermey (1985) prepared ground beef patties by IR broiling in a conveyorized broiler. The results showed that due to high temperatures and short cooking times, the infrared broiler could produce more servings per hour compared to conventional gas heating. In addition, it was found that ground beef patties broiled by tube broiler did not have any adverse effects on the cooking quality (number of samples cooked/min, % shrinkage, number of servings/h) or sensory quality (appearance, flavor, texture, juiciness, and overall acceptability), as compared to conventional gas broiling method. Sakai and Hanzawa (1994) reported on the performance of infrared-based systems with conventional ovens for baking rice crackers and for roasting fish pastes. The comparative study indicated energy savings of 45% to 70% with infrared heating. Abdul-Kadir *et al.* (1990) conducted imbibitions studies and cooking tests to evaluate the effect of IR heating on pinto beans (*Phaseolus vulgaris*) heated to 99 and 107°C. IR-heating was found to improve rehydration rate and degree of swelling of pinto beans; however, cooking time of pinto beans significantly increased. Studies on color development during IR roasting of hazelnuts were reported by Ozdemir and Devres (2000). Olsson *et al.* (2005) found that infrared radiation and jet impingement, as compared with heating in a conventional household oven, increased the rate of color development of the crust and shortened the heating time of par-baked baguettes during post baking. Furthermore, the fastest color development was obtained by combining infrared and impingement heating. The rate of water loss increased due to a higher heat transfer rate, but the total water loss was reduced because of a shorter heating time. In general, the formed crust was thinner for IR-treated baguettes.

Tables 2 : Characteristics of commercially used infrared heat source

	Tungsten Filament		Nickel Chrome Resistance Wire			Wide Area panels	
Infrared source	Glass Bulb	T3 Quartz Lamp	Quartz Tube	Metal Sheath	Ceramic	Ceramic Coated	Quartz Face
Source Temperature (⁰F)	3000-4000⁰F	3000-4000⁰F	Up to 1600⁰F	Up to 1500⁰F	Up to 1600⁰F	200-1600⁰F	Up to 1700⁰F
Brightness	Intense White	Intense White	Bright red to Dull Orange	Dull to Bright Red	Dark to Dull Red	Dark to Cherry Red	Dark to Cherry Red
Typical Configuration	G-30 Lamp	3/8″ Dia. Tube	3/8 or ½″ tube	3/8 or ½″ tube	Various shapes	flat panels	flat panels
Type of Source	Point	Line	Line	Line	Small Area	Wide Area	Wide Area
Peak Wavelength (microns)	1.16	1.16	2.55	2.68	3-4	2.25-7.9	2.5-6
Maximum Power Density	1 kW/ft^2	3.9 kW/ft^2	1.3-1.75 kW/ft^2	3.66 kW/ft^2	Up to 3.6 kW/ft^2	3.6 kW/ft^2	5.76 kW/ft^2
Watts per Linear Inch	N/A	100	34-45	45-55	N/A	N/A	N/A
Conversion Efficiency Infrared Energy	86%	86%	40-62%	45-56%	45-50%	45-55%	45-55%
Response Time Heat /Cool	Seconds	Seconds	1-2 Minutes	2-4 Minutes	5-7 Minutes	5-8 Minutes	6-10 Minutes
Color Sensitivity	High	High	Medium	Medium	Medium	Low to Medium	Low to Medium
Thermal Shock Resistance	Poor	Excellent	Excellent	Excellent	Good	Good	Good
Mechanical Ruggedness	Poor	Fair	Good	Excellent	Good	Good	Fair
Chromalox Model	-	QR	QRT	RAD, URAD	RCH	CPL, CPLI, CPH	CPHI

Selective heating by infrared radiation

Very few attempts have been made to study selective heating in the food industry as well as in nonfood research areas. Certain studies have been found in the literature applied to electronics (Bischorf 1990; Sakuyama *et al.,* 1995). These studies on electronics showed the accessibility of selective heating based on the relation between the optical properties of objects and the spectral distribution of the radiative source. However, the studies did not elaborate on the details or its implementation. Most infrared heaters consist of lamps emitting the spectrum with one specific peak wavelength corresponding to a fixed surface temperature. The type of infrared emitter and control of the accurate wavelength should be considered for optimization of the process. In practice, the IR source emits radiation covering a very wide range. Hence, it is a challenge to cut off the entire spectral distribution to obtain a specific bandwidth.

Efficiency of infrared heaters

Electrically-heated infrared heaters radiate up to 86% of their input as radiant energy. Nearly all the electrical energy input is converted into heat in the filament. Efficiency of the infrared heater depends on matching the emitted wavelength and the absorption spectrum of the material to be heated. Absorption spectrum for water has its peak at around 3000 nm. This means that emission from medium-wave or carbon infrared heaters are much better absorbed by water and water-based coatings than NIR or short-wave infrared radiation. PVC or polyethylene had peak absorption at around 3500 nm. Some metals absorb only in the short-wave range and show a strong reflectivity in the medium and far infrared. This makes a careful selection of the right infrared heater type important for energy efficiency in the heating process. Ceramic elements operate in the temperature of 300°C to 700°C producing infrared wavelengths in the 2 -10 micron range. Most plastics and many other materials absorb infrared best in this range.

Infrared emitters & source temperatures

The amount of radiant energy emitted from a heat source is proportional to the surface temperature and the emissivity of the material. This is described by the Stefan-Boltzmann Law which states that radiant output of an ideal black body is proportional to the fourth power of its absolute temperature. The higher the temperature, the greater the output and more efficient the source. Absorption - Once the infrared energy is converted into heat at the surface, the heat travels into the work by conduction. Materials such as metals have high thermal conductivity and will quickly distribute the heat uniformly throughout. Conversely, plastics, wood and other materials have low thermal conductivity and may develop high surface temperatures long before internal temperatures increase appreciably. This can be an advantage when using infrared heating for drying paint, curing coatings or evaporating solvents on non-metal substrates. Only the energy absorbed is usable in heating the work product. In an unenclosed application, losses from reflection and re-radiation can be excessive. Enclosing the work product in an oven or a tunnel with high reflective surfaces will cause the reflected and re-radiated energy to be reflected back to the work product, eventually converting most of the original infrared energy to useful heat on the work product.

Infrared drying

Most dried vegetable products are prepared conventionally using a hot-air dryer. However, this method is inappropriate when dried vegetables are used as ingredients of instant foods because of low rehydration rate of the vegetables. Freeze-drying technique is a competitive alternative; however, it is comparatively expensive. Application of FIR drying in the food industry is expected to represent a new process for the production of high-quality dried foods at low cost. The use of IR radiation technology for dehydrating foods has numerous advantages including reduction in drying time, alternate energy source, increased energy efficiency, uniform temperature in the product while drying, better-quality finished products, a reduced necessity for air flow across the product, high degree of process control parameters, and space saving along with clean working environment. IR drying operations have been successfully applied in recent years for drying of fruit and vegetable products such as potatoes (Masamura *et al.* 1988; Afzal and Abe 1998), sweet potatoes (Sawai *et al.*, 2004), onions (Mongpreneet *et al.*, 2002; Sharma *et al.*, 2005), kiwifruit (Fenton and Kennedy 1998), and apples (Nowak and Levicki 2004; Togrul 2005). Drying of seaweed, vegetables, fish flakes, and pasta is also done in tunnel infrared dryers. Infrared drying has found its application in food analysis to measure water content in food products (Hagen and Drawert 1986; Anonymous 1995).

Integrated drying technologies: IR and convective drying

Even though IR drying is a promising novel method, it is not a alternate for all drying processes. It appeals, because it is fast and produces heating inside the material being dried, but its penetrating powers are limited. Prolonged exposure of a biological material to IR heat results in swelling and ultimately fracturing of the material. IR heating of legume seeds to 140°C caused cracking on the surface (Fasina *et al.*, 1997). However, a combination of intermittent infrared heating and continuous convection drying of thick porous material resulted in better product quality and energy efficiency (Dostie *et al.*, 1989). IR radiation can be considered as surface treatment similar to other radiation technologies. Application of combined electromagnetic radiation and conventional convective heating is considered to be more efficient over radiation or convective heating alone, as it gives a synergistic effect. Afzal *et al.* (1999) reported that during the combined convective and IR heating process of barley, the total energy required was reduced by about 156%, 238%, and 245% as compared with convection drying alone at 40, 55, or 70°C, respectively. Development of a continuous drying apparatus equipped with FIR heaters, NIR heaters, and hot air blast can reduce the economic costs, drying time, and operating temperature. Hebbar *et al.* (2004) developed a continuous combined infrared and convective dryer for vegetables. The synergistic effect of infrared and hot air led to rapid heating of the materials, resulting in a higher rate of mass transfer. The evaporation of water took 48% less time and 63% less energy consumption in combined mode drying as compared to convective drying. The concept of FIR heating immediately after convective drying (approximately 40°C) for drying of paddy has been utilized in the paddy industry in Japan (Bekki 1991). A combination of IR heating with freeze-drying in sweet potatoes could reduce the processing time by less than half (Lin *et al.* 2005).

Table 3 : The infrared absorption bands for chemical groups and relevant food components (Rosenthal 1992)

Chemical group	Absorption wavelength (μm)	Relevant food component
Hydroxyl group (O-H)	2.7 to 3.3	Water, sugars
Aliphatic carbon-hydrogen bond	3.25 to 3.7	Lipid, sugars, proteins
Carbonyl group (C=O) (ester)	5.71 to 5.76	Lipids
Carbonyl group (C=O) (amide)	5.92	Proteins
Nitrogen-hydrogen group (-NH-)	2.83 to 3.33	Proteins
Carbon -carbon double bond (C=C)	4.44 to 4.76	Unsaturated lipids

Infra red processing of cereals

Advantages

- High digestibility
- Improved palatability, enhanced flavor
- Attractive appearance, color retention
- Energy saving

- Increased nutritional value
- Micronisation help in developing puffed, expanded finished products without deep fat frying.

Applications

- Breakfast cereals/flakes
- Specialty breads
- Soup mixes
- Infant foods
- Brewery adjuncts
- Snack foods
- Instantization of rice and wheat

IR processing of legumes

Advantages

- Consistent inactivation of toxic enzymes
- Development of an agreeable toasted flavor
- Increased oil availability to give higher metabolizable energy values
- Successfully remove beany flavour of full fat soya
- Micronising process offer a highly efficient and economic means of separating the cocoa bean shell from the nib

Applications

- Quick cooking pulses/legumes
- Soup/sauce bases/mixes
- Vegetable burgers
- Toasted products
- Specialty breads
- Help in development of tasty chocolate and high quality chocolate product

IR heating is a unique process; however, presently, the application and understanding of IR heating in food processing is still in its infancy. IR heating offers many advantages over convection heating, including greater energy efficiency, heat transfer rate, and heat flux that results in time-saving as well as increased production line speed. IR heating is attractive primarily for surface heating applications. Combination of IR heating with microwave and other common conductive and convective modes of heating holds great potential. IR heating in the food processing sector will augment in the near future, especially in the area of drying and minimal processing. Research need for upcoming years may include: process modeling, selective IR heating, IR interactions with food components, application of catalytic infrared heating systems.

The infrared processing is emerging field and finding use in advance processing of food products:

- Instant and quick cooking of cereals and pulses.
- Pelletized animal feed and to improve digestibility for better biological value.
- Processing of cocoa beans to separate coca bean shell from the nib with high rate of separation and to give tasty high quality chocolate products.
- Preparation of cereal adjuncts for brewing industry in place of conventional malted cereals.
- Processing of cashew nuts to toast and remove shell with better colour and yield and reduce brokens.
- Processing of soybean to reduce anti-nutrients factors.
- Processing of oil seeds for higher oil extraction
- However, research is required in establishing the IR processing for Indian traditional foods with respect to baking, roasting and frying.

References

Abdul-Kadir, Bargman T, Rupnow J (1990). Effect of infrared heat processing on rehydration rate and cooking of Phaseolus vulgaris (var. Pinto). J Food Sci, 55(5): 1472–1473.

Afzal TM, Abe T, Hikida Y (1999). Energy and quality aspects of combined FIR-convection drying of barley. J Food Eng, 42: 177–82.

Afzal TM, Abe T (1998). Diffusion in potato during far infrared radiation drying. J Food Eng, 37(4): 353–65.

Anonymous (1995). Determination of moisture content in Finnish honey using an infrared apple pomace using an infrared drying technique. New Zealand J Crop Horti Sci, 26: 35–8.

Bekki E (1991). Rough rice drying with a far-infrared panel heater. J Jap Soc Agric Machinery, 53(1): 55–63.

Bischorf H (1990). The answer is electrical infrared. J Microwave Power Electromagnet Ener, 25(1): 47–52.

Dostie M, Seguin JN, Maure D, Tonthat QA, Chatingy R (1989). Preliminary measurements on the drying of thick porous materials by combinations of intermittent infrared and continuous convection heating. In: Mujumdar AS, Roques MA, editors.

Fasina OO, Tyler B, Pickard M (1997). Infrared heating of legume seeds effect on physical and mechanical properties. ASAE Paper No: 976013.

Fenton GA, Kennedy MJ (1998). Rapid dry weight determination of kiwifruit pomace and and apple pomace using an infrared drying technique. New Zealand J Crop Horti Sci, 26: 35-43.

Ginzburg AS (1969). Application of infrared radiation in food processing, chemical and process engineering series. London, U.K.: Leonard Hill.

Hagen W, Drawert F (1986). Determination of water content by infrared. Monatsschrift Brauwissenschaft, 40(6): 240–6.

Hashimoto A, Sawai J, Igarashi H, Shimizu M (1993). Irradiation power effect on pasteurization below lethal temperature of bacteria. J Chem Eng Japan, 26(3): 331-3.

Hashimoto A, Takahashi M, Honda T, Shimizu M,Watanabe A (1990). Penetration of infrared radiation energy into sweet potato. Nippon Shokuhin Kogyo Gakkaishi, 37(11): 876–93.

Hashimoto A, Yamazaki Y, Shimizu M, Oshita S (1994). Drying characteristics of gelatinous materials irradiated by infrared radiation. Drying Technol, 12: 1029-52.

Hebbar HU, Vishwanathan KH, Ramesh MN (2004). Development of combined infrared and hot air dryer for vegetables. J Food Eng, 65: 557–63.

Hung JY, Wimberger RJ, Mujumdar AS (1995). Drying of coated webs. In: Mujumdar AS, editor. Handbook of industrial drying. 2nd ed. New York: Marcel Dekker Inc. p, 1007-1038.

Khan MA, Vandermey PA (1985). Quality assessment of ground beef patties after infrared heat processing in a conveyorized tube broiler for foodservice use. J Food Sci, 50: 707–709.

Lin YP, Tsen JH, King An-Erl V (2005). Effects of far-infrared radiation on the freeze-drying of sweet potato. J Food Eng, 68: 249–55.

Masamura A, Sado H, Nabetani H, Nakajima M (1988). Drying of potato by far-infrared using response surface methodology. J Food Eng, 45: 17–24.

Mongpreneet S, Abe T, Tsurusaki T (2002). Accelerated drying of welsh onion by far infrared radiation under vacuum conditions. J Food Eng, 55: 147–56.

Nowak D, Levicki PP (2004). Infrared drying of apple slices. Innov Food Sci Emerg Technol, 5: 353–360.

Olsson EEM, Triagardh AC, Ahrn´e LM (2005). Effect of near-infrared radiation and jet impingement heat transfer on crust formation of bread. J Food Sci, 70(8): E484–491.

Ozdemir M, Devres O (2000). Analysis of color development during roasting of hazelnuts radiation. Nippon Shokuhin Kogyo Gakkaishi, 35(5): 309–314.

Rosenthal I (1992). Electromagnetic radiations in food science. Berlin, Heidelberg: Springer-Verlag.

Sakai N, Hanzawa T (1994). Applications and advances in far-infrared heating in Japan. Trends Food Sci Technol, 5: 357–362.

Sakuyama S, Uchida H,Watanabe I, Natori K, Sato T (1995). Reflow soldering using selective infrared radiation. Proceedings of the IEEE/CPMT Intl. Electronic Manufacturing Technology (IEMT) Symposium. p 393–396.

Sawai J, Nakai T, Hashimoto A, Shimizu M (2004). A comparison of the hydrolysis of sweet potato starch with b-amylase and infrared radiation allows prediction of reducing sugar production. Int J Food Sci Technol, 39: 967–974.

Sharma GP, Verma RC, Pathare PB (2005). Thin-layer infrared radiation drying of onion slices. J Food Engg, 67: 361-367.

Sheridan P, Shilton N (1999). Application of far-infrared radiation to cooking of meat products. J Food Eng, 41: 203–208.

Togrul H (2005). Simple modeling of infrared drying of fresh apple slices. J Food Eng, 71: 311–23.

□□□

24

Application of Cryogenics in Grinding of Spices

K.K. SINGH AND P. BARNWAL

The spices are an essential part of Indian food, imparting taste and flavour to food preparations. It is being consumed everyday in households, restaurants, hostels, catering services, food processing industries and roadside eateries etc. The main processed product of spices is ground powder.

India produces almost all spices due to its diversity in climate i.e. from tropical to sub-tropical to temperate. Almost all the states and union territories of the country grow one or the other spices. There is a significant increase in production of spice from 3.03 to 4.66 million tonnes from year 2002-03 to 2006-07, respectively. India commands an important position in the world spice trade with 48% share in volume and 44% in value. Although, with the liberalization of the Indian economy, imports of spices are also steadily growing, spices exports have registered substantial growth during the last decade, registering an annual average growth rate of 13.1% in value and 9% in volume. During the year 2009-10, spices export from India has registered an all time high both in terms of quantity and value. In 2009-10 the export of spices from India has been 502,750 tonnes valued Rs. 5560.50 crores US $ 1173.75 million) as against 470,520 tonnes valued Rs. 5300.25 crores

(US $ 1168.40) in 2008-09, registering an increase of 7% in volume and 5% in rupee value. During the year 2008-09, Indian spices and spice products reached more than 135 countries in the world. The leading among them are USA (21%), Malaysia (7%), UAE (6%), China (6%) and UK (5%) (Anon, 2010).

Spices are valued exclusively for their aroma and flavour, which constitute the major attraction for spices. The important flavouring compounds in some spices are presented in Table 1 (Anon, 2010).

Table 1 : Important flavouring compounds in spices

Name of spice	Important flavouring compounds/ components
Black pepper	Piperine,S-3-Carene, ß-caryophyllene
Cardamom	α-terpinyl acetate, 1-8-cineole, linalool
Cinnamon, cassia	Cinnamaldehyde,eugenol
Chilli	Capsaicin,dihydro capsaicin
Clove	Eugenol, eugeneyl acetate
Coriander	d-linalool, C10-C14-2-alkenals
Cumin	Cuminaldehyde, p-l,3-mentha-dienal
Ginger	Gingerol,Shogaol, neral,geranial
Turmeric	Turmerone,Zingeberene, 1,8-cineole
Basil, Sweet	Methylchavicol,linalool, methyl eugenol
Rosemary	Verbenone,1-8-cineole, camphor, linanool
Peppermint	1-menthol, menthone,menthfuran
Spear mint	1-carvone,carvone derivatives

Need for cryogenic grinding of spices

The aim of spice grinding is to obtain smaller particle size with good product quality in terms of flavour and/or color. Due to high fat content in spices, in the normal/ambient grinding process, heat is generated when energy is used to fracture a particle into a smaller size. This generated heat causes temperature rise in the grinder to the extent of 95°C which is responsible for a loss of volatile oil in the tune of about 30%, which is mainly responsible for flavour, and also produces dark color powder. Generally, a continuous operation of grinder is not possible in normal grinding process due to melting of fat and sticking of powder on the grinding surfaces. Thus, normal / ambient grinding of spices produces inferior quality of the product with many operational problems. Also, the spice powder produced by normal / ambient grinding does not conform to the international quality standards due to its poor quality, as a result either fetches lower price or not accepted by the importer countries. The temperature rise of the product can be minimized to some extent by circulating cold air or water around the grinder. But this technique is not sufficient enough to significantly reduce temperature rise of the product.

The application of cryogenic technology for grinding of spices has been scientifically proved to be a suitable technique with less loss of volatile content and improved colour and grinding operation. The loss of volatile can be significantly reduced by cryogenic grinding technique using liquid nitrogen or liquid carbon dioxide that provides the refrigeration needed to pre-cool the spices and maintain the desired low temperature by absorbing the heat generated during the grinding operation. It is better than conventional grinding in terms of higher retention of volatiles and flavouring components, color, particle size distribution of ground powder, free and continuous operation of the grinder without any choking, less energy requirement in grinding, etc.

Cryogenics and its applications

The terminology 'cryogenics' is related with a Dutch physicist, Kamerlingh Onnes, who wanted to produce a gas in his laboratory that could be refrigerated and become a low temperature boiling liquid. The word 'cryogen' was coined to describe a low temperature boiling liquid.

The National Bureau of standards, UK has defined cryogenic temperature as -150°C and below. According to this definition, liquid nitrogen (boiling point -195.6°C) falls in cryogenic range, whereas carbon dioxide (boiling point -78.5°C) does not. However, in general, cryogenics is defined as a branch of engineering specializing in technical operations at very low temperature, generally below -50°C and to create such a low temperature, cryogenic liquids are used. Cryogenic liquids are those which boil at cryogenic temperatures at atmosphere pressure. Liquid forms of hydrogen, helium, nitrogen, oxygen, inert gases, air, methane, carbon dioxide, etc are common cryogens. Liquid nitrogen (LN_2) and carbon dioxide (CO_2), in liquid or solid form, are the two major cryogens used for food applications. The thermo-physical properties of these two cryogenic liquids are given in Table 2.

The use of cryogens in industry, defence and space programmes has simulated the emergence of the new field of cryogenic engineering. As a refrigerant, liquid nitrogen is being used for many decades. But in recent years, it is finding widespread applications in almost every branch of science, vacuum technology, electronics, space technology, biology, medicine, agriculture, etc. Nevertheless, the scope for application of cryogenic technology in the area of food processing is no less important. The cryogenic technology has found application for production of high value frozen products like, fruits and vegetables, meat products, marine products, fish, prawn, etc. Cryogenic freezing offers a wide range of unique benefits including high cooling rates, high throughput, flexibility in terms of acceptability to different products, low capital cost, low dehydration loss and better quality. One of the important applications of cryogenic technology is in grinding of spices, medicinal and aromatic plants.

Table 2 : Some thermo-physical properties of liquid nitrogen (LN_2) and liquid Carbon Dioxide (LCO_2)

Properties	LN_2	LCO_2
Density, kg/m^3	808	464
Boiling point, °C	-195.6	-78.5
Thermal conductivity, W/mK	0.14	0.19
Specific heat of liquid, kJ/kg K	2.05	2.26
Latent heat of evaporation, kJ/kg	199	352
Total usable refrigeration effect, kJ/kg	690	565

Cryogenic technology has also limited applications in peeling of fruits and vegetables where these are immersed in liquid nitrogen for few seconds and then thawed in warm water to loosen the peel. Packaging and homogenization of biological tissues have also limited applications of cryogenic technology.

Cryogenic grinding of food materials

The food materials are ground to reduce size in order to make it more palatable for human consumption. Grinding of materials in the presence of a cryogenic fluid is referred as 'Cryogenic grinding'.The grinding of spices, herbs and other food commodities under the cryogenic conditions causes less change in its natural composition than ambient grinding. In cryogenic grinding, evaporation of liquid nitrogen quickly chills both the material and the grinder. Fibrous materials become brittle and crisp making size reduction easy. Grinding of spices under cryogenic conditions have advantages of improved size reduction, reduced particle size and fine powder, increased production rate, reduced heat build up. The cryogenic grinding is applicable to a wide variety of food materials, viz., spices, herbs, medicinal and aromatic plants, cocoa and chocolate, vanilla-sugar, coconut, coffee, tea, dehydrated meats, etc.

The processes of cereal grinding are well established. However, in normal grinding process, heat is generated due to the use of energy to fracture a particle into a smaller size. Generally, the generated heat results some loss of flavour and quality and the loss can be minimized to some extent by employing some type of cooling to compensate for the heat produced in the grinding process. The use of circulating cold air, cooling water etc. is made to reduce the amount of heat generation to some extent in standard grinding mills. In recent years, the adaptation of cryogenic liquids, such as liquid nitrogen or liquid carbon dioxide to the grinding process has given impetus to the milling industries in order to retain the quality of the food being processed. The extremely low temperature liquid provides the refrigeration needed to compensate for the generated heat in the mill.

Theory of the cryogenic grinding

Grinding, that is conversion of bulk solid to powder, is generally a mechanical process. The large particles are subjected to contact forces which deform the particles and if strong enough, generate internal stresses and cause fractures or cracks. The amount and intensity of fractures determine the size and characteristic surfaces of the particles.

There are many theories which explain how mechanical forces can bring about the individual particle to fracture. The brittle material deforms elastically before developing cracks. Within elastic limit, under a tensile stress an elastic material elongates and then returns to its original shape when the stress is removed. If the material is stretched beyond its elastic limit, it fractures. The characteristics make brittle materials fail suddenly when these are subjected to impact forces.

The strength of materials depends upon their microscopic structure. A crack in brittle material reduces its tensile strength and leads to instability. If the material is subjected to sufficient stress the built-up strain energy is expanded rapidly in creating multiple failures. If food materials are exposed to a cryogenic medium, internal stress is built-up due to fast freezing rate and this internal stress may lead to cracking or shattering under such conditions which could be due to volumetric changes associated with liquid solid phase transition and non-uniform contraction of void spaces perceived through micro-structures. Food materials exposed to cryogenic temperatures become more brittle and crisp, and as such, grinding is rendered easier.

After freezing to its brittle point, if the material is removed from the cryogenic environment, its temperature begins to rise. Therefore, it is essential that in cryogenic grinding the material be fragmented in the embrittled state as quickly as possible so that grinding is completed below its brittle point. Brittle materials can be fragmented by subjecting them to an impact force.

Grinding can be achieved by different mechanisms and sometimes with combinations of two or more mechanisms. These mechanisms may be compression, cutting, impact, shear, rubbing etc. A particular grinding mechanism is employed depending primarily upon the characteristics of the material.

Cryogenic grinding of spices

Spice grinding is an ancient industry like cereal milling industry with the difference that in spice grinding there is the additional problem of natural volatile flavouring components and essential oils getting lost during grinding. Spices are valued for aroma and flavour, these impart to various foods.

The fat content of spices generally poses a problem and is an important consideration in grinding. The other considerations are particle size, product yield, product uniformity, freedom from contamination, economy and dust free operation.

During grinding, spices lose a significant fraction of their volatile oils or flavouring components due to the heat generated. There are considerable losses of oil and moisture from different spices during normal grinding. For different spices the product temperature ranged between 42 and 95°C during grinding.

The quality of spices could be retained by the technique of cryogenic grinding. In cryogenic grinding liquid nitrogen is used to reduce the temperature of the product prior to and during the grinding operation in order to minimize the loss of flavouring volatile components of the product. Thus the flavour strength per unit mass of the resultant ground product is significantly higher than that in the conventionally ground product.

With cryogenic grinding, the temperature of the product can be as low as -195.6°C. But generally such low temperatures are not used for all spices. In practice, temperatures are regulated anywhere from -195.6°C to a few degrees below the ambient temperature.

Cryogenic grinding is accomplished by the controlled injection of liquid nitrogen directly into the mill's grinding zone. The instantaneous evaporation of liquid nitrogen quickly chills both the spice and the mill. It also absorbs the frictional heat of grinding. Thus, the temperatures in the grinding zone generally are well below -73°C.

The extremely low temperature in the grinder solidifies oils so that the spices become brittle, they crumble easily permitting grinding to a finer and more consistent size. The high quality ground product would have domestic as well as international market.

Cryogenic grinding system for spices

In order to obtain high quality ground spices products, a cryogenic grinding system was designed and developed to cool the spices before feeding to the grinder and also to maintain the cryogenic temperature in the grinding zone as shown in Fig.1 (Singh and Goswami, 1999a). The main components of the cryogenic grinding system are (i) pre-cooling unit (cryogenic pre-cooler) and (ii) grinding unit (commercially available grinder, a pin mill and a hammer mill).

The function of the cryogenic pre-cooler is to remove the heat from the material before it enters the grinder. The pre-cooling unit (a cooling device) consists of a screw conveyor assembly, an air compressor, a liquid nitrogen dewar, a power transmission arrangement and control panels. The cryogenic pre-cooler is made up of a screw conveyor enclosed in a properly insulated barrel and a system to introduce liquid nitrogen into the barrel, thereby providing refrigeration (liquid and cold gas) within the system. The particle temperature must be low enough to absorb the heat generated in the grinder and still fracture. Cryogenic pre-coolers, therefore, must have the ability to reduce the temperature of the seed below its brittle point as well as the freezing point of its oil, before it enters the grinder.

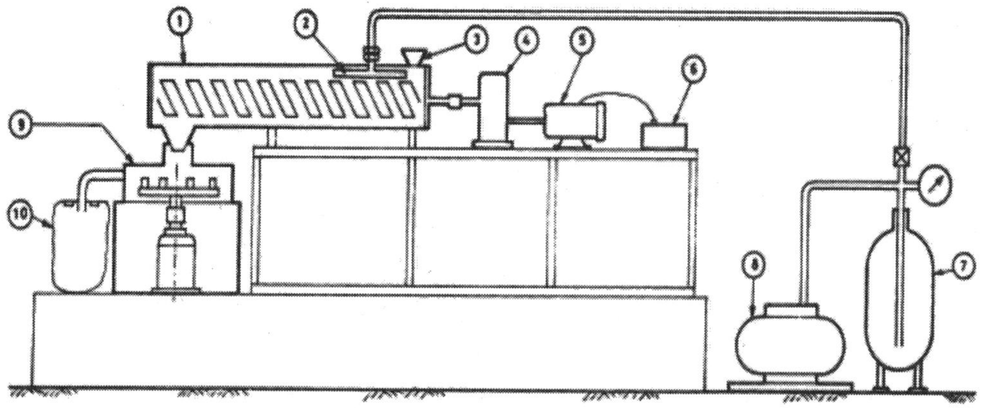

Fig. 1 : Schematic view of cryogenic grinding system
Legend : 1. Screw convevor ; 2. LN_2 distributor ; 3. Hopper ; 4. Reduction gear box ; 5. Motor ; 6. Rectifier ; 7. LN_2 Cylinder ; 8. Compressor ; 9. Grinder ; 10. Collecting bag.

There must be provision to control the temperature of the pre-cooler and the feed rate to the grinder for the obvious purpose of controlling the grinding process. Consumption of liquid nitrogen and the operating cost are important considerations and matters of concern for a cryogenic pre-cooling system. The liquid nitrogen losses can be minimized to a great extent by proper consideration of the design and insulation of the pre-cooler. The design of the pre-cooling unit is to prevent the material from being heated up during grinding. The unit would pre-cool the material before the actual starting of the grinding operation. Thus, the pre-cooling unit is being designed to match with a commercially available grinder (a pin mill and a hammer mill) that could withstand low temperature operations.

The main engineering considerations in the design of a cryogenic pre-cooler are as below:

1. Retention time of the seed in the liquid nitrogen and gaseous zone should be accurately proportioned so that the available refrigeration could be utilised at its optimum level.
2. Appropriate insulation should be used such that losses to the ambient could be minimised.
3. Various components of the pre-cooler should be arranged in such a manner that dismantling and cleaning could be easier.
4. Cooldown losses should be reduced by keeping the machine size and structural components to a minimum.

In addition to the above, other major factors which affect the design of a cryogenic pre-cooler are heat transfer coefficient, freezing or retention time and design of screw conveyor etc. (Singh and Goswami, 1999a).

The screw conveyor was made of aluminium. The shaft is supported by two aluminium bushes fixed with flanges at both ends. The screw shaft is enclosed in aluminium barrel. Both ends of the barrel are covered with flanges which are pressed between barrel ends and bushes. The barrel is insulated from outside with asbestos rope and asbestos powder throughout the surface to minimize heat gain from the surrounding. The thickness of the insulation is 10 mm.

At the upper portion of the barrel, a distributor made of copper tube is fixed. The distributor has a number of perforations to spray liquid nitrogen over the material being conveyed through the barrel. There are three rows of the perforations in a zig-zag manner, one at centre and two at 45° angle from the centre.

Both ends of the tube are closed. The distributor is fixed at 10 mm distance from the inlet and 380 mm from the outlet to allow thorough mixing of the liquid and vapour nitrogen with the material, and to freeze the oil present in the material. The screw shaft and cage are mounted on a frame made of mild steel angle iron.

An air compressor is used to supply compressed air to the liquid nitrogen (LN_2) dewar (Fig. 1). The resulting pressure helps the liquid to flow out of the dewar. It is driven by a 550 W single phase electric motor. The compressor has provision for automatic cut-off of the power supply at a particular air pressure. The air flow at the outlet of the compressor can be varied with the help of a valve.

A double walled, vacuum insulated dewar having capacity of 55 litre is used for storage and transfer of LN_2. An LN_2 transfer assembly is fixed over the dewar to regulate the flow rate of the liquid nitrogen passing to distributor attached to the conveyor assembly. The transfer assembly contained a pressure gauge (range 0-1 kg/cm^2 and least count 0.02 kg/cm^2) to indicate the pressure inside the LN_2 dewar, a relief valve for protection against excessive pressure and a manual shut-off valve to control the LN_2 flow. A 10 mm diameter teflon tube is used to connect the LN_2 dewar and distributor of the screw conveyor assembly.

The electric supply is given through a variable speed DC motor to the speed reduction unit. The ratio of the gear reduction unit is 36:1. The shaft of the gear reduction unit is coupled with the screw shaft through a coupling made of Hylam (bakelite reinforced fabric). The material Hylam insulates the cold screw shaft from the warm motor shaft. The speed of the screw shaft could be varied by regulating the input voltage to the motor through a transformer. The feed rate and retention time of the grinding material in the pre-cooler will be controlled to the desired levels by controlling the speed of the shaft.

The operation of grinding would be performed by impact and attrition. The impact is achieved by the material being struck with rotor ribs, whereas the attrition is achieved while the seeds were present between the stationery sieve ring and the fast moving rotor. The speed of the rotor could be controlled through an in-built control mechanism. The ground powder is collected in the collector pan from an outlet. A nylon bag is attached to the outlet with a locking clamp. The nylon bag retained the particles and allowed the nitrogen vapour to escape from the grinder. The grinder is operated by a 800 W, single phase electric motor.

A single phase wattmeter (range 0-750 W, least count 5 W) may be connected with the machine to measure the power consumed and ultimately to measure the energy required in grinding. The specific energy consumed in grinding may be expressed as:

$$\text{Specific energy consumption} = \frac{\text{Power consumed } (W) \times 3.6}{\text{Feed rate } (kg/h)}$$

Influence of cryogenic grinding on quality of spices

The experiments on grinding of cumin seed were conducted at different temperatures from -160°C to -10°C. The sieve choking, particle size distribution, specific energy consumption and quality of the final product were observed using 8 number of grinder's rotor ribs in the grinder (Singh and Goswami, 1999a).

The choking of sieves was observed by taking photographs of the sieve either at the completion of grinding or when the machine was stopped due to sieve choking. Above -70°C the grinding experiments could not be completed successfully, because the sieve perforations were blocked soon after running the machine and it stopped due to overloading with the incoming material. The accumulation of powder on the sieve surface might be due to the fact that during grinding at a temperature higher than the brittle point of the seed and freezing point of its oil, the seed was soft and behaved like glue in the grinder. Also, during grinding the oil might have come out of the cells, which had sticky characteristics and formed a layer over the sieve surface. The powder was deposited above this layer and formed a thick layer on the sieve surface which created hindrance in grinding. At the same time, the incoming raw material overloaded the grinding surface and stopped the grinder.

It was observed that at the grinding temperature of -70°C, the deposition of cumin powder was minimum. Below -70°C, grinding was smooth without any deposition and the sieve was very clear. This is so because at temperatures lower than the brittle point of cumin seed and freezing point of cumin oil, the oil got solidified and the grinding operation was smoother. Thus, cumin seed could be successfully ground below the temperature of -70°C and at above this temperature, sieve clogging took place.

From Fig. 2, it may be observed that as the grinding temperature increased from -160 to -70°C, there was an increase in the particle size for the same cumulative volume fraction for both the rotors. Thus, increase in grinding temperature from -160 to -70°C resulted in a significant increase in particle size of the product. The volume mean diameter increased from 153.2 to 215 µm with increasing grinding temperature from -160 to -70°C.

The specific energy consumption increased from 55 to 98 kJ/kg with increasing grinding temperature from -160 to -70°C. This is so because at low grinding temperature, degree of brittleness of cumin seed increased due to which it required less energy in grinding. Thus, increase in grinding temperature from -160 to -70°C resulted in specific energy consumption in grinding. The volatile oil content decreased from 3.30 to 3.26 ml/100 g with increasing grinding temperature from -160 to -70°C, but this variation was found to be non-significant at 5% level.

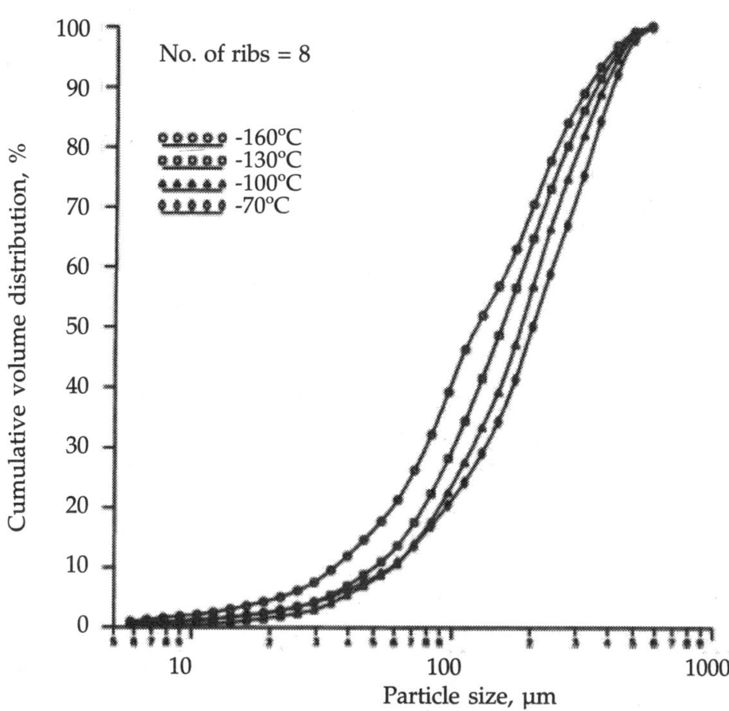

Fig. 2 : Particle size distribution of cumin powder ground at different temperatures

The grinding of cumin seed at various cryogenic and ambient temperatures were also carried out using 12 number of rotor ribs in the grinder to observe its influence on volatile oil content and its components, particle size distribution, volume mean diameter and specific energy consumption (Singh and Goswami, 1999b). It was observed that an increase in temperature in the cryogenic range (-160 to -70°C) had no significant effect on volatile oil content, whereas increase in temperature in the ambient range (40 to 85°C) significantly reduced the volatile oil content from 2.86 to 2.26 mL/100g. The volatile oil components, -pinene, β-pinene, -terpinene, p-cymene and cuminaldehyde were not significantly affected by grinding temperature in the cryogenic temperature range. However, these components decreased significantly with increase in grinding temperature under ambient condition. With increase in temperature from -160 to -70°C for 12 number of rotor ribs, the volume mean diameter of cumin powder increased quadratically from 129 to 164 μm and the specific energy consumption increased from 72 to 108 kJ/kg.

Sridhar and Manohar (2004) reported effect of some operational and machine parameters on volumetric efficiency of cryo-feeder in grinding of black pepper. Further, Murthy and Bhattacharya (2008) reported a study on cryogenic grinding of black pepper at different temperatures and feed rates. The comparison was made between ambient grinding and cryogenic grinding. It was concluded that in cryogenic grinding of black pepper 50% more volatiles were retained in comparison to the ambient grinding.

The study conducted on cryogenic grinding of spices is significant contribution for characterization of grinding parameters vis-à-vis quality of the product. However, the earlier developed laboratory scale cryogenic grinding system (Singh and Goswami, 1999a) can not be directly scaled up for higher capacity because the phenomena of gas circulation, heat and mass transfer, control systems for gas flow, type of grinder would be quite different in the large grinding systems.

Though India is one of the leading producers of the spices in the world but it depends upon traditional grinding technique available in the country. The cold or cryogenic grinding practice is almost inaccessible in India because of non-availability of indigenous technology for cryogenic grinding. Few firms in Germany and Japan manufacture cryogenic grinders for application in rubber and plastic grinding which are basically hammer mills or pin mills with pre-cooling and liquid nitrogen or liquid carbon dioxide injection systems. These mills may also be used for grinding of spices. But its design methodology is not readily available for adoption in India for grinding of spices and other high value agricultural materials. For removing the required heat from a particle prior to its feeding into the grinder, cryogenic pre-coolers are used. The cryogenic pre-coolers can be combined with impact, attrition, or air swept mills. It is ensured that the particle during grinding is at or below its brittle point. Provisions are made to control the pre-cooler temperature and feed rate to the mill.

In view of the absence of indigenous technology for cryo-grinding, some efforts have been made by R & D institutions for development of facility for cryogenic grinding of spices. The purpose is to develop this technology in India. The availability of indigenous cryogenic grinding technology would ensure production of better quality spices within the country and export the processed products to different countries. The foreign earnings could also be enhanced by exporting the cryogenic grinding technology to other neighbouring spice producing countries.

References

Anon (2010). http://www.indianspices.com/

Murthy CT and Bhattacharya S (2008). Cryogenic grinding of black pepper. Journal of Food Engineering, 85(1): 18-28.

Singh KK and Goswami TK (1999a). Design of a cryogenic grinding system for spices. Journal of Food Engineering, 39: 359-368.

Singh KK and Goswami TK (1999b). Studies on cryogenic grinding of cumin seed. Journal of Food Process Engineering, 22: 175-190.

Sridhar BS and Manohar B (2004). Solid rheological characteristics of black pepper (*Piper nigrum* L.) during cryogenic grinding. Journal of Food Science and Technology, 41(6): 630-636.

□□□

25

Microencapsulation for Immobilization of Food Ingredients, Microorganisms and Enzymes

K. NARSAIAH

Microencapsulation is a process of entrapping solid particles, liquid droplets or gases in thin polymeric coatings. The outer film (wall) protects the encapsulated material (core) and allows for its controlled release. Application of this technique is wide spread in several industrial fields, including the manufacture of foods, pharmaceuticals, cosmetics and pesticides.

Common food ingredients subjected to microencapsulation include probiotic bacteria, acidulents, flavours, colours, sweetners, minerals and vitamins. A number of coating materials are employed in production of microcapsules, such as gums, starch, and starch derivatives, cellulosic materials, lipids, proteins and inorganic materials.

Actually the development of the formation of micro-encapsulates was partially inspired by the nature. Eggs, seeds, grains, nuts, fruits etc. are typical examples of the food encapsulation and natures own way of conserving and protecting natural products against its own deteriorating elements. Taking clue from this, man adopted this technology of protection of some of the foodstuffs as sauces, puddings etc.

Microencapsulation can improve the convenience of food. The shell provides a barrier between reactive components (for instance, delaying the release of leavening agents for fluffier bread products or protecting oxygen-sensitive materials during processing and storage). Microcapsules can help fragile and sensitive materials survive processing and packaging conditions and stabilize the shelf-life of the active ingredient. Karel (1990) described variuos mehods of encapsulation of food ingredients.

In addition, microencapsulation can simplify the food manufacturing process by converting liquids to solid powder, decreasing production costs by allowing batch processing using low-cost, powder-handling equipment. To improve food safety, the technology can be used to indicate product tampering, thermal spoilage, and freeze-thaw cycles.

Reasons for microencapsulation

The reasons for microencapsulation are countless. In some cases, the core must be isolated from its surroundings, as in isolating vitamins from the deteriorating effects of oxygen, retarding evaporation of a volatile core, improving the handling properties of a sticky material, or isolating a reactive core from chemical attack. In other cases, the objective is not to isolate the core completely but to control the rate at which it leaves the microcapsule, as in the controlled release of drugs or pesticides. The problem may be as simple as masking the taste or odor of the core, or as complex as increasing the selectivity of an adsorption or extraction process.

Microencapsulation offers food companies a viable means of penetrating this lucrative growth sector because it has the ability to mask the undesirable tastes associated with some of these ingredients.

This also implies significant opportunities in the highly profitable children's market. While consumers are becoming more health-conscious and are demanding more nutritious products, they are unwilling to compromise on taste.

Microencapsulated ingredients can be used to

- Mask undesirable flavours
- Prevent chemical reactions
- Improve delivery, control delay
- Increase stability of finished product

- Reduce production losses
- Increased quality of finished product
- Increase plant capacity without capital expense
- Improve production yield
- Increase saleable products and increase profit.

Although microencapsulation technology was long considered far too expensive for use in the food industry, manufacturers are increasingly adopting this technology to add value to their product range. As food manufacturers consistently track consumer trends and tastes, food ingredients companies must continually look to develop innovative products that help meet end-user needs. Microencapsulation is progressively attracting the interest of food ingredient manufacturers as a way of achieving much-needed differentiation and enhancing product value.

Physical methods to manufacture microcapsules

Pan coating

The pan coating process, widely used in the pharmaceutical industry, is among the oldest industrial procedures for forming small, coated particles or tablets. The particles are tumbled in a pan or other device while the coating material is applied slowly.

Air-suspension coating

Air-suspension coating of particles by solutions or melts gives better control and flexibility. The particles are coated while suspended in an upward-moving air stream. They are supported by a perforated plate having different patterns of holes inside and outside a cylindrical insert. Just sufficient air is permitted to rise through the outer annular space to fluidize the settling particles. Most of the rising air (usually heated) flows inside the cylinder, causing the particles to rise rapidly. At the top, as the air stream diverges and slows, they settle back onto the outer bed and move downward to repeat the cycle. The particles pass through the inner cylinder many times in a few minutes.

Centrifugal extrusion

Liquids are encapsulated using a rotating extrusion head containing concentric nozzles. In this process, a jet of core liquid is surrounded by a sheath of wall solution or melt. As the jet moves through the air it breaks, owing to Rayleigh instability into droplets of core, each coated with the wall solution. While the droplets are in flight, a molten wall may be hardened or a solvent may be evaporated from the wall solution. Since most of the droplets are within ± 10% of the mean diameter, they land in a narrow ring around the spray nozzle. Hence, if needed, the capsules can be hardened after formation by catching them in a ring-shaped hardening bath.This process is excellent for forming particles 400-2000 μm in diameter. Since the drops are formed by the breakup of a liquid jet, the process is only suitable for liquid or slurry. A high production rate can be achieved, i.e., up to 22.5 kg of microcapsules can be produced per nozzle per hour per head. Heads containing 16 nozzles are available.

Vibrational nozzle

Core-Shell encapsulation or Microgranulation (matrix-encapsulation) can be done using a laminar flow through a nozzle and an addtional vibration of the nozzle or the liquid. The vibration has to be done in resonance of the Rayleigh instability and leads to very uniform droplets. The liquid can consists of any liquids with limited viscosities (0-10000 mPas have been shown to work), e.g. solutions, emulsions, suspensions, melts etc. The soldification can be done according to the used gelation system with an internal gelation (e.g. sol-gel processing, melt) or an external (additional binder system, e.g. in a slurry). The process works very well for generating droplets between 100-5000 micrometer, applications for smaller and larger droplets are known. The units are deployed in industries and research mostly with capacities of 1-10000 kg/h at working temperatures of 20-1500°C (room temperature up to molten silicon). Nozzles heads are available from one up to several hundred thousand are available.

Spray–drying

Spray drying serves as a microencapsulation technique when an active material is dissolved or suspended in a melt or polymer solution and becomes trapped in the dried particle. The main advantages is the ability to handle labile materials because of the short contact time in the dryer, in addition, the operation is economical. In modern spray dryers the viscosity of the solutions to be sprayed can be as high as 300mPa.s.

Coaxial air flow bead generator

An easy way for production of small alginate beads in a controllable manner is the use of a coaxial-air-flow bead generator. The basic principle of the instrument is the use of a coaxial air stream which pulls droplets from a needle tip into the gelling bath (Anonymous, 2006).

Simple droplet generator system for encapsulation of the pancreatic islets employing chitosan-alginate matrix was developed by Hardikar *et al.* (1999). The droplet generator system comprised of a needle assembly, a 3-way valve with extended rubber tubing and a filtration unit connected to a pressure pump.

Chemical methods to manufacture microcapsules

Interfacial polymerization

In Interfacial polymerization, the two reactants in a polycondensation meet at an interface and react rapidly. The basis of this method is the classical Schotten Baumann reaction between an acid chloride and a compound containing an active hydrogen atom, such as an amine or alcohol, polyesters, polyurea, polyurethane. Under the right conditions, thin flexible walls form rapidly at the interface. A solution of the pesticide and a diacid chloride are emulsified in water and an aqueous solution containing an amine and a polyfunctional isocyanate is added. Base is present to neutralize the acid formed during the reaction. Condensed polymer walls form instantaneously at the interface of the emulsion droplets.

In-situ polymerization

In a few microencapsulation processes, the direct polymerization of a single monomer is carried out on the particle surface. In one process, e.g. cellulose fibers are encapsulated in polyethylene while immersed in dry toluene. Usual deposition rates are about 0.5 μm/min. Coating thickness ranges 0.2-75 μm. The coating is uniform, even over sharp projections.

Matrix polymerization

In a number of processes, a core material is embedded in a polymeric matrix during formation of the particles. A simple method of this type is spray-drying, in which the particle is formed by evaporation of the solvent from the matrix material. However, the solidification of the matrix also can be caused by a chemical change.

Application of micro encapsulation technology in food industry

Microencapsulation of enzymes

Kailasapathy *et al.* (2006) reported a technique for microencapsulating flavourzyme (a protease-peptidase complex) with sodium alginate and Hi-Maize starch in varying concentrations and proportions for maximizing encapsulation efficiency(EE). Seventy two percent EE was achieved for immobilising Flavourzyme using 1.8% (w/v) sodium alginate and 1% Hi-Maize starch with 30 min hardening of capsules in 0.1 M calcium chloride solution. Hi-Maize TM starch as a filler material improved EE and reduced the enzyme release during capsular hardening in calcium chloride solution, Increasing the cocentration of sodium alginate solution (1.2 to 1.8%,w/v) and calcium chloride concentration (0.05 M to 1.0 M) increased the EE. The optimum hardening time for the microcapsules in calcium chloride solution was 30 min. Shrinkage of microcapsules during hardening in calcium chloride was less with higher concentration of alginate. More spherical were produced with up to 1.8% (w/v) sodium alginate and 1% Hi-Maize TM starch solutions.

Chill proofing of beer (Prem, 2004) is a process by which a proteolytic enzyme is added to beer after fermentation, in order to prevent haze formation upon cooling. The most commonly used enzyme for chill proofing is papain. Its immobilization would decrease the cost of beer processing and furthermore, it could solve the problem of residual enzyme association with the free enzyme method.

Microencapsulation of fats and oils

Prem (2004) studied about microencapsulation of orange oil by using an aqueous gelatin emulsion. Odourless and tasteless bloom of gelatin is used in the form of emulsion in distilled or de-ionized water. Gelatin is dispersed in water at about 60° C along with an antioxidant and surfactant, 'orange oil is introduced and the mixture is stirred/ mixed whereby an emulsion is formed. The step is process of forming a gel capsule in the process liquid. Dehydration is done with anhydrous alcohol. Dried capsules are separated from the two liquids, i.e., process and dehydration liquids either by filtering, decanting or centrifuging. Capsules are washed with a suitable active material. Not more than 35% citrus oil can be encapsulated in gelatin medium. Since encapsulation is carried out in an inert atmosphere, the aromatic oils are not exposed to oxidation. Properly encased citrus oil will withstand rancidity for 18 to 12 months.

Lee *et al.* (2002) prepared microcapsules containing fragrant oil (Foral oil) via the *in situ* polymerization method using melamine-formaldehyde (M-F) as a wall material. The encapsulation efficiency and other physical properties were analysed with varying formaldehyde/melamine (F/M) mole ratio and pH of emulsion medium. The pH of the reaction medium was varied from 5.0-6.0 and the F/M molar ratio, 2.3 ~ 5.5. Microcapsules containing fragrant oil were synthesized successfully and their particle sizes ranged from 12-15 μm. Encapsulation efficiency of fragrant oil varied from 67-81%. It was found that both pH and F/M molar ratio have an effect on the separation of M-F prepolymer, consequently the morphology of the surface of the microcapsule was changed as well as encapsulation efficiency.

Spray dried encapsulated milk fat (Prem, 2004) containing 50% anhydrous butter oil encapsulated in flour or sucrose can be used as shortening replacer in biscuits, sugar snap cookies and muffins. Preliminary results in organoleptic tests indicated that the use of flour encapsulated milk fat significantly improved the flavour and texture of muffins.

Bustos *et al.* (2003) studied about stability of ù-3 polyunsaturated fatty acids (PUFAs) and astaxanthin pigments in an aqueous suspension of krill oil/chitosan microcapsules. Physically stable and homogeneous microcapsules were obtained with chitosan of different molecular weight. PUFAs were stable during the first 15 days of the study retaining over 95% of the eicosapentaenoic acid and docosahexaenoic acid originally present in the oil. Astaxanthin and its ethers were not as stable as the PUFAs, however low molecular weight chitosan showed a better protective effect than high molecular weight chitosan. The potential application of this microparticulate system for PUFAs and liposoluble vitamin supplementation of processed human food and aquafeed is discussed.

Kapusniak and Tomasik (2006) studied about Lipid microencapsulation in starch. Short microwave heating of granular potato, waxy corn and tapioca starches with such lipids as *cis*-9-octadecenoic acid (oleic acid), *cis, cis*-9,12-octadecadienoic acid (linoleic acid), octadecanoic acid (stearic acid), ethyl *cis*-9-octadecenoate, ethyl *cis,cis*-9,12-octadecadienoate and methyl octadecanoate provided microcapsules in which encapsulated guest molecules did not interact with starch microcapsules. On the formation of microcapsules, the lipid guest molecules did not react to starches. The encapsulation yield varied between 11–94%.

Minemoto *et al.* (2002) studied about emulsification of linoleic acid with gum arabic or maltodextrin at various weight ratios of the acid to the polysaccharide in the presence or absence of a small-molecule emulsifier. The emulsions were spray-dried to produce microcapsules. Emulsions prepared with gum arabic were smaller in droplet size and more stable than those prepared with maltodextrin, and linoleic acid in a gum arabic-based microcapsule was also most resistant to oxidation than that in a maltodextrin-based microcapsule. Although the oil droplet size in the emulsion with maltodextrin decreased and the emulsion stability was improved by addition of a small-molecule emulsifier to linoleic acid, the oxidative stability of the encapsulated linoleic acid was not significantly improved. Encapsulated linoleic acid of small droplet size oxidized more slowly than that of large droplet size.

Bocanegra *et al.* (2005) described a method for the production of cocoa butter microcapsules containing an aqueous solution of an oil-in-water emulsion via electrified coaxial jets. These coaxial jets are generated by electro hydrodynamic forces with diameters that range from hundreds of nanometers to tens of microns. The controlled break-up of the jets gives rise to an aerosol of compound droplets in which an outer liquid coats an inner one. This technique has turned out to be an effective method to encapsulate aqueous-based food ingredients such as flavors, aromas, enzymes, salts, minerals, and vitamins in the micrometric range with relatively high monodispersity and controllable geometrical parameters of the capsules such as thickness of the shell or number of inner cores in the capsules.

Microencapsulation of flavours and pigments

Shu *et al.* (2005) prepared lycopene microcapsules by a spray-drying method using a wall system consisting of gelatin and sucrose. The effects of technological parameters including the ratio of core and wall materials, ratio of gelatin and sucrose, homogenization pressure, inlet temperature, feed temperature, and lycopene purity on encapsulation yield (EY), encapsulation efficiency (EE) were investigated. Results showed that EY and EE were significantly affected by the ratio of core and wall materials, ratio of gelatin and sucrose, homogenization pressure, inlet temperature, feed temperature, and lycopene purity. The optimal condition was determined as follows: the ratio of gelatin/sucrose of 3/7 and the ratio of core and wall material of 1/4, feed temperature of 55°C, inlet temperature of 190°C, homogenization pressure of 40 MPa, and lycopene purity of not less than 52%, at which the micro-encapsulated lycopene showed some isomerization, but a good storage stability. Scanning electron microscopy (SEM) and X-ray diffraction (XRD) analysis showed that lycopene microcapsules had a regular spherical shape with a particle size distribution (PSD) of 2–15 μm, mean particle size (PS) of 5 μm, smooth outer surface, "bee net" -like interior structure, and lycopene was embedded in the wall system consisting of gelatin and sucrose.

Korus *et al.* (2003) described a method for preparation of microcapsules from granular potato starch by its prolonged (up to 48 h) soaking in water is proposed. The effects of temperature and size of granules is studied. Such treatment removes the amorphous part of the granule interior, forming empty domains inside granules. Material evacuated from the granules was identified as amylopectin together with amylose. The application of such pre-treated granules for microencapsulation of various fragrant compounds (angelicalactone, diacetyl, dibenzyl ether, 2,6-lutidine and myrcene) from their vapours and from their liquid state is described. Depending on pre-treatment of starch and the microencapsulation method applied, the amount of trapped guest molecules is up to 30 wt%.

Soottitantawat *et al.* (2005) studied about microencapsulation of *l*-menthol by spray drying, using gum arabic (GA) and modified starch (CAPSUL, HI-CAP 100) as capsule materials. The results showed a higher retention of *l-menthol* with the increasing of initial solid concentration. HI-CAP 100, showed a higher retention than the other capsule materials. However, it also showed a higher residue of *l-menthol* on the surface of powder especially at the high concentration of *l-menthol* in the feed emulsion. That might be from the interaction between the wall materials and *l*-menthol which can re-crystallize to form whisker after the spray drying. Furthermore, the release characteristics of *l*-menthol were also investigated. The release rate increased upon elevation of relative humidity and temperature. The activation energies of the release of *l*-menthol from GA wall matrices at 75 and 83%RH were 140 and 48 kJ/mol, respectively.

Shaikh *et al.* (2006) reported a method for microencapsulation of black pepper oleoresin by spray-drying, using gum arabic and modified starch as wall materials. The microcapsules were evaluated for the content and stability of volatiles, non-volatiles, total piperine and entrapped piperine for six weeks. Gum arabic offered greater protection to the pepper oleoresin than modified starch, as seen from the $t_{1/2}$, time required for a constituent to be reduced to 50% of its initial value.

Prem (2004) reported a method for coffee aroma encapsulation. Liquefied coffee aroma is passed in to the emulsifying tank that contains the wall material, i.e., an aqueous gelatin material and a modifier, e.g., non -ionic surface reactive agent, such as tween. The mixture is homogenized and an emulsion of the coating substance and aroma. The emulsion is poured in to the spinning tank, which has white mineral oil, such as Nujol. With proper stirring and cooking of the content in the spinning tank, droplets or globules formation occurs in spinning tank. Droplets are formed and remain suspended in the process oil. Globules formation occurs in spinning tank primarily because of aqueous gelation emulsion is insoluble in process liquid. Suspended globules are dried by anhydrous alcohol in the dehydration tank. Dried globules become capsules and are suspended from the process liquid and dehydrating medium by centrifugation or filtration. The capsules are washed free of the process liquid with the aid of a solvent and the solvent is recovered. Last traces of solvent are removed. Encapsulated coffee aroma can be retained for a year.

Prem (2004) stated that chewing gum composition has been found to have prolonged sourness, flavour and juiciness through incorporation of a delivery system comprising a food acid (e.g.malic acid) encapsulated in low molecular weight polyvinyl acetate material. Other process relates to the hydrophobic encapsulation of active material into the chewing gum. Controlled release of flavour together with aspartame gives a long lasting sweetened flavour perception during chewing of the gum.

Ersus and Yurdagel (2007) described a method for Microencapsulation of anthocyanin pigments of black carrot (*Daucus carota* L.) by spray drier. The acidified ethanol extracts of black carrots which has a high anthocyanin content (125 ± 17.22 mg/100 g) were spray dried using a range of maltodextrins [Stardri 10 (10DE), Glucodry 210 (20–23DE) and MDX 29

(28–31 DE)] as a carrier and coating agents, at 3 different inlet/outlet air temperatures with constant feed solid content (20%). Higher inlet/outlet air temperatures caused greater anthocyanin loss during spray drying. The quality attributes of the powders which were produced at optimum drying temperatures (160 °C) were characterized by their anthocyanin content, antioxidant capacity, L^*, a^*, b^*, C^* and $H°$ values, dry matter content and hygroscopicity. The best dried pigment containing powder was found where the Glucodry 210 was used as wall material. Scanning electron microscope was used for monitoring the structures and size (3–20 μm) of the powders. For determination the stability and half-life period of microencapsulated pigments, samples were stored under different storage temperatures (4°C and 25°C) and light illumination (3000 lx).

Souza *et al.* (2005) studied about Chitosan microspheres containing the natural urucum pigment. An increasing trend in the food and pharmaceutical industries is toward replacing synthetic additives with natural products. Encapsulation has become an important process to protect natural pigments. This paper reports on the encapsulation of the natural urucum pigment with chitosan using different techniques and its release under different pH conditions. The material loaded with pigment was evaluated by infrared spectroscopy, scanning electron microscopy and thermal analysis. Chitosan was found to be an effective encapsulating agent for urucum pigment. No investigations have previously been reported on the relation of chitosan to the stability of encapsulated natural pigments.

Microencapsulation of microorganisms

Picot and Lacroix (2003) described a simple and easily scaled-up technology for microencapsulation of dry probiotic cultures. The microcapsules, which contained micronized skim milk powder dispersed in milk fat droplets surrounded by an enteric coating, were produced using a continuous emulsification/spray drying process.

Chen, *et al.* (2005) described a technique for probiotic microencapsulation using prebiotics and modern optimization techniques to determine optimal processing conditions, performance, and survival rates. Prebiotics (fructooligosaccharides or isomaltooligosaccharides), growth promoter (peptide), and sodium alginate were incorporated as coating materials to microencapsulate 4 probiotics (*Lactobacillus acidophilus, Lacto-bacillus casei, Bifidobacterium bifidum,* and *Bifidobacterium longum*). The proportion of the prebiotics, peptide, and sodium alginate was optimized using response surface methodology (RSM) to first construct a surface model, with sequential quadratic programming (SQP) subsequently adopted to optimize the model and evaluate the survival of microencapsulated probiotics under simulated gastric fluid test. Optimization results indicated that 1% sodium alginate mixed with 1% peptide and 3% fructooligosaccharides as coating materials would produce the highest survival in terms of probiotic count. The verification experiment yielded a result close to the predicted values, with no significant difference ($P > 0.05$). The storage results also demonstrated that addition of prebiotics in the walls of probiotic microcapsules provided improved protection for the active organisms. These probiotic counts remained at 10^6 to 10^7 colony-forming units (CFU)/g for microcapsules stored for 1 month and then treated in simulated gastric fluid test and bile salt test.

The encapsulation of microorganisms such as bacteria and fungi in biopolymers is currently being evaluated as delivery systems in many fields. Information about the viability and morphology of the organisms in the microparticle is often required to ascertain the longevity of the systems. A *Bifidobacterium infantis* strain was microencapsulated within a film-forming protein-carbohydrate-oil emulsion (Crittenden *et al.*, 2006). Microencapsulation protected the viability of probiotic bacteria when they were stored in an open container at 25°C and 50% relative humidity as well as during simulated gastrointestinal transit.

Song, *et al.* (2003) developed a method for microencapsulation of *Lactobacilluscasei* YIT 9018. Several process parameters of membrane emulsification were investigated for producing a stable emulsion. The droplet diameter in the emulsion depended upon the membrane pore size. The monodispersed emulsion obtained by this technique resulted in well-formed microcapsules with a narrow particle size distribution. For artificial gastric acid and bile, the viable count of encapsulated cells was constant through the incubation time, while the count of non encapsulated cells was significantly decreased. A storage stability test at different temperatures resulted in a viability of encapsulated cells 3 to 5 log cycles higher than the viability of non encapsulated cells.

Mercier *et al.* (2001) described a new method for encapsulation of living cells. PC12 rat adrenal pheochromocytoma cells, which have been shown to synthesize, store and release dopamine were employed. The particles are made first and the cells then incorporated in a gentle mechanical procedure. The morphology (by light and electron microscopic observation), stability, rheology, texture and permeability of these microcapsules provided by Kappa Biotech were investigated. Membrane permeability studies demonstrated exclusion of 69000 Da human serum albumin, but equilibrium of D-glucose and inulin was within 24 h, indicating a molecular weight cut-off in the 5000-70000 Da range. The viability and the function of the encapsulated cells were evaluated by measuring the spontaneous release of dopamine by high performance liquid chromatography with electrochemical detection. The results show that dopamine- secreting cells can be sequestered in a semi-permeable capsule and still display good viability and proliferation for at least 1 month.

Koyama and Seki (2004) studied about cultivation of yeast and plant cells entrapped in the low-viscous liquid-core of an alginate membrane capsule prepared using polyethylene glycol. A liquid-core alginate- membrane was prepared by a novel method using polyethylene glycol as a thickener and the cells of *Saccharomyces cerevisiae* were encapsulated in its core and cultured. After 2hr of cultivation, the cell concentration in the capsule core-liquid reached 222 g/mm^3 on a dry weight basis, which was 1.4 times as large as that in the core of double-layered alginate beads, i.e., alginate coated alginate gel beads. The diameter increase of the capsule prepared by the proposed method using immobilized cell growth was suppressed compared to those using the double layer method and simple alginate- gel bead entrapment, most likely because of the mobility of the entrapped cells in the capsule. They also confirmed that this encapsulation method is applicable for the cultivation of cultured cells of the plant *Fragaria ananassa.*

Alginate microencapsulated genetically engineered cells (Leon *et al.*, 2002) can be used for *in vitro* recombinant DNA derived protein production as well as for *in vivo* slow release system of a therapeutic polypeptide. Both applications require the optimization of alginate beads generation to reproducibly obtain uniformly shaped and sized particles that contain viable producing cells. Such particles allow *in vitro* high-density cell culture and are also suitable for *in vivo* intra-tissue injection.

Microencapsulation of vitamins and proteins

Desai, *et al.* (2005) reports the properties of vitamin C encapsulated sodium alginate beads prepared by an alternative approach. The alternative encapsulation process mainly involves immobilization of vitamin C in hydrated zinc oxide layers and encapsulation of prepared immobilized particles in sodium alginate bead. The immobilization of vitamin C in hydrated zinc oxide layers was achieved by a coprecipitation process. Fourier transform infrared (FTIR) spectroscopy showed that the vitamin C was found to be stable after its immobilization. X-ray diffraction (XRD) studies revealed that anionic vitamin C molecules are adsorbed between positively charged zinc hydroxide layers with a 1:1 layer sequence, since well-defined change in basal spacing was observed. Well-defined change in surface morphology was observed by scanning electron microscopy (SEM) when vitamin C immobilized particles are encapsulated in sodium alginate bead. The biological activity of vitamin C was retained, even after its immobilization which was confirmed by 4-dihydroxy-L-phenylalanine (L-DOPA) oxidase inhibition and free radical scavenging activity studies. The release rate of vitamin C from immobilized particles and beads was sustained through an ion exchange process. A higher amount of stable vitamin C was recovered from the bead when compared to neat vitamin C itself.

Katti (1999) described preparation of albumin microspheres by an improved process. In this study microspheres were prepared by the suspension cross linking method for the first time in the absence of any surface active agent, using paraffin oil as the dispersion medium and formaldehyde as the cross linking agent. The microspheres thus obtained were characterized using a Scanning Electron Microscope and found to be spherical and having a particle size distribution in the range 50-400 microns. A preliminary drug release study of chlorothiazide invitro indicated a diffusion controlled release of the drug. This method, being simple and cost effective, could be a promising technique for the large scale manufacture of microspheres.

Taste and odour masking (Prem, 2004) is yet another application of microencapsulation. High protein concentrate from casein, soy protein, fish protein are made palatable by microencapsulation with fats having melting point above 70°F. To overcome the objection, encapsulated garlic oil is strongly recommended, specially in food plant, as it also helps in preventing the cross contamination of the flavour.

Patents on microencapsulation

Most patents or patent applications in microencapsulation relate to fields of application in the pharmaceutical industry. The major roles are taken here by the capsule materials used (generally in combination with specific activities or effects), the controlled release of substances via the application of the encapsulating technique and the stabilization of substances. In the food sector there are far fewer patents, which, however, are essentially determined by these three directions. In the case of ***microencapsulation*** of cells or cell free extracts (CFE), physiological stability in combination with their use as a pharmaceutical product play the major roles.

Patented method of encapsulation involved immobilization of microbial cells in calcium alginate (Chang *et al.*, 1998). Whole microbial cells are mixed with a $CaCl_2$ solution, a small amount of xanthan gum is added, the resultant mixture is added drop-wise while stirring to a Na-alginate solution containing a small amount of surfactant which is preferably polyoxyethylenesorbitan monolaurate to obtain the capsules containing the cells. The capsules may be washed with distilled water, hardened by stirring in a $CaCl_2$ solution and incubated in a growth medium.

Ramasubramanian (2006) patented method and device for micro-encapsulation of cells. It consists of two chambers and an aperture plate.

Jager *et al.* (2005) described the process of microencapsulation of dietary fiber and bioactive ingredients. The capsules were produced by spray drying. Seifert (1992) patented a method for immobilization of bioactive cells by suspending bioactive cells in an aqueous solution of a salt-free osmolarity adjuster, alginate and polyethylene glycol, dividing the suspension into bead-sized globules, contacting the globules with a solution of divalent cations to gel the alginate to form beads containing polyethylene glycol-filled areas and removing the polyethylene glycol to form voids in the beads entrapping the bioactive cells. By perfusing the beads with a nutrient medium, the cells can proliferate in the voids to produce high cell densities and produce a product which can be separated from the medium.

Ganan and Miguel (2003) described the production of particles for introduction into food using a stable microjet and a monodisperse aerosol formed when the microjet dissociates. A variety of devices and methods were disclosed which allow for the formation of a stream of a first fluid (e.g. a liquid) characterized by forming a stable capillary microjet over a portion of the stream wherein the microjet portion of the stream is formed by a second fluid (e.g. a gas).

References

Anonymous (2006). Product details of VAR J1 of Nisco Engg. A.G. URL: http://www.nisco.ch/30.htm.

Bocanegra R, Anil Kumar G, Gaonkar, Barrero A, Ignaico G, Lascertales Pechak D and Marquez M (2005). Production of Cocoa Butter Microcapsules Using an Electrospray Process.Journal of Food Science, 70(8): E492-E497.

Bustos R, Romol L, Yanez K, Diaz G and Romo C (2003). Oxidative stability of carotenoid pigments and polyunsaturated fatty acids in microparticulate diets containing krill oil for nutrition of marine fish larvae. Journal of Food Engineering, 56(3): 289-293.

Chang, HN, Seong, GH, Yoo, IK, Park, JK and Seo, JH (1998). Method for immobi-lization of whole microbial cells in calcium alginate capsules. US Patent No. 5: 766,907.

Chen K, Chen M, Liu J, Lin C and Chiu J (2005). Optimization of Incorporated Prebiotics as Coating Materials for Probiotic Microencapsulation .Joornal of food science, 70(5): M260-M266.

Crittenden, R, Weerakkody, R, Sanguansri, L and Augustin MA (2006). Synbiotic Microcapsules that enhance microbial viability during nonrefrigerated storage and gastrointestinal transit. Applied and Environmental Microbiology, 72 (3), p. 2280–2282.

Desai KGH, Liu C, and Park HJ (2005). Characteristics of vitamin C immobilized particles and sodium alginate beads containing immobilized particles. Journal of Microencapsulation, 22(4): 363-376.

Ersus, S and Yudragel U (2007). Microencapsulation of anthocyanin pigments of black carrot (Daucus carota L.) by spray dryer. Journal of food engineering, 80: 805-812.

Fuchs M, Turchiulic C, Bobin M, Cuvelier ME, Ordonnaud C, Peryrat-Mailard MN and Dumoulin E (2006). Encapsulation of oil in powder using spray drying and fluidized bed agglomeration. Journal of food engineering, 75: 27-35.

Ganan, C and Miguel, A (2003). Enhanced food products. US Patent No. 6: 589,579.

Hardikar, AA, Risbud, MV and Bhonde, RR (1999). A simple microcapsule generator for islet encapsulation. Journal of biosciences, 24(3): 371-376.

Jager, M, Haber, B Kunz, B Strater, S, Weissbrodt, J, Bollinger, H, Brendle, HG and Bache, G (2005). Encapsulated multifunctional biologically active food component, process for its production and its use. US patent No. 6: 841,181.

Kailasapathy K, Perera C, and Philips M (2006). Evaluation of alginate starch polymers for preparation of enzyme microcapsules. International Journal of food Engineering, 97(2): 1-10.

Kapusniak J and Tomasik P (2006). Lipid microencapsulation in starch. Journal of Microencapsulation, 23(3): 341-348.

Karel M (1990). Encapsulation and controlled release of food components. In: Biotechnology and Food Process Engineering Eds. G Schwartzberg and MA Rao, Marcel Dekker, NY, pages: 277-294.

Katti D (1999). Preparation of albumin microspheres by an improved process. Journal of microencapsulation, 16(2): 231-242.

Korus J, Tomasik P and Lii CY (2003). Microcapsules from starch granules. Journal of Microencapsulation, 20(1): 47-56.

Koyama K and Seki M (2004). Cultivation of yeast and plant cells Entrapped in the Low-viscous Liquid-core of an Alginate Membrane Capsule prepared Using Polyethylene Glycol. Journal of bioscience and bio engineering, 97(2): 111-117.

Lee HY, Lee SJ, Cheong IW, and Kim JH, (2002). Microencapsulation of fragrant oil via in situ polymerization: effects of pH and melamine-formaldehyde molar ratio. Journal of Microencapsulation, 19(5): 559-569.

Leon FG, Luis KA, Maria EF and Clandio GG (2002). A Laboratory scale device for microencapsulation of genetically engineered cells in to alginate beads. Electronic journal of Biotechnology, 5(3).

Mercier, Fernandez F, Tortosa F, Bagheri H, Duplan H, Tafani M, Bes JC, Bastide R, Lazorthes Y, and Sallerin B, (2001). A new method for encapsulation of living cells: preliminary results with PC12 cell line. Journal of Microencapsulation, 18(3): 323-334.

Minemoto Y, Hakamata K, Adachi S, and Matsuno R, (2002). Oxidation of linoleic acid encapsulated with gum arabic or maltodextrin by spray-drying. Journal of Microencapsulation, 19(2): M181-189.

Picot A and Lacroix A (2003). Production of multiphase water-insoluble microcapsules for cell microencapsulation using an emulsification/spray-drying technology. Journal of food science, 68(9): 2693-2701.

Prem PD (2004). Application of microencapsulation technology in Food Industry. Indian Food Industry, 23(4): 52-58.

Ramasubramanian (2006). Methods and devices for microencapsulation of cells. Intl. Patent No. WO/2005/071060.

Shaikh J, Bhosale R and Singhal R (2006). Microencapsulation of black pepper oleoresin. Food chemistry, 94(1): 105-110.

Shi G, Rao L, Yu H, Xiang H, Pen G, Long S and Yang C (2006). Yeast-cell-based microencapsulation of chlorogenic acid as a water soluble antioxidant.

Shu B, Yu W, Zhao Y and Liu X (2005). Study on microencapsulation of lycopene by spray drying. Journal of food engineering, 76: 664-670.

Song H, Cho YH and Park J (2003). Microencapsulation of Lactobacillus casei YIT 9018 using a microporus glass membrane emulsification system.Journal of food science, 68(1): 195-200.

Soottitantawat A, Takayama K, Okamura K, Muranaka D, Hidefumi Y, Furuta T, Ohkawara M and Linko P (2005). Microencapsulation of of l-menthol by spray drying and its release characteristics. Innovative Food Science and Emerging technologies, 6(2): 163-170.

Souza TCR, Parize AL, Brighente IMC, Fávere VT, and Laranjeira MCM (2005). Chitosan microspheres containing the natural urucum pigment. Journal of Micro-encapsulation, 22(5): 511- 520.

□□□

26

High Hydrostatic Pressure : An Emerging Technology for Better Food Processing

R.K. VISHWAKARMA, R.K. GUPTA AND V.E. NAMBI

Consumer trends and therefore the food markets are changing continuously and this change will continue with increasing awareness. Foods with more fresh-like attributes and high quality are in demand, consequently less extreme treatments with fewer additives are need of the present as well as future consumers (Palou *et al.*, 1999). Some identified food characteristics that must be attained in response to new consumer demands are less heat and chill damage, more freshness, less acid, salts sugars and fat (Gould, 1992). Moreover, food safety is an aspect of increasing importance, and improvement in microbial control must be attained. On the other hand, the use of Good Manufacturing Practices in the food industry and maintenance of hygiene standards in food service establishments are essential to control the food borne diseases (Patterson *et al.*, 1995). However, these procedures alone may not be sufficient to ensure microbial safety, making food preservation methods necessary.

Traditionally, foods are processed and preserved using thermal processing. Heating the foods inactivates spoilage microorganisms and ensure prolong shelf life and food safety (Pothakmury *et al.*, 1995). However, the uses of high temperatures cause detrimental changes on the processed products. These undesirable changes affect nutritional as well as sensory attributes. These changes result in products that are far from original fresh products. The claim for freshness in food products has, since 1994, been at the top of the list of the most desired claims by consumers (Sloan, 1999). This view sets a challenge for the food industry, as new processes need to be developed and adopted to satisfy consumer's demands. As a result novel non-thermal technologies are currently under extensive research.

One emerging technology receiving great attention is the application of high hydrostatic pressure (HHP). The widest application of HHP process within the food industry is mainly for extending the shelf life of food products, although as research progresses other uses like solute diffusion process (Lambert *et al.*, 1999), assisted freezing thawing and modifications of functional properties of macromolecules (Chevalier *et al.*, 1999) are foreseen. This article describes the HHP process, equipment, its effect on microorganisms and food components.

High hydrostatic pressure: The process and machine

Studies examining the effects of high pressure on foods started in the end of nineteenth century, but renewed research and commercialization efforts worldwide could place high pressure treated foods in market in the end of twentieth century (Pothakmury *et al.*, 1995). In April 1990, the first high pressure product, a high acid jam, was introduced in the Japanese retail market. In 1991, number of food products were introduced in Japanese market and two Japanese fruit juice processors installed semi-continuous high pressure machines for bulk processing of citrus juice (Williams, 1994).

High pressure means that the phase transition occurs under constant pressure that is accompanied by a decrease in volume, which is based on Le Chatelier principle. High pressure processing (HPP) at refrigeration, ambient or moderate heating temperature allows inactivation of pathogenic and spoilage microorganisms in foods with fewer changes in texture, color and flavor as compared to conventional technologies (Ledward, 1995).

The term "high pressure" may be understood with the fact that at the deepest point in the oceans the pressure is about 100 MPa and at the centre of the earth it is 360 GPa. In commercial applications the highest pressure used is around 5-6 GPa; which is applied for artificial diamond grit production (Crossland, 1995). In the food industry, high hydrostatic pressure

(400-700 MPa) is used to preserve food. It is the application of pressure uniformly throughout a product. The process is also known as high pressure processing (HPP) and ultra high pressure processing (UHP).

The pressurization is carried out for the duration of treatment in a closed confined space (Pressure Vessel) containing a pressure transmitting fluid (usually water). Since the pressure is applied isostatistically, it allows solid foods to retain their original shape (San Martin *et al.,* 2002). The pressure is held for the desired time and then released. Isostatic compression is advantageous over conventional thermal treatment is the sense that the process is independent of size and geometry of the product.

The key HPP equipment technologies are the pressure vessels and the high hydrostatic pressure generating pumps. Pressurizing liquid at 20 MPa is fed on the high pressure side of the main pump piston having an area ratio of 30:1 with respect to the high pressure fluid piston displacing into the high pressure vessel a food-grade contact fluid, typically purified water at 600 MPa as shown in Fig. 1 (Torres & Velazquez, 2005). When the main piston reaches the end of its displacement, the system is reversed and high pressure oil is then fed to the other side of the main piston and the high pressure fluid exits on the other pump side.

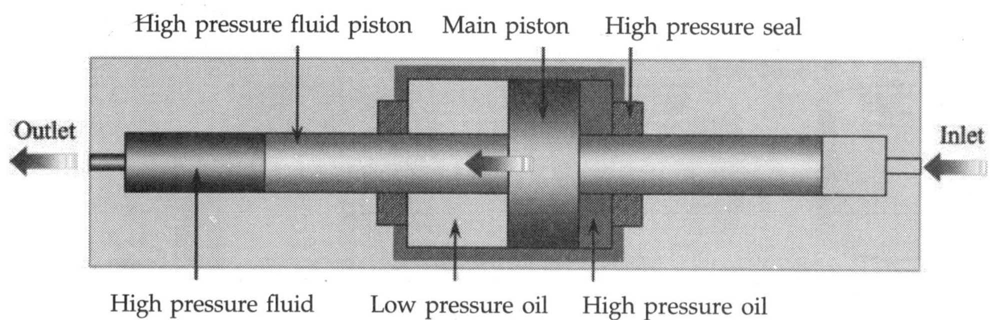

Fig. 1 : High hydrostatic pressure pump

High pressure machine

A high pressure processing system consists of a high-pressure vessel and its closure, pressure generating pump, temperature control system, and material handling devices. Once the vessel is loaded with product and closed, the vessel is filled with a pressure-transmitting medium. Air is removed from the vessel by means of low-pressure fast-fill and drain pump (Palou *et al.,* 1999).

Traditionally, pressure vessels were designed for small volume and at low frequency cycles. The casting limitation of pressure vessel construction from a single block limits them to 25 liter for operating pressures in excess of 400 MPa (Torres & Velazquez, 2005). However, to process large volumes, the industry requires high-frequency cycles, over a given time, and, in addition, safety issues as well as ease of operation must be considered.

Prestressed wire-wound vessels are needed for safe, durable and reliable operation for larger volumes at higher pressures (Torres & Velazquez, 2005). Typically, the same technology is used for the yoke holding the top and bottom seals (Fig. 2). A replaceable liner of steel is inserted inside the cylinder so as to cause a leak before failure.

Wire winding increases equipment costs leading to the present definition of low cost operations such as oyster shucking requiring 200-400 MPa separated by a technology barrier at 400 MPa from higher cost operations such as guacamole. A second technology barrier exists at 680 MPa and above this pressure level there are no vessels available for commercial applications. However, the next generation of equipment is expected to reach 700 MPa and operate at temperatures 100°C to inactivate bacterial spores (Torres & Velazquez, 2005).

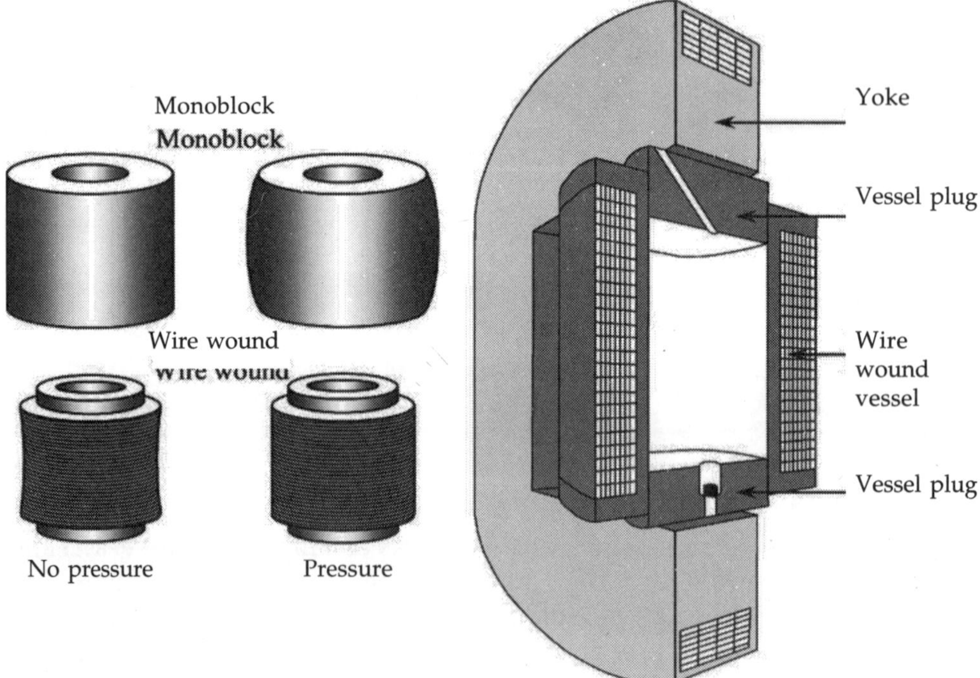

Fig. 2 : Monoblock casting technology for moderate pressure/size vessels and wire-winding technology for vessels and yoke used in larger size and higher pressure applications.

If the processed product is to be in direct contact with the pressure vessel, sanitary materials must be used in construction. When products to be treated are previously packaged, other metal alloys may be used. The generation of pressure inside the vessel may be achieved by any of the three means as described below (San Martin *et al.,* 2002).

Direct compression is generated by pressurizing a medium with the small diameter end of a piston (Fig. 1). The large diameter end of the piston is operated by a low-pressure pump. The direct compression method allows very fast generation of required pressure. But the high-pressure dynamic seal between the piston and the vessel internal surface restricts its use to small diameter laboratory and pilot plant systems.

Indirect compression process consists of a high-pressure intensifier to pump a pressure medium from a reservoir into the closed pressure vessel till the desired pressure is achieved. In industrial scale machine this method is used.

Heating of the pressure medium utilizes expansion of the pressure medium with increasing temperature to generate the pressure inside the vessel. Heating of the pressure medium is therefore used when high pressure is applied in combination with high temperature. The process requires accurate temperature control within the entire internal volume of the pressure vessel.

HHP systems that allow for the temperature control are varied and the simple electric heating to complex heat-exchanging systems may be used. Operations temperatures will depend on the particular type of system but the range may vary from -20°C to 80°C.

Handling systems consist of baskets that are filled with the packed products to be processed that are then loaded into the pressure vessel (San Martin *et al.,* 2002). The baskets may be transported by conveying belt (Horizontal Loading) or by cranes (top loading).

Effect of HHP processing

The successful application of HHP in the food processing results from this fact that high pressure treatment cause on microorganisms and on some enzymes, which are the most important factors contributing to food safety and stability.

The water present in food plays important role during high pressure processing application. The properties of water can change if the volume is changed over significant ranges as constant temperatures. The application of high pressure to water and ice results in more compact structures with distorted hydrogen-bond angles and a close approach between the non hydrogen bonded neighbors (Jonas, 1997). The characteristic behavior of phase transition of water under high pressure can be used to freeze or thaw products that will have a better quality.

Effect on biomembranes

Biomembranes are the main site affected by high pressure (Hoover *et al.*, 1989). The membrane is composed by a bilayer of phospholipids with embedded functional proteins that, among others, play an important role in transporting ions and other substances across the membranes. The lipid bilayer undergo phase transitions under pressure and fluidity of the membrane is decreased. It may result in breakage of the membrane bound – proteins causing a functional disorder of the proteins. Thus, the HHP is capable of altering membrane permeability in fruit tissues which can enhance mass transfer during osmotic dehydration, fluidized bed drying and freezing.

Effect on microorganisms

HHP inactivates microorganisms by interrupting cellular functions responsible for reproduction and survival. HPP damages bacterial membranes and thus affects transport phenomena involved in nutrient uptake and disposal of cell waste. Intracellular fluid compounds have been found in the cell suspending fluid after pressure treatment demonstrating that leaks occur while cells are held under pressure (Shimada *et al.*, 1993). Membrane damage has been shown to occur later than cell death, suggesting that dye exclusion measurements used to assess pressure effect may be used for characterization of microbial pressure inactivation (Ulmer, Gaenzle, & Vogel, 2000).

HPP treatments (345 MPa for 5 min at 25°C) altered *L. mesenteroides* cell walls and made cell membranes permeable (Kalchayanand, *et al.*, 2002). This damage reduced the potential gradient across the membrane, preventing cells from synthesizing ATP, which activated an autolytic enzyme degradation of cell walls. No growth was observed after 48 h of culture on plate count agar for cells treated at 400 MPa for 10 min in pH 5.6 citrate buffer.

The sensitivity of different microorganisms to HHP treatment is strongly dependent of the strain and under the same treatment the response may be greatly different. The shape, age, medium and growth conditions of the microorganisms influence the effect of HPP (San Martin *et al.*, 2002). In addition to inactivation, high pressure also induces morphological changes. These include reversible changes such as elongation and cessation of the motion due to structural changes and irreversible changes such as separation of the cell wall from the cell membranes as well as decrease in number of ribosomes (Hoover *et al.*, 1989). The factors influencing inactivation of microorganisms are given in Table 1.

Table 1 : Factors contributing to inactivation of microorganisms

S.No	Contributing factor	Type
1	Pressure applied	
2	Type of Pressure Treatment	Cyclic, continuous
3	Temperature	
4	Time of pressure application	
5	Species and strain of microorganisms	Shape, Gram type
6	Growth, stage and age of culture	
7	Medium composition and properties	Ionic strength type of ions, water activity, pH

One of the major drawbacks of HHP is the inability to inactivate spores by pressure alone. Spores are highly pressure resistant and capable to survive pressure upto 1200 MPa (Cheftel, 1995). However, it has been demonstrated that low pressure treatments (60-100 MPa) can induce spore germination and with higher temperatures leading to higher percentage of germinated spores. The germinated spores can then be inactivated by treatments at higher pressures (Cheftel, 1995). However, the time required to achieve spore germination were long, in order of 30 min. The combination of HHP with other treatments has been suggested as a way of reducing spore number. However, more research is required in this area.

Effect on food components

Enzymes: The response of enzymes subjected to HHP treatments is varied and depends on the origin of the enzyme, substrate nature, pressure, temperature and processing time. In some cases pressurization increases enzyme activity, like polyphenol oxidase in potato, mushroom, apples increased browning rate after a pressure treatment of 400 MPa, while at 800 MPa, complete inactivation was achieved. However the pressure treatment of purified enzyme did not cause as enhancement in activity.

Protein gelation: Influence of pressure on gel formation by proteins was observed early as in 1914. High pressure can induce gelation at low temperatures, resulting in gels with properties different from gels obtained with heat.

Vitamins and volatile profiles: The high pressure treatment does noı have a significant effect on water soluble vitamins and it contributes to preservation of nutritional quality of foods. High pressure treatment does not affect the volatile compounds and flavor of the product remain stable during low temperature storage. It shows that fresh – like product can be achieved by HHP treatment while extending their shelf life.

Niche opportunities for HPP foods

The successful introduction of a new technology demands competitive advantages over existing ones. In the case of HPP, an additional constraint is the large capital investment which is overcome by operating HPP plants at full capacity. Processing of seasonal commodities requires identifying a product mix achieving maximum utilization of the equipment investment.

The classical example of satisfying a consumer demand for a fresh product is the pressure processing of avocado. The Avomex Inc initiated HPP plant operations in 1996 with a 25-l high pressure vessel for batch processing (Torres & Velazquez, 2005) to satisfy the need of a local chain of Mexican-style restaurants, adding later a second 25-l vessel and then a wire-wound 50-l vessel when product demand expanded rapidly. In 1999, Avomex acquired a semi-continuous unit and in 2000 a 215-l wire-wound batch processing vessel. The treatment required is under 90,000 psi for 1 min (Torres & Velazquez, 2005). HPP avocado is now in distribution in whole USA because there was an unsatisfied consumer demand for products with acceptable shelf-life, convenient to use and free from chemical additives.

The classical example of unique pressure effect is oyster shucking by HPP. In 2002, California banned the sale of untreated Gulf of Mexico oysters harvested between April and October costing local producers an estimated $20 million/year loss. The process discovered in 1997 places live oysters under pressures of 240–350 MPa for 3 min. These moderate pressure treatments denature the abductor muscle and oysters can be opened without knife damage. Replacing the laborious and costly hand-shucking is the most significant development in the oyster industry in the past 100 years (Torres & Velazquez, 2005). Most importantly, the HPP treatment eliminates a high safety risk to production workers, extends refrigerated shelf-life to three weeks and reduces the microbial risk to consumers when consuming raw oysters by inactivating *V. parahaemolyticus, Vibrio vulnificus, Vibrio cholerae, Vibrio cholerae* non-O:1, *Vibrio hollisae* and *Vibrio mimicus* (Berlin *et al.,* 1999).

Refrigerated fresh-cut fruit salads that consumers demand for healthiness, convenience and labor-savings reasons are another example (Pao *et al.*, 1997). During cutting and packaging, these products may be contaminated with *Salmonella, Escherichia coli* and other pathogens of concern. These organisms can be inactivated by HPP without altering product "freshness."

The demand for juices with no thermal treatment still exists and high pressure processors seek to satisfy it by products that meet the new pasteurization requirement and that could be labeled "fresh" because they are undistinguishable from fresh-squeezed juice.

Inactivation of enzymes and spoilage microorganisms in juice have been extensively studied (Goodner *et al.*, 1998; Parish, 1998). Non-carbonated and carbonated juices subjected to conditions ranging from 200 to 600 MPa pressure, 30 to 300 s dwell time and 15 to 50 °C final processing temperature showed that CO_2(g)-assisted HPP increased the rate but not the extent of PME (pectin-methylesterase) inactivation. Determinations of color and antioxidant levels expressed as vitamin C determinations showed no significant differences between pressure-treated and control samples. Shellhammer *et al.* (2003) concluded that pressure-treated juices are safe and similar to fresh-squeezed samples.

The unique physical, textural and sensory properties of food processed by high pressure technology offer new chances for food product development such as long shelf- life convenience foods like fresh and natural colors, new type of food gels, minimally processed meat and fish and frozen foods with improved quality. Commercial feasibility is the most challenging problem for high pressure technology for India. Two important questions need to be answered: Will the consumer accept a high-pressure processed product, and are they prepared to pay an extra cost for a high pressure- processed product? The commercial application of high- pressure technology in Indian Food Industry will depend largely on the economical feasibility of the process. The capital cost associated with the equipment purchase, particularly pressure vessel, and installation is an important hurdle for its commercial implementation. It is unlikely that pressure processing will replace canning or freezing, nevertheless it could find applications for expensive foods with short shelf lives and high value ingredients such as flavors, vitamins and functional biopolymers that are heat sensitive. High hydrostatic pressure is not a cheap technology, and a systematic approach must be taken to search for processing options to ensure that this technology can be successfully and economically applied to a wide range of products.

References

Berlin, DL, Herson, DS, Hicks, DT, & Hoover, DG (1999). Response of pathogenic Vibrio species to high hydrostatic pressure. Applied Environmental Microbiology, 65: 2776-2780.

Cheftel, JC (1995). Review: high-pressure, microbial inactivation and food preservation. Food Science and Technology International, 1: 75-90.

Chevalier, D, Le Bail, A & Chourot, JM (1999). High pressure thawing of fish (Whiting): influence of the process parameters on drip losses. LWT, 32: 25-31.

Crossland, B (1995). The development of high pressure processing. In: High Pressure Processing of Foods (D.A. Ledwar, D.E. Johnston, R.G. Earnshaw & A.P.M. Hasting eds.). Nottingham University Press, Nottingham, 7.

Goodner, JK, Braddock, RJ, & Parish, ME (1998). Inactivation of pectinesterase in orange and grapefruit juices by high pressure. Journal of Agriculture and Food Chemistry, 46: 1997-2000.

Gould, GW (1992). Ecosystem approaches to food preservation. Journal of Applied Bacteriology, 73-85.

Hoovar, DG, Metrick, C, Papineau, AM, Farkas, DF & Knorr, D (1989). Biological effects of high hydrostatic pressure on microorganisms. Food Technology, 43(3): 99-107.

Jonas, J (1997). Cold denaturation of proteins. ACS Symposium Series, 676: 310-323.

Kalchayanand, N, Frethem, C, Dunne, P, Sikes, A, & Ray, B (2002). Hydrostatic pressure and bacteriocin-triggered cell wall lysis of Leuconostoc mesenteroides. Innovative Food Science and Emerging Technologies, 3: 33–40.

Lambert, Y, Demazeau, G, Largeteau, A & Bouvier, JM (1999). Changes in aromatic volatile composition of strawberry after high pressure treatment. Food Chemistry, 76: 7-16.

Ledward, DA (1995). High pressure processing-the potential. In: High Pressure Processing of Foods (D.A. Ledwar, D.E. Johnston, R.G. Earnshaw & A.P.M. Hasting eds.). Nottingham University Press, Nottingham, 1.

Palou, E, Lopez-Malo, A, Barbosa-Canovas, V & Swanson, BG (1999). High-pressure treatment in food preservation. In: Handbook of Food Preservation, Marcel Dekker, Inc, New York , pp 533-576.

Pao, S., Petracek, P.D., & Ismail, M.A. 1997. Advances in preparing peeled fresh-cut citrus. Food Technology International Europe, pp 39-40, 42.

Parish, ME (1998). High pressure inactivation of Saccharomyces cerevisiae, endogenousmicroflora and pectinmethylesterase in orange juice. Journal of Food Safety, 18: 57-65.

Patterson, MF, Quinn, M & Gilmour, A (1995). Effects of high pressure on vegetative pathogens. In: High Pressure Processing of Foods (D.A. Ledwar, D.E. Johnston, R.G. Earnshaw & A.P.M. Hasting eds.), Nottingham University Press, Nottingham, 47.

Pothakamury, UR, Barbosa-Canovas, V & Swanson, BG (1995). The pressure builds for better food processing. Chemical Engineering Progress, pp 45-53.

San Martin, MF, Barbosa-Canovas, GV & Swanson, BG (2002). Food processing by high hydrostatic pressure. Critical Reviews in Food Science and Nutrition, 42(6): 627-645.

Shellhammer, TH, Aleman, GD, McDaniel, MR, & Torres, JA (2003). A comparison of the sensory and chemical properties of orange and apple juices treated with and without high pressure. In IFT Annual Meeting, Chicago, IL.

Shimada, S, Andou, M, Naito, N, Yamada, N, Osumi, M, & Hayashi, R (1993). Effects of hydrostatic pressure on the ultrastructure and leakage of internal substances in the yeast Saccharomyces cerevisiae. Applied Microbiology and Biotechnology, 40: 123-131.

Sloan, AE (1999). Top ten trends to watch and work on the millennium. Food Technology, 53 (8): 40-58.

Torres, JA & Velazquez, G (2005). Commercial opportunities and research challenges in the high pressure processing of foods. Journal of Food Engineering, 67: 95-112.

Ulmer, HM, Gaenzle, MG, & Vogel, RF (2000). Effects of high pressure on survival and metabolic activity of Lactobacillus plantarum TW1.460. Applied Environmental Microbiology, 66: 3966–3973.

Williams, A (1994). New technologies in food preservation and processing. Nutrition and Food Science, 1: 16.

□□□

27

Application of Biodegradable Plastics for Food Packaging

S.K. NANDA

Food packaging is an essential technique for preserving food quality and minimizing food wastage. The package serves the important function of containing the food, protect it against chemical and physical damage and provide product information to the consumer while allowing convenient handling and efficient distribution at the same time. In order to preserve food quality, it has to act as a barrier to moisture and gas, microbial infection, insect infestation, other chemical and biological contamination and also light which, in some cases, can cause product deterioration. As food wastes increase where packaging is not appropriate, or does not exist, the packaging can reduce the overall cost of goods.

A wide variety of polymers are used in the production of plastics packaging materials, the most prevalent being:

Polyethylenes (HDPE, LDPE, LLDPE)

Polypropylene (PP)

Ethylene copolymers (with propylene, vinyl alcohol, vinyl acetate)

Polyestrenes, mainly polyethylene terephthalate (PET)

Polyamides (mainly nylon 6)

Polyvinyl Chloride (PVC)

Packaging is one of the largest consuming sectors of plastics all over the world. Despite all the positive functions fulfilled, packaging waste has increased along with the consumption of consumer goods. The visibility of packaging in the domestic wastebin has contributed to the common perception that packaging is a significant waste stream. Packaging can have numerous environmental impacts, direct and indirect, over its life cycle related to the use of resources and emission of pollutants to water, air and land. In fact, concern over the disposal of enormous tonnage of plastics packaging materials has led to the search for biodegradable plastics.

Recycling is an option to reduce the problem of waste disposal, besides conserving limited natural resources. Polymer composites and multilayers pose considerable difficulties during recycling process (collection, sorting, washing, grinding, refinement). Among the uses of post-consumer recycled (PCR) plastics, the commercial use of chemically recycled PET in food applications has been pioneered by Coca-cola and Pepsi. PCR plants should not transfer their constituents to foodstuffs in qualities that might endanger human health or bring about unacceptable change in the composition and/or organoleptic characteristics of the food stuff. Using less material in each packaging (known as light-weighting or down-gauging) and distribution of product in minimal packaging to be transferred into reusable containers at home are other means of packaging waste prevention or minimization.

Biobased packaging materials

Traditionally, packaging materials for foods, with the exception of paper-based materials, have been based on non-renewable materials. However, due to an increasing focus on sustainability, major efforts have gone into development of renewable materials for production of packaging materials. At present, the market is based on biopolymers such as starch, cellulose, proteins, and lactic acid. So far, most biopolymer-based plastics are used for different purposes such as composting (waste bags), catering (cutlery, drinking cups, plates), agriculture (mulch films, pots), hygiene utensils (diapers, napkins) and very few applications of biobased food packaging materials (Chandra and Rustgi, 1998; Guilbert, 2000).

Application of these novel materials for packaging of foods is challenging, since the demands made upon the packaging materials by the foods are often complex due to specific requirements in terms of oxygen and water vapour permeability, mechanical properties, and safety issues. The commercial use of these materials for packaging of foods is still in its infancy, although this situation is changing along with the increasing efforts put into research and development as well as growing environmental consumer awareness.

Types of biodegradable materials

Biobased packaging materials have been divided into three types, which illustrate their historical development (Nanda, 1996). First-generation materials consist of synthetic polymers and 5-10% starch fillers. These materials do not biodegrade after use, but will biofragment, i.e. they will break into smaller molecules. Second-generation materials consist of a mixture of synthetic polymers and 40-75% starch. Some of these materials are fully biodegradable. Finally, third-generation materials consist of fully biobased and biodegradable materials. Materials from all three categories either are already used for packaging or have considerable potential for this application.

The most commonly available biobased polymers are extracted from agricultural or forest plants and trees. Examples include cellulose, starch, pectins, proteins and also chitosan derived from chitin from the exoskeleton of invertebrates. An interesting property of chitosan in relation to food packaging is its antimicrobial properties (Chandra and Rustgi, 1998). All polysaccharide-based polymers are by nature hydrophilic and are somewhat crystalline, both factors causing processing and performance problems, especially in relation to packaging of moist foods. On the other hand, these polymers constitute materials with medium to excellent gas barriers (Petersen *et al.*, 1999). Today, these biobased packaging materials are produced commercially, (Table 1). Due to the hydrophilic nature of many of these polymers, many commercially available packaging materials are not fully biobased, but are blended with more hydrophobic polymers (first and second generation materials). An example is the starch-based packaging materials, which for many applications are mixed with the synthetic, but biodegradable, polymer such as polycaprolactone (PCL).

Among the polymers synthesized from biomass to date, poly lactic acid (PLA) has the highest commercial potential for large-scale production of renewable packaging materials. PLA is a bio-polyester polymerised from lactic acid monomers. The monomers may be produced by fermentation of carbohydrate feedstock such as corn, wheat, or waste products, e.g. whey, molasses. PLA has a high potential for packaging applications and offers numerous opportunities for tailoring the properties of the final packaging solution. Furthermore, PLA packaging materials can be produced without blending with other polymers. PLA is the first biobased material produced on a larger scale, as evidenced by the production capacity of 180 000 tons PLA per year in 2002 by Cargill Dow.

Table 1 : Selected Manufacturers of biobased packaging materials

Materials	Suppliers	Trade names
Starch and starch-based materials	Novamont (Italy)	Mater-Bi
	EarthShell (USA)	EarthShell
	National Starch (numerous countries)	Eco-FOAM
	Avebe (The Netherlands)	PARAGON
	BIOP Biopolymer (Germany)	BIOPar ®
	Biopac (Austria)	
	Biotec@ (Germany)	BIOPLAST ®
	Plastiroll (Finland)	Bioska
	Apack (Germany)	
	Biologische Verpackung systeme (Austria)	
	Cabot Plastics International (Belgium)	
	Rodenburg Biopolymers (The Netherlands)	Solanyl
Polylactate and polylactate - based materials	Cargill Dow (USA)	Nature Works PLA
	Hycail bv (The Netherlands)	
	Galactic (Belgium)	Galactid
	Trespaphan (Belgium)	Biophan
	Mitsui Chemicals (Japan)	Lacea ® PLA
	Biomer (Germany)	
Poly (hydroxybutyrate) based Materials	Biomer (Germany)	
	Monsanto Metabolix (USA)	Biopol ™
	Zeneca Bio Products (USA)	
	Procter & Gamble (USA)	NODAX ™

Polymers produced directly by natural or genetically modified organisms, consist mainly of the poly hydroxyalkanoates (PHAs) of which poly (hydroxybutyrate) (PHB) is the most common. PHAs are accumulated by a large number of bacteria as energy and carbon reserves. PHAs have a promising potential for food packaging applications, since they possess properties close to those of conventional packaging materials. However, due to exorbitant production costs, few suppliers exist in the market. From 1996 to 1999, PHB was produced by Monsanto under the trade name Biopol ™. However, the production folded, as prices were prohibitive (ten times higher than those of conventional plastics). Production technology and the Biopol ™ trade name were subsequently acquired by Metabolix (USA).

Requirements for biodegradable packaging materials

Protection, which is twofold, is often regarded as the primary function of the package. The package must first protect its contents from outside environmental effects such as water, water vapour, gases, odours, microorganisms, dust, shock, vibrations, compressive forces, etc., and second must impede product spill into the environment (Robertson, 1993). Thus, packaging requirements of foods are complex. Unlike inert packaged commodities, foods often constitute dynamic systems with limited shelf-life and specific packaging requirements. In addition, since foods are consumed to sustain life, the need to guarantee safety is a critical dimension of food packaging requirements. Biobased packaging materials must meet the food criteria applying to conventional packaging materials. Consequently, these materials must remain unchanged and function properly until time of disposal. These criteria relate to mechanical and barrier properties (oxygen, carbon dioxide, water, light, aroma). Furthermore, safety aspects (migration), resistance properties (temperature and chemical resistance), production requirements (welding and moulding properties), convenience, optical properties (transparency and haze) and marketing requirements (communication, marking, and printing properties) are important criteria of food packaging materials (Robertson, 1993; Haugaard *et al.*, 2001a).

Contrary to packaging materials based on non-renewable materials, future requirements for biobased materials should include full post-use compostability. Therefore, environmental conditions conducive to biodegradation must be avoided during product shelf-life, whereas optimised conditions for biodegradation are desirable after discarding. The most important parameters for controlling stability of the biobased packaging materials include appropriate water activity, pH, nutrients, oxygen, storage time, and temperature. Another important aspect is the potential risk of undesirable mould and bacteria growth on the packaging, since some organisms may utilise the biobased materials as an energy source. Furthermore, exterior microbial migration may occur, thereby contaminating the food.

Before using biobased materials for food packaging, the aspects listed should be evaluated in order to meet quality and safety requirements of foods and to comply with legal standards. In this context, testing of shelf-life under realistic storage conditions and evaluation of durability and migration aspects of the packaging materials are of vital importance (Haugaard *et al.,* 2001a).

Barrier properties

The proper gas barriers depend on the foods to be packaged. For instance, a high gas barrier is required for packaging of foods such as meat products, whereas a relatively low gas barrier is essential when packaging fruits and vegetables, since these produce carbon dioxide during respiration. This carbon dioxide must subsequently exit from the packages to avoid internal anaerobic conditions. Additionally, a high water vapour barrier is required when packaging dry products in order to avoid water absorption by the foods, and in the case of high-moisture products to prevent moisture loss.

Packaging films based on PLA, blends of wheat starch and PCL, as well as blends of cornstarch and PCL, possess oxygen transmission rates in the range of 630-2640 ml/m^2/24 h, which is lower than those of LDPE and HDPE films measured at similar conditions and thicknesses, being 8800-9200 ml / m^2/24 h (Petersen *et al.,* 2001). Thus, these biobased packaging films provide a better protection against oxygen than LDPE and HDPE. Oxygen permeabilities of cups based on PLA and PHB are lower than those of polypropylene (PP), polystyrene (PS), and polyethylene (PE) (Petersen *et al.,* 2001). In general, the PLA properties resemble those of PS. Furthermore, PLA can be plasticised in many ways, resulting in properties which will match those of polymers such as LDPE, linear low-density polyethylene (LLDPE), and PS. Thus, PLA seems to be a very promising food packaging polymer.

A major challenge is the hydrophilic behaviour of many biobased polymers, as many foods demand materials resistant to moist conditions. However, when comparing the water vapour permeability of various biobased packaging materials to synthetic plastics , it becomes evident that if a high water vapour barrier is required, *few* biobased materials apply. Development efforts are currently directed at this challenge, and hence, future biobased materials must mimic the water vapour barriers of conventional materials used today.

Increasing the amounts of starch in the packaging materials will result in increased water vapour permeability (Nanda, 1996). The water vapour permeabilities of blends of starch and oil-based films and cups have been compared to those *of* LDPE and HDPE films as well as PP, FE, and PS cups of identical thickness. Results showed that the water vapour permeabilities of the starch-based films were 4-6 times greater than those of conventional films. The starch-based cups had water vapour permeabilities of 100-300 times greater than those of PP and PE cups, and 5-9 times greater than those of PS cups.

PLA provided a better water vapour barrier than did the conventional materials evaluated, i.e. *four* times higher than the conventional films, two times higher than PS cups, and 40-60 times higher than PP and PE cups (Petersen *et al.*, 2001). Petersen *et al.* (2001) found water vapour permeabilities of PHB cups in the range of starch-based cups.

Aroma barrier is an important food packaging parameter. However, data is lacking on the aroma permeability of most packaging materials (Robertson, 1993). Furthermore, light barrier properties are important, but for most polymers these properties can be modified to match the requirements of the foods. With respect to the resistance of the packaging materials towards UV light, a study has shown that the resulting decrease in physical integrity and degradation of the polymer were much lower for PLA than for PE (Ho and Pometto III, 1999).

Mechanical properties

Most biobased polymer materials have mechanical properties similar to conventional polymers. Thus, mechanical properties can be altered by the processing method and selection of raw materials. This indicates that PS and polyethylene terephthalate (PET)-like polymers (relatively stiff materials) and PE-like polymers (relatively flexible polymers) can be found among the available biobased polymers (Van Tuil *et al.*, 2000). Tensile strength, percentage elongation, and tear strength of starch/PCL films and PLA films were found to be lower than for LOPE and HOPE films (Petersen *et al.*, 2001). PHB resembles isotactic PP with respect to mechanical behaviour (Van Tuil *et al.*, 2000). A negative effect on the mechanical properties was pinpointed when the starch content was increased in LDPE/starch blends (Nanda, 1996).

Thermal resistance

The usage temperature range of biobased materials with respect to food packaging is limited. For instance, stability of PLA is poor at elevated temperatures and a supplier of the two commercial products, PLA and PHB, states that the usage temperatures range from -100C to 50°C and from -30°C to 120°C, respectively (Biomer, 2002). Starch/PE blends are relatively more stable (Nanda, 1996).

Migration

Migration is an important aspect to consider when employing packaging materials for foods, and the overall migration may not exceed the 10 mg/dm^2 limit. In general, because of the natural origin of the polymers and the fact that some of the raw materials are present in the foods, i.e. starch and lactic acid, many of the biobased packaging materials are considered safe for food packaging purposes. However, it is imperative that the final food packaging material be safe.

Using biobased biodegradable packaging with particular foods

Usage of biobased biodegradable packaging materials within the food industry remains limited. Some examples of their commercialization are listed in Table 2 below.

Table 2 : Examples of commercial use of biobased packaging materials for foods

Packaging materials	Products	Example of producers	Example of end-users
Polylactate (PLA) cups	Organic yoghurt Fresh salads		Danone (Germany) McDonald's (Sweden)
PLA-coated paper cups and lids	Hot and cold beverages	Huhtamaki Van Leer (Finland)	Coca-Cola Co.: Olympic Games, Salt Lake City (USA)
Clear PLA cups and lids	Cold beverages	Autobar (France) Bartling GmbH (Germany) Fabri-Kal Corporation (USA) Fardis (Belgium) Rexam Thin Wall Plastics (Sweden) Termoplast S.r.L. (Italy) Biocorp (USA)	
PLA punnets	Fruit and produce	Autobar (Spain)	
PLA trays and films	Fresh foods and pasta		Supermarket chain: IPER (Italy)
PLA trays	Mushrooms		Modellprojekt Kassel, e.g. Tegut supermarket chain (Germany)
PLA in general	Plates, cutlery and straws	Biocorp (USA)	

Contd...

PLA films	Bell peppers and cookies	Trespaphan (Belgium)	Modellprojekt Kassel, e.g. Tegut supermarket chain (Germany)
Starch and starch-based materials	Hamburgers and sandwiches Dry products	Apack (Germany) Biologische Verpackungsysteme (Austria)	McDonald's (USA)
Starch, limestone, and fibre mixtures	Plates, bowls, food service wraps and salad containers Dry pasta	Apack (Germany)	Used in Italy
Starch and paper/pulp mixtures	Meat products French fries and chips	Apack (Germany)	Used in Belgium
Starch films laminated on paper	Cutlery, cups and plates		McDonald's (Sweden and Austria) Fast food restaurants Canteens
Starch and fibre-based trays	Organic fruits and vegetables	Apack (Germany)	Sainsbury supermarket chain (UK)
Reinforced nets of starch blends	Onions	Novamont (Italy)	Modellprojekt Kassel, e.g. Tegut supermarket chain (Germany)

Future trends

Biobased materials are not expected to replace conventional materials on a short-term basis. However, due to their renewable origin, these are indeed the materials of the future (Weber, 2002). In the long term, the biobased materials are expected to compete with the conventional ones, when the properties and costs resemble conventional materials.

Packaging of high-quality products such as organic products, where extra material costs can be justified, may be a starting point. In the short term, biobased materials will most likely find usage for foods requiring short-term chill storage due to the biodegradability of the materials. Examples include fruits and vegetables, since biobased polymers facilitate production of films with variable carbon dioxide, oxygen, and moisture permeabilities (Haugaard *et al.,* 2001a). The high carbon dioxide to oxygen ratio of selected biobased packaging materials is interesting in relation to respiring foods. Cheeses, fruits, and vegetables produce carbon dioxide, which accumulates in the package headspace, resulting in inflation of the package. Therefore, using bio-based materials for such products may result in improved quality and longer shelf-life. Biobased packaging for foods may also provide new possibilities for modified atmosphere packaging. Biobased packaging may create new atmosphere conditions, which improve the quality of specific food products during shelf-life, e.g. vegetables (Petersen *et al.,* 1999).

High-value foods such as mushrooms, minimally processed salads, unpasteurised orange juice, organic foods with short shelf-life, etc. constitute a potential food group for biopackaging, since these foods typically are expensive and targeted at consumers who are willing to pay a premium for such products.

With biodegradability being a main issue, potential applications include packaging for short shelf-life foods such as fast food packaging or egg boxes, fresh or minimally processed fruits and vegetables, dairy products, organically grown products, etc. (Guilbert, 2000).

There is great potential for using PLA for food packaging. It is expected that PLA will be suitable for dairy products (other than yoghurt) to include milk, sour cream, and fresh and ripened cheeses. PLA, PHB, and paperboard cartons coated with PLA or PHB have been suggested for dairy products with short shelf-life (Haugaard *et al.,* 2001a). Cargill Dow recommends the use of PLA for packaging of candy, fast food products, fresh foods, ice cream, cereals, coffee, and snack foods.

The novel hydrophilic, starch-based materials would be best suited for packaging of dry or frozen foods, since the materials do not come into contact with moisture. Other suggestions include starch-based trays for packaging of fresh meats and trays and films for fruits and vegetables (Haugaard *et al.,* 2001a). Furthermore, it is now possible to produce starch based foams, which for some applications could replace expanded PS (Bastioli, 2001).

Extensive research and development on biopolymers is required before the materials become competitive alternatives with respect to food packaging.

References

Andersen PJ, Kumar A and Hodson SK (1999). Inorganically Filled Starch Based Fiber Reinforced Composite Foam Materials for Food Packaging', Mat Res Lnnovat, 3: 2-8.

Barron C, Varoquaux P, Guilbert S, Gontard N and Gouble B (2001). Modified Atmosphere Packaging of Cultivated Mushroom (Agaricus Bisporus L) With Hydrophilic Films', J Food Sci, 66 (8): 251-255.

Bastioli C (2001). Global Status of the Production of Biobased Packaging Materials', Starch/ Starke, 53: 351-355.

Blomer (2002). Http://Www.Biomer.Delmechdate.Html (Accessed November 2002).

Chandra R, and Rustgi R (1998). Biodegradable Polymers', Prog Polym Sci, 23:1273-1335.

Frederiksen CS, Haugaard VK, Poll L and Miquel Becker E (2003). Light-Induced Quality Changes In Plain Yoghurt Packed In Polylactate And Polystyrene', Eur Food Res Technol Dol 10.1007/S 00217-003-0722-3.

Guilbert S (2000). Edible Films And Coatings And Biodegradable Packaging', Bulllnt Dairy Fed, 346: 10-16.

Haugaard VK and Festersen RM (2000). Biobased Packaging Materials for Foods', In Weber CJ, Conference Proceedings. The Food Biopack Conference, Copenhagen, 27-29 August, 119-120.

Haugaard VK, Udsen AM, Mortensen G, Hoegh L, Petersen K and Monahan F (2001a). Potential Food Applications of Biobased Materials. An EU-Concerted Action Project', Starch/Starke, 53: 189-200.

Haugaard VI, Festersen RM and Bertelsen G (2001b). Light Induced Changes In Orange Juice', In Oestergaaid S, Conference Proceedings. 2nd Nordic Foodpack Conference, Stavanger, September 5-7. Taastrup, Danish Technological Institute.

Ho KLG, and Pometio III AL (1999). Effects of Electron-Beam Irradiation and Ultraviolet Light (365 DID) on Polylactic Acid Plastic Films', J Environ Polym Degrad, 7(2): 93-100.

Holton EE, ASP EH and Zoitola EA (1994). Corn-Starch-Containing Polyethylene Film Used As Food Packaging', Cereal Foods, World, 39: 237-241.

Krochta JM and DE Mulder-Johnston C (1997). Edible and Biodegradable Polymer Films: Challenges and Opportunities', Food Technol, 51(2): 61-74.

Miller KS and Krochta JM (1997). Oxygen And Aroma Barrier Properties of Edible Films: A Review', Trends Food Sci Technol, 8(7): 228-237.

Nanda SK (1996). Biodegradability And Starch Based Plastics. Trends In Carbohydrate Chemistry, 3: 99-109.

Petersen K, Nielsen PV, Bertelsen G, Lawther M, Olsen MB and Mortensen G (1999). Potential of biobased Materials for Food Packaging', Trends Food Sci Technol, 10: 52-68.

Petersen K, Nielsen PV and Olsen MB (2001). Physical and Mechanical Properties of Biobased Materials', Starch/Starke, 53: 356-361.

Robertson GL (1993). Food Packaging. Principles and Practice. New York, Marcel Dekker.

Sodergard A (2000). Lactic Acid Based Polymers for Packaging Materials for the Food Industry', In Weber CJ, Conference Proceedings. The Food Biopack Conference, Copenhagen, 27-29 August, 14-19.

Sodergard A and Stolz M (2002). Properties of Lactic Acid Based Polymers and their Correlation with Composition', Prog Polym Sci, 27: 1123-1163.

Van Tuil R (2000). Converting Biobased Polymers into Food Packaging', In Weber CJ, Conference Proceedings. The Food Biopack Conference, Copenhagen, 27-29 August, 28-30.

Weber CJ, Haugaard V, Festersen R and Bertelsen G (2002). Production and Applications of Biobased Packaging Materials for the Food Industry', Food Addit Contam, 19, Supplement, 172-177.

□□□

28

Application of Nanotechnology in Packaging of Food Products

M.R. MANIKANTAN

Food packaging is a critical technology addressing the ever-increasing demands for convenience, freshness, ease, shelf-life, safety and security of food products. The main purpose of food packaging is to protect the food from microbial and chemical contamination, oxygen, water vapor and light. The type of packaging used therefore has an important role in determining the shelf-life of the food. Nanotechnology can be intervened to achieve the purpose of food packaging and it will become one of the most powerful forces for innovation in the food packaging industry.

Nanoscience and nanotechnology are new approaches to research and development that concern the study of phenomena and manipulation of materials at atomic scales, where material properties differ significantly from those at a larger scale. The research focuses on understanding, observing and controlling the properties of matter with lengths between 1 and 100 nanometers (1 nanometer = 10^{-9} meter).

Nanotechnology already formed the basis for a number of commercially available products. The technology has been widely used by the electronics industry for many years. Even agricultural production and food industries have witnessed the use of nanotechnology that it can be integrated into number of food systems and food packaging products (Manikantan and Varadharaju, 2007). The use of nanotechnology in Food Science and Technology is shown in Fig. 1.

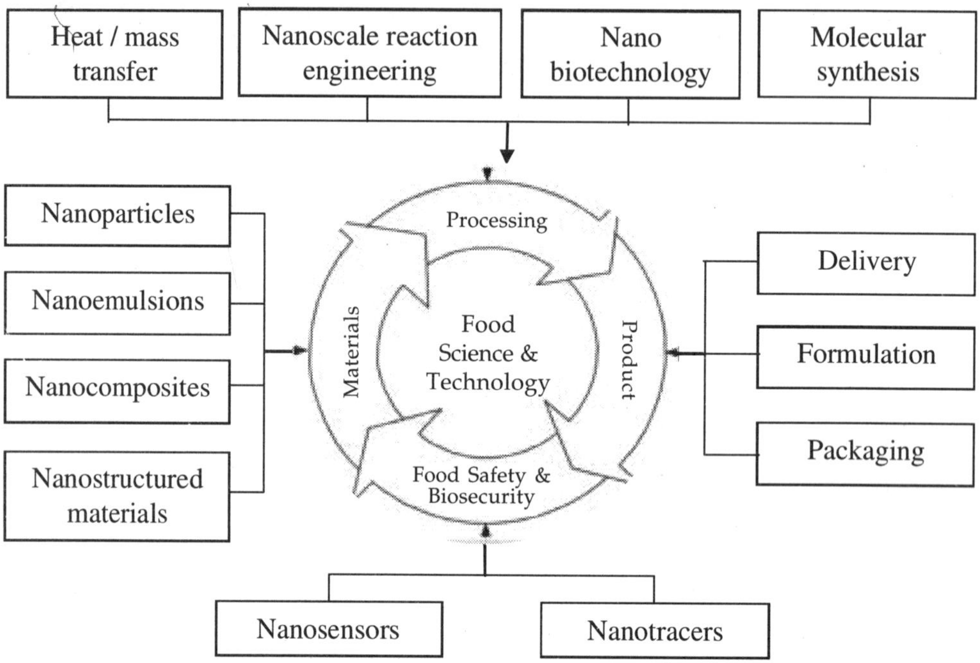

Fig. 1 : Use of nanotechnology in Food Science and Technology

In the food packaging arena, nano materials are being developed with enhanced mechanical and thermal properties to ensure better protection of food from exterior mechanical, thermal, chemical or microbiological effects. This would endow packaged food with an additional level of safety and functionality. Also it would offer advantages along the supply chain and potentially increase the shelf-life of food. Some of the potential uses of nanotechnology in food packaging include modifying the permeation behaviour of films, increasing barrier properties, improving mechanical and heat resistance properties, developing active anti-microbial and antifungal surfaces, sensing and signaling microbiological and biochemical changes. Materials exhibiting anti microbial properties by nano particles of silver have already entered the market.

Nanotechnology enables the designers to alter the structure of the packaging materials on the molecular scale, to give the materials desired properties. With different nanostructure, the plastics can obtain various gas/ water vapor permeability to fit the requirements of preserving fruit, vegetable, beverage, wine and other food. By adding nanoparticles, people can also produce bottles and packages with more light- and fire-resistance, stronger mechanical and thermal performance and less gas absorption. These properties can significantly increase the shelf life, efficiently preserve flavour and colour, and facilitate transportation and usage. Further, nanostructured film can effectively prevent the food from the invasion of bacteria and microorganism and ensure the food safety.

About 400-500 nano-packaging products are estimated to be in commercial use at the moment, while nanotechnology is predicted to be used in the manufacture of 25 percent of all food packaging within the next decade. The contributing factors to these developments include significant benefits in terms of lightweight but strong packaging materials and prolonged shelf life of packaged foodstuffs, and the likely low risk to the consumer attributable to the fixed or embedded nature of engineered nano materials (ENMs) in plastic polymers. A number of nanotechnology-derived food contact materials (FCMs) are currently available worldwide, the main areas of application of which fall into the following broad categories:

- FCMs incorporating nanomaterials for improved packaging properties (flexibility, gas barrier properties, temperature/ moisture stability);
- Biodegradable polymer–nanomaterial composites, with enhanced mechanical and functional properties.
- "Active" FCMs incorporating nanoparticles with antimicrobial or oxygen scavenging properties;
- "Intelligent" and "Smart" food packaging, which incorporates nanosensors to monitor and report the condition of the food.

Nanoparticles reinforced materials for improved packaging properties

The most prominent products in the food industry's research and development pipeline include new polymer nanocomposites for packaging and wrapping. Nanocomposite technology has been described as the next great frontier of material science in packaging. This technology is developed to improve barrier performance pertaining to gases such as oxygen and carbon dioxide. It also enhances the barrier performance to ultraviolet rays, as well as adding strength, stiffness, dimensional stability, and heat resistance. (Manikantan *et al.*, 2009).

Polymer composites with nanoclay

These are among the first nanocomposites to emerge on the market as improved materials for packaging (including food packaging). Nanoclay has a natural nanoscaled layer structure, which when incorporated into polymer composite restricts the permeation of gases. Nanoclay–polymer composites have been made from a thermoset or thermoplastic polymer reinforced with nanoparticles of clay. These include polyamides (PA), nylons, polyolefins, polystyrene (PS), ethylene-vinylacetate (EVA) copolymer, epoxy resins, polyurethane, polyimides and polyethyleneterephthalate (PET).

Polymer nanocomposites consist of resins (either thermoset or thermoplastics) that have fillers added with at least one dimension measured in nanometres. Because the nanoparticles are so small and their aspect ratios (largest dimension/smallest dimension) are very high, even at such low loadings, certain polymer properties can be greatly improved without the detrimental impact on density, transparency, and processability associated with conventional reinforcements like talc or glass. New plastics created with this technology demonstrate an increased shelf life and are less likely to shatter. Once perfected, these plastics will offer the improved characteristics at competitive price and more attractive for use in food and beverage packaging and pharmaceutical applications.

Preparations methods of nanocomposite are solution included intercalation, in-situ polymerization and melt processing. Melt processing is done simultaneously when the polymer is being processed through an extruder, injection molder, or other processing machine. The polymer pellets and clay are pressed together using shear forces to help with exfoliation and dispersion. With in-situ polymerization, clay is added directly to the liquid monomer during the polymerization stage. In the solution included intercalation method, clay is added to a polymer solution using solvents to integrate the polymer and clay molecules.

There are a number of nanoclay–polymer composites available commercially. Table 1 gives the details of commercial application of polymer clay nanocomposites. Known applications of nanoclay in multilayer film packaging include bottles for beer, carbonated drinks and thermoformed containers. Some large breweries are reported to be using the technology already in their beer bottles.

Chemical giant Bayer produces a transparent plastic film (called Durethan) containing nanoparticles of clay. The nanoparticles are dispersed throughout the plastic and are able to block oxygen, carbon dioxide and moisture from reaching fresh meats or other foods. The nanoclay also makes the plastic lighter, stronger and more heat-resistant.

Table 1: Commercial application of Polymer-Clay Nanocomposites

Supplier (Trade Name)	Base Resin	Reinforcement	Commercial Applications
Bayer AG, Germany (Durethan LDPU)	Nylon 6	Organo-clay	Barrier films
Clariant Technologies	Polypropylene (PP)	Organo-clay	Packaging
General Motors R&D, Basell Polyolefins and Southern clay product	Thermoplastic Polyolefin (TPO)	Organo-clay	Auto body-side claddings,step assist for 2002 General Motors Safari and Astro vans
Ube Industries, Japan	Nylon 6 and Nylon 6/66 copolymer	Organo-clay	Barrier films, Timing belt cover for Toyota Motors
Honeywell (Aegis)	Nylon 6	Organo-clay	Bottles and film
Mitsubishi Motors	Nylon	Organo-clay	Engine cover on its GDI models
Kabelwerk Eupen of Belgium	Ethylene vinyl acetate copolymer	Organo-clay	Wire and cable
Nanocor (Imperm)	Nylon 6	Organo-clay	Molding, PET beer bottles
RTP	Nylon 6	Organo-clay	Multi-purpose

Until recently, industry's quest to package beer in plastic bottles (for cheaper transport) was unsuccessful because of spoilage and flavour problems. Nanocor, a subsidiary of Amcol International Corp., is a world leader in the production of nanocomposites, which are polymers bonded with nanocrystals to provide the materials with enhanced properties. The company is currently producing nanocomposites for use in plastic beer bottles that give the contents a six-month shelf- life. The material works by introducing nanocrystals into the plastic that essentially create a maze from which oxygen molecules find it difficult to escape (Fig. 2). Nanocor and another company, Southern Clay, are working on improvements expected to increase shelf life up to 18 months.

In addition to addressing spoilage and flavour issues, nanocomposites also offer other benefits like lighter weight and better recyclability, and suggest that producers may be able to significantly reduce transportation and production costs by reducing the amount of material used to package items.

Nanocomposites, at the present time, are more expensive than most plastic resins but the price continues to drop and it is expected that nanocomposites materials will be very cost competitive in the near future.

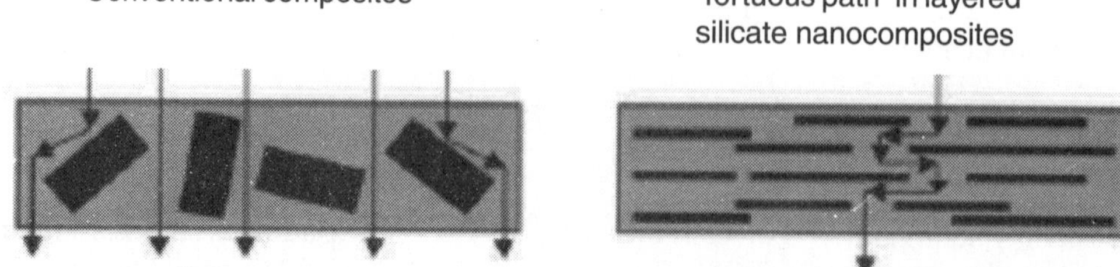

Fig. 2: Formation of tortuous path in polymer clay nanocomposite

Nano biodegradable packaging

The use of nanomaterials to strengthen bio plastics (plant-based plastics) may enable bio based nanocomposite plastics to be used instead of fossil-fuel based plastics for food packaging and carry bags. Bio based nanocomposites are a new class of materials in food packaging industry with improved barrier and mechanical properties as compared to those of neat biopolymers. They are biodegradable and they are also produced from renewable resources. So, these make them environment friendly. It should be noted that barrier properties and especially mechanical properties of bio based nanocomposite films are stronger than edible films and synthetic polymeric films. Unlike edible films, they could not have been consumed as a part of food.

Bio based nanocomposites are produced from incorporation of nanoclay into biopolymers (or Edible films). Advantages of bio based nanocomposites are numerous and possibilities for application in the packaging industry are endless. Bio based nanocomposites can be used to extend the shelf-life of the fresh products such as fruits and vegetables by controlling respiratory exchange. Also it can improve the quality of fresh, frozen, and processed meat, poultry, and seafood products by retarding moisture loss, reducing lipid oxidation and discoloration, enhancing product appearance, and reducing oil uptake by battered and breaded products during frying.

Bio based nanocomposites are composed of biopolymer, nanoclay and usually compatibilizing agents. Major component of bio based nanocomposites is biopolymers. Biopolymers have great commercial potential for bio plastic and edible films, but some of the properties such as brittleness, low heat distortion temperature; high gas permeability and low melt viscosity for further processing restrict their use in a wide range of applications (Ray and Bousmina, 2005). As mentioned earlier, modification of biopolymers with nanotechnology is an effective way to improve their properties. Biopolymers derived from renewable resources are broadly classified according to method of production. This gives the following three main categories (Petersen *et al.*, 1999):

(i) Biopolymers directly extracted/removed from natural materials (mainly plants) such as hydrocolloids (polysaccharides and proteins). The most frequently utilized polysaccharides were cellulose and starch (and their derivatives), chitosan, seaweed extracts (carrageenans and alginates), exudates (arabic gum), seed (guar gum), xanthan and gellan gum and pectin. Proteins include collagen, gelatin, casein, whey proteins, corn zein, wheat gluten and soy proteins.

(ii) Biopolymers produced by classical chemical synthesis from renewable bio derived monomers like polylactate.

(iii) Biopolymers produced by microorganisms or genetically transformed bacteria like Polyhydroxyalkanoates.

Hence, biopolymers which can be used in bio based nanocomposites formulation are numerous.

The utilization of special compatibilizing agents (modifier) between the two basic materials (biopolymer and nanoclay) for the preparation of biobased nanocomposite is necessary. Layered silicates are characterized by a periodic stacking of mineral sheets with a weak interaction between the layers and a strong interaction within the layer. The space in-between the layers is occupied by cations. By cation exchange reactions between the clay and organic cations (such as alkyl ammonium salts), the layered silicate can be transformed into organically modified clay. The inter-layer distance will increase by using voluminous modifiers. If this modifier is compatible with biopolymer as well, a homogeneously and nanoscaled distribution (exfoliation) of the clay sheets can be effected in the polymer matrix. The pure clay shows an interlayer distance of 1.26 nm. It has been proven by X-ray diffraction (XRD) analysis that most of the layers are indeed *swollen* after the modification reaction. The inter-layer distance changes to 2.34 nm, an increase of nearly 100% compared to the pure clay.

Biopolymers like starch present some drawbacks, such as the strong hydrophilic behavior (poor moisture barrier) and poorer mechanical properties than the conventional non-biodegradable plastic films used in the food packaging industries (Huang *et al.*, 2006; McGlashan and Halley, 2003; Park *et al.*, 2003; Park *et al.*, 2002). So, incorporation of nanoclay in biopolymers like starch can improve its properties such as barrier and mechanical properties (Vaia *et al.*, 2000). The most commonly used nanoclays include montmorillonite, a 2:1 phyllosilicate (Chiou *et al.*, 2005).

Antimicrobial nanopackaging materials

The incorporation of antimicrobial compounds into food packaging materials has received considerable attention. Films with antimicrobial activity could help control the growth of pathogenic and spoilage microorganisms. An antimicrobial nanocomposite film is particularly desirable due to its acceptable structural integrity and barrier properties imparted by the nanocomposite matrix, and the antimicrobial properties contributed by the natural antimicrobial agents impregnated within. Materials in the nanoscale range have a higher surface-to-volume ratio when compared with their microscale counterparts. This allows nanomaterials to be able to attach more copies of biological molecules, which confers greater efficiency. Nanoscale materials have been investigated for antimicrobial activity so that they can be used as growth inhibitors, killing agents or antibiotic carriers.

Polymer composites with nano-metals or metal oxides

Polymer nanocomposites incorporating metal or metal oxide nanoparticles are utilized mainly for their antimicrobial action, abrasion resistance, UV absorption, and strength. Some nano materials have been used to develop active packaging that can absorb oxygen and therefore keep food fresh. Other nano materials have been incorporated as UV absorbers to prevent UV degradation in plastics such as PS, PE and PVC. The commercially important nanomaterials in this respect include nanosilver and nanozinc oxide for antimicrobial action, nanotitanium dioxide for UV protection in transparent plastics, nanotitanium nitride for mechanical strength and as a processing aid, and nano silica for surface coating.

The most common nanocomposites used as antimicrobial films for food packaging are based on silver, which is well known for its strong toxicity to a wide range of microorganisms, with high temperature stability and low volatility. Some mechanisms have been proposed for the antimicrobial property of silver nanoparticles (Ag-NPs): adhesion to the cell surface, degrading lipopolysaccharides and forming "pits" in the membranes, largely increasing permeability; penetration inside bacterial cell, damaging DNA; and releasing antimicrobial Ag^+ ions by Ag-NPs dissolution. The latter mechanism is consistent with findings by Kumar and Munstedt (2005), who affirmed that the antimicrobial activity of silver-based polymers depends on releasing of Ag^+, which binds to electron donor groups in biological molecules containing sulphur,oxygen or nitrogen.

It is important to note that the surface biocides, such as nano silver, in packaging materials are not intended to have a preservative effect on the food. Instead, the biocidal agent is intended to help maintain the hygienic condition of the surface by preventing or reducing microbial growth. Where the use of a nano material gives a preservative effect in the packaged product, there would be a requirement for additional regulatory authorization as a direct food additive in most countries. Based on the antimicrobial action of nano silver, a number of "active" FCMs have been developed that are claimed to preserve the food materials by inhibiting the growth of micro-organisms. Examples include food storage containers and plastic storage bags. Nano silver has also been incorporated into the inner surface of some domestic refrigerators to prevent microbial growth and maintain a clean and hygienic environment in the fridge. The discovery of antimicrobial properties of nano zinc oxide and nano magnesium oxide at the University of Leeds may provide more affordable materials for such applications in food packaging (Zhang *et al.*, 2007). A plastic wrap containing nano zinc oxide is also available, which is claimed to sterilize under indoor lighting.

Nanoscale chitosan also has antibacterial activity. One possible antimicrobial mechanism by nanoscale chitosan involves interactions between positively charged chitosan and negatively charged cell membranes, increasing membrane permeability and eventually causing rupture and leakage of intracellular material. Another two antimicrobial mechanisms suggest: chelation of trace metals by chitosan, inhibiting enzyme activities; and in fungal cells, penetration through the cell wall and membranes to bind DNA and inhibit RNA synthesis. Carbon Nano Tubes (CNTs) have been also reported to have antibacterial properties. Direct contact with aggregates of CNTs was demonstrated to be fatal for E. coli, possibly because the long and thin CNTs puncture microbial cells, causing irreversible damages.

Coatings containing nanoparticles

Coatings that contain nanoparticles are used to create antimicrobial, scratch resistant, anti-reflective, or corrosion-resistant surfaces. This involves the coating of nanoparticulate form of a metal, metal oxide or a film resin substance with nanoparticles. Examples of FCMs with nanocoating include antibacterial kitchenware, cutting boards and teapots. High-barrier nanocoatings have also been developed that contain numerous nanodispersed platelets per micron of coating thickness to increase the barrier properties of PET; this enhances the oxygen barrier when used in food and drink applications, ensuring longer shelf life. The coatings have been reported to be very efficient at keeping out oxygen and retaining carbon dioxide and can rival traditional active packaging technologies such as oxygen scavengers (Garland, 2004). Examples include a nanocoating which is an aqueous-based nanocomposite barrier coating that provides an oxygen barrier with a 1–2 micron coating for food packaging use, and plasma arc deposition of amorphous carbon inside PET bottles as a gas barrier.

Waxy coating is used widely for some foods such as apples and cheeses. Recently, nanotechnology has enabled the development of nanoscale edible coatings as thin as 5 nm thick, which are invisible to the human eye. These edible nano-coatings could be used on meats, cheese, fruits, vegetables, confectionery, bakery goods and fast foods. They could provide a barrier to moisture and gas exchange, act as a vehicle to deliver colours, flavours, antioxidants, enzymes and anti-browning agents and could also increase the shelf life of manufactured foods, even after the packaging is opened. The U.S. Company Sono-Tec Corporation announced in early 2007 that it has developed an edible antibacterial nano-coating, which can be applied directly to bakery goods. It is currently testing the process with its clients.

Another trend in this respect is the chemical release nano-packaging. This technique enables food packaging to interact with the food. The exchange can be processed in both directions. Packaging can release nanoscale antimicrobials, antioxidants, flavours, fragrances or nutraceuticals into the food or beverages to extend its shelf life or to improve its taste or smell.

In many instances chemical release packaging also incorporates surveillance elements, where the release of nano-chemicals will occur in response to a particular trigger event. Conversely, nano-packaging using carbon nanotubes are being developed with the ability to pump out carbon dioxide that would otherwise result in food or beverages deterioration. Nano-packaging that absorbs undesirable flavours is also in development stage.

The chemical release packaging is also designed to release biocides in response to the growth of a microbial population, humidity or other changing conditions. Other packaging and food contact materials incorporate antimicrobial nanomaterials, that are designed not to be released, so that the packaging itself acts as an antimicrobial agent. These products commonly use nanoparticles of silver, although some use nanosize zinc oxide or chlorine dioxide and other materials. Nano magnesium oxide, nano copper oxide, nano titanium dioxie and carbon nanotubes are also predicted for future use in antimicrobial food packaging.

Antimicrobial nano-emulsions

Nano-emulsions have been developed for use in the decontamination of food packaging equipment and in the packaging of food. A typical example is a nanomicelle-based product which is claimed to contain natural glycerine and removes pesticide residues from fruits and vegetables, as well as the oil/dirt from cutlery.

Intelligent packaging concepts based on nanosensors

Nanotechnology has also enabled the development of nanosensors that can be applied as labels or coatings to add an intelligent function to food packaging in terms of ensuring the integrity of the package through detection of leaks (for foodstuffs packed under vacuum or inert atmosphere), indications of time-temperature variations (e.g. freeze-thaw-refreezing), or microbial safety (deterioration of foodstuffs). Examples include an indicator that turns from transparent to blue, informing the consumer that air has entered the modified atmosphere of the packaged materials. For this type of application, nanotechnology-derived printable inks have been developed. One example is an oxygen detecting ink containing light-sensitive titanium oxide (TiO2) nanoparticles, which only detect oxygen when they are "switched on" with UV light. Other conductive inks for ink jet printing based on copper nanoparticles have also been developed (Park *et al.,* 2007).

Food safety also requires confirmation of the authenticity of products. This is where application of nanobarcodes incorporated into printing inks or coatings has shown the potential for use in tracing the authenticity of the packaged product (Han *et al.,* 2001). Food quality indicators have also been developed that provide visual indication to the consumer when a packaged foodstuff starts to deteriorate. An example of such food quality indicators is a label based on detection of hydrogen sulphide, which is designed for use on fresh poultry products. The indicator is based on a reaction between hydrogen sulphide and a nanolayer of silver (Smolander *et al.,* 2004). The nanosilver layer is opaque light brown, but when processed meat starts to deteriorate silver sulphide is formed and the layer becomes transparent, indicating that the food may be unsafe to consume.

Other materials developed for potential food packaging applications are based on nanostructured silicon with nanopores. The potential applications include detection of pathogens in food and variations of temperature during food storage. Another relevant development is aimed at providing a basis for intelligent preservative packaging technology that will release a preservative only when a packaged food begins to spoil (ETC Group, 2004).

Based on applied studies of the surface properties of materials, several types of gas sensors have been developed, which translates chemical interactions between particles on the surfaces into a response signal. Metal oxide gas sensors are one of the most popular type of sensors because of their high sensitivity and stability. Conducting polymers (CPs) or electro active conjugated polymers, which can be synthesized either by chemical or electrochemical oxidation, are very important because of their electrical, electronic, magnetic and optical properties, which are related to their conjugated π electron backbones. Polyene and polyaromatic CPs such as polyaniline (PANI), polyacetylene, polypyrrole (PPy) has been widely studied. Electrochemically polymerized CPs have a remarkable ability to switch between conducting oxidized (doped) and insulating reduced (undoped) state, which is the basis of many applications. Food spoilage is caused by microorganisms, whose metabolism produces gases which can be detected by conducting polymer nanocomposites (CPC) or metal oxides, which can be used for quantification and/or identification of microorganisms based on their gas emissions. Sensors based on CPC consist on conducting particles embedded into an insulating polymer matrix. The resistance changes of the sensors produce a pattern that corresponds to the gas under investigation.

Scientists at Kraft, as well as at Rutgers University and the University of Connecticut, are working on nano-particle films and other packaging with embedded sensors that will detect food pathogens. Referred as "electronic tongue" technology, the sensors can detect substances in parts per trillion and would trigger a colour change in the packaging to alert the consumer if a food has become contaminated or if it has begun to spoil.

Researchers in the Netherlands are going on further to develop intelligent packaging that will release a preservative if the food within begins to spoil. This "release on command" preservative packaging is operated by means of a bio switch developed through nanotechnology.

Georgia Tech in the United States used modified carbon nanotube as biosensor to detect microorganisms, toxic substances and spoilage of foods or beverages. Another example, Opel, which makes Opalfilm incorporating 50nm carbon black nano particles, was used as biosensor that can change colour in response to food spoilage.

Nanotechnology is also enabling sensor packaging to incorporate cheap Radio Frequency Identification (RFID) tags. The nano-enabled RFID tags are much smaller, flexible and can be printed on thin labels. This increases the tags versatility and thus enables much cheaper production. Other nano-based track and trace packaging technologies are also being improved. For instance, a United States company Oxonica Inc, has developed nano-barcodes to be used for individual items or pellets, which must be read with a modified microscope. These have been developed primarily for anti- counterfeiting purposes.

Packaging has developed into an essential technology in the handling and commercialization of foodstuffs to provide, by maintaining or even increasing, the required levels of quality and safety. There are high expectations in food and packaging: longer shelf life, safer packaging, better traceability of products and healthier food is only a few of the expected improvements. Nanotechnology may be effectively applied in the various areas of food packaging such as designing the packaging material according to the barrier requirements of the food materials to be stored, anti microbial coatings to ensure the safety of the packaged food, sensing the reactions inside the packaging environment and utilizing the bio materials as packaging materials to reduce the pollution and intelligent packaging by using nanosensors.

References

Chiou, BS, Yee, E, Glenn, GM and Orts, WJ (2005). Rheology of starch–clay nanocomposites. Carbohydrate Polymers, 59: 467–475.

ETC Group (2004). Food Packaging Using Nanotechnology Methods: an Overview of 'Smart Packaging' and 'Active packaging', http://www.azonano.com/Article ID=1317

Garland, A (2004). Nanotechnology in plastics packaging. Commercial applications in nanotechnology (ed). Pira International Limited UK, 14 -63.

Han, M, Gao, X, Su, JZ and Nie, S (2001).Quantum-dot-tagged microbeads for multiplexed optical coding of biomolecules. Nat. Biotechnol., 19: 631–635.

Huang, M, Yu, J and Ma, X (2006). High mechanical performance MMT-urea and formamide-plasticized thermoplastic corn starch biodegradable nanocomposites. Carbohydrate Polymers, 63: 393–399.

Kumar, R and Munstedt, H (2005). Silver ion release from antimicrobial polyamide/ silver composites. Biomaterials, 26: 2081–2088.

Manikantan, MR, Arumuganathan, T, Balakrishnan, M and Varadharaju, N (2009). Role of Nanotechnology in Food Industry, Beverage and Food World, 41-44.

Manikantan, MR and Varadharaju (2007). Nanotechnology in Food Preservation. In: Nanotechnology Applications in Agriculture. (eds.) Chinnamuthu, C.R., Chandrasekaran, B. and Ramasamy, C., TNAU, Coimbatore, 63-76.

McGlashan, SA and Halley, PJ (2003). Preparation and characterization of biodegradable starch-based nanocomposite materials. Polymer International, 52: 1767–1773.

Park, BK, Kim, D, Jeong, S, Moon, J and Kim, JS (2007). Thin solid films, 2007, 7/16. 515 (19): 7706–7711.

Park, HM, Lee, WK, Park, CY, Cho, WJ and Ha, CS (2003). Environmentally friendly polymer hybrids Part 1. Mechanical, thermal, and barrier properties of thermoplastic starch/clay nanocomposites. Journal of Materials Science, 38: 909–915.

Park, HM, Li, X, Jin, CZ, Park, CY, Cho, WJ and Ha, CS (2002). Preparation and properties of biodegradable thermoplastic starch/clay hybrids. Macromolecular Materials and Engineering, 287: 553–558.

Petersen, K, Nielsen, V, Bertelsen, G, Lawther, M, Olsen, MB, Nilsson, NH and Mortensen, G (1999). Potential of biobased materials for food packaging. Trends in Food Science & Technology, 10: 52-68.

Ray, SS and Bousmina, M (2005). Biodegradable polymers and their layered silicate nanocomposites: In greening the 21st century materials world. Progress in Materials Science, 50: 962–1079.

Smolander, M, Hurme, E, Koivisto, M and Kivinen, S (2004). PCT International Patent Application, WO 2004/102185 A1.

Vaia, RA, In Pinnavaia, TJ and Beall, GW (2000). Polymer–clay nanocomposites (Eds.), New York, Wiley, 229–266.

Zhang, L, Jiang, Y, Ding, Y, Povey, M and York, D (2007). Investigation into the antibacterial behaviour of suspensions of ZnO nanoparticles (ZnO nanofluids). J. Nanoparticle Res., 9(3): 479–489.

□□□

29

Nondestructive Methods for Quality Evaluation of Food

S.N. JHA

Quality conscious consumers nowadays want to get assured about various quality attributes of food items before they purchase. Fruits, vegetables and milk are increasing in popularity in the daily diets in both developed and developing countries. Products' quality and its measurement techniques are thus naturally extremely important. The decisions concerning the constituents, level of freshness, ripeness, and many other quality parameters are based mostly on subjective and visual inspection of the foods' external appearance. Several nondestructive techniques for quality evaluation have been developed based on the detection of various physical properties that correlate well with certain factors of a product. The quality of foods including milk and milk products is mostly based on constituents, purity; *i.e.*, levels of adulterants; color, gloss, flavor, firmness, texture, taste and freedom from external as well as internal defects. Numerous techniques for evaluating these parameters are now available commercially, but most of them are destructive in nature. Internal quality factors of fruits such as maturity, sugar content, acidity, oil content, and internal defects, however, are difficult to evaluate. Methods are needed to better predict the internal quality of fruits, vegetables, constituents of foods and level of adulterants, if any, without destroying the sample. Recently, there has been as increasing interest in nondestructive methods of quality evaluation, and a considerable amount of effort has been made in that direction. But the real problem is how these methods are to be exploited practically and what the difficulties are in implementing them. The objective of the present chapter is thus to give exposure of recent nondestructive methods such as nuclear magnetic resonance, x-ray computed tomography, near-infrared spectroscopy and some other important methods to the stakeholders of food industry in India and to evaluate their pros and cons for suitability in commercial application.

Nuclear magnetic resonance (NMR) techniques

The nuclear magnetic resonance technique, often referred as magnetic resonance imaging (MRI), involves resonant magnetic energy absorption by nuclei placed in an alternating magnetic field. The amount of energy absorbed by the nuclei is directly proportional to the number of a particular nucleus in the sample such as the protons in water oil. The theory of NMR is presented in detail elsewhere (Farrar & Becker, 1971). The basic concepts, types of pulsed experiments and the type of information that can be extracted from these experiments are described. Information on experimentation, assembling hardware, conducting laboratory tests and interpreting the results is also available from Fukushima and Roeder (1981). These authors also provided detailed theory for better understanding of what a scientist should seek and what he might expect to find out by using NMR.

There are many applications of NMR in agriculture (Rollwitz, 1984). The simplest among them is the determination of moisture and oil content (Mousseri *et al.*, 1974, Leung *et al.*, 1976; Miller *et al.*, 1980; Brosio *et al.*, 1978; Rollwitz and Persyn, 1971). But the NMR response many times is not clear and poses problems especially when constituents other than water are present in the material (Steinberg & Richardson, 1986). Besides the established relationship between the moisture and output of NMR experiments, various other facts helpful in determining the quality of food materials without destroying them are available in the literature: Selections of chocolate confectionary products can be made non-invasively by three-dimensional magnetic resonance imaging (Miquel *et al.*, 1998); using a spin echo pulse sequence, 128x64x64 data sets were acquired with either a 5-or 20-ms echo time, 500-ms repetition time and signal averages, in total 2-h scan time. Such images localize and distinguish between the constituents, and visualize both the internal and external structure of matter.

Most perishable food products are now marketed in packaged form. To increase the marketability longer shelf life is needed and this is achieved by freezing and secondary processing of the food. During freezing it is natural that ice will form within the food that may change its characteristics. Ice formation during food freezing can be examined using the NMRI method as the formation of ice has been seen to reduce the spatially located NMR signal. The characteristics of a food can be better controlled as MRI can serve to assess freezing times and the food structure during the freezing process (Kerr *et al.*, 1998). The secondary processing changes almost all characteristics of a food, such as physical and aerodynamic (Jha & Kachru, 1998), thermal and hygroscopic properties (Jha & Prasad, 1993; Jha 1999), which in turn, change its key acceptability factors, i.e. sensory texture and taste. The sensory texture of cooked food such as potatoes has been predicted using the NMRI technique (Thybo *et al.*, 2000). In addition, NMR image intensity, the ratio of the oil and water resonance peaks of the one-dimensional NMR spectrum, and both the spin-lattice relaxation time and spin-spin relaxation time of water in the fruit are correlated with maturity of a fruit like avocado before harvesting (Chen *et al.*, 1993). This important finding has desirable features for high speed sorting using a surface-coil NMR probe that determine the oil/water resonance peak ratio of the signal from one region in an intact fruit.

An on-line nuclear magnetic resonance quality evaluation sensor has recently been designed, constructed and tested (Kim *et al.*, 1999). The device consists of a super-conducting magnet with a 20mm diameter surface coil and a 150 mm diameter imaging coil coupled to a conveyor system. These spectra were used to measure the oil/water ration in avocados and this ratio correlated to percent dry weight. One dimensional magnetic resonance images of cherries were later used to detect the presence of pits inside.

X-ray and computerized tomography (CT)

X-ray imaging is an established technique to detect strongly attenuating materials and has been applied to a number of inspection applications within the agricultural and food industries. In particular, there are many applications within the biological sciences where we wish to detect weakly attenuating materials against similar background material.

X-ray computed tomography (CT) has been used to image interior regions of apples with varying moisture and, to a limited extent, density states (Tollner *et al.*, 1992). The images were actually maps of x-ray absorption of fruit cross sections. X-ray absorption properties were evaluated using normal apples alternatively canned and sequentially freeze-dried, fruit affected by water core disorder, and normal apples freeze-dried to varying levels. The results suggested that internal differences in x-ray absorption within scans of fruit cross-sections are largely associated with differences in volumetric water content. Similarly, the physiological constituents have been monitored in peaches by CT methods in which x-ray absorbed by the peaches is expressed in CT number and used as an index for measuring the changes in internal quality of the fruit (Barcelon *et al.*, 1999). Relationships between the CT number and the physiological contents were determined and it was concluded that x-ray CT imaging could be an effective tool in the evaluation of peach internal quality. In another study, the potential for Compton scattered x-rays in food inspection was evaluated by imaging the density variation across a food material by measuring the Compton scatter profile across a food material by measuring the Compton scatter profile across polystyenespheres with internal voids (MacFarlane *et al.*, 2000). In this study particular attention was paid to simulate the obscuring influence of multiple scatter. The simulated result was found to be in close agreement with the experimental observation. Some experimental test sample of a Perspex block with various embedded soft materials showed that care should be taken to ensure that the transmission image is taken with x-ray within an appropriate energy range (Zwiggelaar *et al.*, 1997). For low Z materials the contrasts between the materials became more pronounced at lower x-ray energies. If more than one soft material has to be distinguished from the surrounding area it may be advantageous to image over a range of x-ray energies.

Visual spectroscopy and colour measurements

Colour measurement is now little bit old technique to check the quality of any items in terms of appearance. It has also been tested for assessing the ripeness of fruits and measurement of aesthetic appearances of dairy products. Recently many works have been reported to correlate the internal quality such as total soluble solids contents, maturity of fruits in tree and sweetness of intact fruits using Hunter colour values and reflectance spectra in visual range of wavelengths (Jha *et al*, 2005). This in fact is possible through rigorous analysis of data and modeling for a huge number of samples of varied nature.

Near-infrared spectroscopy

The use of near-infrared spectroscopy as rapid and often nondestructive technique for measuring the composition of biological materials has been demonstrated for many commodities. This method is no longer new; as it started in early 1970 in Japan (Kawano, 1998), Just after some reports from America. Even an official method to determine the protein content of wheat is available (AACC, 1983). The National Food Research Institute (NFRI), Tsukuba has since become a leading institute in NIR research in Japan and has played a pivotal role in expanding near-infrared spectroscopy technology all over the country (Iwamoto *et al.*, 1995). In Japan, NIR as a nondestructive method for quality evaluation was started for the determination of sugar content in intact peaches, Satsuma orange and similar other soluble solids (Kawano, 1994).

To determine the solid content of cantaloupe Dull *et al.* (1989) used NIR light at 884 nm and 913 nm. Initially the correlation of their findings was poor mainly due to light losses. Later, Dull and Birth (1989) modified the earlier method and applied it to honey-dew melons; the improved methods showed better correlation. Similarly, a nondestructive optical method for determining the internal quality of intact peaches and nectarines was investigated (Slaughter, 1995). Based upon visible and near-infrared spectrophotometer techniques, the method was capable of simultaneously predicting the soluble solid content, sucrose content, sorbitol content, etc. of intact peaches and nectarines was investigated (Slaughter, 1995). Based upon visible and near-infrared spectrophotometer techniques, the method was capable of simultaneously predicting the soluble solid content, sucrose content, sorbitol content, etc. of intact peaches and nectarines, and required no sample preparation.

Now various NIR spectrometers are available and are being used commercially. Some modifications in these available spectrometers, especially for holding the intact samples, are reported (Kawano *et al.*, 1992; 1993). In the same sample holding a test tube for holding liquid food such as milk was also used to determine fat content (Chen *et al.*, 1999). Recently a low cost NIR spectrometer has been used to estimate the soluble solids and dry matter content of kiwifruit (Osborne & Kunnemeyer, 1999). Errors are within the permissible limit and the time requires for obtaining data has been reduced. The influence of sample temperature on the NIR calibration equation was also evaluated and a compensation curve for the sample temperature was developed (Kawano *et al.*, 1995) to rectify the result.

Now detection of almost all adulterants in milk in single stroke (Jha and matsuoka, 2004) and composition of milk and effect of somatic cell count on determination of milk constituents are very accurately determined (Tsenkova *et al.*, 2001). Similarly taste of tomato juice in terms of acid-brix ratio can be determined with high accuracy (Jha and Matsuoka 2004). NIR spectroscopy in fact is the most suited technique for nondestructive analysis of dairy products.

Miscellaneous techniques

Quality attributes such as invisible surface bruises, color, gloss, firmness, density, volume expansion of processed food etc are also important (Jha & Prasad, 1996). Often consumers select food materials, particularly fruits and vegetables by judging these parameters visually. Multiple efforts have been made to determine these parameters visually. A fluorescence technique was used to detect invisible surface bruises on Satsuma mandarins (Uozumi *et al.*, 1987). The authors have also tested this method successfully to know the freshness of cucumbers and eggs and found it very useful for detecting the freshness of agricultural produce.

Matsuoka *et al.* (1995) measured the gloss of eggplant by a spectral radiometer system and found to be a viable parameter for determining freshness. They observed remarkable change in relative spectral reflectance values after 48 h. Later, they compares their evaluation by eye in a sorting house with the integrated results of relative spectral reflectance in the visible range and found that the gloss on the surface differs with light and is caused by round and adhesives substances on the epidermal cells (Matsuoka *et al.*, 1996). A unique gloss meter for measuring the gloss of curved surfaces was used in parallel with a conventional, flat surface gloss meter to measure peel gloss of ripening banana (Ward & Nussinovitch, 1996). Usually banana ripeness is judged by the color of the peel. The new gloss meter is able to measure the peel correctly which helps in predicting the correct time and level of ripening. This is also able to measure the gloss of other fruits and vegetables such as green bell pepper, orange, tomato, eggplant and onion (Nussinovitch *et al.*, 1996).

Glossiness and color, in fact, are the only visual attributes for measuring the quality of fruits and vegetables. Another property that helps a consumer in deciding the quality is firmness. Takao (1988) developed a fruit hardness tester that can measure the firmness of kiwifruit nondestructively. The tester is called a 'HIT counter' after the three words, hardness, immaturity and texture. By just setting the sample in the tester, the amount of change in shape is measured and a digital reading within a few seconds indicates about the freshness. Based on the same principal another on-line prototype HIT counter, fruit hardness sorting machine has also been developed (Takao & Omori, 1991). The relationship between density and internal quality of watermelon can also be determined. An optimum range of density was first determined and then a new automatic density sorting system was develops and then a new automatic density sorting system was developed to measure the hollowness of a watermelons with cavities or deteriorated porous flesh to be removed and permits estimation of the soluble solid content of this fruits. Using gloss and other physical parameters such as stiffness and density, Jha and Matsuoka, 2002 have also determined the freshness of eggplants and have correlated it very easily with the day to day price in vegetable *mandis*.

Neural networks have lately gained in popularity as an alternative to regression models to regression models to characterize the biological process. Their decision-making capabilities can best to used in image analysis of biological products where the shape and size classification is not governed by any mathematical function. Many neural network classifiers have been used and evaluated for classifying agricultural products, but multi-layer neural network classifiers perform such tasks best. Recently one scientist used a gamma- absorption techniques combined with a scanning device for continuous non-destructive crop mass and growth measurement in the field (Gutezeit, 2000). Most important in this study was the accuracy of the measurement, which was found to be in agreement with the direct weighting system. This method has made it possible to assess the reaction of plants and their dependence on environmental factors by growth analysis.

Determination of quality of any food material including milk and milk products is actually a complex problem that requires a variety of specific sensor, more than an accumulation of simple sensor. Various techniques are being tried. IMR, x-ray CT and NIR techniques may be useful for a large volume of work in agriculture, especially for evaluation of qualities such as maturity, internal quality of fruit and conditions of food materials after processing, level of adulterants and useful constituents. These techniques, although give a correct picture and precise measurement of parameters, are not convenient for small business except NIR and visual spectroscopy. Their high cost restricts application to large entrepreneurs and developed countries only.

Two examples of the use of x-ray imaging relevant to the agricultural and food industries have been given, notably in the inspection of vegetables and food materials using low energy x-ray imaging and in the inspection and control of dynamic processes. The x-ray imaging results have been compared with the full three-dimensional information obtained by computer tomography. The CT results show more detail in the test sample than the single transmission image and detail in the inspection of materials of variable shape usually encountered in the agricultural and food industry. The imaging techniques MRI and x-ray CT are able to show only the internal structure of the material, not the composition of nutritional details, whereas NIR and visual spectroscopy techniques are very successfully being used to determine the compositional quality of a food and can be used even at farm. However, it is not yet possible to produce an image of the internal physical quality of fruits and vegetables. All techniques are costly because most of the expertise is imported. Central Institute of Post-harvest Engineering and Technology (CIPHET), Ludhiana has taken the lead by initiating R & D works in the country about four years ago. Dairy and food processing industries and other research organizations should also work together to develop such type of instrumentation indigenously.

References

American Association of Cereal Chemists (1983). Approved methods of the Am. Assoc. of Cereal Chemists, 8th Ed. (March). St. Paul, Minn.

Barcelon, EG, Tojo, S and watanable, K (1999). X-ray computed tomography for internal quality evaluation of peaches. J. Agric. Eng. Res., 73: 323-330.

Brosio, EF, Conti, CL and Sykara, S (1978). Moisture determination in starch-rich food products by pulsed nuclear magnetic resonance. J. Food Technol., 13: 107-116.

Chen, JY, Iyo, C and Kawano, S (1999). Development of calibration with sample cell compensation for determining the fat content of unhomogenised raw milk by a simple near infrared transmittance method. J. Near Infrared Spectosc., 7: 265-273.

Chen, P, McCarthy, MJ, Kauten E, Sarig, Y and Han, S (1993). Maturity evaluation of avocados by NMR methods. J. Agric. Eng. Res., 55: 177-187.

Dull, GG and Birth, GS (1989). Nondestructive evaluation of fruit quality: Use of near infrared spectrophotometry to measure solube solids in intact honeydew melons. Hortscience, 24: 754.

Dull, GG and Birth, GS Smittle, DA and Leffler, RG (1989). Near infrared analysis of soluble solids in intact cantaloupe. J. Food Sci., 54: 393-395.

Farrar, TC and Becker, ED (1971). Pulse and Fourier Transform NMR: Introduction to Theory and Methods. Academic Press, New York.

Fukushima, E and Roeder, SBW (1981). Experimental Pulse NMR. Addison-Wasley Publishing Company, Reading, M.A.

Gutezeit, B (2000). Non-destructive measurement of fresh plant mass by the gamma-scanning technique applied to broccoli. J. Agric. Eng. Res., 75: 251-255.

Iwamoto, M, Kawano, S and Yukihiro, O (1995). An overview of research and development of near infrared spectroscopy in Japan. J. Near Infrared Spectrosc., 3: 179-189.

Jha, SN and Matsuoka T (2002). Development of freshness index of eggplant. Applied Engineering in Agriculture, ASAE, 18 (5): 57-60.

Jha, SN and Matsuoka, T (2004). Detection of adulterants in milk using near infrared spectroscopy. Journal of Food Science and Technology, 41(3): 313-316.

Jha, SN and Matsuoka, T (2004). Nondestructive determination of acid brix ratio (ABR) of tomato juice using near infrared (NIR) spectroscopy. International Journal of Food Science and Technology, 39(4): 425-430.

Jha, SN, Chopra, S and Kingsly, ARP (2005). Determination of sweetness of intact mango using spectral analysis. Biosystems Engineering, 91(2): 157-161.

Jha, SN, Chopra S, and Kingsly, ARP (2007). Modeling of color values for nondestructive evaluation of maturity of mango. Journal of Food Engineering, 78(1): 22-26.

Jha, SN (1999). Physical and hygroscopic properties of makhana. J. Agric. Eng. Res., 72: 145-150.

Jha, SN and Prasad, S (1993). Physical and thermal properties of gorgon nut. J. Food Process Eng., 16: 237-245.

Jha, SN and Prasad, S (1996). Determination of processing conditions of gorgon nut (Euryale ferox). J. Agric. Engg. Res., 63: 103-112.

Jha, SN Matsuoka T (2000). Review: Nondestructive Techniques for quality evaluation of intact fruits and vegetables. Food Science and Technology Research, 6(4): 248-251.

Jha. SN and Kachru, RP (1998). Physical and aerodynamic properties of makhana. J.Food Process.Eng., 79: 301-316.

Kato, K (1997). Electrical density sorting and estimation of soluble solids contents of watermelon. J. Agric. Engg. Res., 67: 161-170.

Kawano, S (1994). Nondestructive near infrared quality evaluation of fruits and vegetables in Japan. NIR News, 5: 10-12.

Kawano, S (1998). New application of nondestructive methods for quality evaluation of fruits and vegetables in Japan. J. Jpn. Soc. Hort. Sci., 67: 1176-1179.

Kawano, S, Abe, H and Iwamoto, M (1995). Development of a calibration equation with temperature compensation for determining the brix value in intact peaches. J. Near infrared Spectrosc., 3: 211-218.

Kawano, S, Fujiwara, T, and Iwamoto, MC (1993). Nondestructive determination of sugar content in satsuma maddarin using NIR transmittance. J. Jpn. Soc. Hort. Sci., 62: 465-470.

Kawano, S, Watanabe, H and Iwamoto, M (1992). Determination of sugar content in intact peaches by near infrared spectroscopy with fibre optics in interactance mode. J. Jpn. Soc. Hort. Sci., 61: 445-451.

Kerr, WL, Kauten, RJ, McCarthy, MJ and Reid, DS (1998). Monitoring the formation of ice during food freezing by magnetic resonance sensors. J. Agric. Eng. Res., 74: 293-301.

Kim, SM, Chen, P, McCarthy, MJ and Zoin, B (1999). Fruit internal quality evaluation using on-line nuclear magnetic resonance sensors. J. Agric. Eng. Res., 74: 293-301.

Leung, HK, Steinberg, MP Wei, LS and Nelson, AI (1976). Water binding of macromolecules determined by pulsed NMR. J. Food Sci., 41: 297-300.

Macfarlane, NJB, Bull, CR, Tillett, RD, Speller, RD, Royle, GJ and Johnson, KRA (2000). The potential for compton scattered x-rays in food inspection: The effect of multiple scatter and sample inhomogeneity. J. Agric. Eng. Res., 75: 265-274.

Matsuoka, T, Miyauchi, K and Sun, D (1995). Basic studies on the quality evaluation of agricultural products-Quantification of gloss of eggplants by spectral characteristics. J. Jpn. Sci. Agric Mach., 57: 33-40.

Matsuoka, T, Miyauchi, K and Yano, T (1996). Basic studies on the quality evaluation of agricultural products (Part 2). The evaluation of colour and gloss decrease on the surface of eggplants. J. Jpn. Sci. Agric. Mach., 58: 69-77.

Miller, BS, Lee, MS, Hughes, JW and Pomernz, Y (1980). Measuring high moisture content of cereal grains by pulsed nuclear magnetic resonance. Cerreal. Cereal Chem., 57: 126-129.

Miquel, ME, Evans, SD and Hall LD (1998). Three dimensional imaging of chocolate confectionary by magnetic resonance methods. Food Sci. technol., 31: 339-343.

Mousseri, J, Steinberg, MP, Nelson, AI and Wei, LS (1974). Bound water capacity of corn starch and its derivatives by NMR, J. Food Sci., 39: 114-116.

Nussinovitch, A, Ward, G and Mey-ial, E (1996). Gloss of fruits and vegetables. Food Sci. Technol., 29: 184-186.

Osborne, SD and Kunnemeyer, R (1999). A low cost system for the grading of kiwifruit. J. Near Infrared Spectocs., 7: 9-15.

Rollwitz, WL (1984). Radio frequency spectroscopy: Versatile sensors for agriculture applications. ASAE Paper No. 84-1590, ASAE. St. Joseph, MI 49085.

Rollwitz, WL and Persyn, GA (1971). On-stream NMR Measurements and control. Am. Oil Chem. Soc. J., 48: 59-66.

Slaughter, DC (1995). Nondestructive determination of internal quality in peaches and nectarines. Trans, ASAE, 38: 617-623.

Tsenkova, R, Atanassova, S, Ozaki, Y, Toyoda, K, and Itoh, K (2001). Near-Infrared spectroscopy for biomonitoring; influence of somatic cell count on cow's milk composition analysis. International Dairy Journal, 11 : 779-783.

Steinberg, MP and Richardson, SJ (1986). Application of nuclear magnetic resonance. Paper No.11. Presented at the International Union Food Science and Technology Symposium on Water Activity: Theory and Application. June 13-14, Dallas, TX.

Takao, H (1988). HIT counter. Noryu Giken Kaihou, 180: 7-9 (in Japanese).

Takao, H and Omori, S (1991). Quality evaluation of fruits and vegetables using light transmittance. Noryu Giken Kaihou, 145: 14-16 (in Japanese).

Thybo, AK, Bechmann, IE, Martens, M and Engelsen, SB (2000). Prediction of sensory texture of cooked potatoes using uniaxial compression, near infrared spectroscopy and low field H NMR Spectroscopy. Food Sci. Technol., 33: 103-111.

Tollner, EW, Hung, YC, Upchurch, BL and Prussia, SE (1992). Relating x-ray absorption to density and water content in apples. Trans. ASAE, 35: 1921-1928.

Uozumi, J, Kawano, S, Iwamoto, M and Nishnari, K (1987). Spectrometric system for the quality evaluation of unevenly couloured food. Nippon Shokuhin Kagyo Gakkaishi, 34: 163-170 (in Japanese).

Ward, G and Nussinovitch, A (1996). Peel as a indicator for banana ripeness. Food Sci. Techol., 29: 289-294.

Zwiggelaar, R, Bull, CR, Mooney, MJ and Czarnes, S (1997). The detection of soft materials by selective energy x-ray transmission imaging and computer tomography. J. Agric. Eng. Res., 66: 203- 21.

□□□

30

Evaluation of Physico-Mechanical Properties of Food Products

M.R. MANIKANTAN

Food grains are one of the essential ingredients of human life. The grains are produced, processed, preserved and utilized for the human consumption. The grain is tested at various stages for its utility. In general, the grains are tested by various methods such as physical, chemical and biological tests. The grain is tested for its physical appearance, chemical composition and microbiological toxicity. The physico-chemical test for grain quality includes its spatial dimensions, shape factors (roundness and sphericity), density factors (bulk density, true density and porosity), texture, colour, water activity and moisture content. The marketability of grain depends on the above parameters.

In a broad sense, the physico-chemical properties of foods may be defined as those properties that lend themselves to description and quantification by physical and chemical means. Their importance stretches from product handling to processing to consumer acceptance. Much of this knowledge has been developed against the objective of determining and quantifying quality factors that govern consumer acceptance, suitability of the product for specific uses and thus its economic value.

The knowledge of important physical properties such as size, shape, volume, surface area, density, porosity, texture, color, etc. of different grains is necessary for the design of various separating and handling, storing and drying systems. The density and specific gravity values are also used for the calculation of thermal diffusivity and Reynolds number.

Various types of cleaning, grading and separation equipment are designed on the basis of physical properties of seeds such as size, shape, specific gravity, surface roughness, colour etc. For designing an air screen grain cleaner, the shape and size of the grain determine the shape and size of screen openings, angle of inclination and vibration amplitude and frequency of screens. The density of the grain decides the size of the screening surface. The frontal area and related diameters and density are essential for determination of terminal velocity of the grain. Terminal velocity is necessary to decide about the winnowing velocity of air blast for the separation of lighter materials in air screen grain cleaners.

The shape of product is an important parameter which affects conveying characteristics of solid materials by air or water. The shape is also considered in the calculation of various cooling and heating loads of food materials.

The frontal area and the related diameters are essential for determination of terminal velocity, Reynold's number and drag coefficient. The density and specific gravity are needed for calculating the thermal diffusivity in heat transfer operations, in determining Reynold's number, in pneumatic and hydraulic handling of the agricultural materials.

The surface characteristics, colour and appearance are exploited for selective separation and storage of fruits and vegetables. Some of the physical test parameters for analyzing grain quality are illustrated below.

Moisture content

Moisture content of a substance is defined as the amount of water present in it. It is an index of the probable keeping quality of the product. The amount of moisture in a product is designated on the basis of the weight of water and is usually expressed in per cent. There are two methods of designating the moisture content, viz; wet basis and dry basis. The moisture content on a wet basis (wb) is obtained by dividing the weight of water present in the material by the total weight of the material.

Moisture content (% wb) = $(W_w \times 100) / (W_w + W_d)$

Where,

W_w = weight of water, g and

W_d = weight of dry matter, g

The moisture content on a dry basis (db) is determined by dividing the weight of water by the weight of dry matter.

Moisture content (% db) = $(W_w / W_d) \times 100$

The moisture content on a wet basis is used for commercial designation whereas the dry basis is mainly used by research workers.

Hot air oven method - For determining the moisture content of grain as prescribed in BIS: 3579-1966 is normally used. For this, 5 ± 0.5 g of the sample is weighed accurately in an electronic weighing balance and the sample is kept in a moisture dish which has been dried previously and cooled in the desiccators. The dish is then placed in the air oven for approximately 24 hours at 102 ± 1°C. Then, the dish is removed from the oven and cooled in the desiccators to room temperature and weighed. This procedure is repeated until the difference between the two successive weighing did not exceed one mg or remains constant. Then the moisture content of the seed is determined using the equation as mentioned above. By this method moisture content of seed can be determined on wet basis.

Sample conditioning and preparation

The sample of desired moisture content has to be prepared for various experiments. Initially, all the foreign matters and immature grains have to be removed from the raw material. For this, cleaner of having suitable sieves could be used. The clean grains can be graded if required. Conditioning of cleaned and graded seed for particular moisture content can be achieved by adding pre determined quantity of water (as per the formula given below) and periodically stirred in airtight plastic bags (Double layer LDPE bag of 70μm thickness each having the dimension of 26.5 cm X 17.8 cm).

$$Q = \frac{A(b-a)}{(100-b)}$$

where: 'A' is initial mass of the sample in kilo grams, 'a' is the initial moisture content of the sample in per cent (wet basis) and 'b' is the desired moisture content of the sample in per cent (wet basis).

The treated sample needs to be preserved in a refrigerator at 5°C for 5 days to ensure uniform distribution of moisture. Before starting a test, the required quantity of the grain has to be taken out from the refrigerator and allowed to warm up to the room temperature.

Spatial dimensions

The spatial dimensions of a grain can be measured using micrometers/ vernier calipers with suitable least count. It can also be obtained by measurement of the total length/width of randomly chosen 10 seeds arranged in a line tip to tip/touching along with width or maximum diameter. This measurement could then be divided by 10 to obtain an average seed length/ width. The longest dimension, 'L' is called length, second longest dimension 'B' perpendicular to 'L' is called breadth and the third longest dimension 'T' perpendicular to both is called thickness of an object.

Size or equivalent diameter is the geometric mean of the three dimensions, viz., length, breadth and thickness. Size can be calculated using the following expression:

Size = (length x breadth x thickness) $^{1/3}$

Shape factors

Roundness

It is a measure of the sharpness of the solid material. It can also be used to measure the bluntness of the corners of an object. The most widely accepted method for determining the roundness of irregular particle is given below.

Roundness = A_p/A_c

Where, A_p is the largest projected area of object in natural rest position and A_c is the area of the smallest circumscribing circle. The object area is obtained either by projection or tracing.

Sphericity

It is defined as the ratio of the surface area of sphere having same volume as that of the grain to the surface are of the grain.

Sphericity of the ellipsoid grain = Geometric mean diameter / Major diameter

Sphericity = Di / Dc

Where,

Di = Diameter of the largest inscribing circle

Dc = Diameter of the smallest circumscribing circle

Thousand grain weight

It can be determined using the method described in Indian Standard IS:4333 (Part IV) - 1968. The method is as follows:

Take approximately 500 grains from the sample randomly. Sort out the whole grains and weigh. Subsequently, count the whole grains.

1000 grains weight on 'as-is' basis = (a/b) x 1000

Where,
a = weight of the whole grains, g and
b = number of whole grains in the sample weighed.

Density, specific gravity and porosity

Quality of grains can be assessed by measuring their densities. The density and specific gravity values of grains and other commodities are used in design of storage bins and silos, separation of desirable materials from impurities, cleaning and grading, evaluation of the grain maturity, texture and softness of fruits, quality evaluation of the products etc. The density of any material is expressed as,

Density = Weight of the material / Volume of the material

Bulk density can be determined by measuring the weight of the known volume of the sample. Bulk density is the density of a material when packed or stacked in bulk. Bulk density of particulate solid is measured by allowing the sample to pour into a container of known dimensions. Special care should be taken since the method of filling and the container dimensions can affect the measurement.

True density can be determined by toluene displacement method. Initially, the known weight of the grain is filled with the known volume of toluene. The change in volume due to the filling of the grain sample is noted and the true density is calculated by dividing the weight of grain by the change in volume of the toluene.

Volume and specific gravity can also be determined by using a general purpose reagent (toluene rectified) with the help of measuring cylinder and using following equation:

Volume = Weight of displaced liquid / Weight density of liquid

Sp gravity = (Weight in air x Sp gravity of toluene) / weight of displaced liquid

Porosity is an important physical property characterizing the texture and the quality of dry and immediate moisture foods. Porosity data is required in modeling and design of various heat and mass transfer processes such as drying, frying, baking, heating, cooling, and extrusion. It is an important parameter in predicting diffusional properties of cellular foods. Porosity can be determined by using the following equation.

Porosity = (True density - Bulk density) x 100 / True density

Water activity

The water activity (a_w) of a food or food ingredient is often more directly related to quality in terms of (a) shelf life stability as affected by chemical, enzymatic, and microbiological changes and (b) compatibility with other foods, formulation, and packaging requirements than is the moisture content of the food itself. The a_w directly or indirectly affects or is related to texture, appearance, aroma, taste, freeze-thaw stability, and the microbiological, chemical, and many other objective and subjective characteristics of food.

Water activity is defined as P/P_o, where P is the partial pressure of water over the food sample and P_o is the vapor pressure of pure water at the same temperature. Water activity may also be defined as the Equilibrium Relative Humidity (ERH) divided by 100.

The concept of water activity is important in determining product quality and safety in a particular environment. An automated instrument, water activity meter is used to determine the water activity of fresh and stored grain samples.

Water activity is measured by equilibrating the liquid phase water in the sample with the vapor phase water in the headspace and measuring the relative humidity of the headspace. The water activity meter is provided with a sample cup, which is sealed against a sensor block. Inside the sensor block, there is a fan, a dew point sensor, a temperature sensor and an infrared thermometer. The dew point sensor measures the dew point temperature of the air and the infrared thermometer measures the sample temperature. From these measurements the relative humidity of the headspace is computed as the ratio of saturation vapor pressure of air to the saturation vapor pressure of the sample. The measurement of the headspace humidity gives the water activity of the sample. The purpose of the fan is to speedup the equilibrium process and to control the boundary layer conductance of the dew point sensor.

Water activity is measured by placing the sample in a container and placing the container in a sealed chamber and then the knob is closed to read the water activity of the sample. The water activity of the grain sample is recorded with respect to atmospheric temperature (28 ± 2°C).

Frictional properties

Frictional properties such as coefficient of friction and angle of repose are important in designing of storage bins, hoppers, chutes, pneumatic conveying system, screw conveyors, threshers etc. In mechanical and pneumatic conveying systems, the material generally moves or slides in direct contact with the trough, casing and other components of the machine. Thus various parameters affect the power requirement to drive the machine. Among these parameters, the frictional losses are one of the factors which must be overcome by providing additional power to the machine. Hence, knowledge of frictional properties of the agricultural materials is necessary.

Angle of repose

The angle of repose is an indicator of the product's ability to flow. It is the angle between the base and the slope of the cone formed on a free vertical fall of the granular material to the horizontal plane. The size, shape, moisture content and orientation of the grains affect the angle of repose.

Coefficient of friction

The coefficient of friction may also be given as the tangent of the angle of the inclined surface upon which the friction force tangential to the surface and the component of the weight normal to the surfaces are acting. If F is the force of friction, and W is the force normal to the surface of contact, then the coefficient of friction 'f' is given by the relationship $f = F/W$.

Aero and hydrodynamic properties

The aero and hydrodynamic properties such as terminal velocity of agricultural products are important and required for designing of air and water conveying systems and the separation equipment. The physical properties such as density, shape and size, drag coefficient etc., are required for calculating the terminal velocity of the material. For example, in pneumatic conveying and separation processes, the material is lifted only when the air velocity is greater than its terminal velocity. The terminal velocity of a particle may be defined as equal to the air velocity which a particle remains in suspended state in a vertical pipe.

Mechanical properties

Mechanical properties may be defined as those which affect the behaviour of the agricultural material under the applied force. The mechanical properties such as hardness, compressive strength, impact and shear resistance and the rheological properties affect the various operations of agricultural processing. Data on these properties are useful for application in designing equipment for milling, handling, storage, transportation, food processing etc.

The mechanical damage to grain and seed in threshing and handling operations causes reduction in germination power and viability of seeds, increases the chances of insect and pest infestation and also affect the quality of final product. The hardness of grain effects the milling characteristics and it is also useful to live stock feeders and plant breeders. The impact and shear resistance are important for size reduction of food grains. This information is useful in determination of the appropriate methods of crushing, breaking or grinding the grains. These properties also play important roles towards seed resistance to cracking under harvesting and threshing conditions.

Texture

Texture is defined as the sensory manifestation of the structure of a food and the manner in which that structure reacts to applied force. It is also an important attribute of food acceptance by consumers. It can also be defined as all the mechanical attributes of a food product perceptible by means of mechanical, tactile and where appropriate, visual and auditory receptors. The major mechanical and texture attributes can be measured by *Texture Analyzer.*

Texture analyzer is an instrument that measures the response of a sample to an applied force. It has a test bed and a moving arm. The load cell is installed in the moving arm and the probe is attached with the load cell. The sample is placed on the test bed. During the test the probe moves from the start position toward the sample, as soon as it touches the sample and measures the trigger force, the test starts and probe moves to a specified distance or force in the sample and curve between force and time or force and distance is plotted. The basic methods of measuring texture are tension, compression, bending and penetration. The parameters can be measured using texture analyzer are hardness, firmness, springiness, compression, shear strength, chewiness, cohesiveness, gumminess, consistency, fracturability, toughness, tenderness, brittleness, break strength, extensibility etc.,

Hardness value is the peak force of the first compression of the product. The hardness need not occur at the point of deepest compression, although it typically does for most products. *Fracturability* Not all products fracture; but when they do fracture the Fracturability point occurs where the plot has its first significant peak (where the force falls off) during the probe's first compression of the product.

Cohesiveness is how well the product withstands a second deformation relative to how it behaved under the first deformation. It is measured as the area of work during the second compression divided by the area of work during the first compression. (Refer to Area 2/Area 1 in the graph).

Springiness is how well a product physically springs back after it has been deformed during the first compression. The springback is measured at the downstroke of the second compression, so the wait time between two strokes can be relatively important. In some cases an excessively long wait time will allow a product to springback more than it might under the conditions being researched (eg you would not wait 60 seconds between chews).

Springiness is measured several ways, but most typically, by the distance of the detected height of the product on the second compression (Length 2 on the below graph), as divided by the original compression distance (Length 1). The original definition of springiness used the Length 2 only, and the units were in mm or other units of distance. We do not subscribe to that original description of springiness since the springiness value can only be compared among products which are identical in their original shape & height. Many TPA users compress their products a % strain, and for those applications a pure distance value (rather than a ratio) is too heavily influenced by the height of the sample. By expressing spriniginess as a ratio of its original height, comparisons can be made between a more broad set of samples and products.

Following graph illustrates the Texture Profile Analysis studies for food products, typical experimental procedure and defines the parameters for reading the graph.

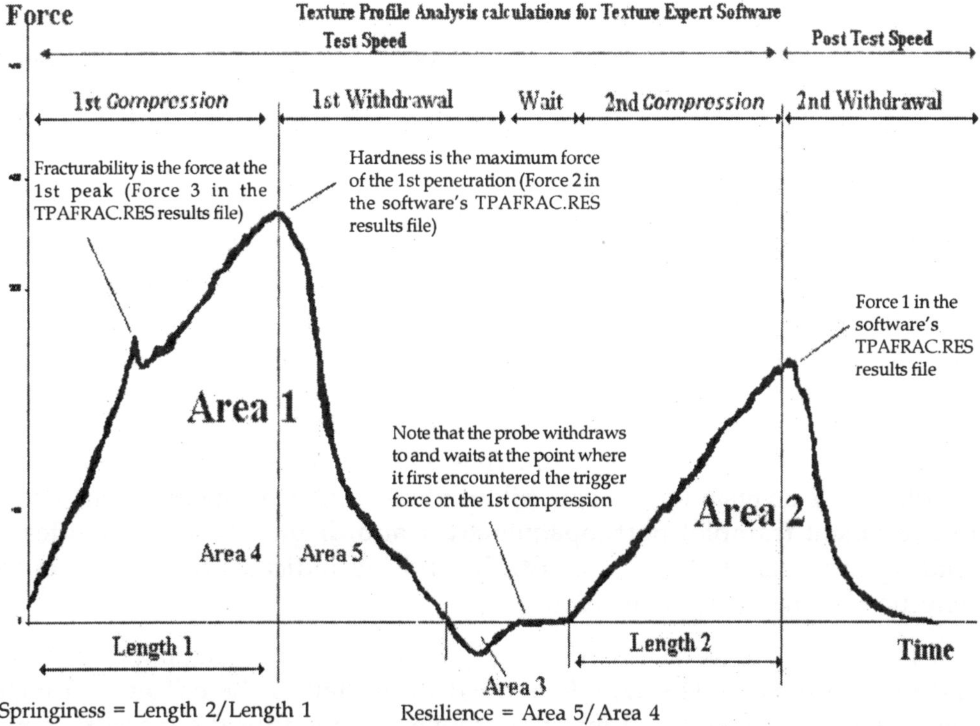

Chewiness only applies for solid products and is calculated as Gumminess*Springiness (which is Length1/Length2). Chewiness is mutually exclusive with Gumminess since a product would not be both a solid and a semi-solid at the same time.

Gumminess only applies to semi-solid products and is Hardness *Cohesiveness (which is Area 2/Area1). Gumminess is mutually exclusive with Chewiness since a product would not be both a semi-solid and a solid at the same time.

Resilience is how well a product "fights to regain its original position". You can think of it as instant springiness, since resilience is measured on the withdrawal of the first penetration, before the waiting period is started. The calculation is the area during the withdrawal of the first compression, divided by the area of the first compression. (Area 5/Area4 on the below graph). Resilience is not always measured with TPA calculations, and was not a direct part of the original TPA work. Resilience can be measured with a single compression; however, the withdrawal speed must be the same as the compression speed.

Colour

Colour is an important physical property which determines the acceptance of a food by consumer. A variety of systems have been developed to describe colours. Color order systems are three-dimensional arrangements of colour according to appearance. There are five colour order systems used for foods: Munsell, CIE, CIE L* a* b* (CIELAB), Hunter Lab, and Lovibond.

The color of the homogenized grain sample can be measured with a Hunter's colorimeter (Hunter Lab Colorimeter, MiniScan XE Plus Hunter Associates Lab inc., Reston, Virginia, USA), calibrated with a white tile having standard values of X=81.0, Y=85.8 and Z=91.2 with corresponding L*, a* and b* values of 67.81, 19.56 and 58.16, respectively as provided by the manufacturer. Similarly, it is calibrated with a black tile. It works on the principle of collecting and measuring the energy from the samples reflected across the entire visible spectrum. The Hunter lab's colour flex spectro colorimeter is a versatile, compact colour measuring instrument consisting of measurement (sample) port, opaque cover and display unit. It provides a reading in terms of L*, a*, -a*, b*, -b* and DE indicates the deviation in reflection *ie.* change in total colour.

The colour can be measured using CIELAB scale at 10° standard observer and D65 illuminant. The sample is filled up to half of the circular 2.5 inch transparent glass cell, fitted into the groove provided for the measurement. An opaque sample cup is covered and sample is measured at three different locations. The 'L*' coordinate measures the value or luminance of a colour and ranges from black at 0 to white at 100. The 'a*' coordinate measures red when positive and green when negative and 'b*' measures yellow when positive and blue when negative.

The change in colour of the sample is expressed as the total colour difference (DE). The DE is determined by using L, a, b colour co-ordinates as defined by the equation.

$$\Delta E = \sqrt{\Delta L^2 + \Delta a^2 + \Delta b^2}$$

where,

DL = $L^*_{smp} - L^*_{std}$ (if, + ΔL sample is lighter than standard, -ΔL sample is darker than standard)

Da = $a^*_{smp} - a^*_{std}$ (if, + Δa sample is redder than standard, -Δa sample is greener than standard)

Db = $b^*_{smp} - b^*_{std}$ (if, + Δb sample is yellower than standard, -Δb sample is bluer than standard)

The L*, a*, b* values of the fresh samples are taken as the standard values and these values were used for determining the total colour difference (DE).

References

Chakraverty, A (1995). Post Harvest Technology of cereals, pulses and oilseeds. Oxford & IBH Publishing Co., Pvt. Ltd., New Delhi.

Kachru, RP, Gupta, RK and Alam, A (1994). Physico-chemical constituents and engineering properties of food crops. Scientific Publishers, Jodhpur.

Serpil Sahin and Servet Gulum Sumuu (2006). Physical properties of foods. Food Science Text Series, Springer, USA.

Sahay, KM and Singh, KK (1994). Unit Operations of Agricultural Processing. Vikas Publishing Pvt. Ltd., New Delhi.

□□□

31

Concept of Colour Measurement and Sampling Techniques for Quality Evaluation of Food

S.N. JHA

The analysis of colour is frequently an important consideration when determining the efficacy of variety of postharvest treatments. Consumers can easily be influenced by preconceived ideas of how a particular fruit or vegetable should appear, and marketers often attempt to improve upon what nature has painted. Recently colour measurements have also been used as quality parameters and indicator of some inner constituents of the material. In spite of the significance of colour in food industries, many researchers continue to analyze it inappropriately. This chapter thus tries to remove unnecessary confusions and through lights on various aspects of colour measurements, its basic units and spectra acquisition for extracting various hidden information through analysis.

Light and colour

Among the properties widely used for analytical evaluation of materials, colour is unique in several aspects. While every material can be said to possess a specific property such as mass, no material is actually coloured as such.

Colour is primarily an appearance property attributed to the spectral distribution of light and, in a way, is related to some source of radiant energy (the illuminant), to the object to which the colour is ascribed, and to the eye of the observer. Without light or the illuminant, colour does not exist. Therefore, several factors that influence the radiation subsequently affect the exact colour that an individual perceives:

- Spectral energy distribution of light
- Conditions under which the colour is viewed
- Spectral characteristics of the object with respect to absorption, reflection, and transmission
- Sensitivity of the eye

Thus, in reality, colour is in the eye of the observer, rather than in the "colored" object. The property of an object that gives it a characteristic colour is its light-absorptive capacity.

Color specification

There are three characteristics of light by which a colour may be specified: hue, saturation, and brightness. Hue is an attribute associated with the dominant wavelength in a mixture of light waves, i.e., it represents the dominant colour as perceived by an observer. Saturation refers to relative purity or the amount of white light mixed with a hue. Brightness is a subjective term, which embodies the chromatic notion of intensity. Hue and saturation taken together are called chromaticity. Therefore, a colour may be characterized by brightness and chromaticity.

CIE system

The Commission de Internationale de l'Eclairage (CIE) defined a system of describing the colour of an object based on three primary stimuli: red (700 nm), green (546.1 nm), and blue (435.8 nm). Because of the structure of the human eye, all colours appear as different combinations of these. The amounts of red, green, and blue needed to form any given colour are called the' 'tristimulus" values, X, Y, and Z, respectively. Using the X, Y, and Z values, a colour is represented by a set of chromaticity coordinates or trichromatic coefficients, x, y, and z, as defined below:

$$x = \frac{X}{X+Y+Z} \quad y = \frac{Y}{X+Y+Z} \quad z = \frac{Z}{X+Y+Z}$$

It is obvious from the equations above that x + y + z = 1. The tristimulus values for any wavelength can be obtained from either standard tables or figures. A plot that represents all colours in x (red)-y (green) coordinates is known as a chromaticity diagram. For a given set of x and y, z is calculated from the above equations. Therefore, colours are generally specified in terms of Y, x, and y.

There are a number of colour metrics based on the CIE system. They include CIE Lightness, CIELUV, CIELAB, etc. In the food industry, the CIELAB system has been popular. For example, objective measurements of colour using the CIELAB colour parameters such as L* (lightness), a* (redness), and hue angle have been used to evaluate pork quality on-line in an industrial context.

Other colour models, such as the RGB, CMY, and HSI, etc., are very similar to the CIE system, and numerical representation of a colour in one system can be converted into another.

Munsell system and atlas

The Munsell colour-order system is a way of precisely specifying colours and showing the relationships among colours. Every colour has three qualities or attributes: hue, value, and chroma. A set of numerical scales with visually uniform steps for each of these attributes has been established. The Munsell Book of colour displays a collection of coloured chips arranged according to these scales. Each chip is identified numerically using these scales. The colour of any surface can be identified by comparing it to the chips under proper illumination and viewing conditions. The colour is then identified by its hue, value, and chroma. These attributes are given the symbols H, V, and C and are written in a form H V/C, which is called the Munsell notations. Using Munsell notations, each colour has a logical relationship to all other colours. This opens up endless creative possibilities in colour choices, as well as the ability to communicate those colour choices precisely. The Munsell system is the colour order system most widely quoted in food industry literature. Food products for which the U.S. Department of Agriculture (USDA) recommends matching Munsell discs to be used include dairy products such as milk and cheese, egg yolks, beef, several fruits, vegetables, and fruit juices.

Other colour atlases and charts are available for use in the food industry, such as the Natural colour System and Atlas, Royal Horticultural Society Charts, etc. These atlases and charts are used for visual comparison of a product colour with that of a standard colour diagram, which is still commonly practiced in the food industry. The evaluation of potato chip colour is a very good example.

Machine reading of colour

A digital colour is usually represented by red, green and blue *(RGB)* tristimullus values. This approximation is a consequence of the fact that human perception of colour is mediated by the response of three different types of photoreceptors in the retina called cones (Wright 1964). A colour standard like the CIE (Commission Internationale de1′ Eclairage) 1931 standard as mentioned above defines colour based on its psycho-physical properties. A useful feature of this system is that it enables the tristimullus values be transformed into lower dimension quantities. In this case they are the chromaticity coefficients expressed in terms of x and y values. An object's colour can be assessed by simply plotting x and y in rectangular coordinates, producing a chromaticity diagram (Fig. 1).

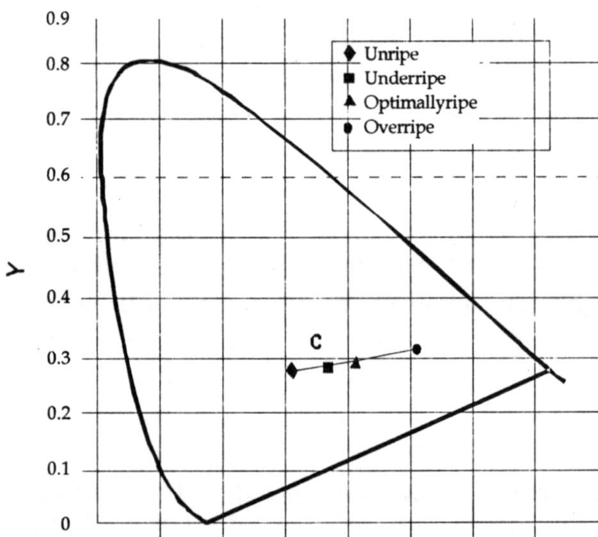

Fig. 1 : Chromaticity diagram showing the ripeness locus of oil palm and also the location of white in illuminant "c"

To facilitate discussion, the locus is superimposed, with the reference horseshoe curve obtained from a standard monochromatic light. It can be seen from this figure that the chromaticity of unripe fruits falls near the center of the CIE diagram. The position marked "C" in this diagram represents colour which is biochromatically achromatic or hueless. Oil palm actually appears reddish black when unripe and this agrees well with colourimeter since this equipment treats both pure white and black as hueless. As the fruit starts to ripen; the locus moves from the hueless zone to the reddish red zone and ends at a point bordering the reddish orange zone. It can be seen from this diagram that the difference in chromaticity between unripe and underripe is relatively small compared to the difference in chromaticity between optimally ripe and overripe, indicating that there is a small degree of change in colour at the early stage of ripening. The distance between unripe and overripe is 0.202 compared to slightly over 0.03 between unripe and underripe. Hence, distinguishing unripe from underripe samples or vice versa may be difficult chromatically.

Even though the chromaticity diagrams are useful in specifying colour, they do not accurately represent colour closest to that of human perception. For proper quantification of tristimulus and efficient colour processing, this variable is usually represented on cylindrical coordinates of L^* a^* and b^* or simply the *CIELab*, a uniform colour space, values. These values later can be transformed to through simple trigonometric functions (Hunter, 1942) comprising psychometric lightness (L^*), hue (h^o) and chroma (C^*) (Eqns 1 - 3). A colour wheel subtends 360°, with red-purple traditionally placed at the far right (or at an angle of 0°), yellow, bluish-green, and blue follow counterclockwise at 90°, 180°, 270°, respectively (Fig. 2).

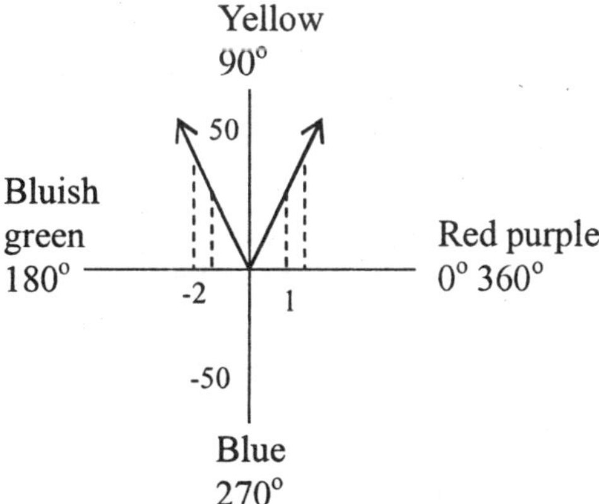

Fig. 2 : Representation of peel hue affected by heat treatments of grapefruits. CIE LAB a* and b* values are plotted on horizontal and vertical axes respectively

$$L^* = L^* \quad \text{... (1)}$$

$$h^0 = \tan-1\left(b/a\right) \quad \text{... (2)}$$

$$C^* = \sqrt{\left(a^{*2} + b^{*2}\right)} \quad \text{... (3)}$$

Arctangent, however, assumes positive values in the first and third and negative values in the second and fourth quadrants. For a useful interpretation, h^o should remain positive between 0° to 360° of the colour shed.

Fig. 3 shows the variation of *CIELab* values calculated from the oil palm image. It can be seen that both hue and chroma increase in curvilinear fashion with ripeness. The small hue and chroma values for unripe class (approximately 7.6° and 2.62, respectively) pushed the psychromatic point nearer to the origin or the achromatic zone of colour. These values increased to approximately 48° in hue and 72.1 in chroma for overripe case. This location is equivalent to reddish orange colour on *CIELab* space. Hence, the hue-moves further away from the origin and in the upward direction as the oil palm ripens. These observations are consistent with human vision and match strongly with the trend of the ripeness locus shown in Fig. 1. Thus hue provides a much better discrimination compared to either *RGB* or CIExy values when specifying colours of food materials. Because of this reason usually hue is chosen for colour inspection by machine.

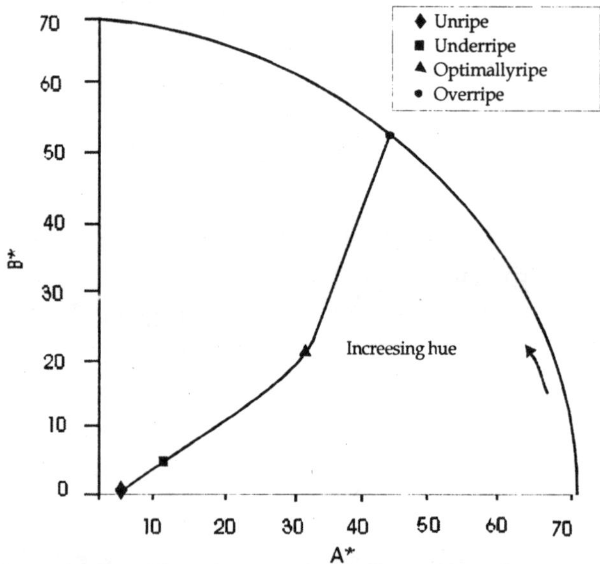

Fig. 3 : Colourimetric plot of oil palm showing the change in hue and chroma during ripening

Unlike colourimeter, the calculation of hue using machine vision system is mathematically involved since it requires colour conversion from *RGB* to *HSl* (Hue, Saturation and Intensity) space. One way of achieving this is by firstly I establishing a new coordinate system, *YIQ*. The relationship between the two coordinate systems is:

$$\begin{bmatrix} Y \\ I \\ Q \end{bmatrix} = \begin{bmatrix} 0.30 & 0.59 & 0.11 \\ 0.60 & -0.28 & -0.32 \\ 0.21 & -00.52 & 0.31 \end{bmatrix} \begin{bmatrix} R \\ G \\ B \end{bmatrix} \quad \text{... (4)}$$

Secondly, *h* is the rotational angle around the Q, I plane and therefore can be written as:

$$h^o = \tan-1\left[\frac{I}{Q}\right] \qquad \dots (5)$$

Equations (4) & (5) are theoretically valid and they can be found in almost any textbook on colour and image processing. For practical reasons, h° was calculated according to the Munsell's colour system, which is given by:

$$h^o = \left\{360^o - \cos^{-1}\left(\frac{-0.5\left[(R-G)+(R-B)\right]}{\sqrt{(R-G)^2+(R-B)(G-B)}}\right)\right\}\times\frac{255}{360} \quad \text{if } B \geq G \qquad \dots (6)$$

or

$$h^o = \left\{\cos^{-1}\left(\frac{-0.5\left[(R-G)+(R-B)\right]}{\sqrt{(R-G)^2+(R-B)(G-B)}}\right)\right\}\times\frac{255}{360} \quad \text{if } B < G \qquad \dots (7)$$

The above equation transforms *RGB* information from three-dimensional space to one-dimensional *h* space. In order to speed-up analysis only *h* values may be processed. The hue values shown in this figure are normalized to 255 for the 8-bit machine vision system. A different approach is needed to solve this type of problem. The method investigated for this application was to treat hue distributions as features and apply multivariate discriminant technique to establish classification.

Example of transformation of CIE/Hunter L a b values to chroma (C) and hue (h°)

Assume a practical data given in Table 1 for analysis of grapefruit colour after three heat treatments for quarantine control.

Treatment	Colour characteristics				
	L	a	b	C	h°
1	76.6	-2.0	56.0	56.0	92.0
2	74.4	2.0	56.0	56.0	88.0
3	63.0	1.2	34.0	34.0	88

Following subprogram was used to compute the h° and C for above data

```
Data colour

READ (*,*) L, a, b

C=SQRT((a*b)+(b*b))
```

```
THETA=(ATAN(b/a)/6.2832)*360
IF a>0 AND b>=0 THEN h = THETA
IF a<0 AND b>= THEN h=180+THETA
IF a<0 AND b<0 THEN h=180+THETA
IF a>0 AND b<0 THEN h=360+THETA
WRITE (*,*) a, b, THETA, h
STOP
END
```

After converting L, a, and b values to hue and chroma as shown in Table 1 can be correlated to any desirable attributes of food. But sometimes these values may give poor correlation with internal attributes such as total soluble solids, dry matter content etc. In order to search for better correlation one should investigate whole range of spectral data of available wavelengths.

Spectra of light

Spectra are curves drawn as a continuous function of relative reflectance or absorbance data with respect to wavelengths. Since light is the basic stimulus of colours, it is important to consider the electromagnetic spectrum. Several optical methods have been developed based on radiation from different regions of this spectrum.

Radiation is one of the basic physical processes by which energy is transferred from one place to another. Propagation of radiation through free space is the same for the entire electromagnetic spectrum, i.e., radiation of all wavelengths-from the shortest gamma rays to the longest radio waves-travels with the same speed in vacuum. Visible light forms only a small part of the electromagnetic spectrum, with a spectral range from approximately 390 nm (violet) to 750 nm (red). The sensitivity of the eye varies even within this narrow visible range. Under conditions of moderate-to-strong illumination, the eye is most sensitive to yellow-green light of about 550 nm.

If the spectral distribution throughout the visible region is unequal, then the sensation of colour is evoked by radiant energy reaching the eye's retina. An equal spectral distribution makes the light appear as white. The unequal distribution responsible for colour sensation may be characteristic of the source itself; such as flame spectra composed of one or more monochromatic wavelengths, or may result from selective absorption by the system, which appears coloured. The latter includes several systems that show selective absorption for light and exhibit colour as a result of reflection or transmission of unabsorbed incident radiant energy (Fig. 4). The three basic factors required in colour sensation include the radiator or illuminant, the object, and the observer. The radiant energy emitted by the radiator is characterized by its spectral quality, angular distribution, and intensity.

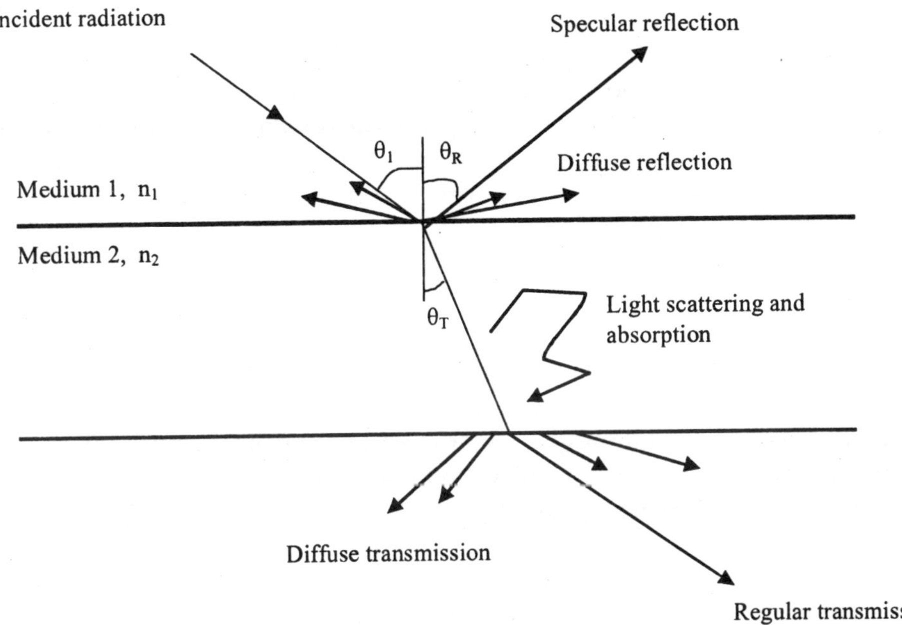

Fig. 4 : Schematic representation of interaction of light with matter, θ_1 = angle of incidence, θ_R = angle of reflectance, θ_T = angle of transmittance, n_1 n_2 = refractive index of medium 1 and 2, respectively

The following material properties and lighting of the scene as affecting the total appearance of the object:

Material properties

Optical properties (spectral, reflectance, transmission)

Physical form (shape, size, surface texture)

Temporal aspects (movement, gesture, rhythm)

Lighting of the scene

Illumination type (primary, secondary, tertiary)

Spectral and intensity properties; directions and distributions

Color-rendering properties

Interaction of light with matter

A. Physical laws

When light falls on an object, it may be reflected, transmitted, or absorbed (Fig. 4.) Reflected light is the part of the incident energy that is bounced off the object surface, transmitted light passes through the object, and absorbed light constitutes the part of the incident radiant energy absorbed within the material. The degree to which these phenomena take place depends on the nature of the material and on the particular wavelength of the electromagnetic spectrum being used. Commonly, optical properties of a material can be defined by the relative magnitudes of reflected, transmitted, and absorbed energy at each wavelength. Conservation of energy requires that sum of the reflected (I_R), transmitted (I_T), and absorbed (I_A) radiation equals the total incident radiation (I). Thus,

$$I = I_R + I_T + I_A \quad \text{... (8)}$$

According to its transmittance properties, an object may be transparent, opaque, or translucent. Almost all food and biological products may be considered to be opaque, although most transmit light to some extent at certain wavelengths. The direction of a transmitted ray after meeting a plane interface between any two nonabsorbing media can be predicted based on Snell's law:

$$n_2 \sin \theta_T - n_1 \sin \theta \quad \text{... (9)}$$

The attenuation of the transmitted ray in a homogeneous, nondiffusing, absorbing medium is defined by Beer-Lambert's law:

$$\log (I_T/I) = abc \quad \text{... (10)}$$

The ratio I_T/I is known as the transmittance T and is related to absorbance A as:

$$A = \log (I/T) \quad \text{... (11)}$$

From Eqs. (10) and (11), absorbance A can also be written as:

$$A = abc \quad \text{... (12)}$$

where a is called the absorptivity. [if c is expressed in mol/L and b in cm, a is replaced by the molar absorptivity, ε (L/mol.cm).]

Various constituents of food products can absorb a certain amount of this radiation. Absorption varies with the constituents, wavelength, and path length of the light. Reflection is a complex action involving several physical phenomena. Depending on how light is reflected back after striking an object, reflection may be defined as regular or specular reflection and diffused reflection (Fig. 5.). Reflection from a smooth, polished surface is called" specular" or "regular". It mainly produces the gloss or shine of the material.

The basic law of specular reflection states that the angle at which a ray is incident to a surface must equal the angle at which it is reflected off the surface. Fresnel equations define the phenomenon of specular reflection. The intensity of parallel R_{pl} and perpendicular R_{pr} components of the reflected light are:

$$R_{pl} = \left[\frac{(n_2/n_1)^2\cos\theta_1 - [(n_2/n_1)^2 - \sin_2\theta_1]^{1/2}}{(n_2/n_1)^2\cos\theta_1 + [(n_2/n_1)^2 - \sin^2\theta_1]^{1/2}}\right]^2 \quad \text{... (13)}$$

$$R_{pr} = \left[\frac{\cos\theta_1 - [(n_2/n_1)^2 - \sin_2\theta_1]^{1/2}}{\cos\theta_1 + [(n_2/n_1)^2 - \sin^2\theta_1]^{1/2}}\right]^2 \quad \text{... (14)}$$

The regular reflectance $R = R_{pl}^2 + R_{pr}^2$ and for normal incidence ($\theta = 0^o$), $R_{pl} = R_{pr,}$ and hence.

$$R = \left[\frac{n_2 - n_1}{n_2 + n_1}\right]^2 \quad \text{... (15)}$$

where n_1 and n_2 are reflective index of the medium and object, respectively: and θ_1 is the incident angle (Fig. 5.). If the material is absorbing, the reflective index is a complex number n (1-ik), where n is the real part of the complex number and k is an absorption constant, and the regular reflectance is written as:

$$R = \left[\frac{(n_2 - n_1)^2 + (n_2 k)^2}{(n_2 + n_1)^2 + (n_2 k)^2}\right] \quad \text{... (16)}$$

When the incident light is reflected from a surface evenly at all angles, the object appears to have a flat or dull finish termed "*diffuse reflection*" No rigorous theory has been developed for diffuse reflectance, but several phenomenological theories have been proposed, the most popular being the Kubelka-Munk theory. The Kubelka-Munk model relates sample concentration to the way Beer-Lambert's law relates band intensities to concentration to the intensity of the measured spectrum in a manner analogous to the way Beer-Lambert's law relates band intensities to concentration for transmission measurements. The Kubelka-Munk function f(R∞) is generally expressed as:

$$f(R_\infty) = \frac{(1 - R_\infty)^2}{2R_\infty} = \frac{k}{s} \quad \text{... (17)}$$

where R = absolute reflectance of an infinitely thick layer, k = absorption coefficient, and s = scattering coefficient.

Kubelka-Munk theory predicts a linear relationship between spectral data and sample concentration under conditions of constant scattering coefficient and infinite sample dilution in a nonabsorbing matrix such as KBr (potassium bromide). Hence, the relationship can only be applied to highly diluted samples in a nonabsorbing matrix. In addition, the scattering coefficient is a function of particle size, so samples must be prepared to a uniform fine size of quantitatively valid measurements are desired.

It is not easy to quantify diffuse reflectance measurements since sample transmission, scattering, absorption, and reflection all contribute to the overall effect. By reducing particle size and dilution in appropriate matrices, surface reflection that can give strong inverted bands is reduced and the spectra more closely resemble transmission measurements. Typically, quantitative diffuse reflectance measurements are presented in log (I/R) units, analogous to absorbance log (I/T) units for transmission measurements. Bands increase logarithmically with changes in the reflectance values. By comparison, bands in spectra displayed in Kubelka-Munk units vary as a function of the square of reflectance. This difference emplasizes strong absorbance bands relative to weaker bands.

The diffuse reflectance may be measured with respect to a nonabsorbing standards and converted to produce a nearly relationship with concentration c as follows:

$$\log(R'/R) = \log(I/R) + \log(R') \cong ac/s \qquad \text{... (18)}$$

where R′ and R – reflectance of the standard and the sample (R′ > R), a = absorptivity, c = concentration, and s = scattering coefficient. For monochromatic radiation, log is constant and may be ignored, and Eq. (18) may be writes as (17):

$$c = k + (s/a)\log(I/R) \qquad \text{... (19)}$$

where k = absorption coefficient. It should be noted that s is not a constant but depends on a number of properties of the sample such as particle size (s is inversely proportional to particle size) and moisture content. In food materials, the primary factors that influences light reflection is a phenomenon known as scattering or diffusion. If the surface of incidence is rough, incident light will be scattered in all directions. Since the incident rays strike a rough surface more than once before being reflected, they would be expected to have a lower total reflectance than those reflected from a smooth surface.

In classical optics, diffuse reflection was thought to be responsible for colour. It was also commonly believed that colour of natural objects, such as foods and plant foliage, are seen by means of light reflected off their surfaces.

It is also known that the light must be transmitted through pigment within the cells in order to produce a coloured appearance. Since most food materials are optically nonhomogeneous, light entering such material is scattered in all directions. Only about 4-5% of the incident radiation is reflected off the surface of these materials as regular reflectance. The remaining radiation transmits through the surface and encounters small interfaces radiation from within the material is scattered back to the surface through the initial interface. This type of reflection is termed as "body reflectance". The body reflectance is nearly always diffuse and is the most significant form of reflectance for foods. Some part of the transmitted light diffuse deeper in to the material and may eventually reach the surface some distance away from the incident point.

B. Factors affecting diffuse reflectance spectral data

Diffuse reflectance spectroscopy offers exceptional versatility in sample analysis. This versatility results from both its sensitivity and optical characteristics. Classically, diffuse reflectance has been used to analyze powered solids in a nonabsorbing matrix of an alkali halide such as KBr. The sample is typically analysed at low concentrations, permitting quantitative presentation of the data in Kubelka-Munk units. This technique yields spectra that are qualitatively similar to those produced by conventional transmittance or pellet methods. However, they exhibit higher sensitivity for quantification and are less subject to scattering effects the cause slopping baselines in pellet measurements.

Several factors determine band shape and relative/absolute intensity in diffuse reflectance spectroscopy through their effect on the reflection/ absorbance phenomena specific to the sample. These include:

- Refractive index of the sample
- Particle size
- Sample homogeneity
- Concentration

1. Refractive index

Refractive index affects the results via specular reflectance contributions to diffuse reflectance spectra. With organic samples, the spectra display pronounced changes in band shape and relative peak intensities, resulting in nonlinearity in the relationship between band intensity and sample concentration. For some inorganic samples, strong specular reflection contributions can even result in complete band inversions. Diluting the sample in nonabsorbing matrix can minimize this overlay of diffuse reflectance and specular reflectance spectra, as well as the resulting spectral distortions. In addition, accessory design can help reduce specular reflectance contributions.

2. Particle size

Particle size is a major consideration when performing diffuse reflectance measurements of solids. The bandwidth is decreased and relative intensities are dramatically altered as particle size decreases. These effects are even more pronounced in spectra of highly absorbing inorganic materials with high refractive indices. For these samples, specular contributions can dominate the final spectra if the particle size is too large. To acquire a true diffuse reflectance spectrum, it is necessary to uniformly grind the sample and dilute it in a fine, nonabsorbing matrix. Similar preparation must be applied to the nonabsorbing matrix material in order to provide and "ideal" diffuse reflector for background analysis and as a support matrix for the samples.

3. Sample homogeneity

The Kubelka- Munk model for diffuse reflectance is derived for a homogeneous sample of infinite thickness. However, some sample analysis methods, especially those designed for liquid sample (e.g., deposition of sample onto a powdered supporting matrix) can result in a higher concentration of sample near the analysis surface. In these circumstances, variations in relative peak intensities may be noticed. In particular, more weakly absorbing wavelengths tend to be attenuated at higher sample concentrations. To avoid these peak intensity variations it is necessary to distribute the analyte as uniformly possible within the nonabsorbing background matrix.

4. Concentration

One particularly important advantage of diffuse reflectance spectroscopy, especially in comparison to transmittance measures, is its extremely broad sample-analyzing range. While it is theoretically possible to acquire usable diffuse reflectance spectra on samples of wide-ranging concentrations, practical considerations often complicate the analysis process. With high concentration samples, especially those with a high refractive index, one can expect a dramatic increase in the specular contribution to the spectral data. As a result, some sample data may be uninterpretable without adequate sample dilution. Even when samples can be measured satisfactorily at high concentrations, it is advisable to grind the sample to a very uniform and fine particle size to minimize both specular reflectance and sample scattering effects, which adversely affect quantitative precision.

Samples and sample preparation

Sampling for measurement of colour and prediction of internal attributes affecting quality of materials requires considerable attention towards the factors described above.

Availability of wide range of techniques for measurement of quality parameters, necessitates to know the suitability of characteristics of samples, i.e. whether it is liquid, solid, paste, semisolid, transparent, opaque or translucent etc for a particular technique to be employed. Complete history of the source of sample and a careful attention to proper instrument operation and consistent sample handling is required particularly for colour measurement.

Preparing samples for measurement

When measuring samples it is important to select samples appropriately, use an established measurement method, and handle all samples in a consistent manner.

Selecting samples

Samples representative of the entire batch should be selected for measurement. Always one should try to collect as number of varied materials as possible from different sources whose colour is to be measured and

1. Choose samples that are truly representative of those materials collected from various sources,
2. Prepare samples in exactly the same manner each time they are measured. Follow standard method, if they exist such as ASTM, BIS etc, and
3. Present the samples to the instrument in a standard, repeatable manner. Results obtained depend on the condition of the samples and their presentation. For established procedure, make a checklist so that laboratory personnel may simply check each step. The checklist will also help in training of new workers.

The sample must also be representative of attributes that are of interest. If samples are non-representative of the batch or are spoiled, damaged, or irregular, then the sample may be biased. When choosing a sample, select in random fashion and examine the sample to avoid biased results. If sampling procedures are adequate, a different sample selected from the same batch should result in comparable measured values.

Sample handling and measurement methods

If method of measurement is established so that same procedure is used each time for specific samples or types of samples, results may be validated for comapsrision purposes. This also insures repeatability of results when measuring the same sample.

There are a variety of techniques that can be used in handling various forms of objects and materials so that the most valid and repeatable measurement of their appearance results. Consideration must be given to the conditions for sample preparation that are dependent upon the type of measurement to be made. For example, when measuring the colour of sample that might pillow into the viewing aperture, one should hold the surface flat by using a cover glass taped over the aperture window. Other materials being measured for colour may be chopped up and placed in a glass specimen cell or made into paste and applied to a glass plate. Sheets and films should be flattened by tension or by a vacuum, if necessary.

Directional samples

Averaging several measurements with rotation of the sample between readings can minimize directionality. Examination of the standard deviation displayed with the average function can guide in selecting the appropriate number of readings to average.

Non-opaque samples

Non-opaque samples must have a consistent backing. A white uncalibrated tile is recommended. If the sample is such that it can be folded to give multiple layers, such as fruit leather, the number of layers for each sample should be noted.

Translucent samples

Light trapped in a translucent sample can distort the colour. The thickness of the sample presented should be chosen to maximize the haze or colour difference.

Granular, powdery and liquid samples

These foods in required quantity may be taken into a petty dish of known composition and characteristics and covered by other complete transparent and flat petty dish of known properties. Thickness or depth of the sample should be so maintained that it presents an opaque mass. Colour readings may be taken keeping the flat portion of nosecone of the colourmeter on the surface of the top petty dish ensuring that light thrown by the instrument neither goes out of the nosecone nor passes through the sample. Part of the light is absorbed by the sample and remaining portion (reflected from the sample) again comes back to the nosecone of the instrument for measurement and interpretation. If samples cannot be prepared to make it opaque (in case of transparent liquid sample) instruments such as tintometer, photospectrometer etc should be used in visual range of wavelength.

References

Birth GS (1978). The light scattering properties of foods. J. Food Sci, 43: 915.

Hutchings JB (1999). Food colour and Appearance. 2nd edition, Gaithersburg. MD: Aspen Publishers Inc.

Jha, SN (2004). Non-destructive methods for quality evaluation of foods. Indian Food Industries, 23(5): 21–26.

Jha, SN (1999). Physical and hygroscopic properties of makhana. J. Agric. Eng. Res., 72: 145-150.

Jha, SN and Prasad, S (1993). Physical and thermal properties of gorgon nut. J. Food Process Eng., 16: 237-245.

Jha, SN and Prasad, S (1996). Determination of processing conditions of gorgon nut (Euryale ferox). J. Agric. Engg. Res., 63: 103-112.

Jha, SN Matsuoka T (2000). Review: Nondestructive Techniques for quality evaluation of intact fruits and vegetables. Food Science and Technology Research - Review, 6(4): 248-251.

Jha. SN and Kachru, RP (1998). Physical and aerodynamic properties of makhana. J. Food Process Eng., 79: 301-316.

Kortum G (1969). Refectance spectroscopy: Priciples, Methods, Applications. New York:Springer-Verlag.

Osborne BG, Fearn T (1986). Near-Infrared spectroscopiy in Food Analysis. Longman Scientific and Technical Avon, England.

□□□

32

Nutritional Quality Analysis of Food and Food Products

MRIDULA D.

Food material specially grains, nuts, fruits and vegetables contain mainly carbohydrates, proteins and lipids. They also contain 2-5% minerals and small proportions of vitamins and other minor but important constituents. Cereal grains contain more of carbohydrates while oilseeds contain more of oil/ fats. Oilseeds and nuts may contain 30-50% protein while cereal grains contain 6-12% protein. The starch contains of oilseeds may be less than 10% while oil content of cereal grains may range from 1-5%. The actual content and the composition of these groups of compound really account for the quality as well as commercial value. Further the composition and quality of food and food products vary with respect to their genetic history, location or soil under which crops are grown, environment or the season, maturity or harvest stage of crops and post harvest processing and storage, etc. Here only a few and most important procedures relating to food materials have been presented. The analyst can find more details and methods of evaluation of other constituents in the standard laboratory analytical manuals.

Preparation of sample

Sample preparation is an important step of the quality analysis of food products. Thorough mixing and accurate weighing are two important steps before analysis. If proper care is not taken, the results obtained may show wide variations. For quality analysis of food grains, sample should be drawn from the whole lot in such a way that the sample drawn represents the whole lot. Sample should also be free from sand, stone and other extraneous material and ground. Use of iron grinder should be avoided when estimation of iron content is needed. It is preferable to use stainless steel trays, knives and scissors for the preparation of the sample.

Fruits and vegetables - Fruits and vegetables, which are small in size, can be taken as such but only edible portion for sample preparation. Bigger size fruits and vegetables should be cut and portions (½ to ¼) of the edible material are used. Samples for the analysis are taken from a mixture of these pieces.

Flesh foods-either the whole tissue or a portion of it is minced and then blended with double its weight of distilled water.

Liquid foods- Samples are mixed thoroughly and filtered through a muslin cloth. Aliquots are taken from the representative samples for analysis.

Moisture

Moisture content is one of the simplest to determine but at the same time is one of the most important constituent of the food or food product.

According to the standard procedure of AOAC, 1990, weighed sample (5 g) of finely ground material is kept in a dried and pre-weighed perti dish and dried in a hot air oven at 105°C and cooled in a desiccator. The process of heating and cooling is repeated till a constant weight. Cooled petri dish with dried material is then weighed. Usually this process may take 5-6 hours or more depending on the moisture content in the sample.

Moisture % = (Loss in weight/ weight of sample) × 100

Crude protein

The protein content of a food product is obtained by estimating the nitrogen content of the material and multiplying the nitrogen value by 6.25 (a factor of 5.7 is used for wheat, 5.75 for rice). This is referred to as crude protein content since the non-protein nitrogen (NPN) present in the material is not taken into consideration. The true protein nitrogen can be determined by subtracting NPN from the total nitrogen. The estimation of nitrogen is done by Kjeldhal method (AOAC, 1980).

Reagents

- Digestion mixture: 1 part CuSO4 and 9 parts of K_2SO_4
- Concentrated Sulphuric acid (H_2SO_4)
- Boric acid 4%: 40 g of boric acid was mixed in 600 ml of hot distilled water. After cooling to room temperature 10 ml of bromocresol green and 7 ml of methyl red indicator was added and final volume was made to 1000 ml.
- Sodium Hydroxide Solution (NaOH) 40%: It is prepared by dissolving 400 g of sodium hydroxide pellets in 1000 ml distilled water.
- 0.1N H_2SO_4 for titration.

Digestion of sample: The purpose of digestion is to break the structure and chemical bonds to simple chemical and ionic structures. Weigh 0.5 g of food samples and transfer them into kjeldhal digestion tube and label them. Add 3g of digestion mixture and 10 ml conc. Sulphuric acid to each digestion tube. Digest the samples using Digestion system at 420°C till it turns clear bluish green and all nitrogen present is converted to $(NH_4)_2SO_4$. the digested samples are then removed from the digestion system/ heater and allowed to cool at ambient temperature for 3-4 hours followed by distillation using distillation system. A blank is also need to be run with the sample.

Distillation of samples: distillation involves separation of ammonia-nitrogen from the digested sample. This accomplished by raising the pH with sodium hydroxide. Carry out distillation for 8 min using 50 ml of 40% NaOH using Distillation system. Collect the distillate in a conical flask containing 25 ml of 4% boric acid.

Titration: Titrate the distillate against 0.1 N H_2SO_4 to a pink end point.

Calculation

N% = {(Volume of 0.1N H_2SO_4 used × 0.0014)/weight of sample}× 100

Crude Protein = %N × 6.25

Note : Vol of 0.1 H_2SO_4 used is taken after subtraction of blank sample. Factor of 6.25 is based on the fact that most proteins contain 16% N and thus 100/16 = 6.25.

Crude fat

Fat is estimated as crude ether extract of the dry material.

Solvent: Petroleum ether (40-60°C)

Procedure: The dry sample (1-3 g) is weighed accurately into a thimble and plugged properly with cotton. The thimble is then placed in previously weighed empty soxlet beakers. Pour 80 ml of anhydrous ether in beakers. Now place these beakers to the fat extraction system. Run the system at 80°C for about 1 hr hours. After boiling for 1 hr, evaporate the solvent at 70°C, cooled in a desiccator (AOAC 1980). Now weigh beakers along with the extract. Calculate fat percentage as per the given formula.

Crude fat% = (Weight of fat, g/weight of sample, g) × 100

Crude fibre

Reagents: 0.255 N Sulphuric Acid (1.25%W/V), 0.313 N Sodium Hydroxide (1.25% NaOH)

Procedure: Place the weighed (2-4 g) amount of fat free sample in Fibra crucibles and fit them to the adopter of Fibra Plus system (fibre estimation system). Pour 150 ml of warm 0.255 N Sulphuric Acid into the extractors from the top and boil at 400°C for 30 min, followed by draining the acid and washing of sample thrice with hot distilled water. Then, pour 150 ml of warm NaOH into the extractor from the top and boil for 30 min. at 400°C, followed by draining the alkali and washing of sample thrice with hot distilled water. After this, dry crucibles in hot air oven till constant weight, cool and weigh. Now place the dried crucible in muffle furnace for ashing at 550°C for 2-3 hours (AOAC 1980). Calculate the crude fibre content as per the following formula.

Crude fibre % = {(Weight of residue after drying – weight of ash)/ weight of sample)} × 100

Total ash

About 3-5 g of the sample is weighed accurately into a tarred porcelain crucible which has previously been heated to about 600°C and cooled. Ignite samples on a hot plate or low flame till all the material is completely charred. Then place them in a muffle furnace for ashing at 550°C to 600°C for 5-6 hrs. It is then cooled in a desiccator and weighed. To ensure completion of ashing, the crucible is again heated in the muffle furnace for 1 h, cooled and weighed. This is repeated till two consecutive weighs are the same and the ash is almost white or grayish white in clour (AOAC 1980). Weigh the crucible along with the residue left in the crucible after cooling and calculate the ash content using the following formula.

Ash% = (Weight of ash, g/weight of sample, g) × 100

Estimation of sugar

Extraction: Mix 1 g of ground sample (cereal or pulses) with 10 ml of 80% ethanol in 100 ml conical flasks fitted with loose condensers. The flasks are heated on boiling water bath for 20-30 min. Extract is decanted into another beaker and residue is again used. Sugars are extracted twice with 80% ethanol followed by complete extraction with 70% ethanol once. The pooled alcoholic extract of sample is concentrated on water bath at 60°C till 6-8 ml of extract is left. To this, 1 ml of saturated lead acetate solution is added to remove protein and volume made to 100 ml. The contents are kept overnight for proper precipitation of protein. The contents are filtered through watman No.1 filter paper and excess lead ions from filtrate is precipitated with sodium oxalate crystals (may be one pinch) and filtered off after 30 minutes. The clear alcohol free extract so obtained is used for the estimation of free sugars. Sugar free residue obtained is analysed for the starch content.

Estimation of sugars (Phenol-sulphuric acid method)

Reagents

- 5% phenol in glass distilled water
- 95.5% Sulphuric acid.
- Standard glucose solution -100 mg glucose in 100 ml of water.

Working standard solution: 10 ml of standard glucose solution is diluted to 100 ml with distilled water (100 μg/ml).

Method

Pipetted out 0.1 ml of sample solution extract in two separate test tube. Add water to each test tube to make the final volume 1 ml. Blank was set with 1 ml of distilled water. Now add 1 ml of 5% phenol solution to each test tube. The tubes are then placed in ice cold water and 5 ml of 95.5% sulphuric acid is to be added to each test tube. The content is mixed equilibrated at to room temperature. Read the colour at 490 nm (Dubois et al, 1956).

Standard curve: Pipetted out 0.2, 0.4, 0.6, 0.8 and 1 ml of working standard solution into a series of test tubes, develop the colour in the same way as in case of samples. The concentration of total sugars in the test extract is then calculated as glucose from the standard curve and multiplying with dilution factor.

Estimation of reducing sugars (Nelson, 1944).

Reagents

I. *Reagent A:* It is prepared by dissolving anhydrous sodium carbonate (25 g), sodium potassium tartarate (25 g), sodium bicarbonate (20 g) and anhydrous sodium sulphate (200 g) in 800 ml distilled water and make the volume to 1 L.

II. *Reagent B:* Dissolve 15 g copper sulphate in distilled water and make the volume to 100 ml. To this add 2-3 drops of concentrated sulphuric acid.

III. *Reagent C*: It is prepared by mixning 25 parts of reagent A and one part of reagent B, just before use.

IV. *Nelson's arsenomolybdate reagent*: Dissolve 25 g ammonium molybdate in 450 ml distilled water. To this add 21 ml of concentrated sulphuric acid slowly with stirring. Separately 3 g of crystalline sodium arsenate is dissolved in 25 ml of distilled water and it is added to the above ammonium molybdate solution with stirring. Incubate the solution at 37°C for 48 h and store in an amber coloured bottle.

Procedure: Take 0.2 ml of each of the test extract in test tubes and make the volume to 1 ml. To each test tube add 1 ml of freshly prepared reagent C and keep tubes in a boiling water bath for 20 min. Add 1 ml of Nelson's arsenomolybdate reagent to each tube after bringing them to room temperature. The contents are then diluted with water to 10 ml and intensity of the colour is read at 620 nm against reagent blank as zero. Simultaneously glucose standards (1-6 μg) are also run taking the standard working solution and develop the colour in the same way as in case of samples. The concentration of total reducing sugars in the test extract is then calculated from the standard curve of glucose.

The concentration of non-reducing sugars is determined as the difference in the concentration of total sugars and reducing sugars.

Carbohydrates

Add the values of moisture, crude protein, crude fat, crude fibre and ash and subtract from 100. The difference will give the values of available carbohydrates.

Estimation of minerals

Processing of samples

Ash solution

Food sample (5g) is taken in duplicate in crucibles. These are charred on a hot plate and these are placed in a muffle furnace at 600°C for ashing. After ashing and cooling the crucible to room temperature, the ash in the crucible is dissolved in 25 ml of dilute hydrochloric acid on a water bath. The contents are transferred to a 250 ml volumetric flask and the volume is made to the mark with distilled water. The content is filtered through Whatman filter paper No. 42 and the filterate is used for the determination of calcium, phosphorus and iron.

Calcium estimation from food

Calcium is precipitated as calcium oxalate. The precipitate is dissolved in sulphuric acid and the solution is titrated with standard potassium permanaganate.

Reagents

- Saturated ammonium oxalate solution
- Dilute hydrochloric acid (1 part HCI + 4 part water)
- Methyl red indicator (0.5% in absolute alcohol)
- 0.1 N potassium permanaganate solution
- Dilute ammonium hydroxide (1 part NH_4OH + 1 part water)
- Dilute sulphuric acid (10%)
- 01N Oxalic acid solution: Sodium oxalate is dried in an oven at 100°C for 12 h. Exactly 0.67 g is dissolved in distilled water, 5 ml of conc. H_2SO_4 is added and solution made up to one L. After it has cooled down.

Standardization of potassium permanganate solution: 25 ml of 0.01 N oxalic acid solutions is transferred to a conical flask. One ml conc. H_2SO_4 is added, warmed to about 70°C titrated against $KMnO_4$ solution, till the faint pink colour remains.

Normality of $KMnO_4$ solution = 25x0.01/titre value (ml)

Procedure: Take 50 ml of clear sample filterate prepared from ash into a beaker. Add 10 ml of saturated ammonium oxalate solution. Boil and add two drops of methyl red indicator. The contents are neutralized with dilute ammonium hydroxide and boil the content again to have coarse crystalline precipitate. Add a few drops of dilute hydrochloric acid until the colour is adjusted to a faint pink. The solution is allowed to stand overnight. The precipitates are filtered through Whatman filter paper No. 42 and washed thoroughly with hot distilled water till the precipitates are free of oxalates (tested with silver nitrate solution). The precipitates along the filter paper are added in the original beaker and dissolved in 20 ml of 10% sulphuric acid. The contents are heated to about 70°C and titrated against 0.01 N potassium permanganate solution to a definite pink colour persisting for at least 1 min. A blank is also run using similar procedure (AOAC 1980).

Calculation: 1 ml of 0.01 N $KMnO_4$ used = 0.2004 mg Ca

$$\text{Calcium, mg/100g} = \text{ml of 0.01 N KM}nO_4 \text{ used} \times 0.2004 \; x \frac{A}{B} x \; \frac{100}{\text{Wt. of sample (g)}}$$

Where A= Total volume of the extract prepared from the food sample, B= ml of the sample filterate taken for Ca precipitation.

Total phosphorus

The phosphorus containing solution is treated with ammonium molybdate to produce phosphomolybdic acid. This is reduced by the addition of hydroquinone reagent to give blue colour which is proportional to the amount of phosphate present.

Reagents

- *Ammonium molybdate solution*: Ammonium molybdate 25g is dissolved in 300 ml water. To this 75 ml of conc. H_2SO_4 in 200 ml distilled water is added slowly and contents are mixed.
- *Hydroquinone solution*: Hydroquinone (0.5g) is dissolved in distilled water and volume is made to 100 ml. One drop of conc. H_2SO_4 is added to retard oxidation.
- *Sodium sulphite solution*: Dissolve 200g sodium sulphite in water, make the volume to one L and filter.
- Conc. hydrochloric acid
- *Standard solution*: Potassium dihydrogen orthophosphate (0.439 g) is dissolved in glass distilled water and volume is made to 1 L. Dilulte 50 ml of the stock solution to 200 ml with water (25 µg P/ml).

Procedure: The ash solution prepared from the food sample is used for the determination of phosphorus. For the development of colour 0.2 ml aliquot is taken in a test tube and diluted to 1 ml with distilled water. 2 ml of ammonium molybdate, 1 ml of hydroquinone and 1 ml of sodium sulphite solutions are added and mix the contents. The volume is made to 10 ml with distilled water, mix the contents and blue colour is measured after 30 min in spectrophotometer at 660 nm. For the blank, instead of test solution, 1 ml of glass distilled water is taken and rest of procedure is same as for the sample (Fisher, 1971).

Standard curve: Take 0.2, 0.4, 0.6, 0.8 and 1.0 ml of standard solution in the test tubes and develop the colour in the same way as for the sample. These solutions contain 5, 10, 15, 20 and 25 μg phosphorus, respectively and a standard curve is prepared against O.D. The concentration of phosphorus in unknown sample is calculated from the standard curve and the results are expressed as mg P/100g sample.

Iron estimation (food sample)

Reagents

- Conc. HCl
- Ortho-phenanthroline Solution. Dissolved 0.1 g of ortho- phenanthroline in 80 ml glass distilled water at 80°C, cool and volume made to 100 ml.
- 10% hydroxylamine hydrochloride.
- Acetate buffer solution: Dissolved 8.3 g anhydrous sodium acetate in glass distilled water. Added 12 ml acetic acid and diluted to 100 ml.
- Standard Iron solution: Dissolve 702.2 mg of ferrous ammonium sulphate in 100 ml of glass distilled water. Add 5 ml of 1:1 HCl (HCl: distilled water). Make the volume 1 L with glass distilled water and mix thoroughly (0.1 mg iron/ ml). The standard solution is prepared once in six months. Diluted 10 ml of this solution to 1 liter.

Working standard solution (10 μg iron/ml) is prepared by diluting the 10 ml of above solution to 100 ml.

Procedure: Ten ml of ash solution was pipetted in the 25 ml of volumetric flask. To this add 1 ml of hydroxylamine hydrochloride solution. In a few minutes, 5 ml of buffer solution was added, followed by the addition of 1 ml of ortho- phenanthroline solution. The contents were mixed and volume was made up to the mark. The intensity of colour developed was measured at 540 nm(AOAC, 1980).

Blank: 2 ml of conc. HCl was diluted to 100 ml and 10 ml of this solution was treated as per sample.

Standard curve: to develop standard curve, 400 μl, 600 μl, 800 μl and 1000 μl of working standard solution are taken in duplicate and the colour is developed as in test samples. After developing the colour, the intensity is measured at 540 nm. A standard curve is drawn with the readings of working standard solution. The concentration of iron in test sample is calculated from the standard curve and multiplying with dilution factor.

Estimate of vitamin A (as *β-carotene*)

Take a representative fruit sample and crush it in a pestle and mortar. Then weight 25 g of crushed pulp and place it in the oven at 40-45^0 for 3-4 hours to remove excess moisture. Transfer the partially dried residue to the mortar and add 5g of anhydrous sodium sulphate to facilitate further grinding. Yellow pigment (β-carotene) then extracted with 3 percent acetone in petroleum ether in 4-6 times to ensure complete extraction. Combine extraction of all lots and make up the volume to 50 ml. Put this extract in the separating funnel and remove excess acetone by adding distilled water. Repeat this process 3-4 times for complete separation of acetone. Now read its optical density (OD) at 452 nm using petroleum ether as a blank. Read the concentration of β-carotene in test sample from the standard curve.

Determination of fatty acid

The separation of different components is based on a partition coefficient of various components of the mixture between stationery phase (liquid) and mobile phase, which is an inert gas. Fatty acids are made volatile by converting them into methyl/ethyl esters (Appleqvist, 1968). The esters are identified and quantified by injecting into GLC and comparing with a set of standard esters.

Procedure: Mix 50 mg of oil sample thoroughly with 1 ml of petroleum ether and leave it for 30 min. Add 1.5 ml sodium ethylate (0.02 M NaOH in 99.5% ethanol) and cover the tube with cap. Now shake the content for few minutes and leave it for 30 min. Subsequently add 1.5 ml of 8% NaCl (w/v) and mix the content properly. As soon as the two layers get separated, transfer the upper layer (petroleum ether) to another tube and allow evaporating. Finally dissolve the content in 10µl of petroleum ether and inject 2 µl using micro syringe (Hamilton) on gas chromatograph equipped with flame ionization detector fitted with 6% butane diol succinate (BDS) on chromosorb WAW/DWCS column 6 feet in length into ¼ inch outer diameter. The conditions for the separation are: oven temperature 250°C, hydrogen flow 40 ml per min, nitrogen flow 60 ml per min, air flow 300-400 ml per min. Identification of peaks of fatty acids is to be done by comparison of their retention time with those of standard fatty acyl esters.

References

AOAC (1980). Official Method of Analysis, Association of Official Analytical Chemists, 13th Edition, Washington, DC.

AOAC (1990). Official Method of Analysis, Association of Official Analytical Chemists, 15th Edition, Washington, DC.

Appleqvist, LA (1968). Rapid methods of lipid extraction and fatty acid esters preparation for seed and leaf tissue with special remarks on preventing the accumulation of lipid contaminants. Ark Kenci, 28: 351-70.

Dubois M, Gilles KA, Hamilton JK, Rebers PA and Smith F (1956). Colorimetric method for determination of sugars and related substances. Anal. Chem., 28: 350.

Hira CK, Kawatra BL, Sadana B, Sharma KK (1996). A manual of laboratory techniques for nutrition research, Department of Food and Nutrition, Punjab Agrricultural University, Ludhiana, p 10.

❑❑❑

33

Protein Quality - Significance and Assay

KIRAN BAINS

India, like many other developing countries has not, as yet, eliminated problems of undernutrition, especially among children of its poor communities. The diet surveys by National Nutrition Monitoring Bureau (NNMB) during 2001-02 (Brahmam, 2007) revealed that among 1-6 years children, except for roots and tubers, the consumption of various foods like cereals and pulses were less than the recommended levels. The intake of protective foods like green leafy vegetables, milk and milk products, income elastic foods like fats and oils and meat/fish/eggs was very low, often less than 20-30% of Recommended Dietary Intake (RDI) levels. The intake among the adults was relatively better where the requirements of cereals were met but the legumes were still inadequate. Distribution of individuals according to protein calorie adequacy status of individuals revealed that while only 30% of children were having protein calorie adequacy, the proportion among adults was about 80-90%.

Since diets from different regions of India are cereal and legume based, the values for the true digestibility of protein in diets from India were considerably lower compared to that of protein in typical North American diets (57-75 versus 88-94%). The presence of less digestible protein fractions and high levels of insoluble fibre and anti nutritional factors in cereals and legume based Indian diets are responsible for poor digestibility of proteins (Gilani *et al.,* 2005). Moreover, diets in India depend mainly on cereals and secondly on legumes for protein, they insufficiently provide amino acids. In particular, lysine and sulphur containing amino acids likely to be limited in cereals and legumes, respectively. Further, the loss of available lysine, the worst nutritional consequence of maillard reaction during food preparation is particularly significant in cereals, where the amino acid is most limiting (O'Barien & Morrissey, 1989) and to some extent in legumes which are supplementing cereals for lysine in Indian diets. Since legumes contain relatively high amounts of lysine, especially in comparison with cereal protein sources thereby legume preparations in Indian diets are a valued source of lysine.

The dietary intake of amino acids is relevant since essential amino acids stimulate skeletal muscle protein synthesis to a greater extent than non-essential amino acids (Tipton *et al.,* 1999). Lysine intakes are marginal in low socio economic group Indians. There are several lines of reasoning to suggest that lysine intake may influence muscle function. Well nourished Indians showed a trend toward improve muscle function with higher lysine intakes (Kurpad *et al.,* 2003). If protein quality is assessed through the use of protein digestibility amino acid score (PDCAAS), and assuming lysine to be the first limiting amino acid in cereal protein, then it becomes evident, with the new higher daily lysine requirement, that the protein quality of Indian diets is particularly with poor PDCAAS adjusted Protein:Energy (PE) ratios of about 0.08 for average Indian diets. This would mean that between 15-45% (depending on age and gender) of the sedentary Indian population would be at the risk of protein deficiency since a sedentary activity pattern would automatically raise the required PE ratio of the dietary requirement. (Millward and Jackson, 2003).

It has been noted that an emerging Asian Indian phenotypes have high body fat with relatively less BMI, less lean body mass and marked abdominal obesity. It is likely that amino acid imbalance may be a contributing factor. Diets with increased protein and reduced carbohydrates have been shown to improve body composition, lipid and lipoprotein profiles, and glycemic regulations associated with treatment of obesity and weight loss. High-protein, low-carbohydrate diets have been found to have positive effects on reducing risk factors for heart disease, including reducing serum triacylglycerol, increasing HDL cholesterol, increasing LDL particle size, and reducing blood pressure. (Donald *et al.,* 2008).

Traditionally, this has been discussed solely in the context of a protein's ability to provide specific patterns of amino acids to satisfy the demands for synthesis of protein as measured by animal growth or, in humans, nitrogen balance. As understanding of protein's actions expands beyond its role in maintaining body protein mass, the concept of protein quality must expand to incorporate these newly emerging actions of protein into the protein quality concept. New research reveals increasingly complex roles for protein and amino acids in regulation of body composition and bone health, gastrointestinal function and bacterial flora, glucose homeostasis, cell signaling, and satiety. The evidence available to date suggests that quality is important not only at the minimum Recommended Dietary Allowance level but also at higher intakes (Millward *et al.,* 2008).

Ninety per cent of world population is dependent on animal foods for high quality protein (Singh, 2008). The concern for optimum protein quality to prevent metabolic abnormalities holds a significance for large lacto vegetarian population of India which depends on plant foods for obtaining essential amino acids. The traditional vegetarian Indian meals usually take care of including foods from different sources in order to improve their protein quality. Optimum protein quality from meals along with physically active lifestyle may help the people to have right proportion of fat mass and lean mass.

Protein quality is a measure of the essential amino acid content of a protein relative to the essential amino acid needs of the body Proteins are not alike. They vary according to their origin (animal, vegetable), amino acid composition (particularly their relative content of essential amino acids), digestibility and texture. Good quality proteins are those that are readily digestible and contain the essential amino acids in quantities that correspond to human requirements.

Supplementation value: If a protein of diet is deficient in one or more of the essential amino acids, N_2 equilibrium cannot be achieved. But if another protein containing the missing amino acids in adequate amounts is added to the diet, N_2 equilibrium can be achieved. This capacity of the protein to make good one another's deficiency is known as supplementary value. By 1915 Mendal divided protein into two groups, i) Allowed growth ii) Failure to growth

Afterwards they divided protein into animal protein and vegetable protein.

So, to check the protein to be tested can be fed to animals and its capacity to maintain nitrogen balance.

Methods for assessing protein quality

Many attempts have been made to give a numerical value to the quality of both individual dietary proteins and mixtures of proteins present in various human diets.

The methods for protein quality evaluation are grouped under the following headings:

1. Analytical methods for the determination of nitrogen and amino acids in foods
2. Chemical and microbiological assays of protein quality
3. Evaluation of protein quality in experimental animals
4. Clinical methods of evaluation of protein quality

Both qualitative and quantitative evaluation of protein and amino acid content in foods is carried out (Bodwell *et al.,* 1989).

Quantitative / Analytical method for estimation of nitrogen and protein

Nitrogen in food not only comes from amino acids in protein but also exists in additional forms that may or may not be used as a part of the total nitrogen economy of animals and humans. Thus, determination of nitrogen is commonly used to determine the protein content of a sample. Total nitrogen, which is determined by the Kjeldahl procedure, is converted to crude protein by a factor of 6.25. Thus, the conventional measure of 'protein' or 'crude protein' in foods is N x 6.25, and it is recommended that this one factor be used in nutritional studies in which whole diets contain more than one source of nitrogen. The Kjeldahl's method is commonly used in estimation of protein content of foods and has been extensively used for protein estimation of various food stuffs. Although over a period of time many other methods have emerged for determination of organic nitrogen, this method still remains a favourite because it is reliable and has very well standardized procedures.

Qualitative estimation of protein / Amino acid content in foods

The protein and amino acid content of foods helps to interpret nutritional differences among foods in terms of their amino acid make up. This has been obtained only through experiments on animals or human beings. The methods available to determine protein quality are based on parameters such as growth and body weight changes, carcass nitrogen analysis, nitrogen balance, regeneration of blood and liver constituents and availability of amino acids. The most commonly used methods are discussed below:

PER (Protein Efficiency Ratio): This method is developed by Osborne, Mendal and Ferry in 1919 based on the growth of young rats. The diets contain

10% of the protein to be tested and are complete in all other dietary essentials. Group of albino rats (21 days old) are fed for a period of 4 weeks on different diets. Records of the gain in body weight and protein intake of the rats are maintained. The PER is calculated :

$$PER = \frac{\text{Gain in body weight (g)}}{\text{Protein intake (g)}} = \text{gain in weight/g of protein consumed}$$

Table 1 : PER values of egg, whole wheat and gelatin

Diet	Protein intake in 4 wks (g)	Gain in body weight in 4 wks (g)	PER
Egg	28	132	4.7
Whole wheat	26	40	1.5
Gelatin	20	- 4	- 0.2

Digestibility coefficient (DC) : Digestibility coefficient of the protein refers to the percentage of the ingested protein absorbed into the blood stream after the process of digestion is complete. Protein differs in their digestibility eg. some proteins like milk and eggs are easily digested and converted into amino acids while protein of pulses are slowly digested to amino acids since only amino acids are absorbed into the blood stream from the intestine, unhydrolyzed portion of protein is wasted in the faeces.

In the experiment when an animal is fed on N-free diet, certain amount of N is excreted in the faeces. This is derived mainly from the digestive juices. This is called endogenous faecel N. When a portion of food is given, the N_2 formed in faeces consists both of endogenous N_2 and food N_2 lost in digestion.

$$DC = 100 \times \frac{N_2 \text{ Intake- } (N_2 \text{ in faeces- endogenous faecal } N_2)}{N_2 \text{ Intake}}$$

$$= 100 \times In \frac{(Fn\text{- } fe)}{In}$$

When fn- fe is the food N_2 lost in the digestion. Protein digestibility depends on the

1. Presence of indigestible carbohydrates e. g. cellulose, hemicellulose
2. Presence of proteolytic enzyme inhibitors.

Biological value (BV) : A method for determining the biological value of protein was developed by Mitchell in 1925. It measures the quantity dietary protein utilized by the animal for meeting its protein needs for maintenance and growth. Group of albino rats (25 days old) are fed successively on two diets for a period of 10 days, i) protein free diet; ii) Diet containing 10% protein to be tested.

Urine and faeces are collected by keeping the rats in metabolic cage. Records of food intake are maintained. The diet, urine and faeces are analyzed for N_2.

$$BV = \frac{N_2 \text{ digested} - N_2 \text{ lost in metabolism}}{N_2 \text{ digested}} \times 100$$

N_2 digested = N intake (In) – N in faeces (Fn) on the protein diet - N in faeces (Fe) on protein free diet is N_2 digested = In- (Fn- Fe)

N_2 lost in metabolism = N_2 in urine (Un) on protein diet

- N_2 in urine (Ue) on protein free diet

= Un- Ue

$$BV = \frac{In - (Fn- Fe) - (Un- Ue)}{In - (Fn- Fe)} \times 100$$

Net Protein Utilization (NPU): Mitchell in 1922 introduced the term "Net Protein Utilization" which was the product of DC and BV divided by 100.

$$NPU = \frac{DC \times BV}{100}$$

In 1955, Miller and Bender developed a direct method. 25 days old albino rats are used. One group is fed on a non-protein diet while other group is fed on the test diet containing different proteins at 10% level for a period of 10 days. The food intake of the animal is measured. The animals are killed at the end of 10 days and the body N_2 is determined by Kjeldahl method on a sample of dried and powdered carcass.

$$NPU = \frac{\text{Body N of the test group} - \text{Body N of the non protein group} + \text{N consumed by Non protein group}}{\text{N consumed by test group}} \times 100$$

Chemical score : Egg proteins contain all essential amino acids in adequate amount and possess the highest nutritive value among dietary proteins. Block and Mitchell (1946) assigned a chemical score of 100 to eggs protein. The nutritive value of the protein will depend on the essential amino acid most limiting in the dietary proteins. Based on the most limiting amino acid, chemical score is evolved which can be an index of nutritive value of the protein. So chemical score is the ratio between the content of the most limiting amino acid in the test protein to the content of the same amino acid in egg protein expressed as a percentage e.g. in milk most limiting amino acid is sulphur amino acids ie. methionine and cysteine.

$$\text{Chemical score (CS)} = \frac{\text{mg of amino acid in 1 g of test protein}}{\text{mg of amino acid in 1 g of reference protein}} \times 100$$

$$\text{e.g. Chemical score of milk protein} = \frac{3.4}{5.5} x\, 100 = 6.5$$

The chemical score of gelatin and zein is 0

Amino acid score (AAS): A chemical technique considered fast, consistent, and inexpensive. It measures the indispensable amino acids present in a protein and compares the values with a reference protein. The protein is rated based upon the most limiting indispensable amino acid. The amount of different amino acids in a particular food is calculated. It is then compared to the amount of essential amino acids required by a 2-5 year old children (FAO/ WHO/UNU, 1985).

Table 2 : Reference patterns for essential amino acids

Amino Acid	Egg mg/N	SAAP* mg/g pr	SAAP mg/N
Histidine	150	19	119
Isoleucine	410	28	175
Leucine	520	66	413
Lysine	440	58	363
Methionine + Cysteine	350	25	156
Phenylalanine + Tyrosine	610	63	394
Threonine	320	34	213
Tryptophan	90	11	69
Valine	450	35	219

*SAAP- Suggested Amino Acid Pattern for 2-5 year old children ((FAO/WHO/UNU, 1985).

Protein digestibility corrected amino acid score (PDCAAS): PDCAAS is a standard used to know the quality and efficiency of protein in a particular food (Sarwar and McDonough,1990). The PDCAAS evaluation procedure is considered to be superior to other methods. PDCAAS has been adapted by the U.S. Food and Drug Administration (FDA), the United Nations Food & Agricultural Organization (FAO) and the World Health Organization (WHO) as the preferred method for evaluating food protein. PDCAAS measures the quality of a food protein based on the amino acid requirements of 2–5 year olds, who are considered the most protein-requiring age group. The amount of different amino acids in a particular food is calculated. It is then compared to the amount of essential amino acids required by a 2-5 year old. The following is the requirement of different amino acids for a 2-5 year old. The digestibility of the food protein is also taken into account. That is, how much of the protein is absorbed by the body after digestion. The fecal method is used to evaluate digestibility. A food must have all the essential amino acids in required amounts to be considered a good source of protein. So, only the amino acid with the least presence in the food is taken into consideration for getting PDCAAS. The highest possible PDCAAS is 1.0. A score of 1.0 means that after digestion of the food, it provides 100% or more of the recommended amount of essential amino acids per unit of protein. A score above 1.0 is rounded off to 1.0. This is because excess protein is not absorbed by the body. It is either converted into fat or thrown out as waste.

Table 3 : PDCAAS of various foods

Food stuff	PDCAAS value
Whey	1.0
Egg White	1.0
Casein	1.0
Milk	1.0
Soy protein isolate	1.0
Beef	0.92
Soybean	0.91
Tubercles	0.74
Vegetables	0.74
Legumes in general	0.69
Kidney beans	0.68
Rye	0.68
Cereals and derivatives	0.58
Whole wheat	0.54
Lentils	0.52
Peanuts	0.52

Protein-Energy ratio (NDP Cal%) : Platt and his colleagues (1961) introduced the ratio of protein energy to total energy (PE ratio) as a useful measure of dietary quality in human nutrition. To take into account both the quality and concentration of the protein, they introduced the concept of the net dietary protein calories as a percentage of total calories (NDPCal%). The

NDP Cal%, is the net dietary protein value expressed as percent of total calories The NPU or the chemical score could be utilized in calculating the NDP Cal% as both of them are indicative of protein value. The formula used is:

$$\text{NDP Cal\%} = \frac{\text{Protein calories in the diet}}{\text{Total calories}} \times \text{NPU/Chemical Score} \times 100$$

For adults, diet with an NDP Cal% of 5% would be adequate to maintain health. In infants, children, adolescent and pregnant women, growth is supported only by a diet providing an NDP Cal% of 8% or above.

Rat as animal model for determining protein quality

Animal models have been commonly employed for determining protein quality in human. This allows assays to be undertaken routinely unlike those in humans that are more technically and economically challenging. The laboratory rats, usually considered as a suitable animal model, has been widely used, especially as it is easy to raise and to manipulate (Rangaramula *et al.*, 1983). Housing and environmental requirements for rats has been shown in the Table 4.

Table 4 : Housing and Environmental Requirements for Rats

	Rat
Adult weight (range)	200-300 g
Type of housing	Cage
Single floor area per animal (Sq. cm)	100-150
Cage height minimum (cm)	14
Room temperature °C	22-24
Relative humidity (%)	50-60
Suitable bedding material	Paddy husk
Sand/dust	
Nesting material	Paper Cutting
Ventilation	
Air changes per hour	10-12
Light	
Intensity (cd)	300-400
Photocycle (Light:Dark)	12-12

Nutritional requirements of rats : Adequate nutrition is one of the important environmental factors influencing the ability of experimental animals to attain their genetic potential for growth, reproduction, longevity and response to stimuli. Recommended dietary allowances of rats have been given in Table 5.

Table 5 : Recommended dietary allowances for rats

	Nutrient (percentage)
Protein	20
Carbohydrate	60
Fat	5
Fibre	5

Preparation of diets: Control and experimental diets are prepared depending upon the experiment. Protein content of all diets is maintained at 10% level. For the control, skimmed milk or casein is taken as reference protein. The percentage of crude protein, crude fibre and ether extracts of different diets are maintained at a constant level of 10, 5 and 10%, respectively. Vitamin mixture and salt mixture are added at the rate of 1% and 4% in each diet. Ingredients are mixed thoroughly to ensure uniform distribution of vitamins and salt mixtures.

Feeding and management of rats: Disease free weanling albino rats about 28-30 days old are selected for the purpose of protein quality evaluation. Their initial weights are taken. Grouping is done having 50% of female and 50% male rats and average weight of all the groups should be uniform. Each group should have atleast eight rats. The animals are housed in individually in aluminium cages having wire mesh bottoms or in metabolic cages with provision to collect urine and faeces with feeding and drinking cups.

References

Bodwell CE, Carpenter KJ and McDonough FE (1989). A collaborative study of methods of protein evaluation: introductory paper, Plant Foods Hum. Nutr, 39: 3-11.

Brahmam SNV (2007). National Nutrition Monitoring Bureau in India – An Overview. Indian Journal of Community Medicine 32 : 1.

Donald KL, Peter C, Mary CG, Ronald MK and Frank QN (2008). Protein in optimal health: heart disease and type 2 diabetes. Amer J Clin Nutr, 87: 1571-1575.

FAO/WHO/UNU (1985). Energy and protein requirements: Report of a joint FAO/WHO/UNU Expert Consultation WHO Technical Report No. 724 Geneva: WHO.

Gilani GS, Cockell KA and Sepehr E (2005). Effects of antinutritional factors on protein digestibility and amino availability in foods. The Journal of AOAC International, 88: 967:987.

Gilani S, and McDonough FE (1990). Evaluation of protein digestibility-corrected amino acid score method for assessing protein quality of foods. J. Assoc. Off. Anal, Chem, 73: 347-356.

Kurpad AV, Regan MM, Nazareth D, Nagaraj S, Gnanou J and Young VR (2003). Intestinal parasite increase the dietary lysine requirement in chronically undernourished adult Indian subjects. American Journal of Clinical Nutrition, 78: 1145-1151.

Millward DJ, Donald KL, Daniel T and Gertjan S (2008). Protein quality assessment: impact of expanding understanding of protein and amino acid needs for optimal health. Am J Clin Nutr, 87: 1576-1581.

Millward DJ and Jackson AA (2003). Protein/energy ratios of current diets in developed and developing countries compared with a safe/protein energy ratio: implications for recommended protein and amino acid intakes. Public Health Nutrition, 7: 387-405.

O' Brien JO and Morrisey PA (1989). Nutritional and toxicological aspects of Maillard browning reaction in food. Critical Review of Food Science and Nutrition, 28: 211-248.

Rangaramula N, Nair KM and Kalyansundram S (1983). A Manual of Laboratory Techniques. National Institute of Nutrition, Indian Council of Medical Research, Hyderabad.

Singh K (2008). In defense of meat-eaters. The Tribune dated 31.05.08, p3.

Tipton KD, Gurki BE, Matin S and Wolfe RR (1999). Non essential amino acids are not necessary to stimulate net protein synthesis in healthy volunteers. J Nutr Biochem, 10: 89-95.

□□□

34

Evaluation of Microbial Quality from Food Sample

PRANITA JAISWAL

Microbial contamination of food/food products can take place at different stages right from the stage of production to the processing. The various sources can be such as

- Crop Production Water
- Manure and Municipal Bio-solids
- Animal Fecal Contamination
- Worker Health and Hygiene
- Post-harvest Water During Packing
- Field and Harvest Sanitation
- Packing Facilities
- Transportation
- Storage & Distribution
- Processing

Water used in the irrigation process is also one of the sources of microbial contamination (besides the surrounding air and soil), which is generally overlooked. Washing of fruits after harvesting is generally applied to reduce the microbial contamination, but if not done carefully, will result in the increasing the contamination process. Washing of fruits and vegetables are generally done by soaking, which is achieved by agitating in the water. Soaking and washing by agitation will result into distribution of spoilage microbe from damaged to the whole lot. Besides storage with damaged foods, possible contact with decaying products, handling by salespeople, spraying with water etc, all these further add to the contamination. Then in processing plant processed like trimming, mechanical abrasion, cutting may result in further addition of microbes from the equipments involved.

Primary causative agents

- Bacteria
- Fungi

Besides viruses have the capacity to damage animal and plant tissues, they along with *Mycoplasma* are not generally regarded as being important in food spoilage.

Counting microbial population in food: Counting in food microbiology refers to determining the number of colony forming unit (CFU), or viable microbial cells present in a unit volume or weight of a sample. Each colony forming unit represents a bacterium that was present in the diluted sample. The numbers of colony forming units (CFU's) are divided by the product of the dilution factor and the volume of the plated diluted suspension to determine the number of bacteria per ml that were present in the original solution. Microbial counting is useful in the basic sciences and is used determine the number of microbes present for physiological or biochemical studies. For example, if one knows the number of microbes present in a culture then one can calculate the amount of protein or DNA that can be isolated from that population. Microbial enumeration is also routinely used in the areas of public health.

Assessment of microbial load

Assessment of micro-flora from any source requires culturing of the microbes present in them on artificial media. In order to culture this microorganisms, we need to understand the basic requirements for their growth. Factors which affect the growth of microorganisms includes physical factors (*e.g.* pH, temperature, osmotic pressure etc.) and chemical factors *e.g.* availability of a useable carbon source, temperature, useable sources of nitrogen, sulfur & phosphorus, availability of trace elemental nutrients, presence (or absence) of oxygen gas etc. An appropriate medium contains nutrients to support growth of targeted microorganisms, which should be properly sterilized before inoculation with desired sample.

Assessment of microbial load from any food samples involve five basic steps and those are :

- Food Homogenization
- Preparing Dilution
- Plating on suitable media
- Counting
- Examination of colonies

Methods for enumeration of microbes from food sample depend upon the type of food to be analyzed. For example liquid food samples are just manually mixed before analyses while solid foods are mechanically homogenized in a blender or stomacher so as to uniformly distribute the food particles in the diluent. Thereafter the food samples are appropriately diluted. And suitable dilutions are plated on appropriate media which depends upon the qualitative and quantitative expectation of microbial population from food sample. Thereafter the plates are incubated and colonies are counted to determine the microbial population in the sample.

Microbiological media

Culturing of microorganisms requires an appropriate growth medium, with following basic characteristics:

- Material containing all nutrients required for the desired organism to grow
- Can be liquid or solid (i.e., solid agar)
- Media must initially be sterile and can be sterilized by heat or by filtration (i.e., no *live* organisms)
- Growth should only occur following inoculation of the medium with the desired sample

Culture media are commonly described as non-selective, selective, differential or selective differential.

Non-selective media: This is an undefined medium because the amino acid source contains a variety of compounds with the exact composition being unknown. Non selective media are also known as nutrient media; it is nutritionally rich and contains all the elements required for growth of microflora, so they are used for the general cultivation and maintenance purpose in laboratory culture collections.

Selective media are used for the growth of only select microorganisms. For example, if a microorganism is resistant to a certain antibiotic such as ampicillin or tetracyclin, then that antibiotic can be added to the medium in order to prevent other cells, which do not possess the resistance, from growing. In other words it suppresses the growth of competing organisms while still aloowing microorganisms of interest to grow.

Differential media or *indicator media* distinguish one microorganism type from another growing on the same media. This type of media uses the biochemical characteristics of microorganism growing in the presence of specific nutrients or indicators. Example of differential media include: eosin methylene blue (EMB), which is differential for lactose and sucrose fermentation

Selective-differential media is one which contains both selective and differential agent. It provides an efficient means of selection and differentiation of microorganisms of interest when present in complex microbiota.

Common media used for isolation of bacteria and fungus from food samples are standard Plate Count Agar and Mortin Rose Bengal Agar, respectively. The media is prepared by dissolving required quantity of different ingredients in 1000 ml distilled water. Dye is added after adjusting the pH of the medium.

Chemical composition of Plate Count Agar and Mortin Rose Bengal Agar

Ingredients	Quantity,g	
	Standard Plate Count Agar	Mortin Rose Bengal Agar
Peptone/Tryptone	5.0g	5.0g
Yeast Extract	2.5g	–
Beef Extract	2.0g	–
KH_2PO_4	–	1.0g
$MgSO_4.7H_2O$	–	0.5g
Rose Bengal	–	0.035g
Glucose	10.0g	10.0g
Agar	18.0g	18.0g
Distilled water	1 litre	1 litre
pH	7.0	6.5

Counting by serial dilution

Microbial load on any food sample depends on no. of factors (as said earlier). Therefore, before microbiological analyses of any food sample they need to be suitably diluted so that enumeration can be accomplished with reasonable accuracy. The serial dilution of food sample is prepared by mixing one part of the sample with nine part of diluent. The schematic representation of serial dilution is given below:

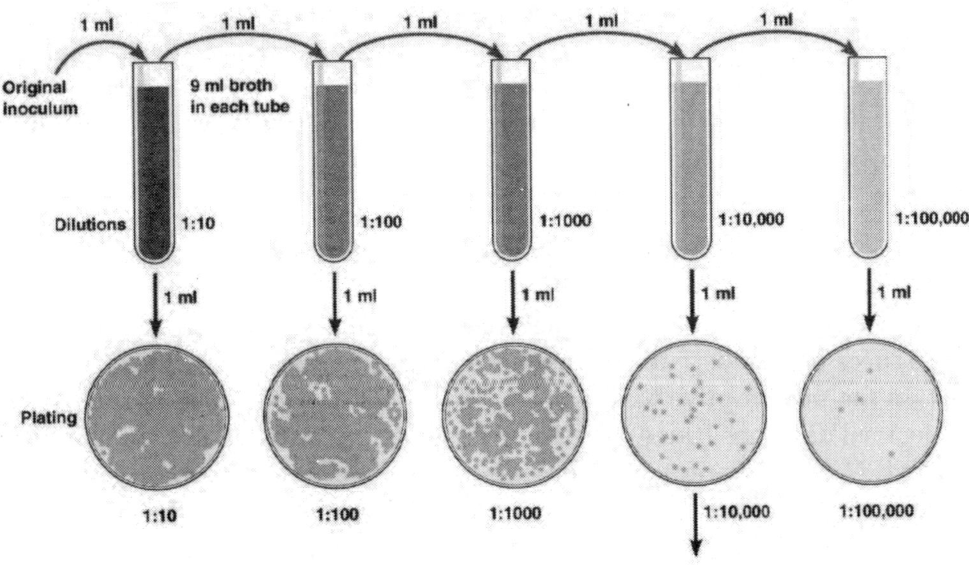

The food samples are diluted in series and each dilution is plated

Methods of plating

There are two basic methods:

(i) Mix 1 ml of culture with molten agar (not too hot, ~45-50°C.) & pour in plate
In this method cells grow both in as well as on the agar surface, but the higher temperature of molten agar may harm the cells.

(ii) Spread small volume of culture (0.1 ml) on solid agar surface.

After plating the plates are incubated at 37°C for 48 h in case of bacteria and at 28°C for in case of fungi (yeast and molds). Following incubation for desired period, the colonies are examined and counted.

The plates with colonies in the range of 30-300 are considered for counting microbial population. Microbial population in a sample can be calculated as:

(# of colonies × dilution factor) / Sample vol. = cell density*

Coliform count in food

Coliform is the name of a test adopted in 1914 by the Public Health Service for several genera of bacteria belonging to the Enterobacteriaceae family. It is the commonly-used bacterial indicator of sanitary quality of foods and water. They are defined as rod-shaped Gram-negative non-spore forming organisms. Some enteron forms can ferment lactose with the production of acid and gas when incubated at 35-37°C. Coliforms are abundant in the feces of warm-blooded animals, but can also be found in the aquatic environment, in soil and on vegetation. In most instances, coliforms themselves are the cause of many nosocomial illnesses, they are easy to culture and their presence is used to indicate that other pathogenic organisms of fecal origin may be present. Typical genera include *Citrobacter, Enterobacter, Escherichia, Klebsiella, Serratia.*

Coliform test-technique multiple tube fermentation technique (MTFT) is a standard method followed all over the world to determine whether the water/food sample is potable or faecally contaminated. The technique involves three successive steps, namely, presumptive test, confirmed test, ad completed test. In the presumptive test, lactose broth tubes are inoculated with a series of decimal dilutions (three) to give an estimate of the most probable number, The MPN confirmation procedure is carried out by transferring positive presumptive tubes (gas production within 24 to 48 h) to 2% Brilliant green bile lactose broth to test for gas production within 48 hr at 35°C for total coliforms and at 44.5°C for fecal coliforms. Positive tubes are used to calculate the most probable number (MPN) value. Presence of coliform bacteria in samples is established in completed tests with EMB.

Detection Methods for Specific Microbes

Microorganisms	Enrichment/ Isolation Media	Differential media	Commercially available facility	
			Company	Methodology
Staphylococcus aureus	Baired Parker medium	BHI, MSA	Taqman Gene Trak TECRA	PCR based NA hybridization Immunoassay
Salmonella	LB, TT, SC, BE, HE, XLD	TSI, LIA	Taqman Gene Trak Warnex Diag. Qualicon, Idaho Technology's	PCR based NA hybridization Real Time PCR PCR based

Contd...

Clostridium perfringens	Tryptose Sulfite Cycloserine	Motility-Nitrate Medium Lactose Gelatin	Probelia	PCR Based
Listeria monocytogenes	UVM1 broth, Fraser broth	PALCAM Agar, MOX, TSAYE, TSA-blood	Qualicon, Idaho Technology's	PCR based
E coli O157 H:7	Tryptone phosphate broth, MacConkey agar	CHROMagar	Qualicon, TECRA, Idaho Technology's	PCR based, Immunoassay, real-time fluorimetery
Yersinia	Irgasan-Ticarcillin-Cholate broth Trypticase Soy Broth Bile-Oxalate-Sorbose broth	Simmon's Citrate agar slants, Christensen's urea agar slants	Gene Trak	Probe
Campylobacter	Modified Campylobacter Charcoal Differential		Gene Trak Biomeriux vitek TECRA	NA hybridization ELISA Immunoassay
Aeromonas	Agar (MCCDA)	Tryptic soy broth, Starch-ampicillin agar, Triple sugar iron (TSI) agar		

References

Brackett RE and Splittstoesser DF (1992). Fruits and vegetables InFP Downes and K. Ito (eds) Compendium of methods of microbiological examination of foods. Ammerican Public Health Association, Washington, DC.

Deak T and beuchat LR (1996). Handbook of food spoilage yeasts. CRC Press, Bocaraton, FL.

Doyle MP (1996). Fecal coliforms in Tea: What is the Problem? Food Technology, 50: 104.

Frazier WC and Westhoff DC (1998). Food Microbiology. Tata McGraw-Hill Publishing company Limited, New Delhi.

Fung DYC (1992). Historical development of rapid methods and automation in Microbiology. Journal of rapid method and automation in microbiology, 1: 1-14.

Pitt JI and Hocking AD (1997). Fungi and Food Spoilage, 2nd ede. Blackie Academic and Professional, Cambrid.

Hamner S, Broadaway SC, Mishra VB, Tripathi A, Mishra RK, Pulcini E, Pyle BH, Ford TE (2007). Isolation of Potentially Pathogenic Escherichia coli O157:H7 from the Ganges River Appl Environ Microbiol, 73(7): 2369–2372.

□□□

35

Sensory Quality Evaluation of Food Products

MRIDULA D.

Food products intended for marketing are the outcome of group of a number of activities such as standardization of formulation; optimization of processing parameters; enhancement of product characteristics; preservation of food quality; and selection of appropriate packaging material. Food quality is multi dimensional comprising physico-mechanical, chemical parameters, microbiological quality and sensory characteristics. In addition, quality in a broader sense encompasses intrinsic factors such as nutritional aspects, which are not immediately or directly comprehensible to a consumer. In today's competitive era, quality is also linked with economic aspect, which is a vital factor both for manufacturer as well as consumer. However, it is a common observation that sensory quality greatly influences the marketing aspects of a product. So, it is desirable for every manufacturer to produce the food products having well identified and highly acceptable sensory characteristics.

Sensory analysis is defined as "A scientific method used to evoke, measure, analyze and interpret the responses to those characteristics of food products as perceived through the senses of sight, smell, taste, touch and hearing" (Stone and Sidel, 1993). Sensory quality of a food product is described in terms of colour and appearance, odour, texture (sensory feeling), mouthfeel, aroma, taste and after taste.

In India, many food manufactures have the misconception that sensory evaluation merely involves tasting and judging the quality of a product in an arbitrary way. But sensory evaluation is not solely taste testing, it encompasses all the 5 senses: vision, audition, taste, smell and touch. It involves measurement and evaluation of sensory characteristics of food; and interpretation of the results. A systematic sensory evaluation provides the link between the manufacturer and the consumer. In order to successful marketing of a product, the manufacturers need to know what is that makes a product acceptable to the buyer. The product should have palatability, should be easy to use and reasonably priced.

Acceptance of food products could be measured in terms of sensorily perceivable characteristics using human subjects as instrument. It is often observed that sensory evaluation is prone to variability in measurement due to individual and psychological factors; even then sensory evaluation is important aspect of product development. Although instrumental analysis is another means of assessing food products quality, it has two main drawbacks namely (1) analysis of individual components do not provide satisfactory estimate of quality and acceptability (2) instrumental methods are not available to integrate the measures of stimuli for texture, taste and aroma to provide data which reflect overall sensory qualities of food products.

Sensory evaluation methods : Sensory evaluation methods are classified into two groups both of which are required for development, testing and marketing of the food products.

(i) *Analytical sensory evaluation*: These can be categorized as an objective method wherein biases are minimized to a greater extent. These methods employ a trained panel for examining the products. It is a quantitative method, the results of which could be reproduced to a greater extent.

(ii) *Affective analysis*: This is a subjective evaluation relating to sensory testing where biasness is not minimized. Untrained large numbers of consumers participate in this type of sensory evaluation.

Physical set-up, product and panel for sensory evaluation

A number of variables must be controlled if the results of a sensory evaluation are to measure the product differences. The variables can be grouped as follows:

(a) *Test controls*: Test room environment, the use of booths or a round table, lighting, room air, preparation area, entry and exit areas.

(b) *Product controls*: The equipment used, the way samples are prepared, screened, coded and served for sensory evaluation.

(c) *Panel controls:* The procedure to be used by a panelist for sensory evaluation.

Physical set up

The physical setting should be designed so as to minimize the subject's biases, maximize their sensitivity and eliminate variables, which do not come from the products themselves. The sensory testing area should be centrally located, easy to reach and free from crowding and confusion, as well as comfortable, quiet, temperature controlled, and above all, it should be free form noise and any type of odour such as production lines, cafeteria kitchens, loading docks etc. The physical setup should include booth area, which is the principal room use for difference test as well as descriptive test, a round table for training and other descriptive tasks, sample preparation area located near by but separate form test room. Office area, sample storage area and data processing area are also required. A signal system is sometime included so that the panel supervisor knows when an assessor is ready for a sample or has a question. Three types of pass-throughs', passing samples to and from panelists are use-sliding door, bread box-type and carousel.

Colour and lighting: The colour and lighting in the booths should be appropriately planned to permit adequate viewing of samples. Walls should be of white; the absence of hues of any colour will prevent unwanted difference in appearance. Fluorescent lighting generates less heat and allows a choice of whiteness. Day light tubes are recommended for food testing. Lights of various colours such as red, green, and/or blue lighting at low intensity obtained through the use of coloured bulbs or special filters. These can be used to mask visual differences between samples in difference tests calling for the subject to determine the taste or by feel which samples are identical.

Air circulation, Temperature and humidity: The sensory evaluation area should be air conditioned at 20 ± 2°C and 45-55% relative humidity (RH).

Sample presentation

The equipment and procedures used for product presentation during the test should be carefully selected to reduce biases and new variables. Serving containers of glass or glazed china is preferable. Proper care should be taken to regulate the amount of sample to be given to each panelist. In general, at least 30 g of a solid food or 15 ml of a beverage should be served (ASTM STP 434, 1968). As part of any test, the order, coding and number of samples presented to each panelist should be monitored. Portions of all the sample should be of same size. When food products consist of a number of small pieces which may differ from piece to piece, panelists should receive a portion large enough that they can evaluate a number of pieces for each characteristic. When beans, for example, are being tested for firmness, panelists should test 3-4 beans before recording their score for the firmness of the sample. After sample is distributed into each serving container and just before serving, the products should be checked to determine if it is at the appropriate temperature. Milk should be served at refrigerator temperature, but bread or cake at room temperature. Some foods require heating to bring out their characteristic odours or flavours. Vegetable oils are often evaluated for odour after being equilibrated at 50^{0}C. The food samples being evaluated may be swallowed or expectorated, however, the panel should be encouraged to develop a consistent technique. Room temperature water is often presented to panelists so that they can rinse their mouths before and between samples. Warm water is particularly helpful when fats or oily foods are being tested. The time between evaluations of each sample may have to be longer than usual if the products being tested have strong flavours. It may also be necessary to restrict to two or three the number of samples presented at one session. When characteristics other than colour are being evaluated it may be necessary to mask colour differences; otherwise they may influence the panelists' judgments of other characteristics. Red, blue, green or yellow light, whichever masks the sample differences most effectively, can be used.

The way, in which a panelist interacts with the environment, the product and test procedure are all potential sources of variations in the test design. Control or regulation of these interactions is essential to minimize the extraneous variable, which may potentially bias the results. Product testing just after meals, tea or coffee may introduce bias and must be avoided. The optimum time for conducting a sensory evaluation is between 10 A.M. to 1 P.M.. It is necessary to prepare panelists for what they expect in the actual test situation, to give them the orientation and time to feel comfortable with the test protocols and to provide them with enough information to respond properly to the variables under study.

Order, coding and number of samples: The order of presentation should be balanced so that each sample appears in a given position and equal number of times. Most sensory analysts rely on three-digit random numbers for product coding. Codes should not be very prominent, either on the product or on the scorecard. The number of samples that can be presented in a given session is a function of both and mental fatigue in the subject.

Panel selection

Panelists are usually from production plant or researchers. General information like name, age, address, occupation, sex, education, family type, income range, ethnic background, food habits, likes and dislikes and sensory experience, etc. are required to record and to be maintained for reference. The following are the general requirements of members in panels:

- Sound health without any defects in sensory perception.
- Average sensitivity.
- Capability of independent judgment.
- Ability to concentrate.
- Intellectual curiosity and interest in quality evaluation.
- Willingness to spend time for evaluation.
- Freedom from prejudices in respect of specific food product types.

Panel screening

Potential panelists are subjected to initial screening process based on

- Discriminating difference between solutions/substances of known chemical composition.
- Ability to recognize flavours and odours.
- Ability to discriminate graded levels of intensity of a given attribute.
- Performance in comparison with other panel members.
- Ability to discriminate differences in samples to be used later in the test.

The sensitivity of panelists needed, will decide the screening limits.

Flavour is the subtle and most complex sensation, which is the major delight one experience in food. It is a combination of the gustatory, olfactory and mouthfeel properties and hence the panel selected is screened and trained thoroughly for these sensory modalities.

Gustatory: Panelist's capacity is assessed in recognizing basic tastes viz., sweet, sour, bitter and salty. Their sensitivity is further judged using different levels of concentrations near threshold. Ageusics (taste blind) subjects are to be eliminated.

Olfactory: Ability to recognize familiar odours and to remember and to recognise unfamiliar odours are of paramount importance. This is ensured by presenting them with a series of odour notes belonging to different food groups. Panelists later on are encouraged to give atleast related description during sessions and then rated on their performance. Anosmics (smell blind) subjects are also to be eliminated.

Mouthfeel: The next step would be to assess the sensitivity of panelists with various components of texture and variations in texture profile arising due to mechanical, geometrical, shape, size and different flow properties of the constituent components of foods. Recognition and evaluation of changes in situations of different interactions arising due to multi- dimensionality of foods are also taken care of. Final selection is made on the performance of all the above tests.

Panel training

Training should be distinguished from experience and its basic aim is to achieve greater homogeneity of responses. Usually training is designed to evaluate a particular product or product types based on the definite problem posed so as to ensure that the panel is adequately oriented in not only identifying the differences but also scaling them. For this, members of the panel are trained sufficiently in various types of different tests with directional ratings and descriptive analysis with scaling techniques etc. Particularly for descriptive analysis, training must be provided towards the use of descriptive languages as well as scoring intensity of different parameters consistently.

Care has to be taken that psychological errors and bias that can affect judgment are avoided to the best degree. For a panel to perform without bias, especially for the selected attribute with certain homogeneity, training is necessary. The ability to communicate the perceived sensation needs constant recalling from memory of sensation and the related descriptor; the training needs to be kept as individual exercise for each class and type of food and has to be specific to attributes to be examined. Therefore panel training has to be rigorous when a panel is taken fresh and also when a more sensitive analysis is to be done. Subsequent training can be planned to keep the panel alert and experienced. The first training for any product should provide the complete range in product type that can occur. Training is necessary in the use of methodology and scale too.

Motivation

A well motivated panelist performs better with the sense of great responsibility. Feedback about their performance will provide much of the motivation for panelists, particularly during training. If there is not sufficient time during the panel sessions to discuss the previous day's results, the data can be posted on a wall chart for the panelists to see at their convenience. However, it is more beneficial if the panel leader personally discusses the results with the panelists, individually or in the group. In addition, a small refreshment at the end of each day's panel session is commonly used for motivating them for the next day's evaluation. In addition to this, sometimes incentives are given to futher boost the sense of involvement.

Performance monitoring

The credibility of sensory analysis and subsequent interpretation depends heavily on panel performance. It cannot be assumed that once the discriminative and descriptive sensory evaluation panel has been screened and trained, it continues to function consistently as an analytical device. The performance evaluation and the selection of panel are constantly monitored with coded controls and the sensitivity can be expressed in terms of repeatability, reproducibility and occasionally with respect to known compounds of taste or odour. The panel leader monitors performance by evaluating the ability of the panel as a whole, and of the individual panelists, to discriminate differences among the samples being tested and to reproduce results consistently. For both types of evaluation, a set of different samples, which the panel leader knows to be different, must be evaluated by each panelist repeatedly on several occasions to provide the necessary data. Statistical analysis (analysis of variance) is used to assess the results. Significant differences among panelists, although not unexpected, may be reduced with further training. Lack of significant differences among samples indicate the need for further training, if the panel leader knows that differences do in fact exist.

Individual panelist's results can also be analyzed. Panelists who are able to distinguish significant differences among the samples with small error mean squares in the analysis should be retained on the panel. If none of the panelists found significant differences among samples for a particular characteristic, additional training for that characteristic is indicated. Panelist performance can also be monitored during the sensory study by comparing replicate judgments. This ensures that panelists continue to perform in a reliable, consistent manner and will indicate when additional re-training may be required or when panelists need further motivation.

Sensory tests

Sensory problems are individualistic, that is sometimes unfruitful following the exact procedures of one worker by another who may have different objectives, products, or situations. The selection of a test will depend upon i) the defined objective of the test, ii) appropriateness to the problem and the situation and iii) the efficiency, simplicity, yet being practical. No one method serves all purposes and is might be wise to investigate several approaches before selecting the one method for your purposes.

Sensitivity tests are often used in laboratory sensory analysis. Common among them, threshold tests are of use in situation where panel of homogeneous sensitivity is to be selected and synergetic or antagonistic effect of odour or taste compounds is to be understood. In quality assurance for prescribing minimum for acceptance through dilution technique', where a correlation can be established between concentration of stimuli with response, threshold tests are very useful.

The concepts of sensitivity is one of immense importance to the researchers and it plays an important role in quality control, assurance, process evaluations, etc. as a valuable tool in fixing tolerance levels of many of the undesirable notes which are generally present, or due to several practical reasons where one ingredient is being replaced by another substitute on economic grounds but costly from the sensory quality point of view. Among sensitivity tests, threshold tests are used for determining the minimum detectable level of concentration of stimuli. Threshold is the point on the stimuli scale where a transition occurs in sensation from a blank or 'white noise'. It is recognized that precise values of threshold for a given stimuli cannot be obtained, being influenced by a number of external (experimental) parameters and internal (individual) parameter.

Sensitivity tests have also been used in the evaluation of ingredients, packaging materials and in the evaluation of odour as a characteristic of polluted water. The dilution type of sensitivity test is used as a means for selecting and maintaining a reference standard for scoring or difference methods. More recently threshold tests are found very useful in bioengineering and biotechnology studies where flavour production is attempted and monitoring of ingredients, nutrients down stream processing are required.

Stimulus threshold or detection threshold: Minimum value of a stimulus needed to give rise to a sensation. The sensation need not be identified.

Recognition threshold: Minimum value of a sensory stimulus permitting identification of the sensation perceived.

Difference threshold: Value of the smallest perceptible difference in the physical intensity of a stimulus.

In Threshold method, each panelist evaluates a series of gradually increasing concentrations of a material or an ingredient. The panel member is asked to record the sample in which he/she first detects a difference or a definite flavour characteristic. The series can be evaluated one level at a time using a difference technique if the test is run in duplicate. It is observed that difference approach is cumbersome and time consuming while the series approach is efficient if bland check samples are served.

Dilution method: The concentration of the test food is varied by the addition of a diluents material. Foods should be homogeneous if this method is to be used. Testing procedure, analysis and questionnaire are the same as those used in the threshold method.

Applications

- Determining sensitivity of panelists to sensory attributes and grouping them into homogeneous panel for sensory analysis.
- Dilution test is useful to establish intensity of sensory responses of food or food components.
- *Scoville value*: Dilution test of spices for pungency and the reciprocal of the highest dilution where recognition threshold is reached is called the scoville value of sample. This test is useful in determining and specifying pungency value.
- Odor threshold to identify production/activity of enzymes in flavour produced by bioengineering, in newer hybrids, alternate chemical reactors etc.

Consumer oriented tests

Preference, acceptance and hedonic (degree of liking) tests are considered to the consumer oriented tests since they should be conducted using untrained consumer panels. Although panelists can be asked to indicate their degree of liking, preferences or acceptance of a product directly, hedonic tests are often used to measure preference or acceptance indirectly.

Preference tests : Preference tests allow consumers to express their choice between samples; one sample is preferred and chosen over another or there is no preference. The paired preference test is the simplest preference test but the category scale and ranking tests are also often used to determine preference. In a paired preference test, a pair of coded sample is presented and panelists are asked to choose the sample which they prefer. Both the sample should be presented in identical containers coded with three digit random numbers. Proforma for paired test for sensory testing of puree sample is given hereunder.

Name : .... Date : ....

Taste the given two puree samples in front of you, starting with the sample on the left. Circle the number of the sample that you prefer. You must choose a sample, even if both samples seem equal.

529	716
(Sample A)	(Sample B)

Proforma for puree paired preference test

Acceptance tests

Acceptance tests are used to determine the degree of consumer acceptance for a product. Category scales, ranking tests and the paired comparison test can all be used to assess product acceptance. Acceptance of a food product usually indicates actual use of the product i.e. eating and purchase. In the acceptance test using ranking, panelists are asked to rank coded samples for acceptance in order from the least acceptable to the most acceptable. Ties where the sample is given equal acceptance ranks are not usually allowed. In this case three or more numbers of samples are presented in coded identical containers. All the samples are simultaneously presented to each panelist in a balanced or random order and re-tasting should be allowed. Proforma for ranking test for energy bar sample is given hereunder.

Name :......................... Date: ....

Please taste each of fruit bar samples. Assign the sample with the most acceptable texture a rank value of 1, the sample with the next most acceptable texture a rank 2 and the third sample with the least acceptable texture a rank value of 3. Do not give the same rank to two samples.

Code	Rank assigned
------	------
------	------
-------	-------

Proforma for energy bar texture acceptability ranking test

Hedonic tests

Hedonic tests are used to measure degree of liking for a product. Category scales, ranging from like extremely, through neither like nor dislike, to dislike extremely, with varying numbers of categories are used (ISI, 1971). Panelists indicate their degree of liking for each sample by choosing the appropriate category. The sample presentation order should be randomized for each panelist or balanced, if possible. Sample may be presented all at once or one at a time. Presenting the sample simultaneously is preferred as it is easier to administer and also allows panelists to re-evaluate the sample if desired and make comparisons between the samples.

Product oriented tests

Product oriented tests are commonly used in food testing laboratories, include difference, ranking for intensity, scoring for intensity, and descriptive analysis tests. These tests are always conducted using a trained panel.

(i) Difference tests

Difference tests can be used to determine whether a noticeable change has occurred in a food's appearance, flavour or texture as a result of storage, of a change in processing methods, or of alteration of an ingredient by simple discrimination. The difference panel is usually small, pre selected and trained. Products which build up flavour factors which carry over from one sample to another do not lend themselves to difference testing. It is inefficient to use difference tests for screening tests. To test for discrimination of differences related to one characteristic, the sample being compared must be identical in all other characteristics. The following are the common difference tests.

Paired comparison test : This test is similar to the paired-preference test except all panelists are asked which of the two samples has the greater intensity of specific characteristics. They indicate whether the sample, are same or different. For example, panelists may be asked 'which sample is sweeter?' or 'which sample is most tender'? but the extent of the difference is not measured using this test. There is a chance probability of one half of placing the samples in a certain order. The number of pairs, which can be served at one session, depends upon the number, which can be examined before 'fatigue" that may interfere with judgment. To avoid order, duplication and bias, all possible combinations of samples must be given.

Triangle test : Three samples, two alike and one odd, are presented simultaneously. The panelist is asked to determine which of the three the odd one is. He/she must give an answer even if it is a guess. The chance probability of selecting the correct odd sample is one third. Statistical analysis can determine if a significant difference exists. This discriminatory method is often used as a tool in quality assurance programme to ensure that samples from different production lots are different or the same as expected. Proforma for sensory testing of energy bar sample to check the storage effect using triangle test is given hereunder.

Name :........................ Date: ....

Out of three given samples of energy bar, two samples are identical and one is different.

Taste the samples listed and place a check beside the code number of the sample that is different.

Code	The different sample is:
-------	-------
-------	-------
-------	-------

Proforma for sensory testing of energy bar sample for storage effect using triangle test

Duo-Trio test : In this test, the panelist is presented with three samples. One sample is labeled as R for reference and the other two samples are coded samples. One of the coded samples is identical to the reference (R) and the other is not. Panelists are asked to taste R first, then the coded samples and identify which of these two coded samples are same as R or different from R. This test indicates differences, but the magnitude of the difference between samples. This method is not as popular as the triangular method, possibly because of the higher chance probability and surmised lower efficiency.

(ii) Ranking for intensity tests

This test is used to determine how several samples differ on the basis of a single characteristic. Panelists have to order three or more samples according to the perceived intensity of a specific characteristic without identification of control from the most intense to the least intense. Panelists are presented all samples simultaneously with code numbers and asked to rank all samples according to the intensity of the specified character and their preference. The panelists are allowed to re-evaluate the samples as necessary to make the required comparisons among them. This test can be used up to six samples. The coding could be done preferably in letters. These are often used to screen one or two "best samples" from a group of samples rather than to thoroughly test all samples. Magnitude of difference between samples is not possible with this test as evaluation of the samples is only on relative performance basis. Proforma for sensory testing of energy bar sample for texture acceptability using ranking test is given hereunder.

Name :........................ Date: ....

Please evaluate the energy bar samples for toughness (the force required to bite with the front teeth).

Evaluate samples and arrange them in order of their toughness. Assign the toughest sample a rank value of 1, the sample with the next toughness a rank value of 2 and 3 accordingly and the sample with the least toughness-assign a rank value of 4. Do not give the same rank to two samples.

Code	Rank assigned
------	------
------	------
------	------
------	------

Proforma for sensory testing of energy bar for texture acceptability using ranking test

(iii) Scoring for intensity tests

Scoring tests for intensity measurements require panelists to score samples, on line scale or category scales, for the perceived intensity of a sensory characteristic. Scoring tests measure the amount of difference between samples, and allow samples to be ordered by increasing or decreasing intensity of a characteristic.

In this test, panelists are asked to score the perceived intensity of the specific characteristic in each coded sample on an interval scale from low intensity to high (of strong) intensity. All samples are simultaneously presented to each panelist in a balance or random order. Panelists are instructed to evaluate each sample independently. In order to minimize inter-comparison of samples, the researcher may present samples one at a time to each panelist, removing each sample after testing and before presenting the next sample. In either case, the panelists are instructed to evaluate each sample and indicate the intensity of the specified characteristic by checking an appropriate category or by making a vertical mark on a line scale. Proforma for sensory testing of energy bar sample for texture acceptability using intensity test is given hereunder.

Name :......................... Date: ....

Please evaluate the energy bar samples for hardness (the force required to bite with the front teeth).

Evaluate samples hardness in the order show on ballot, from top to bottom. Bite down once with the molar teeth. Place a vertical line on the horizontal line scale at the position that indicates the hardness of energy bar sample.

Sample code		
------	Not hard	hard
------	Not hard	hard
-------	Not hard	hard
-------	Not hard	hard

Proforma for hardness scoring test using a line scale
(actual line scale should be 15cm in length)

(iv) Descriptive methods

This method determines the specific difference and how extreme are those differences. This test is similar to scoring for intensity tests except that panelists score the intensity of a number of sample characteristics. Discrimination of qualitative and quantitative traits of a sample is done by use of a small (5-10) trained panel to judge flavor profile, textural profile and QDA (quantitative descriptive analysis). Descriptive analysis requires at least three evaluative processes e.g. discrimination, description and qualification of trait. The steps of discrimination description of traits are qualitative. The language used is developed through careful training and practice with the panel. This requires panelists to develop a common vocabulary, which catalogues various sensory stimuli with appropriate language. Descriptive analysis is a complex cognitive process which requires more mental acuity than sharp taste or olfactory senses. Discrimination among stimuli is only part of the qualitative process. The third step in descriptive analysis is to quantify their traits as to how strong they are. Two products may be quite similar in their qualitative components but differ overall because of the relative intensities of these characteristics. Descriptive analysis is the only sensory method that deals with the total picture or profile of a food product. It may be used to suggest or interpret instrumental methods or other sensory test information or be used as a research guideline tools, or as a method of quality assurance or quality control. Proforma for sensory testing of flavour profile of pineapple squash using descriptive test is given hereunder.

Name :...................... Date: ....

- The samples will be provided to you one by one
- Examine for the presence of each of the attributes listed below
- Indicated intensity of the perceived attributes by cutting the line at an appropriate place and writing the code number
- Cleanse your palate with water in between samples

Colour and appearance

Yellow

------ ....

Light Dark

Clarity

------ ....

Clear Hazy

Aroma

Pineapple aroma

------ ....

Low High

Mouthfeel

Body

------ ....

Thin Thick

Taste

Sweet

------ ....

Low High

Sour

------ ....

Low High

Bitter

------ ....

Low High

Astringent

------ ....

Low High

Overall Quality

------ ....

Very Poor Very Good

Proforma for flavour profile of pineapple squash

Awareness of sensory quality of food particularly for marketing a product and satisfy to the consumers is increasing day by day. Entry of multinationals, higher consumer expectations, changing taste preferences of different age groups, increasing purchasing power are resulting in increased competition amongst the processors. Introducing new products or improving the existing ones has become a continuous ongoing activity of the leading processors. Success in these activities to a greater extent is dependent on effective use of sensory evaluation. Manufacturers are realizing that sensory science is an integral part of food products development and greater application of sensory science in the food industry will provide a competitive edge.

References

ASTM Committee E-18 (1968). Manual on sensory testing methods. STP 434. Am. Soc. For testing and Materials, Philadelphia, Pa.

Indian Standard, Indian Standard Institution, Guide for sensory evaluation of foods. (IS: 6273 Part I&II 1971), Manak Bhawan, New Delhi, 1971.

Stone H and Sidel JL (1993). Sensory evaluation practices, 2nd Edn. San Diego, CA: Academic Press.

□□□

36

An Overview on Quality of Extruded Food Products

MRIDULA D. AND R.T. PATIL

Extrusion cooking is a promising technology to achieve the goal of making the snack food products which are cholesterol free, low in fat, and rich in phytofactors. Extrusion involves forcing food through a hole (or die) to produce strands or other desired shapes. High throughput with continuous operation, lower processing cost, higher quality of heat sensitive nutrients because of high temperature for short time, and low moisture process result in environment friendly process, are some of the key attributes of extrusion technology. The thermal energy generated by viscous dissipation during extrusion cooking with combination of shearing effect cooks the raw mixture quickly so that the properties of the materials are modified by physico-chemical changes of the biopolymers. It results in gelatinization of starch, denaturation of proteins, inactivation of many native enzymes and antinutritional factors, reduction of microbial counts, and improvement in digestibility and biological value of nutrients.

Extrusion cooking offers the possibilities for improving the nutritional status of vulnerable segments of population in under developed nations. Extrusion cooking can produce safe, lightweight and shelf stable foods free from microbial contamination that can be stored for longer duration and can be suitable food for distribution during famines and natural disasters. There is no clear trend for effects of extrusion cooking on nutritional quality of food products. Much of the research on the stability of micronutrients during extrusion processing was conducted in 1970s and 1980s. More recent research has focused on non-nutritive healthful food and novel ingredients blends. Extruder operators often focuses upon screw speed, feed moisture and barrel temperature that determine a secondary set of factors - specific mechanical energy (SME), product or mass temperature (PT) and pressure. These factors influence the viscosity of the food within the extruder barrel, residence time of the material in the extruder and shear applied to the food. These factors in turns bring the qualitative changes of the final product (Table 1).

Table 1 : Effect of extrusion cooking on foods

Important changes	Compound affected
I. Physico-chemical changes	
Change in native structure	Starch (gelatinization), proteins (denaturation)
Volatilization	Flavour
Binding	Flavour
II. Chemical changes	
Recombination of fragments	Dietary fibre, starch
Depolymerization	Proteins, starch, dietary Fibre
Thermal degradation	Amino acids, sugars

I. Effect of extrusion cooking on physical characteristics of products

Water absorption and water solubility index

WAI has been generally attributed to the dispersion of starch in excess water. The dispersion is increased by the degree of starch damage due to gelatinization and extrusion-induced fragmentation, that is, molecular weight reduction of amylose and amylopectin molecules (Yagci and Gogus, 2008). Studies indicated that increasing feed moisture significantly decreased the WAI of extrudate; increasing the level of soybean flour tends to decrease the WAI, might be because of oil in soybean which interfered with water uptake. Singh *et al.* (2007) observed a decrease in WAI with addition of pea grits in extrusion of rice. They explained that a decrease in WAI was due to the dilution of starch in rice pea blends. The WAI measures the volume occupied by the extrudate starch after swelling in excess water, which maintains the integrity of starch in aqueous dispersion (Qing-Bo *et al.*, 2005). Mercier and Feillet (1975) observed that higher amylose results in a higher WAI. Colonna *et al.* (1989) indicated that WAI decreases with the onset of dextrinization. Pelembe *et al.*

(2002), reported increased WAI as the percentage of cowpea increased in sorghum extrudates. It could be expected that water solubility index (WSI) decrease since extrudates of high legume content, contain more starch aggregates or microgels which will be suspended in water (Gomez and Aguilera, 1983). This suggests that water solubility index (WSI) is not only due to starch contents but also due to water-soluble components, like proteins which are present in soybean. The level of dried beetroot and DSF or chickpea *dhal* in the feed material significantly affected the WAI of extrudates. WAI increased significantly ($p<0.001$) with increased level of dried beetroot in extrudates in both type of samples which may probably be due to the presence of higher amount of dietary fibre in dried beetroot (Mridul *et al.,* 2011). The increased WAI may probably be caused by uncovering of hydrophilic groups in extrudates and greater availability and easier penetration of structures by water molecules (Collona *et al.,* 1989, Rayas-Duarte *et al.,* 1998).

WSI is often used as an indicator of degradation of molecular components (Yang *et al.,* 2008), which measures the degree of starch conversion during extrusion which is the amount of soluble polysaccharide released from the starch component after extrusion process (Ding 2005). Mercier and Feillet (1975) reported increase in soluble starch with increasing extrusion temperature and decreasing feed moisture. WSI often is used as an indicator of degradation of molecular components (Kirby *et al.,* 1988). Gelatinization, the conversion of raw starch to a cooked and digestible material by the application of water and heat, is one of the important effects that extrusion cooking has on the starch component of foods (Qing-Bo *et al.,* 2005). Water is absorbed and bound to the starch molecule with resulting change in the starch granule structure. It has been established (Pomeranz, 1991; Wolf and Cowan, 1971) that proteins are the most reactant components in foods and some of their reactions are essential for functionality. Among functional properties, water holding capacity is important because of the hydrogen bonds formed between water and polar residues of protein molecules.

Viscosity

The viscosity of a paste depends on to a large extent on the degree of gelatinization of the starch granules and the rate of molecular breakdown. Studies reported that increase in the amount of soybean flour and increase in feed moisture decreased the apparent viscosity of extrudate. In addition to the effect of extrusion, the reduction in viscosity may be attributed to the high level of oil from the soybean flour which consequently decreased the shear effect as a result of lubrication in the metering zone. Increase in moisture on the other hand, will further lubricate the dough leading to less shearing effect. Low moisture in the feed can possibly increase frictional damage, particularly when the residence time is high due to low screw speed. Viscosity generally depends on solubility and water holding capacity as well as the structure of components in a food system. Viscosity profile can be thought of as a reflection of the granular changes in the starch granule that occur during gelatinization (Thomas and Atwell, 1997).

Extrusion can induce starch dextrinization resulting in reduction of viscosity in gruels and a concomitant increase in caloric and nutrient density (Jansen *et al.*, 1981). Arambula *et al.* (1998) reported decreased apparent viscosity of extruded instant corn flour when temperature was increased. Davidson *et al.* (1984) reported that viscosity over a heating and cooling cycle have been used to characterize the changes in extruded products in numerous studies. This characteristic is affected by both physical modifications of the granule structure as well as changes to the structures of the starch polymers. They further reported that, the characteristics of the paste viscosity curves were significantly altered by extrusion processing with extrudates showing low values. Pelembe *et al.* (2002) reported that, reduced viscosity of protein - rich sorghum - cowpea extrudate could be very beneficial for infant feeding. Hagenimana *et al.* (2006), reported that viscosity values of extruded rice flours were far less than those of their corresponding unprocessed rice flour dispersed in the Micro Visco Amylo Graph (MVAG), indicating that their starches have been partially pregelitinized by extrusion process. They reported that peak viscosity indicated a high positive correlation with hot paste viscosity and cold paste viscosity with $r > 0.70$ ($p < 0.01$). Ozcan and Jackson (2005) reported that during extrusion cooking of corn starches, extruded starch had higher water absorption and water solubility indices, and they had lower rapid viscoamylograph viscosity profiles when compared with raw starch. This can be attributed to the fact that degradation of the starch occurred during extrusion. It is suggested that starch degradation in the extruded product is a likely significant factor associated low viscosity profiles. The mixtures of raw and extruded starches have potential applications in the industry for functional properties. Arambula *et al.* (1998) reported decreased apparent viscosity of extruded instant corn flour when temperature was increased; though the temperature used in this study was not varied, it was however in the high regime. Likimani *et al.* (1991) indicated that the degradation of molecular bonding of starch during extrusion influenced the characteristics of the extruded product and was used to characterize the target parameters (solubility and viscosity).

The high bulk (low nutrient density) of cereal weaning porridges is a major cause of infant malnutrition in Africa, since it limits nutrient intake (Da *et al.*, 1982). Increased screw speed resulted in an increase in input energy which caused stretching and sometimes fracture of protein-protein matrix, thus, making product less viscous when reconstituted, the screw speed from this study influenced viscosity. Linkimani *et al.* (1991) reported that extrusion induced starch dextrinization which resulted in reduction of viscosity in gruels and a concomitant increase in caloric and nutrient density. Adeyemi and Beckley (1986), reported that a high level of damaged starch would reduce peak viscosity of flour or 'ogi'. Starch dextrinization during extrusion cooking, however, occurred mostly under processing conditions at very high temperature and low moisture (Gomez and Aguilera, 1983) where shear effects were significant. General increase in water absorption of sorghum extrudates was reported by Gomez *et al.* (1988).

Expansion

Sectional expansion index (SEI) is a measure of the cross-sectional area expansion of extrudates. Hashimoto and Grossmann (2003) reported a decrease in expansion with increasing extrusion temperature for extrusion of cassava bran and cassava starch blends. It was found out that after a critical temperature, which depends both on the type of starch and moisture content, expansion decreases with temperature, most likely due to excessive softening and potential structural degradation of the starch melt which becomes unable to withstand the high vapour pressure and therefore collapse (Kokini *et al.*, 1992). In addition, Launay and Lisch (1983) suggested that the sectional and longitudinal expansions were dependent on melt elasticity and melt viscosity, respectively. The elasticity loss with increasing temperature would then be one of the possible reason for the decrease of SEI. Colonna *et al.* (1989) also reported that the decrease in expansion at higher temperature has been attributed to increased dextrinization and weakening of structure.

Textural properties

Crispness is typically a textural attribute for snack foods. Ryu and Walker (1995) observed that breaking strength and bulk density decreased over the temperature range of 140 to 160°C in extrusion cooking of wheat flour. Ding *et al.* (2005) reported that increasing temperature would decrease melt visocosity, but it also increases the vapour pressure of water. This favour the bubble growth which is the driving force for expansion that produces low density products and thus decreasing hardness of extrudate. It is reported that progressive increase in temperature resulted in pores in the structure due to formation of air cells and the surface appeared flaky and porous and hence decreased hardness (Bhattacharya and Choudhary, 1994). Fibre content in feed material also affects the textural characteristics of extrudate. Fibre reduces the cell size, probably by causing premature rupture of gas cells, which reduces the overall expansion and results in less porous structure (Yanniotis *et al.*, 2007).

II. Effect of extrusion cooking on nutritional quality of food products

Macronutrients

Proteins

Extrusion cooking denatures proteins thus improves digestibility of proteins. Most proteins such as enzymes and enzyme inhibitors lose activity due to denaturation caused by the processing conditions such as heat and shear within the extruder barrel. Changes in the quality of protein may be overlooked if only total protein is assayed, hence animal feeding trials or *in vitro* protein digestibility testing should be performed. Degree of denaturation is typically assessed as change in protein solubility in dilute salt solutions and in water. These changes are more pronounced under high shear extrusion conditions (Della Velle *et al.*, 1994), although mass temperature and moisture are also important factors. Wheat protein solubility decreased even at the relatively low temperatures used in pasta extrusion (Ummadi *et al.*, 1995a).

High temperatures and low moisture promote Maillard reactions during extrusion cooking. Sucrose and reducing sugars, including those formed during shear of starch can react with lysine, thereby lowering the quality of protein. Lysine is the most limiting essential amino acid in cereals. Depletion of this essential amino acid during extrusion cooking can result in growth retardation in young children, if product is intended for them. Soy protein has been identified as a cholesterol lowering food. Extrusion texturization of soy isolate did not reduce its effects on rat serum cholesterol, excretion of cholesterol and other steroids in feces, or protein nutrition compared with non-extruded soy (Fukui *et al.*, 1993). Excessive maillard browining can result in losses of lysine up to approximately 50% (de la Gueriviere *et al.*, 1985). Extensive loss of lysine can take place when legume or cereal legume blends are extruded at very high temperature (180°C) or shear forces (>100 rpm) at low moisture (15%), especially in the presence of reducing sugars (3% glucose, fructose, maltose, lactose) (Bjorck and Asp, 1983). High feed moisture protects lysine in food products intended for special nutritional needs. Sucrose may be hydrolyzed during extrusion. In a model system ot wheat starch, glucose and lysine, pH had the greatest effect on Maillard reactions, as measured by changes in colour (Bates *et al.*, 1994).

Carbohydrates

Starchy foods like food grains and tubers provide calories and satiation to the consumers. It is usually the major food constituents in extruded foods such as breakfast cereals, snacks and weaning foods. Extrusion cooking is a unique process because gelatinization occurs at much lower moisture content (12-22%) than is necessary in other food operations. Processing parameters like temperature, pressure and shear tend to increase the rate of gelatinization.

Starch molecules can be physically broken into smaller and more digestible fragments during extrusion processing. For example, branches on amylopectin molecules can be sheared off easily in the barrel during extrusion. Reduction in molecular weight of both amylose and amylopectin had been observed during extrusion cooking but large amylopectin molecules were more likely to degrade during extrusion. Politz *et al.* (1994b) found that larger corn amylopectin molecules were subjected to the greatest molecular weight reduction during extrusion cooking. In a related study, wheat flour starch showed greater starch degradation after extrusion, and higher die temperature (185° C) and feed moisture (20%) helped to maintain molecular weight (Politz *et al.*, 1994a). Gautam and Choudhoury 1999 recommended designing screw configuration to minimize or maximize starch breakdown. The researchers observed that the magnitude of starch degradation was greater for wheat than for corn. This molecular degradation may be exploited to produce dextrins and/ or free glucose for syrups or subsequent fermentation process. Starch degradation may reduce radial expansion, which is critically related to product texture. Starches with 50% amylose showed optimal expansion, and barrel temperatures close to 150° C and low feed moistures favour expansion. Irradiation and certain additives were thought to improve expansion, but only sodium chloride proved to be useful. Over a range of amylose content, 1% sodium chloride consistently produced more expansion than did the starch alone. Addition of dietary fibre, sucrose, salts and lipids to starchy foods such as corn meal may also affect the gelatinization, and thus expansion (Jin *et al.* 1994). Extrusion cooking may not result in complete gelatinization, but digestibility is often improved (Wang *et al.*, 1993a).

Sugars provide sweetness and involved in numerous chemical reactions during extrusion. Foods with high digestible starch is desirable for weaning foods and invalids but may not be good for diabetic individuals due to rapid rise in post-prandial blood sugar levels after consumption of such foods. Extrusion offers the ability to reduce the glycemic index of some foods by converting the starch to digestion-resistant starch (RS). Reducing sugars on detached branches may react with other starch molecules, forming indigestible anhydro linkage. Theander and Westerlund (1987) reported such transglycosidation in extruded wheat flour, presumably from attachment of sheared amylopectin branches to other reactive sites, however Politz *et al.* (1994a) did not observed any changes in 2,3-glucose linkage in extruded wheat flour after methylation analysis. Difference in extrusion condition may be responsible for this observed discrepancy. Chiu *et al.* (1994) were awarded a patent for a process that subjected high amylose starch to digestion with pullulanase, followed by extrusion to remove moisture. This process raised the resistant starch as much as 30%, measured as dietary fibre.

Food additives such as citric acid or certain form of starch also affects the resistant starch. Unlu and Faller (1998) demonstrated that adding certain forms of starch or citric acid to corn meal prior to extrusion modified resistant starch plus dietary fibre. Addition of 30% corn, potato or wheat starch did not increase resistant starch (RS) values. RS and fibre values more than doubled when 7.5% citric acid was mixed with cornmeal, and 30% high amylose corn starch with 5 or 7.5% citric acid resulted in values of 14%, compared with slightly more than 2% in 100% cornmeal. It is speculated that polydextrose may have been formed within the extruder. Formation of amylose lipid complex can also reduce starch digestibility. Monoglycerides and free fatty acids are more likely to form complexes than triglycerides when added to high amylose starch (corn starch at 4% level) (Bhatnagar and Hanna, 1994a). Stearic acid mixed with normal corn starch with 25% amylose and extruded at low feed moisture (19%) and low barrel temperatures (110-140° C) induced the most starch-lipid complexes (Bhatnagar and Hanna, 1994b). High viscosity and longer residence time favoured complex formation in small extruders.

Incorporation of dietary fibre to starchy foods during extrusion can also affect the digestibility. Longer cellulose fibres reduced solubility (Chinnaswamy and Hanna, 1991), possibly due to transglycosidation. Studies showed that addition of 20% protein with removal of insoluble dietary fibre from wheat flour resulted in pasta with significantly delayed dextrin release under *in vitro* digestion condition (Fardet *et al.*,1999). Microscopic evaluation showed that fibre removal facilitated starch-protein interaction that may have increased enzyme-resistance. Flatulence causing oligosaccharides viz. raffinose and stachyose also decreased significantly during extrusion cooking of extruded high-starch fractions of pinto beans (Borejszo and Khan 1992) that may improve the consumer acceptability of legume based extruded products.

Dietary fibre

Fibre is a term used to describe many food components. According to American Association of Cereal Chemists (1999), dietary Fibre is the edible parts of plants or analogous carbohydrates that are resistant to digestion and absorption in the human small intestine with complete or partial fermentation in the large intestine. Dietary Fibre includes polysaccharides, oligosaccharides, lignin, and associated plant substances. Dietary Fibres promote beneficial physiological effects including laxation, and/ or blood cholesterol attenuation, and/or blood glucose attenuation. Like starches, branched dietary Fibre molecules are susceptible to shear during extrusion. These smaller molecules may be soluble in water. These smaller molecules could unite to form large insoluble complexes or Maillard compounds that may be analyzed as lignin. Beans (*Phaseolus vulgaris* L) subjected to conditions making them hard-to-cook, and then treated beans were extruded under various conditions in order to make them more functional (Martin Cabrejas *et al.*, 1999). The total fibre values were unaffected by extrusion but insoluble fibre decreased when extruded at 25% moisture content. Soluble fibre increases in those samples and especially in a sample processed at 30% moisture and at 180° C. Extrusion did not affect uronic acids (components of pectin) but insoluble nonstarch polysaccharides (NSP) increased in oatmeal and potato peels (Camire and Flint, 1991). Soluble NSP was higher in oatmeal and potato peels but corn meal Fibre was unaffected by extrusion.

Soluble fibre offers certain health benefits, including binding of bile acids in the small intestine, leading to reduced serum cholesterol levels. However, extrusion induced solubilization may not offer the health benefits. This 'new' soluble fibre is chemically distinct from naturally occurring soluble fibre compounds such as pectin and gums. Extrusion nearly doubled soluble fibre in potato peels, and significantly more cholic acid and deoxycholic acid were bound by peels extruded at the lower temperature (Camire *et al.*, 1993). Viscous gums and other soluble Fibres may reduce cholesterol levels by trapping bile acids; increased excretion of bile eventually depletes body stores of cholesterol, which are tapped to synthesize new bile acids. The viscosity of aqueous suspensions of extruded wheat, oat and barley were higher than unprocessed grains, lending support to the theory that bile acids are trapped in viscous fibre mixtures in the small intestine. Young rats fed extruded oats, barley or wheat had lower total serum and liver cholesterol levels than did rats fed a control diets or feeds containing unextruded grains (Wang and Klopfenstein, 1993). Extruded lemon and orange peels exhibited higher levels of soluble fibre after processing which were correlated with increased *in vitro* viscosity (Gourgue *et al.*, 1994). However, starch digestion and diffusion of glucose were not affected by extrusion. Extrusion cooking decreased sugar beet pectin and hemicellulose molecular weight and viscosity, but tripled the water solubility from 16.6% to 47.5% (Ralet *et al.*, 1991). Beta-glucans in oats and barley are believed to be responsible for the cholesterol -lowering properties of these grains. Solubility of β-glucans in regular (phoenix) and waxy (candle) barley cultivars increased after extrusion at four barrel temperatures and three different moisture levels (Gaosong and Vasanthan, 2000).

Lipids

Extruded products are usually processed from grains which are low in fat content but fat is often added during post extrusion by frying or spraying of fats or oil (lipids) to hold flavouring compounds. Generally food materials containing less than ten percent fat are extruded because greater quantities of fat if present in feed material making extrusion difficult, particularly for expanded products. Lipids actually grease the internal parts of the extruder and increase the slippage within the extruder barrel. Nutritional quality of lipids is not compromised by extrusion cooking. Complex lipids are digestible by monogastric and ruminant animals. Lipid content of extrudates may appear lower than that found in the original material prior to processing. This apparent loss is most likely due to formation of complex compound with amylose or protein. Lower lipid content has primarily been reported for starchy materials. This 'lost' lipid recovery is improved when extruded products are first digested with acid or amylase, followed by solvent extraction with an organic solvent such as ether. Total fat was not significantly changed in extruded whole wheat, but only half of the ether-extractable lipids in extruded product was detected (Wang *et al.*, 1993b). Cornmeal extruded at lower temperature (50-60°C or 85-90°C) had over 75% bound lipids, but extrusion at 120-125°C only bound 70% of the lipids (Guzman *et al.*, 1992). *Trans* fatty acids are also important from consumption point of view for nutritional safety. Maga (1978) found that extrusion of corn and soy resulted in formation of only 1.5% *trans* fatty acids. Essential and omega fatty acids in extruded foods are more susceptible to lipid oxidation, thus reduced nutritional value. Suzuki *et al.* (1988) reported that both docosahexaenoic (DHA) and eicosapentaenoic (EPA) acids were retained in chum salmon muscle extruded with ten per cent wheat flour.

Table 2 : Nutritional quality of dietary fibre rich extruded foods

Food Source	Nutrition assay	Health effect	Reference
Extruded wheat, barley with husks, or oats with husks	Rat feeding study for 6 weeks	Decreased Cholesterol in rats fed extruded grains versus raw grains or casein control	Wang and Klopfenstein, 1993
Extruded potato peels	*In vitro* bile acid binding	Lower serum cholesterol	Camire *et al*, 1993
Extruded wheat with added guar gum	Human glucose tolerance test	Decreased serum glucose postprandial compared to low-fiber cereal	Fairchild *et al.*, 1996
Extruded rice, oat, corn and wheat brans	Hamster feeding study	Extrusion did not affect cholesterol-lowering properties	Kahlon *et al.*, 1998

Lipid oxidation is a major cause of loss of nutritional and sensory quality of foods and feeds. Although we suspect that lipid oxidation does not occur during extrusion due to brief residence time, lipid oxidation can occur during storage. Artz *et al.* (1992) have reviewed factors affecting lipid oxidation in extruded foods. It has been observed that screw wear results in higher concentrations of pro-oxidant minerals. Semwal *et al.* (1994) observed that iron and peroxide values were higher in extruded rice and dhal compared to oven dried and drum dried products.

Expansion is desirable for snacks and breakfast cereals, but increased surface area formed by numerous air cells in expanded products favours oxidation. However, lipolytic enzymes and other enzymes that promote oxidation may be inactivated during extrusion, and starch-lipid complexes formed in the barrel may be more resistant to oxidation. Oat meal containing potato peels showed lower peroxide values than control cookies, and extruded peels had greater antioxidant activity than did non-extruded peels (Arora and Camire, 1994). Vaccum packaging or packaging under nitrogen in opaque containers may further protect extruded foods.

Vitamins

Vitamins are essential for metabolism of macronutrients. Since extruded foods are intended for consumption by children and other groups with high nutritional needs, their vitamin content is of great importance. Due to reduction of vitamins in extrusion cooking, some processors are prompt to apply vitamins post-extrusion as a spray. Marchetti *et al.*, 1999 reported that fat-coated ascorbic acid, menadione, pyridoxide and folic acid were retained well than crystalline forms in extruded fish feed. Although fortification of extruded foods with micronutrients is a popular concept but little research has been oriented towards the interaction of extrusion cooking conditions and stability of vitamins.

Fat soluble vitamins : Many extruded products generally do not have high levels of lipid soluble vitamins but stability of these vitamins is a concern for fortified products. Carotenoids and tocopherols are susceptible to oxidation and thermal degradation during extrusion and storage. Although β-catrotene is not a vitamin, it has vitamin A activity and is an important antioxidant and colouring agent but unstable when exposed to heat and oxygen. Beta-carotene is added to foods to make them more orange in color, but it is unstable when heated. Retention of retinyl palmitate in tapioca snacks mixed with either fish or protein flour was 52% and 73%, respectively after extrusion cooking (Suknark *et al.*, 2001). More than 50% of all added *trans* β-carotene in wheat flour were destroyed when barrel temperature increased 125 to 200°C (Guzman-Tello and Cheftel, 1990). In the same study, extrusion under a nitrogen atmosphere or addition of butylated hydroxytoluene (BHT) provided some protection against colour loss, which was used as an indicator of oxidation. The possible antioxidant for extruded food products are given in Table 3.

Table 3 : Antioxidant for extruded food products

Anioxidant treatment	Advantages	Limitations
Butylated hydroxyanisole (BHA); butylated hydroxytoluene (BHT)	Disperse well in dry feed material	Due to volatile nature, only low levels may available after extrusion.
Tocopherols	Varying vitamin E activity; natural appeal to final products	Unstable to heat; common forms are difficult to measure in feed material
Phenolic acids	Natural	Result in darker colour due to complex formation during extrusion; may give bitter or astringent flavour
Ascorbic acid	Natural; enhances iron bioavailability	Unstable, thermal degradation, higher levels impart acidic flavour
Nitrogen flush in packaging	No direct additives	Oxidation may take place due to trapping of some air in expanded foods, costlier than ordinary packaging

Vitamins D and K are fairly stable during food processing but they are not commonly used in extruded products intended for human consumption. Vitamin E and related tocoherols, however, perform as both vitamin and antioxidant. Less than 20% of vitamin E was retained in extruded and drum-dried wheat flour (Wennermark, 1993). Rice bran tocopherol decreased with increasing extrusion temperature, and bran extruded at 120/140°C lost more tocopherols over a year's storage than did bran extruded at 110°C (Shin *et al.*, 1997). Tocopherol decreased in puffed snacks containing either fish or partially-defatted peanut flour (Suknark, 1998).

Water soluble vitamins: vitamin C is usually decreased in the presence of heat and oxidation. It is found decreased in wheat flour when extruded at higher barrel temperatures at fairly low moisture (10%) content (Andersson and Hedlund, 1990). Blueberry concentrate had been found to protect 1% added ascorbic acid in an extruded breakfast cereal compared to a product containing corn, sucrose, and ascorbic acid (Chaovanalikit, 1999). Addition of ascorbic acid (0.4-1.0%) to cassava starch to increase starch conversion retained over 50% occurred addition (Sriburi and Hill, 2000). The loss of water soluble vitamins may be upto 100% which has been summarized in Table 4.

Table 4: Loss of water soluble vitamins during extrusion cooking (Source: Camire ME. 2001)

Vitamins	Losses (%)
Vitamins B_1	5-100
Vitamins B_2	0-40
Vitamins B_6	4-44
Vitamins B_{12}	1-40
Niacin	0-40
Vitamins C	0-87

Vitamin B complex : Thiamine (vitamin B_1), a water-soluble vitamin is most susceptible to thermal processing. Killeit (1994) summarized thiamine losses as ranging from 5 to 100%. Wheat flour extruded with no added water resulted in large thiamine losses, and increased barrel temperature decreased thiamine only at the slowest feed rate used (250 g/ min versus 500 and 750 g/min) but riboflavin (B_2) and niacin were unaffected (Andersson and Hedlund, 1990). Thiamine retention in potato flakes reduced under extrusion conditions of lower moisture and higher barrel temperature; sulfites may also have contributed to vitamin destruction in the potato flakes (Maga and Sizer, 1978). Lorenz and Jansen (1980) found the retention of over 90% of thiamine, riboflavin, vitamin B_6 and folic acid in corn–soy blends, extruded at 171°C using low-cost single-screw extruders. Since cereals and pulses are an important source of B complex vitamins in our diets, attention should be directed to prevent the losses of these vitamins during extrusion cooking.

Minerals

Minerals are unlikely to lose in the steam distillate at the die during extrusion cooking but entrapment or binding may take place during this process. Abrasive foods like brans rich in dietary Fibre may abrade the interior of the extruder barrel and screws, resulting in increased mineral content particularly iron. Extruded corn, which has low dietary Fibre, showed no change in total, elemental, or soluble iron after extrusion using twin screw extruder (Camire and Dougherty, 1998). Iron content in extruded potato flakes increased with barrel temperature (Maga and Sizer, 1978). Potato peels extruded under higher temperature had as much as 38% more total iron after extrusion (Camire *et al.*, 1993). Rats fed extruded corn or potato utilized the screw wear as well as endogenous iron (Fairweather-Tait *et al.*, 1987).

Extrusion did not reduce iron and zinc absorption from wheat bran and flour in adult human volunteers (Fairweather-Tait *et al.*, 1989). Although total iron increased in all but white beans, dialyzable (available) iron decreased, possibly due to complex formation with phytate. Ummadi *et al.* (1995b) found that low shear extrusion conditions significantly increased dialyzable iron in navy beans, chickpeas, cowpeas and lentils as compared with samples that were boiled or extruded under high shear conditions. The solubility of iron under condition similar to digestion and subsequent ability to dialyze across a membrane is used to assess bio-availability. Extrusion slightly increased iron availability in corn snacks (Hazell and Johnson, 1989). Zinc bioavailability of blends of semolina and soy protein concentrate (85:15) was found unaffected due to extrusion processing (Kang, 1996). Weaning food blends of pearl millet, cowpea and peanut showed higher protein digestibility and iron availability compared to similar foods prepared following roasting, however none of the weaning foods provided adequate iron to meet infant needs (Cisse *et al.*, 1998).

Mineral bio-availability may be affected in foods rich in dietary Fibre and phytate. Extrusion reduced phytate levels in wheat flour (Fairweather-Tait *et al.*, 1989). Inactivation of phytases during extrusion in these studies may be the possible reason of these findings. Although phytic acid was lower under all processing conditions, total phytate was not affected. Legume phytate was also not affected by extrusion (Lombardi-Boccia *et al.*, 1991; Ummadi *et al.*, 1995b). Gualberto *et al.* (1997) reported that varying screw speed had no effect on phytate retention in wheat, rice and oat barns, but insoluble Fibre content lowered in rice and oat bran after extrusion processing. When phytate was removed from these samples, extruded rice and oat brans bound more calcium and zinc *in vitro*, but not copper (Bergman *et al.*, 1997).

Fortification of extruded products with minerals has become a common practice particularly in ready-to-eat products. As certain iron salts react with phenolics and result in unpleasant dark colours, Kapanidis and Lee (1996) recommended the addition of ferrous sulfate heptahydrate in a simulated rice product for maintaining light colour. Addition of calcium hydroxide (0.15-0.35%) to cornmeal extrudates lowered expansion and increased lightness in color (Martinez-Bustos *et al.*, 1998).

Non-nutritive components of foods

Extrusion cooking improves the safety and digestibility of the food products by destroying many naturally occurring toxins and antinutrients present in them (Table 5). Enzyme inhibitors, hormone-like compounds, saponins and other compounds could retard growth in children but these may offer protection against chronic diseases in adults. Allergens and mycotoxins compounds are more resistant to thermal processing but may be susceptible to extrusion processing in combination of chemical treatments. Glucosinolates present in many *Brassica* species of commercial importance, may protect against certain forms of cancer (Van Poppel *et al.*, 1999). Extrusion processing alone has little effect on retention of glucosinolates (Fenwick *et al.*, 1986). Although extrusion processing of rapeseed meals with ammonium carbonated did not result in complete destruction of glucosinolates, but rats feeding study showed improvement in nutritional parameters (Barrett *et al.*, 1997). Canola total glucosinolates were reduced by added ammonia during extrusion cooking (Darroch *et al.*, 1990).

Phenolic compounds in plants protect against oxidation and some degenerative diseases. In potato peels, total free phenolics, of which chlorogenic acid predominates, were lower (unpublished data, Camire and Dougherty). Higher barrel temperature and feed moisture protected free phenolics. Anthocyanin pigments provide colours to the products and are believed to act as antioxidants that protect vision and cardiovascular health (Camire, 2000). Polymerization and browning may also have contributed to anthocyanin losses. Blueberry anthocyanins were significantly decreased during extrusion cooking in breakfast cereals containing cornmeal and sucrose (Chaovanalikit, 1999).

Table 5: Effect of extrusion cooking on antinutrients and toxins (Source: Camire, 2001)

Antinutrients and toxins	Food sources	Factors favoring reduction
Protease inhibitors	Legumes, potato	Higher extrusion temperatures
Allergens	Peanut, soy	Added starch, Increased shear
Mycotoxins	Grains	Added amine sources, lower temperatures, increased mixing
Gossypol	Cottonseed	High feed moisture
Glycoalkaloids	Potato	Addition of thiamine
Glucosinolates	Canola	Addition of ammonia

Isoflavones present in soybean may protect post menopausal women from osteoporosis and heart disease, while men may get protection against prostate cancer and other testosterone-dependent cancers. Extrusion reduced aglycone (genistein) of okara, a co-product of tofu manufacturing, mixed with wheat flour (Rinaldi *et al.*, 2000). In blends of soy protein concentrate with cornmeal (20:80), increasing barrel temperature caused decarboxylation of isoflavones, resulted increased proportions of acetyl derivatives (Mahungu *et al.*, 1999). Total isoflavones also decreased in the soy-corn blends. Singletary *et al.*, 2000 reported that although the content of the biologically-active aglycones was unaffected with extrusion cooking, extruded corn-soy blends were less effective in preventing proliferation of breast cancer cells *in vitro*.

There are enormous possibilities for development of nutritionally balanced value added extruded products and nutrition research. Very little research has been published on the effects of extrusion cooking on physio-chemicals and healthful components of food viz. flavonoids, inulin, lycopene and related carotenoids, omega-3 fatty acids, tannins, caffeine and other stimulants, etc. Scale-up issues in extrusion cooking is also to be considered for valid interpretation of laboratory studies.

References

Adeyemi IA, Beckley O (1986). Effect of period of maize fermentation and souring on Chemical properties and amylograph pasting viscosity of Ogi. J. Cereal Sci., 4: 353.

Andersson Y and Hedlund B (1990). Extruded whey flour: correlation between processing and product quality parameters. Food Quality and Preference, 2(3): 201-16.

Arambula VG, Figueroa JDC, Martinez-Bustos F, Ordorica FCA, Gonzalez-Hernandez J (1998). Milling and processing parameters for Corn Tortillas from Extruded instant Dry Masa flour. J. Food Sci., 63 (2): 338-341.

Arora A and Camire ME (1994). Performance of potato peels in muffins and cookies. Food Res. Intl, 27: 15- 22.

Artz WE, Rao SK and Suer RM (1992). Lipid oxidation in extruded products during storage as affect by extrusion temperature an selected antioxidants, in Kokini J.L., Ho C-T. and Karwe M.V., Food Extrusion Science and Technology, New York, Marcel Dekker, Inc.

Barret JE, Klopfenstein CF and Leipold HW (1997). Detoxification of rapeseed meal by extrusion with an added basic salt. Cereal Chem, 74(2): 168-70.

Bates L, Ames JM and Macdougall DB (1994). The use of a reaction cell to model the development and control of colour in extrusion cooked foods. Lebebsm-Wiss u-Technol, 27(4): 375-9.

Bergman CJ, Gualberto DG and Weber CW (1997). Mineral binding capacity of dephytinized insoluble Fibre from extruded wheat, oat and rice brans. Plant Foods Hum Nutr, 51(4): 295-310.

Bhatnagar S and Hanna MA (1994a). Amylose-lipid complex formation during single-screw extrusion of various corn starches. Cereal Chem, 71(6): 582-7.

Bhatnagar S and Hanna MA (1994b). Extrusion processing condition for amylase-lipid comlexing. Cereal Chem, 71(6): 587-93.

Bhattacharya, S, Choudhury, GS (1994). Twin-screw extrusion of rice flour: effect of extruder length-to-diameter ratio and barrel temperature on extrusion parameters and product characteristics. Journal of Food Processing and Preservation, 18: 389-406.

Bjorck I and Asp NG (1983). The effects of extrusion coking on nutritional value – a literature review. J Food Engin, 2(4): 281-308.

Borejszo Z and Khan K (1992). Reduction of flatulence-causing sugars by high temperature extrusion of pinto bean high starch fractions. J Food Sci, 57(3): 7771-2 & 777.

Camire ME (2000). Bilberries and blueberries as functional foods and pharmaceuticals, in Mazza G. and Oomah D.B., functional foods: Herbs, Botanicals and Teas, Lancaster, PA, Technomic Press.

Camire ME and Dougherty MP (1998). Added phenolic compounds enhance lipid stability in extruded corn. J Food Sci, 63: 516-518.

Camire ME and Flint SL (1991). Thermal processing effects on dietary Fibre composition and hydration capacity in corn meal, oat meal and potato peels. Cereal Chem, 68(6): 645-7.

Camire ME, Zhao J and Violette DA (1993). In vitro binding of bile acids by extruded potato peels. Journal of Agricultural and Food Chemistry, 41(12): 2391–2394.

Chaovanalikit A (1999). Anthocyanin stability during extrusion cooking, M.S. Thesis Univ. of Maine, Orono ME.

Chinnaswamy R and Hanna MA (1991). Physicochemical and macromolecular properties of starch-cellulose Fibre extrudates. Food Structure, 10(3): 229-39.

Chiu CE, Henley M and Altieri P (1994). Process for making amylase resistant starch from high amylose starch United States Patent. Patent No: 5,281,276 Date of Patent: Jan. 25: 1994.

Cisse D, Guiro A Γ, Diaham B, Souane M, Doumbouya NT and Wade S (1998). Effect of food processing on iron availability of African pearl millet weaning foods. Intl J Food Sci Nutr, 49(5): 375-81.

Colonna P, Tayer J, Mercier C (1989). Extrusion cooking of starch and starchy products. In: Mercier C, Linco P, Harper JM (Eds) Extrusion Cooking. St Paul, MN:American Association of Cereal Chemists. pp. 247-319.

Darroch CS, Bell JM and Keith MO (1990). The effects of moist heat and ammonia on the chemical composition and feeding value of extruded canola screenings for mice. Can J Anim Sci, 70: 267-77.

Davidson VJ, Paton D, Diosady LL, Larocque G (1984). Degradation of wheat Starch in a single-screw extruder: Characteristics of extruded starch polymers. J. Food Sci., 49: 453-458.

De La Gueriviere JF, Mercier A and Baudet L (1985). Incidences de la cuisson-extrusion sur certains parameters nutritionnels de produits alimentaires notamment cerealiers. Cah Nutr Diet, 20(3): 201-10.

Della Valle G, Quillien L and Gueguen J (1994). Relationships between processing conditions and starch and protein modifications during extrusion-cooking of pea flour. J Sci Food Agric, 64(4): 509-17.

Ding, QB, Ainsworth, P, Tucker, G & Marson, H (2005). The effect of extrusion condition on the physico-chemical properties and sensory characteristics of rice-expanded snacks. Journal of Food Engineering, 66: 283-289.

Fairweather-Tait SJ, Symss LL, Smith AC and Johnson IT (1987). The effect of extrusion cooking on iron absorption from maize and potato. J. Sci Food Agric, 39(4): 341-8.

Fairweather-Tait SJ, Portwood DE, Symss LL, Eagles J and Minski MJ (1989). Iron and zinc absorption in human subjects from a mixed meal of extruded and non-extruded wheat bran flour. Am J Clin Nutr, 49(1): 151-5.

Fardet A, Abecassis J, Hoebler C, Baldwin PM, Buleon A, Berot S and Barry JL (1999). Influence of technological modifications of the protein network from pasta on in vitro starch degradation. J Cereal Sci, 30(2): 33-145.

Fenwick GR, Spinks EA, Wilkinson AP, Heaney RK and Legoy MA (1986). Effect of processing on the antinutrient content of rapeseed. J Sci Food Agric, 37(8): 735-41.

Fukui K, Aoyama T, Hashimoto Y and Yamamoto T (1993). Effect of extrusion of soy protein isolate on plasma cholesterol level and nutritive value of protein in growing male rats. J Jap Soc Nutr Food Sci, 46(3): 211-16.

Gaosong J and Vasanthan T (2000). Effect of extrusion cooking on the primary structure and water solubility of β–glucans from regular and waxy barley. Cereal Chem, 77(3): 396-400.

Gautam A and Choudhoury GS (1999). Screw configuration effects on starch breakdown during twin screw extrusion of rice flour. J Food Process Preserv, 23(4): 355-75.

Gomez MH, Aguilera, JM (1983). Changes in the starch fraction during extrusion-cooking of corn. J. Food Sci., 48: 378-831.

Gourgue C, Champ M, Guillon F and Delort-Laval J (1994). Effect of extrusion-cooking on the hypoglycaemic properties of citrus fibre: an in vitro study. J Sci Food Agric, 64(4): 493-9.

Gualberto DG, Bergman CJ Kazemzadeh M and Weber CE (1997). 'Effect of extrusion processing on the soluble and insoluble Fibre, and phytic acid contents of cereals bran. Plant Foods Human Nutrition, 51(3): 187-98.

Guzman LB, Lee TC and Chichester CO (1992). Lipid binding during extrusion cooking, in Kokini J.L., Ho C-T and Karwe M.V., Food Extrusion Science and Technology New York, Marcel Dekker, Inc.

Guzman Tello R and Cheftel JC (1990). Colour loss during extrusion cooking of beta-carotene-wheat flour mixes as an indicator of the intensity of thermal and oxidative processing. Intl Food Sci Technol, 25(4): 420-34.

Hagenimana A, Xiaolin-Ding, Tao-Fang (2006). Evaluation of rice flour modified by extrusion cooking. J. Cereal Sci., 43(1): 38-46.

Hashimoto JM and Grossmann Maria Victoria E (2003). Effects of extrusion conditions on quality of cassava bran/cassava starch extrudates. International Journal of Food Science and Technology, 38: 511–517.

Hazell T and Johnson IT (1989). Influence of food processing on iron availability in vitro from extruded maize-based snack food. J Sci Food Agric, 46(3): 365-74.

Jansen GR, O'Deen L, Tribelhorn RE, Harper JM (1981). The Calorie densities of gruels made from extruded corn-soy blends. Food Nutr. Bull., 3(I): 39.

Jin Z, Hsieh F and Huff HE (1994). Extrusion coking of corn meal with soy Fibre, salt and sugar. Cereal Chem, 71(3): 227-34.

Kahlon TS, Edwards RH and Chow FI (1998). Effect of extrusion on hypocholesterolemic properties of rice, oat, corn and wheat bran diets in hamsters. Cereal Chem., 75(6): 897-903.

Kang SY (1996). Zinc bioavailability in a semolina/soy protein mixture was not affected by extrusion processing, M S Thesis, East Lansing, MI, Michigan State University.

Kapanidis AN and Lee TC (1996). Novel method for the production of color-compatible ferrous sulfate-fortified simulated rice through extrusion. J Agric Food Chem, 44(2): 522-5.

Killeit U (1994). Vitamin retention in extrusion cooking. Food Chem, 49(2): 149-55.

Kirby, AR, Ollett, AL, Parker, R and Smith, AC (1988). An experimental study of screw configuration effects in the twin-screw extrusion cooking of maize grits. Journal of Food Engineering, 8: 247-272.

Kivisto B, Andersson H, Cederblad G, Sandberg AS and Sandstrom B (1986). Extrusion cooking of a high-Fibre cereal product 2. effects on apparent absorption of zinc, iron, calcium, magnesium and phosphorus in humans. British J Nutr, 55(2): 255-60.

Kokini, JL, Lai, L, and Chedid, LL (1992). Effect of starch structure on starch rheological properties. Food technol, 46(6): 124-139.

Launay, B and Lisch, JM (1983). Twin-screw extrusion cooking of starches: flow behavior of starch pastes, expansion and mechanical properties of extrudates. Journal of Food Engineering, 2: 259–280.

Likimani TA, Sofos JN, Maga JA, Harper JM (1991). Extrusion cooking of corn/soy bean mix is presence of thermostable a-amylase. J. Food Sci., (56)1: 99-105.

Lombardi-Boccia G, Di Lullo G and Carnovale E (1991). In-vitro iron dialysability from legumes: influence of phytate and extrusion cooking. J Sci food Agric, 55(4): 599-605.

Lorenz K and Jansen GR (1980). Nutrient stability of full-fat soy flour and corn-soy blends produced by low-cost extrusion. Cereal Foods World, 25(4): 161-2, 171-2.

Maga JA (1978). Cis-trans fatty acid ratios as influenced by product and temperature of extrusion cooking. Lebensm-Wiss u- Technol, 11(4): 183-4.

Maga JA and Sizer CE (1978). Ascorbic acid and thiamin retention during extrusion of potato flakes. Lebensm-Wiss u-Technol, 11(4): 192-4.

Mahungu SM, Diaz-Mercado S, Li J, Schwenk M, Singletary K and Faller J (1999). Stability of isoflavones during extrusion processing of corn/soy mixture. J Agric Food Chem, 47(1): 279-84.

Marchetti M, Tossani N, Marchetti S, and Bauce G (1999). Stability of crystalline and coated vitamins during manufacture and storage of fish feeds. Aquaculture Nutr, 5(2): 115-20.

Martin-Cabrejas MA, Jamie L, Karanja C, Downie AJ, Parker ML, Lopez-Andreu FJ, Maina G, Esteban RM, Smith AC and Waldron KW (1999). Modifications to physicochemical and nutritional properties of hard-to-cook beans (Phaseolus vulgaris L.) by extrusion cooking. J Agric Food Chem., 47(3): 1174-82.

Martinez Bustos F, Chang YK, Bannwart AC, Rodriguez ME, Guedes PA and Gaiotti ER (1998). Effects of calcium hydroxide and processing conditions on corn meal extrudates. Cereal Chem, 75(6): 796-801.

Mercier C, Feillet P (1975). Modification of carbohydrate components by extrusion-cooking of cereal products. Cereal Chem., 52: 283-297.

Ozcan S, Jackson DS (2005). Functionality behavior of raw and extruded corn starch mixtures. Cereal Chem. 82(2): 223-227.

Pelembe LAM, Erasmus C, Taylor JRN (2002). Development of a Protein - rich Composite Sorghum - Cowpea Instant Porridge by Extrusion Cooking Process. Lebensm - Wiss. U. - Technol., 35: 120-127.

Politz ML, Timpa JD, White AR and Wasserman BP (1994a). Non-aqueous gel permeation chromatography of wheat starch in dimethylacetamide (DMAC) and LiCI: extrusion-induced fragmentation. Carbohydrate Polymers, 24(2): 91-99.

Pollitz ML, Timpa JD and Wasserman BP (1994b). Quantitative measurement of extrusion-induced starch fragmentation products in maize flour using nonaqueous automatic gel-permeation chromatography. Cereal Chem, 71(6): 532-6.

Pomeranz Y (1991). Functional Properties of Food Components, 2nd Edition. Academic Press, Inc., NY.

Qing–Bo D, Ainsworth P, Tuker G, Marson H (2005). The effect of extrusion conditions on the physicochemical properties and sensory characteristics of rice-based expanded snacks. J. Food Eng, 66: 284-289.

Ralet MC, Thibault JF and Della Valle G (1991). Solubilization of sugar-beet pulp cell wall polysaccharides by extrusion-cooking. Lebensm-Wiss u- Technol, 24(2): 107-12.

Rayas-Duarte, P, Majewska, K & Doetkott, C (1998). Effect of extrusion process parameters on the quality of buckwheat flour mixes. Cereal Chemistry, 75: 338-345.

Rinaldi VEA, Ng PKW and Bennick MR (2000). Effects of extrusion on dietary Fibre and isoflavone contents of wheat extrudates enriched with wet okara. Cereal Chem, 77(2): 237-40.

Camire ME (2001). Extrusion and nutritional quality in Robin Guy (Editor), Extrusion Cooking-Technologies and Application, CRC Press, Newyork, Washington.

Ryu, GH, Walker, CE (1995). The effects of extrusion conditions on the physical properties of wheat flour extrudates. Starch 47 (1): 33–36.

Sandberg AS, Andersson H, Kivisto B and Sandstrom B (1986). Extrusion cooking of a high-Fibre cereal product. 1. Effects on digestibility and absorption of protein, fat, starch, dietary Fibre and phytate in the small intestine. Br J Nutr, 55(2): 245-4.

Semwal AD, Sharma GK and Arya SS (1994). Factors influencing lipid autooxidation in dehydrated precooked rice and bengalgram dhal. J Food Sci Technol, 31(4): 293-7.

Shin TS, Godber JS, Martin DE and Wells JH (1997). Hydrolytic stability and changes in E vitamers and oryzanol of extruded rice bran during storage. J Food Sci, 62(4): 704-8.

Singh B, Sekhon KS, Singh N (2007). Effects of moisture, temperature and level of pea grits on extrusion behavior and product characteristics of rice. Food Chem., 110: 198-202.

Singletary K, Faller J, Li JY and Mahungu S (2000). Effect of extrusion on isoflavone content and antiproliferative bioactivity of soy/corn mixtures. J Agric Food Chem, 48(8): 3566-71.

Sriburi P and Hill SE (2000). Extrusion of cassava starch with either variations in ascorbic acid concentration or pH. Intl J Food Sci Technol, 35(2): 141-54.

Suknark K (1998). Development of extrusion processes for novel snack food production and the physicochemical and nutritional properties of resulting products PhD Dissertation, Griffin GA, University of Georgia.

Suknark K, Lee J, Eitenmiller RR and Phillips RD (2001). Stability of tocopherols and retinyl palmitate in snack extrudates. J Food Sci, 66(6): 897-902.

Suzuki H, Chung BS, Isobe S, Hayawaka S and Wada S (1988). Changes in w(omega)-3 polyunsaturated fatty acids in the chum salmon muscle during spawning migration and extrusion cooking. J Food Sci, 53(6): 1659-61.

Theander O and Westerlund E (1987). Studies on chemical modifications in heat-processed starch and wheat flour. Starch/Starke, 39(3): 88-93.

Thomas DJ, Atwell WA (1997). Starches. Eagan Press Handbook Series. Eagan Press, Minnesota, USA.

Ummadi P, Chenoweth WL and Ng PKW (1995a). Changes in solubility and distribution of semolina proteins due to extrusion processing. Cereal Chem, 72(6): 564-7.

Ummadi P, Chenoweth WL and Uebersax MA (1995b). The influence of extrusion processing on iron dialyzability, phytates and tannins in legumes. J Food Process Preserv, 19(2): 119-31.

Unlu E and Faller JF (1998). Formation of resistant starch by a twin-screw extruder. Cereal Chem, 75(3): 346-50.

Van Poppel G, Verhoeven DT, Verhagen H and Goldbohm RA (1999). Brassica vegetables and cancer prevention. Epidemiology and mechanisms. Adv Exp Med Biol, 472: 159-68.

Wang WM, and Klopfenstein CF (1993). Effect of twin screw extrusion on the nutritional quality of wheat, barley and oats. Cereal Chem, 70(6): 712-15.

Wang S, Casulli J, and Bouvier JM (1993a). Effect of dough ingredients on apparent viscosity and properties of extrudates in twin-screw extrusion cooking. Intl J food Sci Technol, 28(5): 465-79.

Wang WM, Klopfenstein CF and Ponte JG (1993b). Effects of twin-screw extrusion on the physical properties of dietary Fibre and other components of whole wheat and wheat bran and on the baking quality of the wheat bran. Cereal Chem, 70(6): 707-11.

Wennermark B (19[illegible]). Vitamin E retention during processing of cereals, FILDER Thesis, Lund, Lunds [illegible]itet.

Wolf WF, Cowan J[illegible] ([illegible]971). Soybean as a food source. CRC critical Rev. Food Technol., 2(1): 81.

Yagci S, Gogus F (2008). Response Surface Methodology for Evaluation of Physical and Functional Properties of Exruded Snack Foods Developed from Food – by – Products. J. Food Eng., (86): 122–132.

Yang S, Peng J, Lui W, Lin Jenshinn (2008). Effects of adlay species and rice flour ratio on the physicochemical properties and texture characteristic of adlay-based extrudates. J. Food Eng., (84): 489-494.

Yang S, Peng J, Lui W, Lin Jenshinn (2008). Effects of adlay species and rice flour ratio on the physicochemical properties and texture characteristic of adlay-based extrudates. J. Food Eng., (84): 489-494.

Yanniotis, S, Petraki, A, Soumpasi, E (2007). Effect of pectin and wheat Fibres on quality attributes of extruded cornstarch. Journal of Food Engineering, 80: 594–599.

□□□

37

Quality Assurance and Food Safety in Modern Food Industry

HARINDER SINGH OBEROI

Quality Assurance is a programme for the systematic monitoring and evaluation of the various aspects of a project, service, or facility for ensuring the compliance with the set standards. The two important aspects of Quality Assurance (QA) are the suitability of product for intended purpose and the right "first time approach". Quality Assurance includes regulation of the quality of raw materials, assemblies, products and components, services related to production, production and systems and management, production and inspection processes. The different steps for quality assurance process are:

- Testing of previous articles
- Improvement plans
- Design to include improvements and requirements
- Manufacturing using improved design
- Review of new items and improvements
- Testing of new items

Quality Assurance is a proactive or preventive approach which helps in preventing the hazards to occur whereas quality control (QC) emphasizes on product quality testing and management of reporting systems. Total quality management (TQM) deals with management of quality of each unit operation in manufacturing. It deals with the quality of equipment, instruments, testing procedures, record keeping, personnel employed for the job, audit procedures, planning and implementation.

In food industry, quality assurance has a special place as it deals directly with human health and human well being. In food industry, quality management begins with quality of raw material, quality of equipment used for processing, instrumentation and on-line monitoring and control systems, knowledge and skill of the human resource, audit management systems, testing procedures and checks employed at different processing steps. Hazard analysis critical control point (HACCP), good manufacturing practices (GMP), good laboratory practices (GLP) are not a standard by themselves but provide a means to achieve the desired quality standards. Generally, agencies like Codex Alimentarius Commission, International Organization for Standardization (ISO), United States Food and Drug Administration (USFDA), International Commission on Microbiological Specifications for Foods (ICMSF), National Advisory Committee on National e on Microbiological Criteria for Food (NACMCF) are some of the International organizations responsible for setting food standards. Most of the countries or unions such as European Union (EU) have set up their own standards for different foods, but meeting the standards set by ISO or Codex Alimentarius help in export of products from one country to another.

ISO as a data quality management system

The International Organization for Standardization (ISO) series programme provides standards for data documentation and audits as part of a quality management system. Though the ISO series is not designed explicitly for emissions data development, many of the principles may be applied to ensure the production of a quality inventory. Inventory agencies may find these documents useful source material for developing QA/QC for greenhouse gas inventories. Some countries (United Kingdom and the Netherlands) have already applied some elements of the ISO standards for their inventory development process and data management.

The following standards and guidelines published under the ISO series may supplement source category-specific QA/QC procedures for inventory development and provide practical guidance for ensuring data quality and a transparent reporting system.

ISO 9004-1 : General quality guidelines to impiement a quality system.

ISO 9004-4 : Guidelines for implementing continuous quality improvement within the organization, using tools and techniques based on data collection and analysis.

ISO 10005 : Guidance on how to prepare quality plans for the control of specific projects.

ISO 10011-1 : Guidelines for auditing a quality system.

ISO 10011-2 : Guidance on the qualification criteria for quality systems auditors.

ISO 10011-3 : Guidelines for managing quality system audit programmes.

ISO 10012 : Guidelines on calibration systems and statistical controls to ensure that measurements are made with the intended accuracy.

ISO 10013 : Guidelines for developing quality manuals to meet specific needs.

Hazard analysis critical control point

The hazard analysis critical control points (HACCP) strategy has been promoted as a means by which safe foods can be produced while minimizing the need for expensive and often inefficient end product testing. In general, HACCP identifies points (critical control points or CCP) in the production and processing chain at which failure to maintain control of the process could lead to food borne illness or diseases. For example, irrigation could be considered to be a CCP in the production of fruits and vegetables because it is a point at which it is essential to control contamination by irrigation waters. HACCP has been almost universally accepted in the food processing industry, and some have suggested the implementation of HACCP in farming. However, this concept is new to production agriculture, and many aspects of production do not lend themselves to the HACCP system as well as in the food processing industry. In particular, many if not most steps in food production are virtually uncontrollable. That is, it is difficult and impractical to maintain sanitation, temperature control, and other essentials of typical good manufacturing practices. For example, harvested or processed foods can be held in sanitizable containers or areas that can be kept free from outside contamination and at a proper temperature. In contrast, pre harvest fruits and vegetables are grown outside with complete exposure to the elements, blowing soil, wild animals and birds, and uncontrollable fluctuations in temperature. Consequently, any HACCP system for production agriculture would have to be designed with these constraints in mind. Nevertheless, it is possible for a modified HACCP plan to be developed for many products that would identify controllable points and means for control. Although the modified system might not meet the strict criteria for HACCP, it would make great strides in reducing the risk of food borne illness arising at production. Currently, a HACCP plan is not considered to be mandatory for incoming raw products and ingredients. There are seven principles on which HACCP works.

Principle 1 : Conduct a hazard analysis

Conducting a hazard analysis is the process of identifying the hazards that might affect a particular product in a specific processing operation, and then collecting and evaluating information on the hazards and conditions leading to their presence to decide which are significant to food safety and must be addressed by the HACCP plan.

Principle 2 : Determine the critical control points

A critical control point (CCP) is a point, step or procedure in the food manufacturing process at which a control measure can be applied and is essential to prevent, eliminate or reduce a food safety hazard to an acceptable level. Determining the CCPs involves identifying the hazards in the processing operations which could be addressed by the HACCP plan and can be prevented, reduced or eliminated.

Principle 3 : Establish critical limits

Critical limits are criteria that separate safe product from unsafe product. Critical limits must be established for each CCP. Critical limits must be clearly defined and should be quantifiable (measurable).

Principle 4 : Establish monitoring procedures

Monitoring is the process of conducting a planned sequence of observations or measurements to determine if a CCP is under control. For each CCP, monitoring procedures must be implemented and documented to ensure that the critical limit is being met.

Principle 5 : Establish corrective actions

Corrective actions are predetermined activities that are taken when CCP monitoring results indicate that a deviation has occurred and there is a chance that unsafe food has been, or will be produced. For each CCP there must be planned, written corrective actions.

Principle 6 : Establish verification procedures

Verification is the application of methods, procedures, tests and other evaluations, in addition to monitoring, to determine conformance with the HACCP plan. Verification confirms that the HACCP plan is operating effectively and according to written procedures.

Principle 7: Establish record-keeping and documentation procedures

HACCP plans, including all of the items listed above must be documented. The required monitoring and verification records must be complete and accurate.

Criteria of each CCP

Each CCP should have criteria that assure the safety of food. These criteria should be based on scientific studies and should be measurable. The criteria in food industry are upper or lower limits, not upper and lower limits e.g., only lower limits for cooking temperature and sanitizer concentration, and only upper limits for cold storage temperature, blast chilling time/temperature, and time control. The definitions and descriptions in this chapter are to be considered a framework with regard to set criteria in the sense that criteria may be tightened if required by a regulatory agency, by a customer, or by internal policy. All criteria should be compared internally and with the customer's national criteria, since criteria for various items can differ among countries.

Checklists and corrective action

A checklist of each CCP must be designed to cover all criteria but it should also be easy to fill out. Monitoring must be done by the worker who is doing the job; therefore, this system is called a "self-inspection system." Monitoring results are recorded on the spot, on the checklist of each CCP. Immediate corrective action must be taken if deviations from the established standards are identified. They should be highlighted on the record and the section manager should be notified immediately.

Proper corrective action should be clearly described on the checklist; otherwise no action or wrong action may be taken. The person responsible for monitoring and evaluation should be clearly listed on each definition sheet.

Prerequisite for HACCP program

1. Building layout

The unit layout is very important in the production of foods. To prevent cross-contamination, the layout should permit effective segregation of clean and unclean materials/processes and should flow in one direction. To prevent airborne contamination, the preparation of pastry and bread dough, which generate high amounts of dust containing bacteria, yeast, and molds, should be carried out in areas separate from the rest of meal production areas. The walk-in-coolers for storage of processed or semi-processed foods should be separately differentiated for different products like meat, bakery, dairy, soups, fruit and vegetable products.

Example of a CCP Checklist

CHECKLIST CCP. FOOD COLD STORAGE SECTION : ____________

WEEK COMMENCING: Refrigeration/Freezer Number:__________

DAY	TIME	GUAGE TEMP.	FOOD TEMP.	SEGREGATION[a]	DATE MARK[b]	COVER[c]	COMMENTS	SIGNATURE
MON	am							
	pm							
TUE	am							
	pm							
WED	am							
	pm							
THU	am							
	pm							
FRI	am							
	pm							
SAT	am							
	pm							
SUN	am							
	pm							

*Corrective Action: When wrong item, uncovered item, no date marked or outdated item is found, correct it. When the refrigerator/freezer temp. is over 5/ -18°C, call section head and maintenance team.

CHECKED BY: (USER) FILLING TO: HYGIENE CONTROL OFFICE

DATE:

[a]Check the presence of wrong item in that specific cold holding room. [b]Check the presence of datemark missing items or outdated items. [c]Check the presence of uncovered or inadequately covered items.

2. Cleaning and sanitizing

Cleaning in the food processing industry must be accompanied with sanitizing. Food-handling areas of the building should be designed to be easily cleaned and sanitized. For example, walls, floors, and ceiling should be constructed of a material impervious to water, fats, and chemicals. Working surfaces should be of smooth, durable, and readily cleansable materials. A cleaning program must be setup and used which itemizes the area/equipment to be cleaned, the person responsible, and the frequency with which cleaning must occur. The chemicals and the method of cleaning and sanitizing must also be specified. Chemicals employed must be suitable for use in food processing premises and should be stored separately from foods. Staff must be fully trained in cleaning methods and chemical usage.

3. Hygiene facilities

To control flying insects, electric fly killers should be installed and maintained properly at strategic locations. The locations of electric fly killers must be carefully decided for maximizing efficiency and preventing food contamination. Some models of electric fly killers should be not located less than 6 feet from the exposed food because electrocution creates aerosols of insects parts. UV lamps with the sticky board type are especially recommended for food preparation areas.

The air curtain at every outside entrance should have a blower strong enough to prevent flying insects from entering. High pressure hoses, sometimes used for cleaning trolleys, meal carts, or floors, produce aerosol mists that may spread contaminants around the production areas. There can be special problems when such hoses are used in areas of heavy contamination sources, such as drains. Unrestricted use of high pressure hoses should be avoided. Drains should be opened for regular cleaning and then flushed with a bactericidal agent. Cleaning equipment (mops, squeegees, etc.) should be kept in sets, dedicated to specific areas. Implements should be cleaned after use and kept in separate storage areas, where they can be allowed to dry.

4. Temperature measurement

Liquid-filled glass thermometers are not allowed for taking food temperatures because of the danger of breakage. Bimetallic dial thermometers are not suitable because of poor accuracy and slow response (28). They are also not appropriate for thin foods. These thermometers can be used for refrigerators and freezers with a wall mount, however. A wide variety of electronic instruments are available, such as simple pen-type digital thermometers (thermistors) and probe-exchangeable, handheld digital thermometers (thermocouples). The thermocouple and the thermistor can quickly measure temperature even in this food. Waterproof digital thermometers are more suitable for chefs, since thermometers can easily get wet by water or oil. In addition, the non contact type of infrared (IR) surface thermometer is recommended for some purposes. Measuring equipment should be accurate to, at most, 0.5°C or 1°F. All thermometers should be properly calibrated at certain intervals, as recommended by the manufacturer, and records of this should be kept. Thermometers probes may act as a source of cross-contamination if they remain dirty or are used for both raw and cooked foods. After each use, probes should be wiped and disinfected with a 70% alcohol pad or equivalent. It is important to have on-line monitoring systems for quantitative determination of different parameters such as temperature and pH. Some of the examples of HACCP with identified CPs and CCPs are given below for ready reference.

Example of potential hazards for fresh cut fruits and vegetables

Packaging materials	Microbiological contamination can be brought in on raw or packaging materials.
	Residues of field chemicals on the fruit or vegetable.
Predash	Growth of microbiological contaminants during initial wash step.
	There is also the risk of chemical contamination from the sanitisers
Inspection/ trimming	Microbiological contamination and/ or growth can occur with poor process controls.
	There is also a risk of chemical or physical contamination at this process step.
Peeling/ cutting	Microbiological contamination may occur if this process is not well controlled.
	Risk of chemical or physical contamination at this process step.
Dewatering	Microbiological contamination and/ or growth can occur at this stage.
	There is also a risk of chemical or physical contamination
Storage/ Addition of ingredients	Ingredients may include pre-prepared fresh cut products or other loose/ packaged food products.
	Microbiological or physical contamination can occur at this step.
Blending	Microbiological growth is likely to occur if the product is not chilled to the correct temperature.
	There may also be a risk of microbiological contamination

Contd...

Weighing/ packing	Microbial contamination from product handling, chemical contamination from cleaning chemicals or physical contamination from equipment and staff
Metal detection	Physical (metal) contamination, from the production or processing environment if the metal detector is not functioning adequately.
Finished product storage	Growth of potentially harmful organisms if, the finished product is not stored under temperature control after packaging
Distribution	Growth of potentially harmful organisms if the product is not chilled during distribution.

HACCP is a system or approach to achieve the desired set food standards. HACCP on its own does not give any certification for a process or product; rather it is a system if followed religiously helps in achieving the desired goals. Some of the standards set by different agencies for various food products are given below. These standards have been compiled from different sources. Good laboratory practices (GLP), good manufacturing practices (GMP), HACCP are the approaches to achieve the standards set by different organizations for internal consumption in the country as well as exports. As mentioned previously, these standards for different products vary from country to country. For example, certain specific crops grown in a particular region or country or a processed product made from those specific crops may not find a mention in the standards laid down by Codex Alimentarius Commission (CAC), ICMSF or NACMSF. It is imperative for the modern food processing units to put the QA strategies in place so as to have a final product free from any physical, chemical or biological contamination. Needless to say that the biological contamination caused because of microorganisms constitute about 95% of the total food hazards. Good laboratory practices, GMP and HACCP plans must be implemented in all the food processing units with proper checks and balances in place. Documentation and validation are very important for continuous improvement in food processing industry. The different standards for different products mentioned hereunder are for reference purpose so that the food processors, students, researchers can use them for their specific applications. As the consumer awareness about the food quality is increasing in India and with more and more people employed and keeping in mind the convenience and time, processed food industry in the country is increasing.

Following are the examples of some of the criteria laid down for microbiological standards in foods by different agencies/organizations

Specifications for dried milk powder (Prolacta Inc., USA)

Parameters	Specifications/ Limits
Standard Plate Count	Not more than 10,000 / g
E.coli	Nil
Salmonella	Nil / 25 g
Coliform	Nil
Yeast and mold count	Not more than 50/ g
Clostridium perfringes	Nil/ g
Staphylococcus aureus	Nil/ g
Listeria monocytogenes	Nil/ g

Microbiological limits for fruit and vegetable drink/beverage (Codex Alimentarius Commission)

Type of micro-organism	Limit (number of counts) maximum
Standard Plate Count	1000 cfu / ml
Yeast + Molds	50 cfu/ ml
Coliform	10 cfu/ ml
E.coli	Nil/ 30 g

FPO limits for vegetable sauces

Type of micro-organism	Limit (number of counts) maximum
Standard Plate Count	1×10^8 / ml
Molds	Not more than 40% of the fields examined by Howard mold counter
Yeasts	Not in excess of 125 per 1/8 c.m.m

Food Products Order (FPO) limits for tomato ketchup

Type of micro-organism	Limit (number of counts) maximum
Yeasts + Molds	1×10^8 / ml
Molds	Not more than 40% of the fields examined by Howard mold counter
Yeasts	Not in excess of 125 per 1/60 c.m.m

General microbiological specification – Germany & Netherlands

Parameter	Standard Value	Danger Value
Germany		
Total Aerobic Bacteria	$1x10^5/g$	$1x10^6/g$
E-coli	Absent	Absent
Bacillus cereus	$1x10^4/g$	$1x10^5/g$
Staphylococcus aureus	$1x10^2/g$	$1x10^3/g$
Salmonella		Absent in 25g
Netherlands		
Bacillus Cereus	Absent in 25g	
Escherichia Coli	Absent in 25g	
Clostridum perfringens	Absent in 25g	The danger values similar to that of Germany.
Staphylococus aureus	Absent in 25g	
Salmonella	Absent in 25g	
Yeast and Mold	Absent in 25g	
Coliform	$1x10^3/g$	
	$1x10^2/g$	

FPO specifications for sun dried mango slices

Type of micro-organism	Limit (number of counts) maximum
Total viable count	$1 \times 10^6 / g$
Molds	$1 \times 10^4 / g$
E.coli	$1 \times 10^3 / g$
Salmonella	Nil or negative / g

Microbiological Limits Specified by Prevention of Food Adultration (PFA) For Different Food Products

Prevention of Food Adulteration (PFA) Rules, 1956 specifies microbiological requirements for pathogens such as *E. coli, Staphylococcus auerus, Salmonella & Shigella, Vibrio cholerae, V. parahaemolyticus, Clostridium perfringens, C. botulinum, Listeria monocytogenes*

Infant milk food, infant formula, milk–cereal based complementary food, processed cereal based complementary food follow up formula

- Bacterial count per gram (not more than): 10,000
- Coliform count absent in 0.1g
- Yeast and mould count absent in 0.1g
- *Salmonella* and *Shigella* absent in 25g

- *E. coli* absent in 0.1 g
- *Staphylococcus aureas* absent in 0.1 g

Mineral water

Yeast and mould counts : absent

- Salmonella and Shigella : absent
- *E. coli* or thermotolerant coliforms: absent in 250 ml
- Total coliform bacteria: absent in 250 ml
- Faecal streptococci and *Staphylococcus aureus* : absent in 250 ml
- *Pseudomonas aeruginosa* : absent in 250 ml
- Sulphite-reducing anaerobes: absent in 50 ml
- *Vibro cholera*e : absent in 50 ml
- *V. paraheamolyticus* : absent in 250 ml

Packaged drinking water (Other than mineral water)

- Yeast and mould counts : absent in 250 ml
- Salmonella and Shigella : absent in 250 ml
- *E. coli* or thermotolerant coliforms: absent in 250 ml
- Total coliform bacteria: absent in 250 ml
- Faecal streptococci and *Staphylococcus aureus* : absent in 250 ml
- *Pseudomonas aeruginosa* : absent in 250 ml
- Sulphite-reducing anaerobes: absent in 50 ml
- *Vibro cholera*e : absent in 250 ml
- *V. paraheamolyticus* : absent in 250 ml
- Aerobic Microbial Count The total viable colony count shall not exceed 100 per ml at 20°C to 22°C in 72 h on agar-agar or on agar-gelatin mixture, and 20 per ml at 37°C in 24 h on agar-agar.

Microbiological Limits for Fruit and Vegetable Products

S.No.	Products	Parameters	Limits
1.	Thermally processed fruits and vegetable products	a) Total plate count b) Incubation at 37°C for 10days and 55°C for 7 days	a) Not more than 50/ml b) No change in pH
2.	i) Dehydrated fruits and vegetable products ii) Soup powders iii) Desiccated coconut powder iv) Table olives v) Raisins vi) Pistachio nuts vii) Dates viii) Dry fruits and nuts	Total plate count	Not more than 40,000/g
3.	Carbonated beverages, ready-to-serve beverages including fruit beverages	a) Total Plate Count b) Yeast and mould count c) Coliform count a) Mould Count b) Yeast and Spores	Not more than 50cfu/ml Not more than 2.0cfu/ml Absent in 100 m Positive in more than 40.0 percent of the field examined Not more than 125 per 1/60 c.m.m.
4.	Tomato Products a) Tomato juices and soups b) Tomato puree and paste c) Tomato ketchup and Tomato sauce	a) Mould Count a) Mould Count b) Yeast and Spores c) Total Plate count	Positive in more than 60.0 percent of the field examined Positive in more than 40.0 percent of the field examined Not more than 125 per 1/60 c.m.m. Not more than 10000/ml

5.	Jam/Marmalade/Fruit Jelly/Fruit Chutney and Sauce	a) Mould Count b) Yeast and Spores	Positive in more than 40.0 percent of the field examined
6.	Other fruits and Vegetables products	Yeast and Mould Count	Not more than 125 per 1/60 c.m.m.
7.	Frozen Fruits and Vegetable products	Total Plate Count	Not more than 40000/g
8.	Preserves	Mould Count	Absent in 25 g/ml
9.	Pickles	Mould Count	Absent in 25 g/ml
10.	Fruit Cereal Flakes	Mould Count	Absent in 25 g/ml
11.	Candied and Crystalised or Glazed Fruit and Peel	Mould Count	Absent in 25 g/ml
12.	i) All Fruits Vegetable Products and Ready to Serve beverages including Fruit Beverages and Synthetic	a)Flat Sour Organisms	Not more than 1000 cfu/g for those products which have pH less than 5.2
	Products such as	b) *Staphyloccus aureus*	Absent in 25 g/ml
	ii) Table olives	c) *Salmonella*	Absent in 25 g/ml
	iii) Raisins	d) *Shigella*	Absent in 25 g/ml
	iv) Pistachio nuts	e) *Clostridium botulinum*	Absent in 25 g/ml
	v) Dates	f) *E.coli*	Absent in 1 g/ml
	vi) Dryfruits and nuts	g) *Vibrio Cholera*	Absent in 25 g/ml
	vii) Vinegars		

Microbiological Criteria for Milk and Milk Products

S. No	Requirement	Ice Cream/ frozen Dessert /Milk Lolly/ Ice Candy/Dried Ice Cream Mix	Cheese/ Processed	Evaporated Milk	Sweetened Condensed Milk	Butter	Butter Oil/Butter Fat and Ghee	Yoghurt /Dahi
1.	Total Plate Count	Not more than 2,50,000/g	Not more than 50,000/g	Not more than 500/g	Not more than 500/g	Not more than 500/g	Not more than 500/g	Not more than 10,00,000/g
2.	Coliform Count	Not more than 10/g	Absent in 0.1g	Absent in 0.1g	Absent in 0.1g	Not more than 5/g	Absent in 0.1g	Not more than 10/g
3.	*E. coli*	Absent in 1 g	Absent in 1 g	Absent in 1 g	Absent in 1 g	Absent in 1 g	Absent in 1 g	Absent in 1 g
4.	*Salmonella*	Absent in 25 g.	Absent in 25g	Absent in 25g	Absent in 25g	Absent in 25g	Absent in 25g	Absent in 25g
5.	*Shigella*	Absent in 25 g.	Absent in 25g	Absent in 25g	Absent in 25g	Absent in 25g	Absent in 25g	Absent in 25g
6.	*Staphyloccus aureus*	Absent in 1 g	Absent in 1g	Not more than 100/g	Not more than 100/g	Absent in 1g	Absent in 1g	Not more than 100/g
7.	Yeast and Mould Count	Absent in 1 g	Absent in 1 g	Absent in 1 g	Not more than 10/g	Not more than 20/g	Absent in 1 g	Not more than 100/g
8.	Anaerobic Spore Count	Absent in 1 g	Absent in 1g	Not more than 5/g	Absent in 1g	Absent in 1g	Absent in 1g	Absent in 1g
9.	*Listeria monocytogenes*	Absent in 1 g	Cheese other than hard Cheese: Absent in 25g	Absent in 1g	Absent in 1g	Absent in 1g	Absent in 1g	Absent in 1g

S. No	Requirement	Milk Powder/ Cream Powder	Edible Casein Products	UHT Milk/UHT Flovoured Milk	Pasteurised milk	Sterilised Milk/Sterilised Flavoured milk	Khoya/ Chhana/ paneer	Chakka/ Srikhand
1.	Total Plate Count	Not more than 50,000/g	Not more than 50,000/g	Nil	Not more than 30,000/g	Nil	Not more than 50,000/g	Not more than 50,000/g
2.	Coliform Count	Not more than 0.1/g	Absent in 0.1 g	Absent in 0.1 g	Absent in 0.1 g	Absent in 0.1 g	Not more than 90/g	Not more than 10/g
3.	*E. coli*	Absent in 0.1 g	Absent in 1 g	Absent in 1 g	Absent in 1 g	Absent in 1 g	Absent in 1 g	Absent in 1g
4.	*Salmonella*	Absent in 25g	Absent in 25 g	Absent in 25 g	Absent in 25 g	Absent in 25 g	Absent in 25g	Absent in 25g
5.	*Shigella*	Absent in 25g	Absent in25 g	Absent in 25 g	Absent in 25 g	Absent in 25 g	Absent in 25g	Absent in 25g
6.	*Staphyloccus aureus*	Absent in 0.1 g	Absent in 1 g	Absent in 1 g	Absent in 1 g	Absent in 1 g	Not more than 100/g	Not more than 100/g
7.	Yeast and Mould Count	Absent in 1g	Absent in 1 g	Absent in 1 g	Absent in 1 g	Absent in 1 g	Not more than 250/g	Chakka: Not more than 10/g
8.	Anaerobic Spore Count	Absent in 1g	Absent in 1 g	Not more than 5/g	Absent in 1 g	Not more than 5/g	Absent in 1 g	Absent in 1g
9.	*Listeria monocytogenes*	Absent in 1g	Absent in 1 g	Absent in 1 g	Absent in 1 g	Absent in 1 g	Absent in 1 g	Absent in 1g

Quality Assurance is a proactive or preventive approach which helps in preventing the hazards to occur and thus is extremely important for the food processing industry which deals directly with the food that we consume. . Total quality management (TQM) deals with management of quality of each unit operation in manufacturing. It deals with the quality of equipment, instruments, testing procedures, record keeping, personnel employed for the job, audit procedures, planning and implementation. Hazard analysis critical control point (HACCP), good laboratory practices (GLP) and good manufacturing practices (GMP) are the strategies to achieve the food standards which are required to get the certifications, such as ISO 14001. Such strategies have to be followed on regular basis to maintain the hygiene standards as well as standards for raw food, semi-processed food, cooked food and stored food. Proper validation and documentation is a mandatory requirement in a modern food industry to ensure compliance to regulations stipulated for various foods. Biological hazards constitute about 95% of total food hazards and thus control over microbial growth and proliferation is the key to meeting the food standard requirement. There are different agencies like United States Food and Drug Administration (USFDA), International Commission on Microbiological Specifications for Foods (ICMSF), National Advisory Committee on National e on Microbiological Criteria for Food (NACMCF) and Codex Alimentarius Commission (CAC) which set standards for different food products for internal use as well as exports. In addition, different countries have set their own standards which they adopt for meeting the food laws and regulations. In India, agencies like Prevention of food adulteration (PFA) play a pivotal role in setting microbiological limits for different foods which have been included in this chapter.

References

Arora KC (1998). Total Quality Management and ISO 14000 (IS, 13967), 1st edn, S.K. Kataria and sons, New Delhi.

Bagchi (TP) (1994). ISO 9000- Concepts, methods and implementation, 2nd edn , Wheeler Publishing Company, New Delhi, India.

Farber JM and Todd ECD (eds) (2000). Safe handling of foods. Marcel Dekker Inc, New York, USA.

Hester RE and RM Harrison (2001). Food safety and food quality. Royal society of Chemistry, Cambridge CB4OWF, UK.

Menon S (1999). Food, Nutrition and Agriculture- Crisis management and technological innovation, Kanishka Publishers, New Delhi.

The prevention of food adulteration act and rules (1954). www.dbtbiosafety.nic.in

□□□

38

Food Safety Issues Related to Horticultural Produce

MAHESH KUMAR

Horticultural produce plays a significant role in human nutrition by supplying vitamins, minerals, dietary fibre and anti-oxidants to the diet. The quality and safety of horticultural produce reaching the consumer hinges upon pre-harvest factors as well as proper post harvest management practices throughout the chain, from the field to the consumer. Each stakeholder along the post-harvest chain- i.e. those involved in harvesting, handling and marketing of fresh produce- has a role to play in assuring the safety and quality of fresh produce. Quality, the degree of excellence or superiority, is a combination of attributes, properties, or characteristics that give each commodity value, in terms of its intended use. The relative importance given to a specific quality attribute varies in accordance with the commodity concerned and with the individual (producer, consumer, and handler) or market concerned with quality assessment. To producers, high yields, good appearance, ease of harvest, and the ability to withstand long-distance shipping to markets are important quality attributes. Appearance, firmness, and shelf-life are important from the point of view of wholesale and retail marketers. Consumers, on the other hand, judge the quality of fresh fruits, ornamentals, and vegetables on the basis of appearance (including 'freshness') at the time of initial purchase. Subsequent purchases depend upon the consumer's satisfaction in terms of flavor (eating) quality of the edible part of produce.

The number one issue among consumers is maintenance and improvement of fresh produce safety from field to fork. With intensive agricultural practices being under taken by the farming community, a number of chemicals are being used to maintain the productivity. The excessive use of poisons and fertilisers has made the people fearful of the harmful influence over their health. Further rising cost of medical treatment and life threatening disease becoming prevalent, people are fearful of consuming even fresh fruits and vegetable. Reducing the risk of food-borne illness from consumption of fresh fruits and vegetables is the responsibility of everyone, including growers, shippers, processors and consumers.

Safety factors

A number of factors threaten the safety of fruits and vegetables. These include naturally-occurring toxicants, such as glycoalkaloids in potatoes; natural contaminants, such as fungal toxins (mycotoxins) and bacterial toxins, and heavy metals (cadmium, lead, mercury); environmental pollutants; pesticide residues; and microbial contamination. While health authorities and scientists regard microbial contamination as the number one safety concern, many consumers rank pesticide residues as their most important safety concern. Unless fertilized with animal and/or human waste or irrigated with water containing such waste, raw fruits and vegetables should normally be free of most human and animal enteric pathogens. Organic fertilizers, such as chicken manure, should be sterilized prior to their application in fruit and vegetable production, so as to avoid the risk of contaminating fresh produce with *Salmonella, Listeria,* and other pathogens. Commodities that touch the soil are more likely to be contaminated than those that do not come in contact with the soil. The best approach to achieve and maintain the safety of fresh fruits and vegetables is to focus on reducing the potential contamination during their growth, harvesting, handling, treatment, packaging and storage. Strict adherence to Good Agricultural Practices, i.e. basic food safety principles associated with minimizing biological, chemical and physical hazards from the field throughout the distribution chain of fresh fruits and vegetables; Good Hygienic Practices, i.e. conformance to sanitation and hygienic practices to the extent necessary to protect against contamination of food from direct or indirect sources, are strongly recommended to minimize microbial contamination. Careful handling and washing of all produce to be consumed raw and the strict observance of proper sanitary measures are strongly recommended to reduce microbial contamination at the food-service, retail, and consumer levels.

Preventing contamination of fresh fruits and vegetables by human pathogens and dangerous levels of chemical residues is the best way to assure that foods are wholesome and safe for human consumption. This chapter will provide information on 1) potential hazards during production of raw produce, 2) sanitary postharvest handling of produce and 3) Hazard Analysis Critical Control Points (HACCP) principles as applied to handling fresh produce.

Food safety risks associated with produce

Physical hazards

Physical hazards which may become imbedded in produce during production handling or storage are fasteners (staples, nails, screws, holts), pieces of glass and wood splinters.

Chemical hazards

Chemical hazards which may contaminate produce during production handling or storage are pesticides, fungicides, herbicides, rodentacides, and machine lubricants from forklifts or packing line equipment, heavy metals (lead, mercury, arsenic), industrial toxins and compounds used to clean and sanitize equipment.

Pathogens

Human Pathogens: There are four main types of human pathogens associated with fresh produce:

Soil associated pathogenic bacteria: Clostridium botulinum, Listeria monocytogenes

Faeces associated pathogenic bacteria: Salmonella spp., Shigella spp., E. coli O157:H7 and others.

Pathogenic parasites: Cryptosporidium Cyclospora.

Pathogenic viruses: Hepatitis, Enterovirus.

Many of these pathogens are spread via human (or domestic animal) to food through human transmission route. Handling of fruits and vegetables by infected fieldworkers or consumers, cross contamination, use of contaminated irrigation water, use of inadequately composted manure or contact with contaminated soil are just a few of the ways that transmission of human pathogens to food can occur.

Food safety hazards during the production of fresh produce

Produce quality can be judged by outward appearance on the criteria like colour, turgidity and aroma but food safety cannot. Management of growing conditions are paramount in preventing the contamination of fresh produce by physical hazards, harmful chemicals and human pathogens.

Land use

The safety of food grown on any given piece of land is not only influenced by the current agricultural practices but also by former land use practices. Heavy metals and pesticide residues may remain in soils for long periods of time. Production land soil should be tested to assure that dangerously high levels of these compounds are not present. Former land use should also be investigated to assure that the production land was not used for hazardous waste disposal or for industrial purposes that may have left behind toxic residues. If production land was previously used for agricultural purposes, pesticide use records should be reviewed to assure that proper pesticide management practices were followed. Production acreage should not have been previously used as a feed lot or for animal grazing since fecal contamination of the soil may be extensive.

Fertilizers

Improperly or non composted organic fel1ilizers are a potential source of human pathogens. Human pathogens may persist in animal manures for weeks or even months. Proper composting (the heat treatment the manure undergoes as it breaks down) will reduce the risk of potential food borne illness. Use of inorganic fel1ilizers which have been certified to be free of heavy metals and other chemical contaminants is recommended.

Irrigation water

Irrigation water is another potential way of contamination of fruits and vegetables. Well water is less likely to be contaminated with human pathogens than surface water supplies, however, all irrigation water sources should be periodically tested for contamination caused by pesticides and human pathogens. Overhead irrigation systems are more likely to spread contamination since contaminated water is applied directly to the edible portions of fruits and vegetables.

Pesticide usage

Pesticide should be use in strict accordance with manufacturer recommendations as well as state ordinances. Monitoring and documentation of proper pesticide usage should be done to prevent unsafe levels or illegal residues contaminating fruits and vegetables. All pesticide applications should be documented and proper records of application as well as tests for residues should be available and reviewed by management on a regular basis.

Harvest operations

During harvesting operations field personnel may contaminate fresh fruits and vegetables by simply touching them with an unclean hand or knife blade. Portable field latrines as well as hand wash stations must be available and used by all harvest crew members. Monitoring and enforcement of field worker personnel hygiene practices such as washing hands after using the latrine are a must to reduce the risk of human pathogens contamination. Produce once harvested should not be placed upon bare soils before being placed in clean and sanitary field containers. Field harvesting tools should be clean, sanitised and not be placed directly in contact with soil. Field containers should be cleaned and sanitized on a regular basis as well as being free of contaminants such as mud, industrial lubricants, metal fasteners or splinters.

Postharvest handling of produce

Depending upon the commodity, produce may be packed in production catchments in containers that will go all the way to the destination market or be temporarily placed in bulk bins, baskets or bags which will be transported to a packing shed. Employees, equipment, cold storage facilities, packaging materials and any water which will be contacting the harvested produce must be kept clean and safe, to prevent contamination.

Employee hygiene

Gloves, hairnets and clean smocks are commonly worn by packing house employees in export oriented packing sheds. The cleanliness and personnel hygiene of employees handling produce at all stages of production and handling must be managed to minimize the risk of contamination. Adequate bathroom facilities and hand wash stations must be provided and used properly to prevent contamination of produce by packing house employees. Shoe or boot cleaning stations may also be in place to reduce the amount of field dirt and contamination which enters the packing shed from field operations. Employee training regarding sanitary food handling practices should be done when an employee is hired and reviewed before they begin work each season.

Equipment

Food contact surfaces on conveyor belts, dump tanks etc. should be cleaned and sanitized on a regular scheduled basis with food contact surface approved cleaning compounds. A 200 parts per million sodium hypochlorite (bleach) solution is an excellent example of a food contact surface sanitizer. Sanitizers should be used only after thorough cleaning with abrasion to remove organic materials Such as dirt or plant material. Use of steam to clean equipment should be avoided since steam may actually cake organic materials and form a bio film, which renders equipment almost impossible to sanitize. Steam may also aerosolize bacteria into the air and actually spread contamination throughout the packing house facility.

Cold storage facilities

Cold storage facilities and in particular refrigeration coils, refrigeration drip pans, (forced air cooling fans, drain tiles, walls and floors should be cleaned and sanitized on a frequent and regular basis. The human pathogen *Listeria monocytogen* can proliferate quite well at refrigerated temperatures and may contaminate cold stored produce if condensations from refrigeration units or ceilings drip on to the produce. Placing warm product with field heat into a cold room with insufficient refrigeration capacity will cause a temperature rise in the room and as the room cools a fog or mist will occur. As the water condenses out of the air and onto walls and ceilings, contamination from walls may end up, dripping onto stored produce.

Packaging materials

All packaging materials should be made of food grade materials to assure that toxic compounds in the packaging materials do not leach out of package and into the produce.

Toxic chemical residues may be present in some packaging materials due to use of recycled base materials. Empty packages Such as boxes and plastic bags should be stored in an enclosed storage area to protect them from insects, rodents, dust dirt and other potential sources of contamination. These actions protect, not only the potential loss of valuable materials but protects the integrity and safety of these materials. Plastic field bins and totes are preferred to wooden containers since plastic surfaces are easier to clean and sanitize, which should be done after every use. If containers are not cleaned and sanitized after every use, they may become contaminated and then contaminate the next products which are placed in the container. Wooden containers or field totes are almost impossible to sanitize since they have a porous surface and wooden or metals fasteners such as nails from wooden containers may accidentally be introduced into produce. Cardboard field bins if reused should be visually inspected for cleanliness and lined with a polymeric plastic bag before reuse to prevent the risk of cross contamination.

Wash and hydrocooling water

All water which comes in contact with produce for washing or hydrocooling must be safe to drink. Water should contain total chlorine to the level of 100-150 PPM and have a pH of between 6 and 7.5. Use of chlorine prevents the possibilities of cross contamination of all produce in the washing or hydrocooling system. It will not sterilize the produce. Scientific studies have demonstrated that washing produce in cold chlorinated water will reduce microbial populations by 100 to 1000 fold. But sterility is never achieved because microorganisms adhere so voraciously to the surface of produce and may be present in microscopic nooks and crannies on the surface of produce. Rinsing produce with cold chlorinated water will significantly reduce the number of microorganisms present on the produce but it will not remove all bacteria. Therefore, human pathogens cannot be completely removed from produce by washing it in cold chlorinated water.

Refrigerated transport

Produce is best shipped in temperature controlled refrigerated trucks. Maintaining perishables below 5°C even while being transported to destination markets will extend shelf-life and significantly reduce the growth rate of microbes including human pathogens. Temperatures used for transporting chilling sensitive produce will not protect against the growth of most pathogens. Trucks used during transportation should be cleaned and sanitized on a regular basis. Trucks which have been used to transport live animals, animal products or toxic materials should never be used to transport produce.

Hazard analysis critical control points (HACCP)

HACCP is a food safety system first developed by the Pillshury Co. to reduce the risk associated with the food eaten by astronauts during space flights. HACCP is systems approach which:

- Identifies potential sources of contamination in food production systems
- Establishes methods for detecting the occurrence of contamination
- Clearly prescribes what corrective actions will be taken to prevent consumption of contaminated food items.

HACCP is a systems approach method to assure the safety of a food product, it is NOT a means of assuring food quality. The seven basic principles of HACCP are:

- Assessment of Hazards
- Determine Critical Control Points (CCP) to Control the Identified Hazards

- Establishment of Limits at each CCP
- Establishment of CCP Monitoring Procedures
- Establishment of Corrective Actions to be Taken When CCP Exceed Set Limits
- Establishment of Record Keeping Systems to Document the HACCP Program
- Establishment of Procedures to Verify that the HACCP is Functioning Properly.

References

Kader AA and Rosa S Rolle (2004). The role of post-harvest management in assuring the quality and safety of horticultural produce. FAO Bulletin no 152.

Kader AA, Kasmire RF, Mitchell FG , Reid MS, Sommer NF and Thompson JF (1985). Postharvest technology of horticultural produce. University of California, Division of Agriculture and Natural Resources.

Ashby, BH (1989). Regulations governing sanitation in vehicles transporting food, USDA Transporation Tips. Office of Transporation, OT-ID No 12.

Beuchat. LR (1996). Pathogenic micro-organisms associated with fresh produce, J. Food Prot, 59: 204-216.

Harris, L. (ed) (1997). Perishables Handling. Special Issue No, 91 : Food Safety.

Rushing, JW (1996). Implementation of a HACC P program in a commercial fresh market tomato.

Packing house: A model for the mdustry. Dairy, Food and Environment Sanitation, 16(9): 549-553.

Suslow, TV (1997). Microbial Food Safety an emerging challenge f~)r small-scale growers. Small Farm News June-July: 7-10.

❏❏❏

39

Energy Efficiency Measures in Food Processing Industry

SANGEETA CHOPRA

The processing techniques that are employed by fruit and vegetable industry are dependent on the variety of products that are manufactured. At any given facility, the choice of individual processes as well as the process sequences that are employed depend on the preservation method used (i.e. canning, freezing, or drying and dehydrating) and on the specific fruits and vegetables that are processed. However, there are many unit processes that are common such as washing, blanching, peeling, conveying, drying and size reduction etc. The energy expended in these operations represents a significant operating cost to the fruit and vegetable processing industry. Typically, the energy cost depends on the processes and the product types and accounts for up to 10-15% of the production cost. Significant savings can be made in the energy bill by implementing the following energy conservation measures as the cost of energy is a controllable factor in the food-processing sector. The high energy-consuming areas in various types of food-processing industries are highlighted in the Table 1.

Table 1: High energy consuming areas in food processing industries

Type of products	Critical areas
Frozen fruits and vegetables	Boiler, refrigeration cold storage blast freezing motors, cooling tower
Tomato ketchup/puree/ juice, sauces, canned fruits and vegetables fruit pulps, juices	Boiler, steam heating, steam traps, motors
Ready-to-cook, ready-to- eat, instant mixes, soup mixes and pickle, spices chutneys	Grinder and pulverizer motors, air compressor lighting
Candies and jellies	Boiler, steam usage, motors

Steam generation and distribution system

In food processing industry, steam is generally used for blanching, peeling, heat sterilization, evaporation, pasteurization, hot water generation, indirect heating, and so on. Energy can be saved both in steam generation and distribution through some simple measures as mentioned below.

- The biomass-based boiler should be adopted wherever possible. It is environment friendly and does not contribute to GHG (greenhouse gas) emissions.
- There should be no fuel oil leakage. Even a small oil leakage of one drop per second can result in wastage of 4000 litres per year of fuel oil.
- The gauges for measuring steam pressure and temperature on boilers and temperature gauge in the flue gas outlet may be provided.
- The flue gas should be regularly analyzed using a portable flue gas analyzer. The parameters to be checked are oxygen, carbon monoxide, and temperature.
- The excess air in the boiler may be optimized. Excess air is the quantity of air in addition to the theoretical quantity required for 100% fuel combustion. The recommended excess air and O_2 levels for various fuels are mentioned in Table 2.

Table 2: Recommended oxygen and excess air levels

Fuel	Recommended oxygen levels in flue gas (%)	Excess air (%)
Diesel	2–3	10–15
Bagasse	5-7	25-35
Wood/biomass	4-5	20-25

- A simple maintenance programm to ensure all components of boiler are operating at peak performance can result in substantial energy savings (Galitsky *et al.,* 2005).
- The colour of the smoke coming out of the boiler may be checked. Brown hazy smoke indicates proper combustion; black colour indicates incomplete combustion; and colorless or white smoke shows high excess air quantity.
- The reduction in excess air quantity by 5% as compared to the recommended excess air percentage increases the boiler efficiency by 1%. Similarly, 1% reduction of residual oxygen in the flue gas reduces flue consumption by 1%. The sensible heat may be recovered from the hot flue gas for heating the boiler water (via economizer) or preheating the combustion air (via air-reheater).
- The condensate should be recovered and returned to the boiler, as it still carries about 15 -20% of the total steam energy.
- The temperature gauges should be installed in all steam heating equipment to avoid overheating of material.
- Each steam-using unit should always have an individual steam trap. The steam trap bypass line may be kept closed always except during start up. If the trap is faulty, the trap should be replaced rather than opening the bypass valve using thermostatic element steam traps can ensure reduced energy use and increased reliability (Alesson 1995).

Compressed air system

Compressed air is generally used in all food processing industry for machine operations, pneumatic controls, and other similar applications. Following recommendations may be followed for an efficient compressed air system. (CAC, 2002)

- The air intake to the compressor should be clean, cool and drawn from a place which is away from the heat sources.
- The compressor discharge pressure should be kept at minimum acceptable level. Increase of 1 kg/cm^2 air discharge pressure (above the desired) from the compressor would result in about 4–5% increase in input power. This will also increase compressed air leakage rates roughly by 10%.
- If compressed air is required at two different pressures, it is better to have two compressors catering to air requirement at different pressures rather than having one large compressor generating compressed air at higher pressure.

- The air intake filters should be cleaned at regular intervals to facilitate clean air intake of compressor and low pressure drop across it. (Radgen and Blaustein, 2001).
- Compressed air should be cooled up to the ambient temperature by using after coolers/air dryers before it enters in the system. This will help in removing the moisture from the system.
- The separators should be provided to get rid of any moisture in the system before the compressed air reaches the pneumatic equipment.
- All the compressed air piping should be laid out in such a way that it minimizes pressure drops during transmission. The pressure drops in a smaller compressed air system should not be more than 0.3 bar.
- A properly sized compressed air storage receiver should be used. It dampens the pulsation from the reciprocating compressor and makes the flow of air smooth. It also acts as a reservoir to meet short time excess demand in the system, condenses moisture, and removes the oil traces (through drain trap) from the compressed air. The size of the receiver should be at least 6–10 seconds capacity of the compressor. It is always beneficial to over design the air receiver. The receiver should be fitted with a pressure gauge, safety valve, and a moisture drain valve.
- The compressed air generation is a costly affair therefore the alternatives such as blowers for low-pressure high quantity applications, hydraulic rather than air cylinders, electric rather than air actuators, and electronic rather than pneumatic controls should be considered.

Electrical distribution system

Electrical system is an integral part of all food processing industry. An efficient electrical distribution system and demand management can reduce the electricity bill significantly. Some of the opportunities to improve efficiency of the distribution network are as given below.

- The non-critical load should be staggered according to the electricity tariff to reduce the electricity expenditure.
- The capacitors should be provided at the load (motor) end to have the benefit of reduced distribution loss (for example, line losses, and cable loading). Any shortfall from the desired power factor can be met by connecting the capacitors at the main panel. The power factor should always be less or equal to unity and never be leading, which may lead to motor burning. The automatic power factor relay should be used for effective power factor management.

- The benefits of higher power factor are reduced demand, better voltage, high system efficiency and rebate from the electricity supplying company. The power factor can be improved by installing capacitors in the electrical system.
- The contactor based or thyristorized power factor correction systems may be used for the power factor correction.
- The transformers are, normally, designed to operate at maximum efficiency between loadings of 32 and 35% of its full load capacity. If the load on transformer increases beyond 80% of the designed capacity, it is better to go for a new or bigger transformer to avoid the sharp rise in transformer losses.

Motors and lighting

Motors are used throughout a typical food-processing unit for various processes (such as mixing, grinding, peeling, cutting, pulping, filling, and packaging) and utility (such as ventilation fans, compressors, and pumps) equipment. Some energy conservation opportunities for small food processing units are given below. The typical industrial plant can reduce its electricity use by around 5-15% by inproving efficiency of its motor driven system (US DOE 2006e)

- The properly sized high efficiency motors should be used as per the load application. Oversized motors can result in unnecessary energy wastage due to decrease in efficiency and power factor
- The new replacements should be done with energy-efficient motors having 3-5% higher efficiency.
- A 20% reduction in motor speed will result into almost 50% power savings
- The old motor which has undergone rewinding 3 times should be replaced as the motor efficiency goes down by 3-5% after each rewinding
- The motor should be rewound by a qualified person. This will minimize the losses in the rewound motor.
- If a motor is continuously running below 45% of its design load, it is better to operate it in star mode by changing to star connection or by installing auto delta-star converter. This will give a handsome saving.
- It is appropriate to use VSD (variable speed drive) with the motor, if the load is of variable nature such as that for pumps, fans and mixers. VSD matches the motor speed as per the load and results in savings. (Hackett *et al.,* 2005)

- The lights should be kept off in cold storage to avoid additional load on refrigeration system due to extra heat load of lighting fixture.
- The maximum use of daylight should be made by providing translucent roof sheets, glass window etc.
- All the incandescent bulbs should be replaced with CFL. The fused tube light should be replaced immediately to avoid choke losses in tube light fitting without any useful lighting
- The conventional 40 W tube lights may be replaced with more efficient 28 W T5 lights. The mercury vapor lamps should be replaced with 70 W low pressure sodium vapor (LPSV) lamp, for the areas where colour rendering is not important. This will result in 40–50% electricity savings

Emerging energy-efficient technologies for fruit and vegetable processing

Ohmic heating : Ohmic heating is a thermal processing method in which an alternating electrical current is passed through food products to generate heat internally. Ohmic heating is said to produce a uniform, inside-out heating pattern that heats foods faster and more evenly than conventional outside-in heating methods. The potential applications for ohmic heating relevant to fruit and vegetable processing include blanching, evaporation, dehydration, fermentation, and extraction. However, ohmic heating parameters such as the frequency of the alternating current, the applied voltage, the temperature to which the sample is heated, and the electrical conductivity of the food can all have a significant effect on the performance of the process.

Infrared drying : In conventional drying methods, substantial amounts of air must be heated and circulated around the product to be dried. In contrast, infrared drying uses infrared radiation to heat only the material that needs to be heated—not the surrounding air—and thus saves energy compared to conventional methods. For drying apple slices, a comparison of infrared drying with convective drying done using equivalent processing parameters showed that energy costs were lower and that the time of the drying process could be shortened by up to 50% using infrared methods.

Pulsed fluid-bed drying : The pulsed fluid-bed dryer is a modification of the conventional fluid-bed dryer (used widely in the dehydration of fruits and vegetables). In pulsed fluid-bed drying, gas pulses cause high-frequency vibrations within the bed of product particles. Reported advantages of the pulsed fluid-bed drying approach include easier fluidization of irregular particle shapes, fluidization with 30 to 50% less air than conventional methods (leading to energy savings in heating and circulating hot air), and reduced channeling of particles. Additionally, pulsed fluid-bed dryers are roughly half the size of conventional conveyor-type dryers. Successful trial applications in the food industry include the drying of carrot cubes and the drying of chopped onions. In the drying of carrot cubes, a pulsed fluid-bed dryer reduced the total drying time by two to three times compared to traditional fluid-bed drying methods while providing a final product that was highly uniform in color and moisture content. Similarly, for chopped onions, the final products were of high color and reconstitution quality and uniform in moisture content.

Pulsed electric field pasteurization : The use of pulsed electric fields to pasteurize liquid food products is showing promise as an emerging technology. Pulsed electric field pasteurization for juices may provide superior taste and freshness compared to juices undergoing conventional heat treatment. In the pulsed electric field process, liquids are exposed to high voltage pulses of electricity to inactivate harmful micro-organisms as well as enzymes that degrade the quality of fruit juices over time. The energy savings associated with pulsed electric field processing arise from the fact that the process operates at lower temperatures than conventional heat-based pasteurization methods and thus the pasteurized juices require less cooling energy.

Magnetically-coupled adjustable-speed drives : Magnetically-coupled adjustable-speed drives (MC-ASDs) are a new type of ASD, in which the physical connection between the motor and the driven load is replaced with a gap of air, and the amount of torque transferred is controlled by varying the air gap distance between rotating plates in the assembly. As compared to existing ASDs, MC-ASDs have several advantages in addition to greater energy efficiency, including greater tolerance for motor misalignment, little impact on power quality, ability to be used with regular duty motors (instead of inverters), expected lower long term maintenance costs and extended motor and equipment lives, due to elimination of vibration and wear on equipment

Energy is an important basic input in the manufacturing process and its saving will have direct impact on the profitability of the food industries. It is expected that by implementing the energy conservation measures mentioned above the food industry can save about 5–10 % of their energy costs.

References

Alesson, T (1995). All Steam Traps are not Equal. Hydrocarbon Processing. Gulf Publishing Company, Houston, Texas.

Compressed Air Challenge (CAC) (2002). Guidelines for Selecting a Compressed Air system service http://www.compressedairchallenge.org.

Energy-Efficiency booklet-NEW.p65, Published by TERI Press, The Energy and Resources Institute, Darbari Seth Block, IHC Complex, Lodhi Road, New Delhi-110 003.

Galitsky, C, E Worrell, A Radspieler, P Healy, and S Zechiel (2005). BEST Winery Guidebook; Benchmarking and Energy and Water Savings Tool for the Wine Industry. Lawrence Berkeley National Laboratory, Berkley, California. LBNL-3184.

Hackett, B, Chow, S, and AR Ganji (2005). Energy Efficiency Opportunities in Fresh Fruit and Vegetable Processing/Cold Storage Facilities. Processings of the 2005 ACEEE Summer Study on Energy Efficiency in Industry, American Council for an Energy-Efficient Economy, Washington, D.C.

Radgen, P and E Blaustein (Eds.) (2001). Compressed Air Systems in the European Union, Energy, Emissions, Savings Potential and Policy Actions. LOG_X Verlag, GmbH, Stuttgart, Germany.

Tutterow, V, D Casada, and A McKane (2000). Profiting from your Pumping System. Proceedings of the 2000 Pump Users Expo, Louisville, Kentucky.

United States Department of Energy (DOE) (2006e). Save Energy Now in Your Industrial Steam Systems. Office of Energy/Efficiency and Renewable Energy, Industrial Technologies Program, Washington, D.C. Report DOE/GO-102006-2276.

Worrell, E, Price, L, and C Galitsky (2004). Emerging Energy-Efficient Technologies in Industry: Case Studies of Selected Technologies. Lawrence Berkeley National Laboratory, Berkeley, California. LBNL-54828.

Zeitz, Ronald A (Ed.) (1997). CIBO Energy Efficiency Handbook. Council of Industrial Boiler Owners, Burke, Virginia. 149.

□□□

40

Optimization and Statistical Analysis for Food Processing Experiments

VINOD K. BHARGAV

Experimental methods are widely used in research as well as in industrial settings, however, sometimes for very different purposes. The primary goal in scientific research is usually to show the statistical significance of an effect that a particular factor exerts on the dependent variable of interest.

In industrial settings, the primary goal is usually to extract the maximum amount of unbiased information regarding the factors affecting a production process from as few (costly) observations as possible. While in the former application (in science) analysis of variance (*ANOVA*) techniques are used to uncover the interactive nature of reality, as manifested in higher-order interactions of factors, in industrial settings interaction effects are often regarded as a "nuisance" (they are often of no interest; they only complicate the process of identifying important factors).

These differences in purpose have a profound effect on the techniques that are used in the two settings. If you review a standard ANOVA text for the sciences, you will find that they will primarily discuss designs with up to, perhaps, five factors (designs with more than six factors are usually impractical.

The focus of these discussions is how to derive valid and robust statistical significance tests.

Mixture designs: Cornell (1990) reports an example of a typical (simple) mixture problem. Specifically, a study was conducted to determine the optimum texture of fish patties as a result of the relative proportions of different types of fish (Mullet, Sheepshead, and Croaker) that made up the patties. Unlike non-mixture experiments, the total sum of the proportions must be equal to a constant, for example, to 100%. The results of such experiments are usually graphically represented in so-called triangular (or ternary) graphs. In general, the overall constraint —that the three components must sum to a constant is reflected in the triangular shape of the graph.

Constrained mixture designs: It is particularly common in mixture designs that the relative amounts of components are further constrained (in addition to the constraint that they must sum to, for example, 100%). For example, suppose we wanted to design the best-tasting fruit punch consisting of a mixture of juices from five fruits. Since the resulting mixture is supposed to be a fruit punch, pure blends consisting of the pure juice of only one fruit are necessarily excluded. Additional constraints may be placed on the "universe" of mixtures due to cost constraints or other considerations, so that one particular fruit cannot, for example, account for more than 30% of the mixtures (otherwise the fruit punch would be too expensive, the shelf-life would be compromised, the punch could not be produced in large enough quantities, etc.). Such so-called constrained experimental regions present numerous problems, which, however, can be addressed with the *Experimental Design* module. In general, under those conditions, one seeks to design an experiment that can potentially extract the maximum amount of information about the respective response function (e.g., taste of the fruit punch) in the experimental region of interest.

Response optimization : A typical problem in product development is to find a set of conditions, or levels of the input variables, that produces the most desirable product in terms of its characteristics, or responses on the output variables. The characteristics of the product in terms of the response variables depend on the ingredients, the *X* variables. The problem is to select the levels for the *X*'s which will maximize the desirability of the responses on the *Y*'s. The solution must take into account the fact that the levels for the *X*'s which maximize one response may not come close to maximizing a different response. Also, for some response variables, the most desirable value is not necessarily the highest value, for example, the most desirable value of elongation falls within a narrow range of the possible values. Such problems can be addressed with the *Experimental Design* module using *Response/desirability profiling* and other options for analyzing prediction profiles and response surfaces.

There are basically two general issues to which Experimental Design is addressed:

1. How to design an optimal experiment, and
2. How to analyze the results of an experiment

With regard to the first question, there are different considerations that enter into the different types of designs. In the most general terms, the goal is always to allow the experimenter to evaluate in an unbiased (or least biased) way, the consequences of changing the settings of a particular factor, that is, regardless of how other factors were set. In more technical terms, you attempt to generate designs where main effects are unconfounded among themselves, and in some cases, even unconfounded with the interaction of factors. With regard to the second question (how to analyze results of experiments), the designs (and how to analyze them) all pertain to the estimation (fitting) of response surfaces, following the general model equation:

$$y = b_0 + b_1 * x_1 + ... + b_k * x_k + b_{12} * x_1 * x_2 + b_{13} * x_1 * x_3 + ... + b_{k-1,k} * x_{k-1} * x_k + b_{11} * x_1^2 + ... + b_{kk} * x_k^2$$

Put into words, one is fitting a model to the observed values of the dependent variable y, that include (1) main effects for factors x_1 , ..., x_k, (2) their interactions (x_1*x_2, x_1*x_3, ... ,$x_{k-1}*x_k$), and (3) their quadratic components (x_1^2, ..., x_k^2). No assumptions are made concerning the "levels" of the factors, and you can analyze any set of continuous values for the factors.

There are some considerations concerning design efficiency and biases, which have led to standard designs that are ordinarily used when attempting to fit these response surfaces. The central composite design analysis options do not make any assumptions about the structure of your data file, that is, the number of distinct factor values, or their combinations across the runs of the experiment, and, hence, these options can be used to analyze any type of design, to fit to the data the general model described above.

Orthogonal designs: One desirable characteristic of any design is that the main effect and interaction estimates of interest are independent of each other. For example, suppose you had a two-factor experiment with both factors at two levels. Your design consists of four runs:

	A	B
Run 1	1	1
Run 2	1	1
Run 3	-1	-1
Run 4	-1	-1

For the first two runs, both factors *A* and *B* are set at their high levels (*+1*). In the last two runs, both are set at their low levels (*-1*). Suppose you wanted to estimate the independent contributions of factors *A* and *B* to the prediction of the dependent variable of interest. Clearly this is a silly design, because there is no way to estimate the *A* main effect and the *B* main effect. One can only estimate one effect – the difference between *Runs 1+2* vs. *Runs 3+4* – which represents the combined effect of *A* and *B*.

The point here is that in order to assess the independent contributions of the two factors, the factor levels in the four runs must be set so that the "columns" in the design (under *A* and *B* in the illustration above) are independent of each other. Another way to express this requirement is to say that the columns of the design matrix (with as many columns as there are main effect and interaction parameters that one wants to estimate) should be orthogonal (this term was first used by Yates, 1933). For example, if the four runs in the design are arranged as follows:

	A	B
Run 1	1	1
Run 2	1	-1
Run 3	-1	1
Run 4	-1	-1

Then the *A* and *B* columns are orthogonal. Now you can estimate the *A* main effect by comparing the high level for *A* within each level of *B*, with the low level for *A* within each level of *B*; the *B* main effect can be estimated in the same way.

Technically, two columns in a design matrix are orthogonal if the sum of the products of their elements within each row is equal to zero. In practice, one often encounters situations, for example due to loss of some data in some runs or other constraints, where the columns of the design matrix are not completely orthogonal. In general, the rule here is that the more orthogonal the columns are, the better the design, that is, the more independent information can be extracted from the design regarding the respective effects of interest. Therefore, one consideration for choosing standard central composite designs is to find designs that are orthogonal or near-orthogonal.

Rotatable designs: The second consideration is related to the first requirement, in that it also has to do with how best to extract the maximum amount of (unbiased) information from the design, or specifically, from the experimental region of interest. It can be shown that the standard error for the prediction of dependent variable values is proportional to:

$$(1 + f(x)' * (X'X)^{-1} * f(x))^{1/2}$$

where f(x) stands for the (coded) factor effects for the respective model *f(x)* is a vector, *f(x)'* is the transpose of that vector), and *X* is the design matrix for the experiment, that is, the matrix of coded factor effects for all runs; $X'X^{-1}$ is the inverse of the crossproduct matrix. Deming and Morgan (1993) refer to this expression as the normalized uncertainty; this function is also related to the *variance function* as defined by Box and Draper (1987). The amount of uncertainty in the prediction of dependent variable values depends on the variability of the design points, and their covariance over the runs.

For the simple 4-run orthogonal experiment shown earlier, the information function is equal to

$$I_x = 4/(1 + x_1^2 + x_2^2)$$

where x_1 and x_2 stand for the factor settings for factors *A* and *B*, respectively. Inspection of this function in a plot shows that it is constant on circles centered at the origin. Thus any kind of rotation of the original design points will generate the same amount of information, that is, generate the same information function. Therefore, the 2-by-2 orthogonal design in 4 runs shown earlier is said to be rotatable.

As pointed out before, in order to estimate the second order, quadratic, or non-linear component of the relationship between a factor and the dependent variable, one needs at least 3 levels for the respective factors. What does the information function look like for a simple 3-by-3 factorial design, for the second-order quadratic model, this function looks more complex, contains "pockets" of high-density information at the edges (which are probably of little particular interest to the experimenter), and clearly it is not constant on circles around the origin. Therefore, it is not rotatable, meaning different rotations of the design points will extract different amounts of information from the experimental region.

Star-points and rotatable second-order designs: It can be shown that by adding so-called star-points to the simple (square or cube) 2-level factorial design points, one can achieve rotatable, and often orthogonal or nearly orthogonal designs. For example, adding to the simple 2-by-2 orthogonal design shown earlier the following points, will produce a rotatable design.

	A	B
Run 1	1	1
Run 2	1	-1
Run 3	-1	1
Run 4	-1	-1
Run 5	-1.414	0
Run 6	1.414	0
Run 7	0	-1.414
Run 8	0	1.414
Run 9	0	0
Run 10	0	0

The first four runs in this design are the previous 2-by-2 factorial design points (or square points or cube points); runs 5 through 8 are the so-called star points or axial points, and runs 9 and 10 are center points. The information function for this design for the second-order (quadratic) model is rotatable, that is, it is constant on the circles around the origin

The two design characteristics orthogonality and rotatability depend on the number of center points in the design and on the so-called axial distance a (*Alpha*), which is the distance of the star points from the center of the design (*i.e., 1.414* in the design shown above). It can be shown that a design is rotatable if:

$$\alpha = (n_c)^{1/4}$$

where nc stands for the number of cube points in the design (i.e., points in the factorial portion of the design).

A central composite design is orthogonal, if one chooses the axial distance so that:

$$\alpha = \{[(n_c + n_s + n_0)^{1/2} - n_c^{1/2}]^2 * n_c/4\}^{1/4}$$

where

n_c is the number of cube points in the design

n_s is the number of star points in the design

n_0 is the number of center points in the design

To make a design both (approximately) orthogonal and rotatable, one would first choose the axial distance for rotatability, and then add center points so that:

$$n_0 \approx 4*n_c^{1/2} + 4 - 2k$$

where k stands for the number of factors in the design.

Finally, if blocking is involved, Box and Draper (1987) give the following formula for computing the axial distance to achieve orthogonal blocking, and in most cases also reasonable information function contours, that is, contours that are close to spherical:

$$\alpha = [k^*(1+n_{s0}/n_s)/(1+n_{c0}/n_c)]^{1/2}$$

where

n_{s0} is the number of center points in the star portion of the design

n_s is the number of non-center star points in the design

n_{c0} is the number of center points in the cube portion of the design

n_c is the number of non-center cube points in the design

The analysis of central composite designs proceeds in much the same way as for the analysis of $3^{(k-p)}$ designs. The program will fit to the data the general model described above; for example, for two variables the program would fit the model:

$$y = b_0 + b_1{}^*x_1 + b_2{}^*x_2 + b_{12}{}^*x_1{}^*x_2 + b_{11}{}^*x_1^2 + b_{22}{}^*x_2^2$$

Of course, the Experimental Design allows you to flexibly "toggle" in and out the different terms in this model

Special issues arise when analyzing mixtures of components that must sum to a constant. For example, if you wanted to optimize the taste of a fruit-punch, consisting of the juices of 5 fruits, then the sum of the proportions of all juices in each mixture must be 100%. Thus, the task of optimizing mixtures commonly occurs in food-processing, refining, or the manufacturing of chemicals. A number of designs have been developed to address specifically the analysis and modeling of mixtures. The common manner in which mixture proportions can be summarized is via triangular (ternary) graphs. For example, suppose you have a mixture that consists of 3 components *A*, *B*, and *C*. Any mixture of the three components can be summarized by a point in the triangular coordinate system defined by the three variables.

For example, take the following 6 different mixtures of the 3 components.

	A	B
1	0	0
0	1	0
0	0	1
0.5	0.5	0
0.5	0	0.5
0	0.5	0.5

The sum for each mixture is 1.0, so the values for the components in each mixture can be interpreted as proportions. If you graph these data in a regular 3D scatterplot, it becomes apparent that the points form a triangle in the 3D space. Only the points inside the triangle where the sum of the component values is equal to 1 are valid mixtures. Therefore, one can simply plot only the triangle to summarize the component values (proportions) for each mixture.

Fitting a response surface to mixture data is, in principle, done in the same manner as fitting surfaces to, for example, data from central composite designs. However, there is the issue that mixture data are constrained, that is, the sum of all component values must be constant.

Consider the simple case of two factors *A* and *B*. One may want to fit the simple linear model:

$$y = b_0 + b_A{*}x_A + b_B{*}x_B$$

Here y stands for the dependent variable values, b_A and b_B stand for the regression coefficients, x_A and x_B stand for the values of the factors. Suppose that x_A and x_B must sum to 1; you can multiply b_0 by $1=(x_A + x_B)$:

$$y = (b_0{*}x_A + b_0{*}x_B) + b_A{*}x_A + b_B{*}x_B$$

or

$$y = b'_A{*}x_A + b'_B{*}x_B$$

where $b'_A = b_0 + b_A$ and $b'_B = b_0 + b_B$. Thus, the estimation of this model comes down to fitting a no-intercept multiple regression model.

The Experimental Design computes the coefficients for the requested model, along with the respective standard errors and confidence intervals.

ANOVA 3 Factor mixture design; Mixture total=1., 14 Runs
Sequential fit of models of increasing complexity

Model	SS Effect	df Effect	MS Effect	SS Error	df Error	MS Error	F	p	R-sqr	R-sqr Adj.
Linear	44.755	2	22.378	46.872	11	4.2611	5.2516	0.0251	0.4884	0.3954
Quadratic	30.558	3	10.186	16.314	8	2.0393	4.9949	0.0307	0.8220	0.7107
Special Cubic	0.719	1	0.719	15.596	7	2.2279	0.3225	0.5878	0.8298	0.6839
Cubic	8.229	3	2.743	7.367	4	1.8417	1.4893	0.3452	0.9196	0.7387
Total Adjusted	91.627	13	7.048							

To decide which of the models of increasing complexity provides a sufficiently good fit to the observed data, one usually compares the models in a hierarchical, stepwise fashion. For example, consider a 3-component mixture to which the full cubic model was fitted. The Experimental Design module computes the following ANOVA.

First, the linear model was fit to the data. Even though this model has 3 parameters, one for each component, this model has only 2 degrees of freedom. This is because of the overall mixture constraint, that the sum of all component values is constant. The simultaneous test for all parameters of this model is statistically significant ($F(2,11)=5.25$; $p<.05$). The addition of the 3 quadratic model parameters ($b_{12}*x_1*x_2$, $b_{13}*x_1*x_3$, $b_{23}*x_2*x_3$) further significantly improves the fit of the model ($F(3,8)=4.99$; $p<.05$). However, adding the parameters for the special cubic and cubic models does not significantly improve the fit of the surface. Thus, one could conclude that the quadratic model provides an adequate fit to the data (of course, pending further examination of the residuals for outliers, etc.).

R-square: The ANOVA spreadsheet will also report the *R-square* values for the respective models. The *R-square* value can be interpreted as the proportion of variability around the mean for the dependent variable, that can be accounted for by the respective model. (Note that for non-intercept models, some multiple regression programs will only compute the *R-square* value pertaining to the proportion of variance around 0 (zero) accounted for by the independent variables.

Pure error and lack of fit: The usefulness of the estimate of pure error for assessing the overall lack of fit was discussed in the context of central composite designs. If some runs in the design were replicated, then one can compute an estimate of error variability based only on the variability between replicated runs. This variability provides a good indication of the unreliability in the measurements, independent of the model that was fit to the data, since it is based on identical factor settings (or blends in this case). One can test the residual variability after fitting the current model against this estimate of pure error. If this test is statistically significant, that is, if the residual variability is significantly larger than the pure error variability, then one can conclude that, most likely, there are additional significant differences between blends that cannot be accounted for by the current model. Thus, there may be an overall lack of fit of the current model. In that case, try a more complex model, perhaps by only adding individual terms of the next higher-order model (e.g., only the $b_{13}*x_1*x_3$ to the linear model).

Surface and contour plots: The respective fitted model can be visualized in triangular surface plots or contour plots, which, optionally, can also include the respective fitted function. Note that the fitted function displayed in the surface and contour plots always pertains to the parameter estimates for the pseudo-components.

Residual plots: Finally, it is important, after deciding on a model, to review the prediction residuals, in order to identify outliers or regions of misfit-fit. For that purpose, by default the 3D surface graphs produced by the Experimental Design show the observed data along with the fitted response (surface). In addition, you should review the standard normal probability plot of residuals and the scatter plot of observed versus predicted values. The multiple regression analysis (i.e., the process of fitting the surface) assumes that the residuals are normally distributed, and one should carefully review the residuals for any apparent outliers.

A wide application of Statistical optimization in food processes was observed in the literature. The effectiveness of this optimization depends upon the range of independent parameters. The maximum (or minimum) value of the response without stationary point could not be the optimum point. In statistical optimization, the aim of the study and selection of the range of independent parameters should be explained clearly. The verification of the predicted equation should also be made. The physical mean of the estimated results should be discussed and suitability of the estimated response to the expected result should be evaluated. Bas and Boyacý (2007a, b) described the usability statistical optimization using RSM and compared with artificial neural network. they also pointed that the statistical optimization must used within the range of the parameters. The statistical optimization is powerful tool for any process however a prior knowledge of process involving the independent parameter are required for utilizing the tool effectively.

References

Bas, D and Boyacý, IH (2007a). Modeling and optimization I: Usability of response surface methodology Journal of Food Engineering, 78 : 836–845.

Bas, D and Boyacý, IH (2007b). Modeling and optimization II: Comparison of estimation capabilities of response surface methodology with artificial neural networks in a biochemical reaction. Journal of Food Engineering, 78: 846–854.

Box, GEP, and Draper, NR (1987). Empirical model-building and response surfaces. New York: Wiley.

Cornell, JA (1990). Experiments with mixtures: designs, models, and the analysis of mixture data (2nd ed.). New York: Wiley.

Deming, SN, and Morgan, SL (1993). Experimental design: A chemometric approach. (2nd ed.). Amsterdam, The Netherlands: Elsevier Science Publishers B.V.

Hill, T and Lewicki, P (2007). STATISTICS Methods and Applications. StatSoft, Tulsa, OK.

Yates, F (1933). The principles of orthogonality and confounding in replicated experiments.Journal of Agricultural Science, 23: 108-145.

□□□

41

Mobilization and Impact Assessment of Self Help Groups to Popularize the Post Harvest Technologies

M.S. MEENA

Agriculture has become more commercial and is bracing up for a crucial role in economic development in the era of globalization. Emerging priority areas of concern in changing scenario include post harvest management of agricultural produce to save the enormous loss of agricultural produce as well as to secure enhanced monetary returns. Though, India has made quantum jump in production and productivity, lack of technology application in post harvest handling and value addition of agricultural produce has eluded the farmers of their due profit. About 10% of food grains and 25-40% of fruits and vegetables produce go waste, as the agricultural produces are not processed in the catchments areas. The value addition in the form of preservation and processing of agricultural produce has tremendous potential to provide employment and additional income to farm families in rural catchments and thus help in checking the migration of people to urban areas in search of livelihood (Patil and Singh, 2007). Agro and food processing industries in India have great importance in terms of employment, income generation, poverty alleviation, export promotion and foreign exchange earning. In 2006, the estimated farmers' population in developing countries was 1.32 billion, which is going to rise with increasing population in these countries (FAOSTAT, 2006). In most of these developing countries, the majority of rural poor, primarily the farmers have small landholdings, limited resources and excess family labour. Creating multiple opportunities, especially for small and marginal farmers and farmwomen, through various agribusiness enterprises are the challenging tasks.

In a large country, like India, meeting of scientists with farmers and visit of the farmers to the research institute is a difficult task. So, self-help groups (SHGs) have another important role to play in the transfer of technology to the user population. SHGs are being recognized, as a reliable and efficient mode of transfer of technology and positive attitude of the members is required to foster the transfer of technology process (Meena *et al.*, 2003). Working together through formation of SHGs, members become self-reliant, confident, and more empowered economically (Nylund, 2000; Mok, 2001; Yadav *et al.*, 2006). The empowerment of SHGs has proved beneficial on the basis of cost-benefit evaluation of food processing and value added enterprises (Eswarappa, 2006). The most effective means of reaching the rural poor is through groups. SHGs provide ample scope for the unemployed youth and landless labourers to go for self-employment in agriculture and agro-based enterprises (Prakash *et al.*, 2004). Meena *et al.* (2006) observed that training programme on food processing and preservation aspects had significant impact on gain in knowledge level, experience and family size of participants had contributed significantly in enhancing the knowledge level of the participants. Building social capital is critical in overall agricultural development strategies aimed at reducing rural poverty (Swanson, 2006).

Concept of self help groups

Self Help Groups (SHGs) means a group of persons who are helping themselves or each other. Usually group consists of voluntary association of 10 to 20 members with a basic aim to increase the income of group members. SHG is voluntarily constituted by people themselves to discuss issues of common interests, common needs to them such as livelihood generation, economic assets creation, watershed management and other issues. The group motivates the members for constructive work for their socio-economic upliftment. The specific objectives of group formation are:

- To encourage the saving behavior of group members
- To encourage people to take loan for viable productive income generation activities

- To increase the socio-economic status of rural poor
- To develop leadership qualities amongst group members.
- To facilitate easy loan availability with low interest rates, and
- To select suitable package of practices/technology and to transfer them among different stakeholders.

How to form self help group?

Farmers themselves can form groups. Locally identified and specially trained group leaders can also facilitate this process. Group leaders assist the group development process and act as intermediaries between the groups and outside service providers. The process of SHG formation can be divided into three phases.

(i) Identification of group members

It is one of the most crucial stages and progress and stability of group largely depends on members of the group. Go slowly to form the healthy groups because it takes time. Encourage small groups so they can learn more quickly than large groups. Ensure a homogeneous group in which members should have similar interests and backgrounds. Homogeneity reduces the dispute and facilitates more efficient learning. Assist the villagers in identifying areas for group action. The information about their resources, occupational skill, poverty, education, credit need etc; are required at the initial stage. Promote voluntary and democratic group formation. Participatory Rural Appraisal (PRA) is an important method to know about the village resources. The participatory methods and approaches also represent an opportunity to build better linkages between the various actors and to increase the learning from each other. Focus on individual profitability and upliftment of the living standard of groups. Assist in setting the group objectives. Frequent meeting must be encouraged in the initial stage of group formation. Ensure leadership development, which avoid the too much dependency on a single individual.

(ii) Group stabilization stage

Group stabilization is the second important phase to form the Self-Help Group. For the stabilization of the group, participation of the group members must be ensured. It is always better to include educated people in the formation of group. It facilitates the members in keeping the proper accounts. All the transactions are transparent and in written form so members are free to take information or to see the accounts. By keeping records, a group is free from all types of group disputes and facilitates smooth organization of meetings. Group has its own rules, which may be modifies as per need of the group members. In the formation of the group, help from other organizations and experienced personnel can be taken. Help the group to choose a group name that is important to build a group's identity and promote member solidarity. Group must have a constitution preferably in written form. Group constitution must have a President, Secretary and Treasurer based on group consensus. These people are responsible for the management and operation of SHG.

President: President must be selected/elected by group consensus. She / he should be wise and capable to take decision. President also plays an important role in sanctioning loan to the group members with transparency.

Secretary: Elect preferably educated member as secretary. The role of secretary is to maintain the records, meeting proceeding etc., which are required for group management.

Treasurer: The role of treasurer is to deposit the group money, maintain the group account (detail about the debit/credit), entry in the passbooks, bank transactions etc.

(iii) Self helping / Withdrawal stage

In self-helping or withdrawal phase, emphasis is given on intervention aspect. It includes intervention in the form of technical guidance for the management and productivity. Responsibility of concerned organization is not only to impart training but also to evaluate the trainees later on. So, their performance could be evaluated and feedback can be taken to improve the future training programmes.

Income generation activities through SHGs

Agri-Based: Production of wheat, rice flour and dal, mushroom production and value addition of fruits and vegetables, preparation of vermi-compost, bee keeping, organic farming etc.

Horticultural Based: Preparation of jam, jellies, marmalade, pickles, ketchup, sauce, candy, beverages, cordials, squash, syrup, juice powder, glazed fruit, chutney, processing of spices and production of essential oils, spice powder, spice oil, spice paste drying and dehydration oil spice, flowers vending, nursery raising etc.

Fisheries Based: Fish hatchery, fish culture, fish pickles, rearing of brooding fish etc.

Animal / Dairy based

Milk and Milk Based Products: whey, ghee, paneer, cheese, butter, cream, sweets etc.

Leather Based Products: shoe, bags, belt, jacket etc.

Ideal attributes for sustaining and mobilization of SHGs

Following thumb rules must be kept in mind for the establishment, encouragement and mobilization of the group action.

- Build rapport and create environment for rural participation and the formation of informal self-help groups
- Rural people should be allowed to organize their own group businesses.
- Listen to each other rather than carrying on with dominance of few and non representation of others.
- Encourage group members to discuss, explore, firm up ideas and develop workable proposals.
- Meet regularly and regulations while conducting the meeting.
- Record proceeding and the important decisions.
- Regular follow up of action proposed to be taken by the group.
- Draw immediate as well as long-term objective of the group
- Select with consensus the office bearers viz. chairperson/leader, secretary and cashier/treasurer.
- Take up activities of common interest.
- Work with farmers to identify their problems and prioritize for solving them.
- Help them to assess their group capacities, their strengths and resources for solving those problems
- Go slowly to form the healthy groups because it takes time
- Work to gain the people's confidence and trust
- Listen the farmers carefully and give respect
- Call meeting and discuss village issues and identify how a group approach might be used to help and solve those problems
- Discuss goals and expectations
- Remind people that the advantages of group action are realized through hard work, self-sacrifice and a clear focus on realistic group objectives
- Focus on individual profitability and upliftment of the living standard.
- Group action must make economic sense to each farmer
- Assess all the benefits and costs of cooperation

- Discuss about the participation and benefits to them
- Encourage small groups so they can learn more quickly
- Show more trust so information may be shared more freely and without hesitation.
- Promote voluntary and democratic group formation.
- Help the group to choose a name for it that is important to build a group's identity and promote member solidarity.
- Assist in setting the group objectives.
- Frequent meeting must be encouraged in the initial stage.
- Ensure leadership development, which avoid the too much dependency on a single individual.
- Highlight the importance of member contributions to encourage the members.
- Encourage for simple record keeping which have details of the income and expenses and also essential for monitoring group business activities.

Case study in the context of punjab state

The transfer of technology unit at Central Institute of Post-Harvest Engineering and Technology (CIPHET), Abohar (India) in its stride towards mobilizing farm families for group action in value addition and processing formed a number of women SHGs and developed their skills through training in food processing and preservation aspects. A study was undertaken to see the changes in the attitude construct as a result of training. The specific objectives of the study were (i) to study the socio-economic profiles of rural women of SHGs (ii) to measure the attitude of rural women towards SHGs, and (iii) to suggest policy interventions.

Research methodology: Attitude strength is an important determinant of the attitude-behavior relationship. It was assumed that changing the attitude of rural women would influence their performance in SHGs activities. It was also strongly felt that the pace of transfer and popularization of post-harvest technologies must be accelerated through formation of SHGs. So, even small and marginal farmers could get benefit of the new technologies. Evaluation of the SHGs was conducted at attitude construct. A Likert-type-scale was developed exclusively for this study (Meena *et al.*, 2008). Data were solicited from thirty trainee participants of SHGs on five-point continuum namely strongly agree, agree, undecided, disagree and strongly disagree.

Results and Discussion: Attitude strength is an important determinant of the attitude-behavior relationship. Strong attitudes are based on past knowledge and may be retrieved, whereas weak attitude is often constructed on the spot.

(i) Demographic characteristics

All the members of the self-help groups were female and a majority of them (53.33 %) were 20 to 29 years old. Only one-third participants (33.33 %) had primary and senior secondary level education in each category. All the participants were of rural background. The participants had obtained fifteen days training from CIPHET, Abohar on food processing and preservation aspects. The training was imparted to develop technical skill for preparation of jam, jelly, pickles etc., of high quality at cottage level. Half of the respondents reported their income from SHG activities in the range of (US $) 42 to (US $) 68 per month. Thirty percent of them earned less than (US $) 68 per month. More than half of the respondents had 3 to 4 years experience in SHGs, and only 26.66 per cent had more than 4 years working experience in the self-help groups. All the participants belonged to labour class family and adopted food processing as a complementary occupation (Table 1).

(ii) Participants' attitudes towards SHG

The mean ratings, standard deviations and *t*-values of the participants' attitude before and after the training program have been presented in Table 2. Participants' ratings and responses were obtained on five-point continuum namely strongly agree, agree, undecided, disagree and strongly disagree with respective weightage of 5, 4, 3, 2 and 1 for the positive statements while 1, 2, 3, 4 and 5 for negative statements. Attitude statements were related to five dimensions namely; (i) socio-economic upliftment (ii) educational and training (iii) marketing and entrepreneurship qualities (iv) technology adoption and participatory research, and (v) banking / credit aspect. The change in attitude was compared employing *t*-test. Low standard deviations implied a close agreement with participants' ratings.

Socio-economic upliftment

A significant change in attitude was observed towards the statement that *self help groups work as a powerful tool for socio-economic empowerment of the poor in the rural areas,* as reflected by the increase in mean value of response from 2.83 before the training to 3.50 after the training. The change in mean value from 2.76 before the training to 3.13 after the training reflected the agreement of the participants that *groups resolve the conflicts among the participants.* The participants further endorsed that the *self-help group was an approach for collective efforts,* as revealed from the increase in mean values of their response from 3.23 to 3.66 during before and after training. That the training led to a significant change in the attitude of the participants towards the statement, *SHGs formation, which could be a way to eradicate the poverty and unemployment,* could be gathered from the change in mean value of their response from 3.43 to 3.90. As they had the experience of associating as well as working in the group and also securing additional income through group action, they visualized that SHGs had tremendous potential to capacitate them in overcoming the prevailing scenario of poverty.

Table 1 : Demographic attributes of SHG participants.

Attributes		*f*	%
Age (years)	Less than 20	8	26.67
	20 to 29	16	53.33
	More than 29	6	20.00
Education	Illiterate	2	6.67
	Primary	10	33.33
	Middle	8	26.67
	Senior secondary	10	33.33
	Graduate and above	-	-
Back ground	Rural	30	100
	Urban	-	-
Gender	Male	-	-
	Female	30	100
Income (US $*)	Low (< 42)	9	30.00
	Medium (42 to 68)	15	50.00
	High (> 68)	6	20.00
Years in group	Less than 3	4	13.33
	3 to 4	16	53.33
	More than 4	8	26.66
Training	Food processing and preservation	30	100.00
Occupation Main	Labour	30	100.00
Complementary	Food processing and preservation	30	100.00

1 US$ = Indian Rs. 45

Educational and training

Training is an essential process of increasing knowledge, changing attitudes and developing skills through instruction, demonstrations and by other techniques, which develop confidence among the participants. It enhances their self-confidence and competencies in job as well as proficiency in communicating the desired knowledge among peers and clients. Participants showed an enhancement in positive thinking implying that *training helped in developing positive attitude for employing new techniques.* The mean value of their responses increased from 3.06 to 3.60 before and after the training. The mean value of their response was higher (3.73) with respect to the statement that e*ducation plays a pivotal role in changing the behavior of the participants* after training. Education being a social process is responsible for developing and cultivating various physical, intellectual, aesthetic and moral qualities as well as values in an individual. Their orientation that *educated persons are more likely to be good entrepreneurs* was significantly changed after the training, as indicated by significant negative *t*-value.

Table 2 : Comparison of SHG Participants' Attitude Before and After the Training

Statements rated	Before training		After training		*t*-value
	M	SD	M	SD	
Socio-Economic Upliftment					
Self-help groups work as a powerful tool for socio-economic empowerment of the poor in rural areas	2.83	0.83	3.50	0.86	2.76*
Self -help groups help to resolve the conflicts among the members	2.76	0.67	3.13	0.89	3.00*
Self- help group is an approach for collective efforts	3.23	0.81	3.66	0.88	3.79*
Group rules and regulations are based on democratic principle	3.60	0.81	3.76	0.89	1.04
Female members can better utilize their spare time in productive activities through groups	3.36	0.71	3.46	0.97	0.45
Self-Help Group formation can be a way to eradicate the poverty and unemployment	3.43	0.85	3.90	0.88	5.03*
Self-help group improves the saving behavior of the members	3.60	0.85	3.50	0.82	-0.48
Educational and Training					
Training helps in developing positive attitude for new techniques	3.06	1.08	3.60	0.93	3.76*
Education plays pivotal role in changing behavior	3.26	0.67	3.73	0.82	3.29*
Educated persons are more likely to be good entrepreneurs	4.16	0.69	3.70	0.74	-4.06*

Contd...

Training is essential to improve the competence, understanding and professional behavior	3.43	0.56	3.93	0.82	3.52*
Marketing and Entrepreneurship Qualities					
Good entrepreneurs are developed by training and experience	3.63	0.66	4.10	0.69	4.64*
Group formations is the democratic approach for entrepreneurship development	3.30	0.65	3.96	0.73	4.26*
Market demand is very important factor to take up any business activity	3.33	0.71	3.56	0.67	1.65
Knowledge of marketing is prerequisite to gain maximum benefit	3.53	1.13	3.80	0.84	1.61
Product preparation through self-help groups promotes the healthy competition among the groups.	2.53	0.86	2.86	1.04	1.50
Risk taking is the important characteristics of a successful entrepreneur.	3.33	1.02	3.53	0.73	0.97
Technology Adoption and Participatory Research					
Mass media is helpful for quick dissemination and popularization of the technologies	3.13	0.81	4.03	0.71	5.83*
Through group approach participatory research becomes easier	4.10	0.60	3.80	0.71	-1.66
Groups are emerging as a very reliable and efficient mode for transfer of technology	3.93	0.94	3.76	0.89	-0.86
Need assessment is essential for the planning of a programme	3.23	0.85	3.93	0.90	4.37*
Banking / Credit					
Self help groups improve coordination among the members	3.23	0.62	3.53	0.93	1.87
Meeting of different groups contribute in exchange of their experiences	3.60	0.72	4.06	0.73	3.12*
Frequent meeting may contribute in exchange of social norms and values among members	3.53	0.57	4.10	0.54	4.26*
Banks are more eager to sanction loan to groups compared to individual	3.66	0.80	3.56	0.67	-0.72
Financial assistance is essential to setup a new venture	2.90	0.66	3.86	0.89	4.96*

Rating Scale: 1=Strongly-Disagree; 2=Disagree; 3=Undecided; 4 = Agree; 5 = Strongly-Agree Note: *Indicate 't' value significant at the.05 level of confidence (df = 29)

Marketing and entrepreneurship qualities

Training is an overt process, a sequence of the experiences and a series of opportunities to learn. Trainees are exposed to learning situations and acquisition of new skills. For the statement that *good entrepreneurs are developed by training and experience*, mean value of their response changed from 3.66 to 4.10. The change in pre-and post-training attitude was also significant in case of the statement that *group formation is the democratic approach for entrepreneurship development.* Participants learnt by interactions in the group that record keeping, group meetings, opening of accounts in banks and operations were fully based on group consensus. Nothing could be imposed on the group process.

Technology adoption and participatory research

Participants were aware about the latest information technology and agreed that *mass media was helpful for quick dissemination and popularization of the technologies*. Mass media like radio, television, Internet, print media and other means, can reach a large number of people with minimum cost and time. During the training, by watching television the participants could get knowledge about agricultural programmes and their timings. The mean response values changed drastically from 3.13 to 4.03, and the change was found significant. The participants comprehended that the *need assessment was essential for the planning of a programme*. The change in participants' pre and post-training mean response values were significant. They understood the importance of need based planning for effective attainment of objectives of groups as well as individuals.

Banking / credit

The participants realized that *meeting of different groups contribute in exchange of their experiences*. Meetings facilitated the group members in exchanging their ideas and feelings, operational problems and conflicts as well as their resolutions. The training led to a significant change in their response with respect to this statement. The group meetings promoted the extension principle of learning by doing. With respect to the statement that *frequent meeting may contribute in exchange of social norms and values among members*, the increase in participants attitude from pre-training mean value of 3.53 to 4.10 after the training was also significant. It is known that people adopt the norms and values that they consider the best. Traditionally in India, women do not have the control over production system. Their accessibility to production assets is dependent upon male of their family. They do not have property in their names. Recently, land rights legislation has come as a solace to their development and empowerment. In want of equal rights over production base and property, they face immense hurdles in harnessing the credit benefits from the banks and in their economic prosperity. The mean response to the statement that *banks are more eager to sanction loans to groups compared to individuals* decreased after training, while the mean response significantly increased for the statement that *financial assistance is essential to setup a new venture*. Participants made their efforts to get the loan to start their business in commercialization of horticultural products.

Policy recommendations

The results of this empirical study indicate significant change in attitude of the participants in the areas of socio-economic upliftment, educational and training, marketing and entrepreneurship qualities, technology adoption and participatory research, and banking aspects. Training also changed the attitude of rural women towards SHGs positively. As attitude becomes more permanent in nature when the training programme facilitates it, efforts should be made more adequate training of individuals for sustained performance of groups and individuals. Groups could be used as an effective mechanism for information dissemination; social and mutual learning; institutionalized process of capacity building and empowerment; and sustainable and equitable development. SHGs can promote participatory extension and development. Considering the grim scenario of abject poverty in countryside, the programmes for rural upliftment must integrate the vital dimensions like formation of social capital; pro-poor financial and credit support system; market-driven and decentralized extension system; diversification towards high-value enterprises, value addition and processing; need-based skill development through training; media-mix for technology transfer; informal education at rural level and strong political will. However, the extension system needs to be re-oriented and revitalized with new knowledge base in emerging technologies and methodologies. Besides effective cooperation and coordination among the stakeholders, what is most essential is to infuse positive and favorable intentions and attitude, self-confidence and capacity for self-determination among the clientele system.

References

FAOSTAT (2006). All longitudinal production and population data. Retrieved on September 4, 2006, from http://faostat.fao.org/

Meena, MS, Bhagwat, VR, Jain, RK and Ilyas, SM (2003). Self-help group: An effective instrument for transfer of technology. CIPHET extension Bulletin-1. Central Institute of post harvest engineering & technology, Ludhiana (India).

Meena, MS, Brar, RS and Meena, HR (2006). Impact assessment of training on food processing and preservation aspects. Indian Journal of Extension Science, 1(1): 128-130.

Meena MS, Jain Dilip and Meena HR (2008). Measurement of attitudes of rural women towards Self Help Groups. The Journal of Agricultural Education and Extension, 14(3): 217-229.

Mok, BH (2001). Cancer self-help groups in China: A study of individual change, perceived benefit and community impact. Small Group Research, 32(2): 115 -132.

Nylund, M (2000). The mixed-based nature of self-help groups in Finland. Group Work, 12(2): 64-85.

Patil, RT and Singh, DB (2007). Agricultural engineering for enhanced productivity and employment: Role of post harvest management and value addition. Paper presented in 41st Annual Convention & Symposium of Indian Society of Agricultural Engineers, January 29-31, College of Agricultural Engineering and Technology, Agricultural University, Junagarh (India). T-19: 19-30.

Prakash, N Pal, PP and Sharma, M (2004). Self Help Groups for mobilizing tribal youth for entrepreneurship:The experience of Meghalaya. MANAGE Extension Research Review, 2: 62-66.

Swanson, BE (2006). Extension Strategies for Poverty Alleviation: Lessons from China and India. The Journal of Agricultural Education and Extension, 12(4): 285-299.

Yadav, L Varma, S and Jayanti, NV (2006). Capacity building of farmwomen: SHGs approach. Paper presented in International Conference on Social Science Perspectives in Agricultural Research and Development. Feb. 15-18, Indian Agricultural Research Institute, New Delhi, p 255.

□□□

42

Marketing Strategies for Novel Food Products

ANIL KUMAR DIXIT

The global food economy has changed substantially within last decade. Researchers summarized this, as due to combination of rising demand in industrialized and developing countries for value added products. The domestic market for processed food is huge and fast growing. The retail boom will create a huge demand for the food processing sector in the coming years. To boost economic returns from farming, we must find ways for farmers to earn a greater share of the product sale revenue after adding value to their own produce, through backward and forward linkages. Processing and value addition of agricultural produce not only provide utility but have strong multiplier effect on income and employment (Srinivas *et. al*, 2009). An efficient marketing system can reduce post-harvest losses; facilitate exports, and promote value addition through processing, packaging services and food safety practices. It generates employment opportunities and provide food and nutritional security. Off course, the marketing strategies and organizational models will vary from product to product and region to region. Hence better marketing strategies require numerous marketing decisions. Here, an attempt has been made to cover various aspects of marketing such as: level of food processing and its present status; market functions and retail supply chain for value added products; and finally setting strategies and suggestions.

I. Value addition to farm produce

a) Level of value addition from consumer perspective

Value addition is often understood in the context of adding value to the product. It is important to understand that how a consumer perceives the value delivered to him through product and services. Hence, consumer orientation and their requirements need to be kept in mind while deciding the level of value addition. There are three ways in which value addition of farm produce is possible, i.e. primary processing (cleaning, grading and packaging), secondary processing (packaging and branding), and high end processing (supply chain management, modern processing technology, branding and marketing, for instance, dehydrated vegetables, canned and carbonated drinks, extruded products etc.). The evidences reveal that around 60 percent of the food units in India are engaged only in primary processing. The production base of secondary and tertiary processed food in India is relatively low, thus resulting low value addition. However, in a shifting paradigm of income and life style of middle class population, the demand for high value added products is expected to be high.

The consumers/ customers (i.e., primary, secondary and tertiary) and their interdependence with distributors, retailers and processors in the supply chain management are shown in Fig. 1.

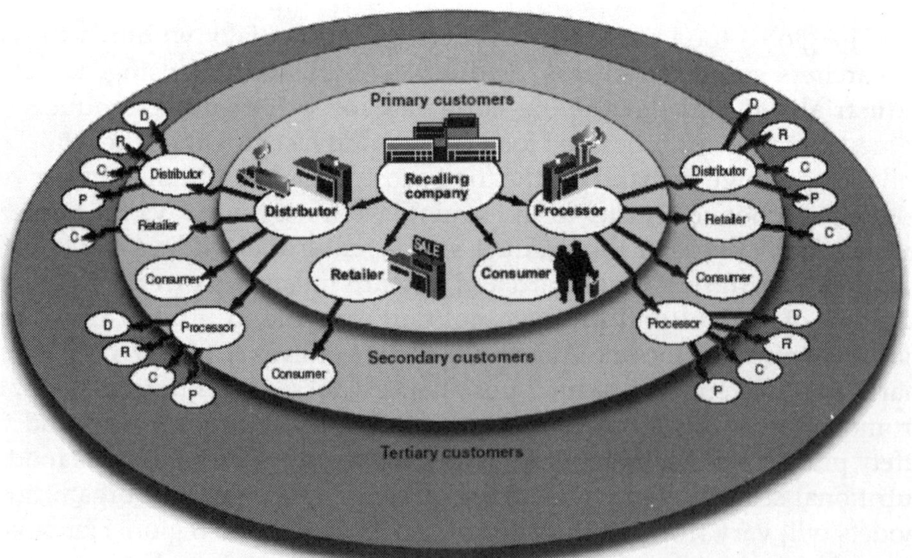

D = Distributor R = Retailer C = Consumer P = Processor

Fig. 1 : Interdependencies in Supply Chain (*Source*: Sporleder, 2005)

a) Present status of Indian food processing sector

The food processing industry is ranked fifth in terms of production, consumption, export and expected growth. Food processing is vital to the Indian economy as it acts as a linkage between manufacturing and agriculture, offering employment to around 13-15 million people, and contributes about 14 percent of agricultural GDP. In 2006-07, food processing sector was estimated 82.6 US $billion (Table 1) (Madras Consultancy Group, 2009). However, when we compare with some of the developed countries, Indian food processing sector is still in nascent stage Sujatha *et. al* (2008). For instance, current level of fruits and vegetables processed is around 2-3 as against 80 percent in USA and Malaysia (Table 2). Further, the level of processing of perishable products in both organized and unorganized sector is furnished in Table 2. The level of processing in milk and milk products was found high in unorganized sector compared to organized sector, whereas the processing of buffalo, meat and poultry products dominated by organized sector.

It was noticed by Sujatha *et al.* (2008) that the growth rates of biscuits, malted food, chacholates, jaggery, wheat flour have decelerated in post WTO period. Whereas, the growth rates of milk powder, guar gum, vanaspathi was high during post WTO period. The demand for import of juices, sauces and jams in 2007-08 was valued at US$16.83 million, US$11.4 million and US$1.55 million, respectively (MCG Compilation). The direction and demand for export and import for valued added products needs to be considered while planning market strategies.

Table 1 : Indian Food Sector, Market Size, 2006-07

Particulars	US $ Billion
Food sector	194.3
Food processing sector	82.6
Organized food processing sector	22.3

Source: Madras Consultancy Group (MCG), 2009

Table 2 : Level of processing of fruits and vegetables in different countries

S. No	Name of the country	Processing (%)
1	USA	80
2	France	70
3	Malaysia	80
4	Thailand	30
5	India	2.2

Source: Sujatha *et al.* (2008)

Table 3 : Level of processing in India (perishable products)

	Organized	Unorganized	Total (%)
Fruits and vegetables	1.3	0.8	2.2
Milk/ milk products	13	22	35
Buffalo meat	21	-	21
Poultry	6	-	6
Marine products	8	15	21

Source: Sujatha *et. al* (2008)

Marketing functions for value added products

Usually, marketers often refer to the "Four *P*s," or the marketing portfolio (Perner, L. http://www.consumerpsychologist.com/).

- *Product*: Firms invest in developing product according to the taste and preferences of consumers by using high quality ingredients and also doing extensive research and development. Product development and refinement is a continuous phenomenon. For instance, McDonald's and Burger King, literally spend millions of dollars to perfect their French fries!

- *Price*: The prices are generally determined by demand and supply forces in the market. The logic here is that consumers will be willing to buy a larger quantity of goods at a lower price than they would at a higher price. As we will see, this assumption is sometimes violated, usually when consumers use price as a cue to quality and assume that a higher priced product is better. Similarly, sellers are generally willing to sell a larger quantity when a higher price is offered. Thus, the equilibrium point is determined by supply and demand curves. However, in case of monopoly situation different strategies may be taken with respect to price. Monopolist can make a profit in two ways- sell a lot and make a small margin on each unit or make a large margin on each unit and settle for lesser volumes.

- *Distribution*: There are many channel/routes such as: i) direct sale (factory outlet or direct home/restaurant delivery), ii) owns sale outlet in cities/ towns, iii) franchise/dealership, iv) wholesalers and retailers-mega mart, v) export (domestic and foreign), and internet, through which the final product reaches to the consumers. The detailed flow chart of the distribution channels expects sale through internet is presented in Fig. 2.

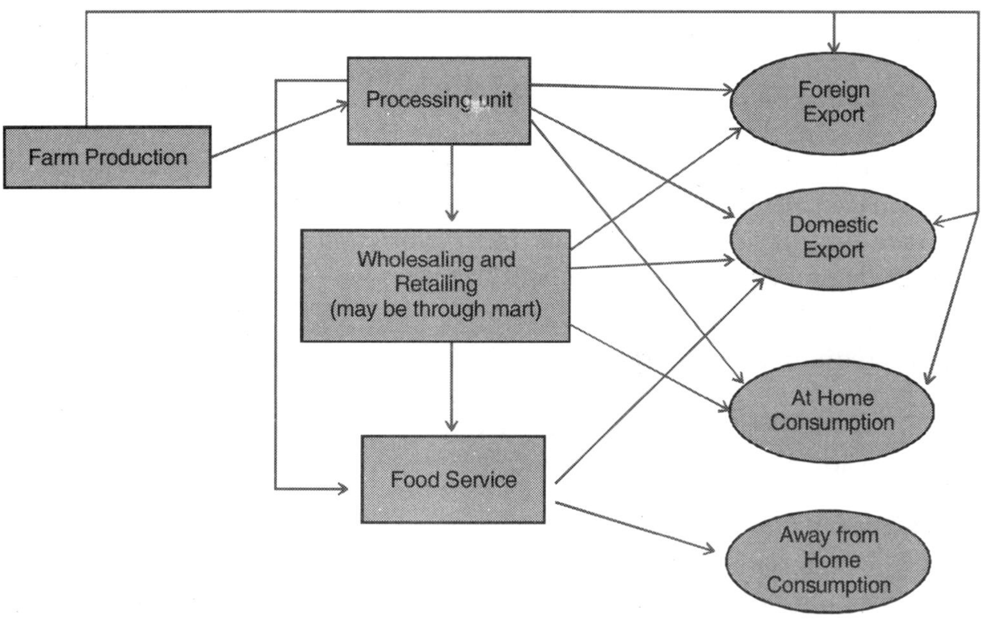

Fig. 2 : Flow chart for distribution of processed products

- *Promotion*: Involves the different tools that firms have to get consumers to buy more of their products, possibly at higher prices. Advertising, demonstration, home delivery, coupons, and discount on sale product are some of the means and method which arouse curiosity of the buyers. Money can be invested in brand building (through advertising and other forms of promotion) to increase either quantity demanded or the price consumers are willing to pay for a product. Manufacturers may be able to leverage their existing brand names by developing new product lines.

Besides, many factors which affect the demand and establish reputation in the market are discussed below:

- Guarantee: It reduces buyer's resistance. Guarantee gain customer confidence and usually customers travel a long way for purchasing a product which might be available in his/her vicinity.
- Delivery: Timely delivery is another important criterion to win customer confidence. However, it is not necessary that it is always free.
- Service: Very important to customers, and a major way to retain customer loyalty. It is difficult to establish credibility for new customers, since 'service' has to be experienced to be truly believed.

- Credit: Can add value to some customers. But make sure that your own finances will not be damaged. Finance companies could be used, and some of them will pay you a commission.
- Packaging and presentation: The more professional and attractive the packaging, the more the perceived value of the product will be. However, the cost of packaging should never exceed 20 percent of the cost of product. More important is that the packaging material should be biodegradable and environment friendly.
- Quality assurance: International standards, such as ISO 9000 accreditation can be of value, but reputation is equally important. Brand loyalty is, of course, a matter of degree. Some consumers will not switch for a moderate discount, but would switch for a large one or will occasionally buy another brand for convenience or variety.

Retail supply chain

No doubt, the supply chain provides an opportunity for many firms to add value to a product but also add cost. These added costs usually result from consumer demand where consumers are willing to pay for additional convenience. Retailing has an important place in sequence of value chain. In India, retail market for processed products is characterized by co-existence of street-side vendors, kirana stores and organized retailers (Table 4). Unorganised and semi-organised retailers like kirana and grocers are characterized by low volume of business and economics of scale are not yet realized. The advantage of this model is that the vendor knows his customers personally and door delivers of the products. It is evident from the Table 4 that organized retail is coming in a big way. These oragnised retail models are competitive through pricing, convenience of shopping and ambience.

Table 4 : Retail models in India

Convenience stores (*kirana* store) and hawkers	Organized retailers
1. Street-side and pushcart vendors (pickles, confectionary items)	1. Hypermarket (Big *Bazar*, Trent)
2. Public distribution system shop (unbranded grocery items)	2. Supermarket (Food world, Spencer's Daily, Reliance Fresh)
3. *Kirana* store (milk and milk products, grocery, beverages, snack foods)	3. Departmental store (Lifestyle)
4. Seven Eleven store (small to medium food retail outlet)	4. Specialty chain (Spar)
	5. Discount chain (*Subhiksha*)

Source : Complied from MCG

Strategies and suggestions

The market strategies are categorized into generalized recommendations and specific suggestions for manufacturer/ entrepreneur

1. General recommendations

- Pricing policies: Linking it with the quality of the produce or a product (for instance, fat content in milk; higher protein quality/quantity in soybean based product).
- Effective linkage: Linkages between farmer- processor on a mutually beneficial contractual agreement (may be through contract farming).
- Cost-effective packaging technology: Ensure safety and prolonged shelf life.
- Encourage farmer to grow processing varieties.
- Encourage Agro Processing Centre (APC) in production catchment: Farmer, entrepreneur and consumer will be benefited.
- Town and Village Enterprises (TVE): Adopt model of China (guided by the market; financed by local financial institutions; availability of factor of production; complete business autonomy with little government interference; solely responsible for profits and losses; adopting flexible management), (Anonymous, 2002) with required modifications.
- Main product in consonance with by-product processing/utilization: Experimental protocols for converting these into usable co-products are available. These need to scale up.
- Decision support system: The system continuous tracking of the demand, supply, consumer needs and prices.
- Basic infrastructure: Public-private investment in road, communication, and other critical market infrastructure, in public-private partnership mode.
- Functional support: Government need to create 'Quality Certification' and establishment of a 'Quality Testing Lab'.
- Policy reform: Incentives to level the playing field and create space for the private sector to boost export. The protection of domestic entrepreneurs/manufactures against low cheap import (in case of dumping off from other countries).
- Institutional arrangements: include contract enforcement, coordinated value chains, grade and standards and market information services.

2. Specific suggestion for manufacturer/ entrepreneur

- *Packing, labeling and bar coding*: Packing should be attractive but least cost. The composition of the ingredients, date for manufacturing, period for safe use and method of use should be labeled. Bar coding is important for maintaining computerized records.
- *Quality control and food safety assurance*: The qualities should be in conformity with claims made in labelling. Food safety concerns assurance that the food will not cause harm to the consumer, when eaten according to its intended use.
- *Estimating consumer willingness- to- pay for product*: Consumers responses may be ascertained pertaining to prices in a hypothetical market. Another method of estimating response is selling a product in certain test stores at different prices and seeing responses.
- *Set market objectives*: A firm would like to reduce costs, improve quality, increase distribution channels, increase awareness, improve consumer perception, and ultimately increase the demand of the product in an integrated manner (Fig. 3).

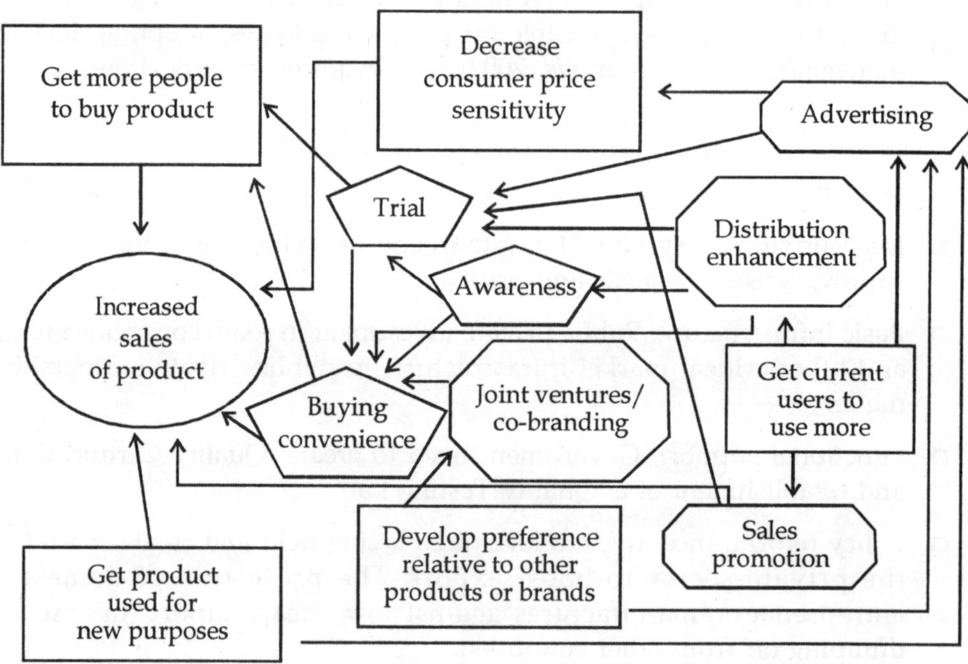

Fig. 3. Strategies for increasing the demand of the product (*Source* : http://www.consumerpsychologist.com/intro.htm)

The huge potential for processed products in the country can be harnessed by strengthening market infrastructure such as market yard, quality control laboratory, efficient and reliable market information along with pre- and post harvest technologies. Nonetheless, institutional support (in terms of better contract agreement and contract enforcement, quality certification), and critical market services (sales promotion and decision support system) are important for the development of value added food product .

References

Anonymous (2002). Agriculture-Industry Interface: Value added farm products. National Academy of Agricultural Sciences, India.

Madras Consultancy Group (2009). Overview of Indian Agriculture and Food Processing Sector.

Perner, L. Food Marketing. University of Southern California Los Angeles, USA.

Sporleder, T (2005). Nature and Scope of Agri-food Supply Chain Dynamics, Paper presented at FAMPS Conference, May 4-5.

Srinivas, K, KP Singh, Ajay Kumar, SRK Singh and KKS Bisht (2009). Agro Processing Centres in Uttarakhand Hills: A Case Study of Enterprise Development, J. Agri. Develop. & Policy, 19 (1) : 44-53.

Sujatha, RV and Y Eswara Prasad (2008). Food Processing Sector in India in Globalization Perspective. Agricultural Situation in India, pp. 303-307, August 2008.

❑❑❑

43

Impact Assessment of Post Harvest Technology

ANIL KUMAR DIXIT AND S.K. NANDA

Post harvest technology has been recognized as a vital tool for bringing rural prosperity through generation of income and employment opportunities by way of minimizing post harvest losses as well as several value added products. It has immense potential for providing food and nutritional security and can ensure inclusive growth of national economy. In recent years, the importance of post harvest research and activities has received greater attention at global level on account of urbanization, contraction of agriculture sector (both by declining a agriculture GDP as a share of total GDP and declining labor force in agriculture) and globalization.

Nonetheless, Indian Council of Agriculture Research has channelized their efforts in this direction by launching the All India Coordinated Research Project (AICRP) on Post Harvest Technology (PHT) in September 1972 and at present there are 38 centres functioning across the country. About 57 technologies developed and translated into use by entrepreneurs and farmers-cum-processors and several other technologies are ready for commercialization. The present chapter highlights the impact (direct and indirect) of some selected technologies developed and commercialized / transferred including Agro Processing Centres established and monitored by the cooperating centres under AICRP on PHT.

Methodology

About 57 technologies (Table 1) generated under AICRP on PHT have been translated into use by entrepreneurs and farmers-cum-processors. As a result of these new technologies, increase in percentage recovery, reduction of post harvest loss, reduction in women drudgery, etc. were observed in many cases, besides, augmenting the income and employment opportunities. It would not be possible to discuss the impact of all these technologies, and hence, for the ease, two technologies (i.e., 1. PKV Mini Dal Mill, 2. Extraction of Apricot Kernel Oil) and Agro Processing centres are considered for discussion in this chapter. Some broad indicators for impact assessment such as income, employment, food and nutritional security and sustainability were identified and translated into survey schedules/questionnaires -'I' and 'II' which were utilized for data collection. The cost economics was worked out for improved technology as well as traditional practices. The basic assumption here is that the technology is the key driver for adoption, though other factors such as infrastructure, policy and extension and institutional support also affecting the adoption rate. Besides, economic surplus model was used to estimate the economic benefits of the technology 'Extraction of Apricot Kernel Oil'. It was assumed for ease of analysis that a parallel research induced parallel shift in supply curve. The assumption of linear supply and demand functions with parallel shift, have been applied in most of the earlier studies on research benefits (Alston *et al.*, 1995, Dixit *et al*, 2010, Ilyas *et al.*, 2004, Mruthyunjaya *et al, 2004*). The data pertaining to annual cost of the technology was computed by considering the salaries of the scientific and other staff proportionate to the time allocated for the development and refinement of the technology, and the contingent and extension expenditure incurred.

$$\text{Change in total surplus} = K_t P_0 Q_0 \left(1 + 0.5\, Z_t \eta\right)$$

where,

$Z_t = K_t\left(\varepsilon / (\varepsilon + \eta)\right)$

K = Vertical shift in supply function

ε = Elasticity of supply

η = Elasticity of demand

P_0 = Base year output price (Average of 2009 and 2010)

Q_0 = Base year output quantity

The data pertaining to recovery advantage, labour cost reduction, cost economics, cost of technology were collected from already established entrepreneurs in this business and AICRP on PHT, Solan centre with the help of schedule/questionnaire 'I' and 'II'. The data on economic variable like elasticity, input prices, and availability of apricot kernel in the target domain were obtained from published sources.

Net present value (NPV), internal rate of return (IRR), and benefit-cost ratio were then computed using the following formula.

$$\text{Net Present Value (NPV)} = \sum\left[(B_t - C_t)/(1 + r)^t\right]$$

$$\text{Internal Rate of Returns (IRR)} = \sum\left[(B_t - C_t)/(1 + IRR)^t\right] = 0$$

$$\text{Benefit-Cost Ratio (BCR)} = \sum\left[(B_t)/\ (1 + r)^t\ \right] / \sum\left[(C_t)\ /\ (1 + r)^t\right]$$

where, B_t denotes the benefits in the year t, C_t is the cost in year t and t is the time period, and r is the discount rate (rate of interest).

Table 1 : List of technologies commercialized / transferred under AICRP on PHT

1. PKV mini dal mill (PDKV Akola)
2. PKV chilli seed extractor (PDKV Akola)
3. Vivek thresher-cum-pearler (VPKAS Almora)
4. Preparation of cane jaggery cubes / blocks (RARS Anakapalle)
5. Process for improvement of setting and quality of jaggery from immature, overaged, scale infested and drought affected cane juice. (RARS Anakapalle)
6. Double grating Anakapalle jaggery furnace for increased heat use efficiency (RARS Anakapalle)
7. 2-in-1 Maize sheller-cum-Sunflower thresher (UAS Bangalore)
8. Arecanut dehusker (UAS Bangalore)
9. 3-in-1 mini Groundnut decorticator, Sunflower thresher and Maize sheller (UAS Bangalore)
10. Pedal operated coconut dehusker (UAS Bangalore)
11. Multi purpose grain mill (CIAE Bhopal)
12. Pedal cum Power operated grain cleaner (CIAE Bhopal)
13. CIAE Dhall mill (CIAE Bhopal)
14. Groundnut decorticator, hand operated (CIAE Bhopal)
15. Solar cabinet dryer for vegetables (CIAE Bhopal)

16. Osmo dehydrated Pineapple slices/rings and tidbits (OUAT Bhubaneswar)
17. House hold insect trap (TNAU Coimbatore)
18. Four Roller Sugarcane Crusher (TNAU Coimbatore)
19. Bottling of Sugarcane juice (TNAU Coimbatore)
20. Household paddy parboiling unit (TNAU Coimbatore)
21. Mini Dahl Mill (TNAU Coimbatore)
22. Value added products from ginger (AAU Jorhat)
23. Feed Block Making Machine (JAU Junagadh)
24. Cleaner- cum-Grader for Cumin (JAU Junagadh)
25. On-Farm Fruit Grader (JAU Junagadh)
26. Agriculture wastes-fired dryer for Red Chillies (JAU Junagadh)
27. Development of Shell fired Copra Dryer (CPCRI Kasargod)
28. Process for production of coconut chips (CPCRI Kasargod)
29. Production of snow ball tender nut (CPCRI Kasargod)
30. Design of solar cum electrical dryer with biofuel as third source of energy (CPCRI Kasargod)
31. Grain puffing machine (IIT Kharagpur)
32. Bamboo Cement Bin (for paddy storage) (IIT Kharagpur)
33. Recirculatory tray dryer (IIT Kharagpur)
34. Dehydrated vegetables (IIT Kharagpur)
35. Ripe Bael drink and unripe mango drink (IIT Kharagpur)
36. Liquid Jaggery Production Process (RS&JRS Kolhapur)
37. Modified Electronic / Digital Thermometer to indicate the striking point temperature (RS&JRS Kolhapur)
38. Churner for defrothing during jaggery making (RS&JRS Kolhapur)
39. Modified Hardness tester for solid jaggery (RS&JRS Kolhapur)
40. Cane Juice Boiling Pan (RS&JRS Kolhapur)
41. Drying-cum-storage bin for jaggery (IISR Lucknow)
42. Single Drum Rotary Screen Grain Pre-cleaner (PAU Ludhiana)
43. Farm level fruit and vegetable washing machine (PAU Ludhiana)
44. Pantnagar Mini Dhal Mill (GBPUAT Pantnagar)
45. Technology for extraction of kernel oil from apricot stones/seeds (YSPUH&F Solan)
46. Hand operated cassava chipping machine (CTCRI Trivandrum)
47. Cassava rasper (CTCRI Trivandrum)
48. Motorised chipping machine (CTCRI Trivandrum)
49. Peeling knife (CTCRI Trivandrum)

50. Maize dehusker sheller (MPUAT Udaipur)
51. Garlic bulb breaker (MPUAT Udaipur)
52. Process technology for Garlic flakes & powder (MPUAT Udaipur)
53. Technology for ginger and turmeric processing (MPUAT Udaipur)
54. Solar heat treatment machine (MPUAT Udaipur)
55. Garlic peeling machine (MPUAT Udaipur)
56. Garlic clove flaking machine (MPUAT Udaipur)
57. Peeler-cum-polisher for ginger and turmeric (MPUAT Udaipur)
58. Agro Processing Centre – APC (multiple centres)

Impact of PKV Mini Dal Mill

Dal milling in villages amounts to a very small share of consumption through burr mills, where recovery seldom exceeds 55%. PKV Mini dal mill (using two horse power single phase electric motor) was developed and commercialized by AICRP on PHT Akola centre with the view to increase the recovery of dal so that the pulses can be processed at village level, and transportation and other logistic cost can also be minimized. Almost all pulses can be dehulled with this machine and the products are quite comparable with that of the commercial dal mills. This plant is commercially manufactured by M/s Shriram Associates Akola and available in the market. The processing capacity of PKV dal mill is 100-125 kg/h for pigeon pea and 125-150 kg/h for green and black gram and the respective recoveries are 72-75% and 82-85% (Borkar *et al.*, 2006). About 538 units of PKV mini dal mill have been sold to various entrepreneurs so far (up to July 2010). The estimated benefit from this dal mill to an entrepreneur is Rs. 3.45 lakh per annum (Table 2). The technology offered rural employment to the tune of 2.58 lakh mandays (direct) & 7.68 lakh mandays (indirect) (Table 2). It is worth to mention here that the society (consumer/customers) were more benefited than entrepreneurs. The benefits were to the tune of Rs. 117.42 crore as a result of enhanced recovery of dal and machine was working mostly on custom hiring basis. The price of broken (Rs. 25 per kg) was adjusted against the market price of dal (Rs. 70 per kg) while calculating the profit to the customers/society. However, the profit to entrepreneurs (538 Nos) was estimated to the tune of Rs 18.53 crore per annum. It is evident from the results that the indirect impact is much stronger than the direct impact as far as enhanced income/saving of final product is concerned. India being a deficit in pulse production, needs to minimize the post harvest losses (particularly processing losses) and this technology is a right step in this direction.

Table 2 : Economic benefits from a PKV Mini Dal Mill

Particular of processing	Economic benefits to the entrepreneur					Employment generation
	Working days/ year	Total quantity processed (q)	Rate of processing (Rs. /q)	Cost of processing (Rs./q)	Net profit (Rs./ year)	(Man-days / year)
A) *Dal* Milling						
Pigeon pea/ Bengal gram	130	1440	250	75	2,52,000	260
Black gram	30	300	200	75	37,500	60
Cowpea/ soyabean	20	200	200	75	25,000	40
B) Cleaning and polishing						
Wheat cleaning	30	600	50	25	15,000	60
Rain touch *mung* cleaning	15	150	100	50	7500	30
Cleaning/ Polishing of *Jowar* for *papad* making	15	150	100	50	7500	30
Total	240	2840	-	-	344500	480

Survey results 2009-10

Table 3 : Economic benefits from PKV Mini Dal Mill (Unit established = 538)

Benefit	Quantity processed* / improved recovery** (Q)	Net profit per annum (Rs. Crore)	Employment generated/ potential employment# (lakh mandays)
Benefits to entrepreneurs	1527920*	18.53	2.58
Benefits to society/ consumer	260930**	117.42	7.68#

* Calculated for 538 units based on the data given in Table 1 and ** Improved recovery @ 25 percent over conventional burr mill and calculated on the basis of data in coloum A of Table 1, *Note: Average prices of dal= Rs.70 per kg and brokens= Rs. 25 per kg*

Impact of apricot kernel oil extraction technology

Oil derived from the kernel of apricot (a stone fruit grown abundantly in the temperate regions) has medicinal uses well known since ages. But the technology for breaking of stone and separation of kernels was not available and it was being done manually, which involved drudgery, particularly for women. Moreover, manual breaking of stones to separate kernel and oil extraction through traditional Kolhu is a tedious, time consuming and unhygienic process, which results in very low yield and poor quality of extracted oil. In this backdrop, AICRP on PHT Solan centre developed and commercialized complete package for extraction of apricot kernel oil which consist of mechanical decortications, kernel separation with gravity method, and extraction of kernel oil with table oil expeller, filtration and then packing of oil in bottles (Gupta and Sharma, 2009). The perusal of data presented in Table 3 clearly indicates that improved technology has added advantage in terms of efficiency of stone decortication, kernel separation over conventional practices. Moreover, the quality of final product (edible oil) was found superior (clear and transparent) in case of improved technology.

Table 4 : Benefits of improvement technology over conventional practices

Particulars	Unit Operation	Improved Technology	Conventional Practice
Stone decortications/ Breaking	-	Mechanical Decorticator	Manual Breaking
Decortications capacity	Kg/ hr	90 (±10.0)	3.9 (±0.3)
Kernel separation		Specific Gravity Separation	Manual Separation
Separation capacity	Kg/hr	6.5 (±0.5)	0.66 (±0.05)
Oil Extraction		Table oil Expeller	Traditional Kohlu
Oil Yield	%	43.5 (±1.5)	32.5 (±1.5)
Filtration	-	Filter Press	Decantation/ Settling
Capacity	Lt/hr	2.5 (±0.5)	-
Oil quality after filtration		Clear and Transparent	Turbid with sediments

Source : Dixit *et al.*, 2010

The comparison of cost economics (Table 4) shows that the fixed cost per unit was higher in case of improved technology but variable cost particularly

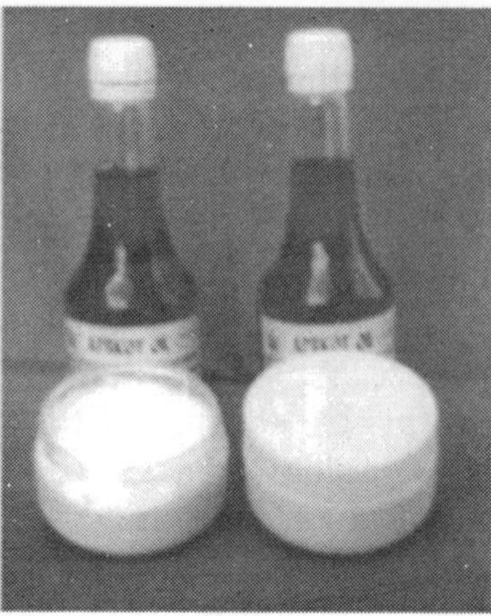

Table 5 : Cost and benefits of improved technology vis-a-vis- conventional practice (n=5)

S. No.	Cost/ benefits	Improved technology (mechanical decorticator, kernel separator, oil extraction and filtration)	Conventional technology (Manual breaking of stones as well as separation of kernels and oil extraction in Kolhu)
1	Annual fixed cost (Rs.)	47500	20500
2	Variable cost (cost of raw material, repairer and maintenance of machine, fuel charges, labour charges, packaging etc.)	259 144	299230
3	Total cost (1+2)	306644	319730
4.	Returns		
a.	Main product Apricot Oil (lt)	1408	1254
b:	Value of apricot oil @ (Rs.300 /lt)	422400	376200
c.	Co-product (Press cake) (kg)	1792	2176
d.	Value of byproduct @ Rs. 10/kg	17920	21760
	Total returns (b+d) in Rs.	440320	397960
5.	Profit (D-C)	133676	78230
6.	Cost of production (Rs/lt.)	192.57	248.20
7.	Reduction in unit cost of production (%)	22.41	
8.	Saving over conventional practice (Rs/annum)	55446	

mechanical breaking of stone and separation of kernel has been substantially reduced by 94.87 and 80 percent, respectively, for processing of equivalent raw material.

Overall, net profit per unit in case of improved technology was estimated to the tune of Rs. 1.34 lakh, which lead to saving of Rs. 55446 over conventional practices. The results derived from economic surplus model in terms of NPV, IRR and benefit-cost ratio of the technology has been found quite attractive. The NPV of incremental benefits was Rs. 486 lakhs, whereas, IRR and B-C ratio were 44% and 21.83, respectively (Dixit *et al.*, 2010). A period of 18 years was considered for economic analysis for developing the cost-benefits stream. The R & D lag was found to be reduced by 10 years.

Further, the opinion of the respondents were ascertained and rated on the scale of 1 to 5 (highly disagreed to highly agreed) to know whether the technology is acceptable, sustainable and environment friendly. The technology was highly acceptable as it helped in reduction in women drudgery, health hazards (pressed fingers and nail during manual breaking of stone), labour friendly (no back pain) and overall mean score (5.0) was recorded (Dixit *et al.*, 2010). The results further experienced that the technology was environmentally safe as it utilized fruit stones which otherwise are thrown as a waste leading to environmental pollution and also contributed towards maintaining hygienic conditions and quality products.

Impact of agro processing centres (APC)

Establishment of new APC and monitoring of existing APC are being undertaken by AICRP on PHT centres (Akola, Almora, Bapatla, Bhubaneswar, Coimbatore, Jorhat, Junagadh, Kasargod, Ludhiana, Pantnagar and Udaipur). APC managed by women entrepreneurs and women workers have been established by Bangalore, Bhubaneswar and Coimbatore centers, an organic APC has been established by Akola centre for export promotion. These APC have been quite effective in post harvest loss reduction, value addition income augmentation as well as employment generation in rural areas.

These APCs are able to tackle effectively the problem of under employment particularly in hilly area by way of establishing forward (with consumer) and backward (with production) linkages. After running the APC in Almora district of Uttarakhand for almost two years, the average operational cost was Rs. 3,50,686/- and total return was Rs. 4,34,913/- per year. The estimated economic indicators such as: IRR (36.22%) and B/C (1.39) and Pay Back Period (2.8 years) of an Agro Processing Centre in Uttarakhand show that the business is on sound footing and can be a very good option for investment (Singh *et al.*, 2007). This can be considered as one of the good alternative for youths in hills region for self-employed and they need not to migrate from hills in search of employment and livelihood security. In Punjab state so far as about 25 APCs were established and monitored by AICRP on PHT, Ludhiana centre. The estimated net benefits from average APC was found to the tune of Rs. 6.0 lakh per annum. It can also be interpreted that the entrepreneurs (25 Nos) were able to generate a sum of Rs. 1.5 crore per annum as a result of establishment of APC, besides employment generation to the tune of 63750 man days per annum (direct). It is worth to mention that a good number of APCs are being inspired from existing APCs, and therefore, the potential for income and employment generation in the state of Punjab through APC need to further explore.

The experiences of APC established by Akola centre highlighted that a group of 280 farmers of Vidharba region with the total land holding of about 4000 acres have come together to grow crops organically under the coordinated management of Sh. Deepak Shinde, progressive farmer-cum-entrepreneur and Chief Executive Organic Farmers, Village Nimkhed. The success story published on ICAR web site (www.icar.org.in). He has established an organic APC in Nimkhed, Amrawati District of Maharashtra during 2004. Sh. Shinde has also been interacting with farmers of developed countries particularly USA, Poland, U.K. for sharing knowledge in the field of organic farming. He has established electronic communication facilities in the rural area for fast communication and to facilitate sale of organic produce. The data on cost and benefits were collected during 2008-09. The annual input cost was found to the tune of Rs. 4.72 crore, whereas, the gross returns were Rs. 5.74 crore, making a net profit of Rs. 1.02 crore per annum.

Post harvest technology has been contributing significantly towards food security of country and also a source of livelihood for entrepreneurs in rural areas. The diverse impact of the technologies of AICRP on PHT is evident from the results. Interestingly, the society (customers/ consumers) were found to be more benefited as a result of increased recovery of dal from PKV Mini Dal Mill. The technology for extraction of apricot kernel oil has found added advantage over conventional practices and also recognized its potential to minimize negative externalities to environment, as it utilizes by product (apricot stone), which otherwise is not properly utilized and thrown as a waste leading to environmental pollution. APCs have also impacted well both in terms of income and employment generation in the production catchment and able to establish backward and forward linkages in hilly regions.

References

Alston, JM, G Norton and P Pandey (1995). Science under Scarcity: Principles and practices for Agricultural Research Evaluation and Priority Setting. Cornell University Press, Ithaca. 585 p.

Borkar, PA, IL Pardeshi, PH Bakane (2006). PKV Mini Dal Mill: Svam Rojgar Ke Sadhan, Dr. Punjabrao Deshmukh Krishi Vidyapeeth, Krishi Nagar, Akola, Bulletin No./ PHT/PKV/06/57.

Dixit, AK, PC Sharma, SK Nanda and SK Aleksha Kudos (2010). Impact of Processing Technology in Hilly Region: A Study on Extraction of Apricot Kernel Oil, Agricultural Economics Research Review, 23 (conference issue): 405-410.

Gupta A and PC Sharma (2009). Technology for extraction of wild apricot oil (Chulli) at cottage scale. Indian Farming Dec, 2009: 12-17.

Ilyas, SM, LS Gangwar, RK Goyal, DK Bhati and Sandeep Kumar (2004). Impact Assessment of Post Harvest Technologies Developed under NATP and AICRP on PHT. Central Institute of Post Harvest Engineering and Technology (CIPHET) Ludhiana, 76 p.

Mruthyunjaya, S Pal, LM Pandey and AK Jha (2004). Impact of Selected Technologies Refined under NATP. Agricultural Economics Research Review, 17 (conference issue): 1-13.

Singh, KP, AK Srivastva, K Srinivas, SRK Singh and HS Gupta (2007). "Entrepreneurship Development in Agriculture through Agro Processing Centre: A Case study of Almora District in NW Himalaya". Invited Overview (International e-journal), IX (2).

□□□

About the Editors

Dr. Mridula D., presently Senior Scientist (Food & Nutrition) at Division of Food Grains & Oilseeds Processing, Central Institute of Postharvest Engineering and Technology (CIPHET), Ludhiana, Punjab under ICAR. She got her Master's and Ph.D. degree from Banaras Hindu University, Varanasi, Uttar Pradesh. Her field of research interest is formulated nutritious foods, functional and designer foods, enriched foods and extrusion processing. In her 13 years of research experience, Dr Mridula has contributed one book, one compendium, 4 technical bulletins, 2 book chapters and 13 technical popular articles. She has so far published 36 research papers in various National and International Journals and 30 presentations in national and international conferences. She is also the Mukhya Sampadak of CIPHET *Rajbhasha Patrika 'Krishi Prashanskaran Darpan'* and so far brought out 10 issues of this Hindi *Patrika*. She is a recipient of the ICAR Award for Team Research in Agriculture and Allied Sciences (Engineering and Technology) for the Biennium 2005-2006. She is also the life member of various professional societies like Indian Science Congress, Nutrition Society of India, Indian Society of Agriculture Engineers and Bioved Research Society.

Dr. R.T. Patil, Principal Scientist (Agricultural Structures and Process Engineering) and former Director of Central Institute of Post Harvest Engineering and Technology (CIPHET) under ICAR which is a nodal institute to undertake lead researches in Post Harvest Management and Value Addition. Dr, Patil has B. Tech degree (Ag. Engg.), M. Tech (Crop Process Engineering) from Indian Institute of Technology, Kharagpur, India and Ph.D. in Bio Process Engineering from University of Saskatchewan, Canada and finished 2 years of Post Doctoral Fellowship at Washington State University, Pullman (USA). He has 32 year of research experience in food process engineering. He has contributed 12 books / bulletins and 22 books chapters. His areas of expertise are drying and dehydration, novel product development, physical properties and advanced extrusion processing. Dr. Patil has so far developed more than 22 research prototypes, 5 process protocols, 3 pilot plants and contributed to two US patents and seven Indian patents with over 100 papers published in national and international journals and over 100 presentations (4 presentation awarded best paper award) in national and international conferences. Dr. Patil is a staunch promoter of food processing entrepreneurship at production catchment and has helped establish over 300 food processing industries. He is a recipient of Best Graduate thesis award of Canadian Society of Agricultural Engineers for his Ph.D. and the team award from Ministry of Education, Government of Poland for Thermal Processing of Bio-materials. He was elected as Director (Food Process Engineering) of ISAE during 1998-2000 and was elected three times as Vice President of Indian Society of Agricultural Engineers and was Vice President, Association of Food Scientists and Technologist, India for 2010. Dr. Patil is a fellow of Indian Society of Agricultural Engineers, Institution of Engineers (India) and National Academy of Agricultural Sciences.

Dr. M.R. Manikantan, is presently Senior Scientist (Agricultural Structure & Process Engineering) at Food Grains & Oilseeds Processing Division, Central Institute of Postharvest Engineering and Technology (CIPHET), Ludhiana, Punjab under ICAR. He has Ph.D. degree in the field of Agricultural Processing from Tamil Nadu Agricultural University, Coimbatore, Tamil Nadu. His fields of specialization are application of nanotechnology in food packaging, oilseeds processing, physical properties, drying and dehydration. He has also obtained Post Graduate Diploma in Operations and Human Resource Management from Indira Gandhi National Open University, New Delhi. He also underwent advance training on "Post Harvest Rice Processing" in Japan. In his 12 years of research experience, he has published 22 research articles in both national and international referred journals, 7 technical popular articles, 5 national and international conference proceedings, 4 technical bulletins, 5 manuals, one compendium and one book chapters. He has so far developed 4 research prototypes, 5 process protocols and one pilot plant. His two research papers have been awarded as best research paper in national conferences. He is a recipient of Jawaharlal Nehru Award for Post Graduate Agricultural Research for the year 2010 for outstanding research in the field of Agricultural Engineering.